CHEMICAL ENGINEERING METHODS AND TECHNOLOGY

PHOTOCHEMISTRY: UV/VIS SPECTROSCOPY, PHOTOCHEMICAL REACTIONS AND PHOTOSYNTHESIS

CHEMICAL ENGINEERING METHODS AND TECHNOLOGY

Additional books in this series can be found on Nova's website
under the Series tab.

Additional E-books in this series can be found on Nova's website
under the E-books tab.

CHEMICAL ENGINEERING METHODS AND TECHNOLOGY

PHOTOCHEMISTRY: UV/VIS SPECTROSCOPY, PHOTOCHEMICAL REACTIONS AND PHOTOSYNTHESIS

KAREN J. MAES
AND
JAIME M. WILLEMS
EDITORS

Nova Science Publishers, Inc.
New York

For permission to use material from this book please contact us:
Telephone 631-231-7269; Fax 631-231-8175
Web Site: http://www.novapublishers.com

Library of Congress Cataloging-in-Publication Data

Photochemistry : UV/VIS spectroscopy, photochemical reactions, and photosynthesis / editors, Karen J. Maes and Jaime M. Willems.
p. cm.
Includes bibliographical references and index.
ISBN 978-1-61209-506-6 (hardcover : alk. paper)
1. Photochemistry. I. Maes, Karen J. II. Willems, Jaime M.
QD714.P47 2011
541'.35--dc22

2011003548

Published by Nova Science Publishers, Inc. † New York

Contents

PREFACE

This new book presents current research in the study of photochemistry, including novel electron-transfer three-component visible light photoinitiating systems; photolabile molecules as light-activated switches to control biomolecular and biomaterial properties; organic photochemistry with computational methods; photoinduced transformation processes in surface waters and photochemical processes in needles of overwintering evergreen conifers.

Chapter 1 - Three-component photoinitiator systems generally include a light-absorbing photosensitizer (PS), an electron donor, and an electron acceptor. To investigate a key factors involved with visible light activated free radical polymerizations involving three-component photoinitiators and 2-ethyl-2-(hydroxymethyl)-1,3-propanediol triacrylate, the authorsused thermodynamic feasibility and kinetic considerations to study photopolymerizations initiated with polymethine dyes as the PS. The Rehm-Weller equation was used to verify the thermodynamic feasibility for the photo-induced electron transfer reaction. The key kinetic factors for efficient visible light activated initiation process are summarized in two ways: (i) to retard back electron transfer process and recombination reaction steps and (ii) to use a secondary reaction step for consuming dye-based radical and regenerating the original PS (dye).

This chapter presents the short review on the influence of a second co-initiator structure on the free radical polymerization ability of photoredox pairs composed of polymethine dye cation acting as an electron acceptor and *n*-butyltriphenylborate anion acting as an electron donor. It is also presented a very efficient concept to raise the photoinitiators activity by introducing of a second co-initiator to the two-component photoinitiating system being the polymethine borate salts. Three-component systems, which contain a light-absorbing species (dye), an electron donor (borate salt), and a third component (N-alkoxypyridinum salt, 1,3,5-triazine derivative, N-methylpicolinium ester, heteroaromatic mercaptan or cyclic acetal), have emerged as efficient, visible light sensitive photoinitiators of free radical polymerization. It was found that three-component photoinitiating systems are more efficient than their two-component counterparts.

The photochemical reactions (primary and secondary) occurring in the three-component photoinitiating system after irradiation with visible light were proposed basing on the laser flash photolysis.

The primary photochemical reaction involves an electron transfer from *n*-butyltriphenylborate anion to an excited singlet state of the dye, followed by the reaction of the second co-initiator with the resulting dye radical to regenerate the original dye. This reaction simultaneously produces a second radical, which undergoes decomposition yielding radicals active in the initiation of free radical polymerization.

The polymethine dyes are able to start a specific chain of an electron transfer reaction involving different additives, giving as a result one photon - two-radicals photochemical response.

Chapter 2 - From the very first use of photoinitiators in the crosslinking of polyethylene, benzophenone has established a reputation as an effective mediator in photoinduced modifications of polymeric materials. Most of the applications are based on the benzophenone's capability to abstract a labile hydrogen atom from the organic substrate when excited by the UVA radiation. These include the photocrosslinking and surface photografting, with the latter method aiming to modify the shallow properties of the substrates. In recent years, there has been an increasing interest in the understanding the benzophenone photophysics at the gas – solid (organic or inorganic) interface or inside the cavities of different types, which might pose some restraints on its mobility. It could be said that in benzophenone chemistry there are general rules, but there are also particular ways of interaction with various kinds of (macro)molecules. This paper summarizes some of the most interesting and successful applications of benzophenone photochemistry in materials science. The photoreactions induced by benzophenone are discussed in relation with the chemistry of the substrate.

Chapter 3 - Some photochemical methods of synthesis of nanomaterials has been proposed as alternative method in the preparation of diverse materials such as semiconductors and ceramics materials. These systems are of easy implementation and management, carried out under mild temperature conditions, utilizing simple precursor compounds showing good results in the photo-deposition of the resultant materials. Among these methods the authors have presented the photo-deposition in solid phase (thin films). This methodology it involves a complex precursor film is spun on to the substrate forming an optical quality film. The film is then irradiated and the precursor fragments generating metals, which remain in the film, and volatile organic by-products. When this process is carried out in the atmosphere air oxidation normally results in the formation of a metal oxide film. This methodology has been employed in the preparation of SnO and ZrO_2 thin films and their preliminary study as chemical sensor and as luminescence device respectively.

Chapter 4 - It is of great interest to develop technologies to dynamically control the properties of biomolecules and materials in order to conduct advanced studies in cell biology and to create new medical devices. Light has been considered an ideal external stimulus to exert such control since it can be readily manipulated spatially and temporally with high precision. To this end, photolabile molecules are often employed as light switches. These photolabile molecules or photocages are light-sensitive compounds which can be covalently bound to biomolecules or materials using a variety of functional groups and can be removed upon exposure to light. A variety of biomolecules such as peptides, proteins, and nucleic acids have been synthesized to incorporate photocages that initially render them inactive. Light exposure can subsequently re-activate these biomolecules at a particular time and area in space. The number of the caged biomolecules and their applications in the life sciences is ever-growing.

More recently, a number of studies have attempted to make creative use of these photolabile molecules by designing novel materials for biological applications that can respond to light in different manners. For example, materials have been created that can degrade upon light exposure in specific locations or patterns. Alternatively, solutions that form gels upon light exposure have been reported. Biomaterials have often been designed to release biologically important molecules or drugs upon implantation with the timing and location of release being of the utmost importance; the design of materials that can release their cargo upon light exposure is now being explored. In another avenue of material development, photolabile molecules have been used to form patterns of biomolecules on solid surfaces towards the invention of new biochips. Yet others have employed photocaged molecules to form patterns of live cells for biological study and tissue engineering applications. The ability to dynamically alter material properties to dictate gel formation, material degradation or the location of proteins and cells will ultimately bring about a new era of highly tunable materials for improved performance in a number of biological applications.

This review will first briefly overview some commonly used photolabile molecules and then examine some of the applications they have found to date in caging biomolecules. This will set the stage for a detailed discussion on how photolabile molecules are incorporated into materials for biological applications and the future potential of these novel strategies in biomedical engineering research and biotechnology.

Chapter 5 - Recent global warming is presumed to reduce productivity of plants near future through deterioration of photosynthetic activity. In particular, photosystem II (PSII) of higher plants is the most sensitive site affected by heat stress. Elevation of temperature causes structural damages in the PSII pigment-protein complexes due to impairment of D1 proteins and release of extrinsic proteins accompanied with degradation of manganese-calcium clusters, resulting in the loss of photosynthetic activities. In contrast, some prokaryotic phototrophs are moderate thermophiles growing at high temperatures over 50°C. Although bacterial photosystems are primitive adapted for ancestral environments, they can provide some implications on stability and tolerance under heat stress condition. In this article, the authors focus on photosynthetic organisms with type II reaction centers (RC), and discuss a relationship between structures and thermal stability and/or tolerance of the pigment-protein complexes, to find possible countermeasures for overcoming the global warming condition.

PSII is evolutionally related to purple bacteria because of similarities in their heterodimeric RC and quinone-mediated electron transport system. The purple bacterial photosystem lacks extrinsic proteins which are ubiquitously residing in the periphery of PSII, but has a circular pigment-protein complex, light-harvesting 1 (LH1), which is closely associating with the RC to maintain structures and functions of the photosystem. Among purple bacteria, *Thermochromatium tepidum* is a unique thermophile. The enhanced thermal stability is attributed to Ca^{2+}-induced structural changes of the LH1 complex at the peripheral region. In cyanobacteria, extrinsic proteins PsbO, PsbU, PsbV at the lumenal side of PSII play significant roles for regulation and stabilization of the water oxidation machinery. Deletion of these proteins resulted in deterioration of oxygen–evolving ability. Also in higher plants, PSII core is stabilized by extrinsic proteins PsbO, PsbP, PsbQ, and liability of interaction between the extrinsic proteins and PSII core under heat stress condition is thought to be a primary cause of heat-derived damage of photosystem. In particular, PsbO is most important for stabilizing the manganese-calcium cluster, and thus, commonly exists in all oxygenic phototrophs. Accumulated knowledge suggests that release of PsbO is critical for loss of oxygen evolution, and biochemical modifications of PsbO by ROS and lipid peroxides enhance its release from PSII core, especially under heated condition in the presence of light.

These findings indicate that stabilizing of RC is crucial for protecting photosystem from heat damage, therefore regulation of light is a possible means to alleviate the heat damage because light is the primary determinant to regulate redox state. In addition, to induce heat tolerance of higher plants, application of artificial chemicals is expected to convenient and effective tools. Several small peroxidized products from lipid peroxidation are shown to induce gene expression involved in heat tolerance. The authors named these chemicals as "environmental elicitor", and malondialdehyde and ethylvinylketone were screened as heat tolerance-inducing environmental elicitors protecting PSII from heat damage. Taken together, improvement of structure of PSII to stabilize RC, control of light condition, and use of environmental elicitors might be potential countermeasures to overcome heat stress.

Chapter 6 - In this chapter a brief description of the well-known wavelength effects in the provitamin D photochemistry which is the first stage of vitamin D synthesis initiated by the UV irradiation is presented. Unusual spectral kinetics revealed under irradiation of provitamin D with XeCl excimer laser at $\lambda = 308$ nm which conflicted with standard neglect of the weak irreversible channel is shown, and the origin of the observed anomaly is described in detail using simplified model and computer simulations. Particular attention is given to the original spectrophotometric analysis of the multi-component mixture of the vitamin D photoisomers which takes into account the irreversible photodegradation. Significant consequences of the effect revealed for the industrial synthesis of vitamin D and for biological UV dosimetry are discussed.

Chapter 7 - Research in photochemistry has substantially changed in the last decades. Some time ago, the focus of research on photochemistry was mainly the effect of light on molecules and the photochemical reactivity or photophysical properties caused by light excitation. More recently, while interest in photochemistry at the molecular level still remains, new and more complex applications to exploit light energy have been developed: photocatalysis, sensing and signaling, photoactivated molecular machines, photoprotection, biomedical and environmental uses and others. The rational design of molecules capable of performing such tasks depends on the knowledge of the mechanisms operating at the molecular level. Thus, much effort has been devoted lately to gather a deeper insight into the way these photoreactions occur.

In recent years, computational chemistry has emerged has an important tool for the detailed investigation of photochemical reactions. With the exponential increase of computer power, together with fundamental breakthroughs in the theoretical descriptions of photochemical reactions the scientific community now has the tools to explore and understand those organic photoreactions.

In this contribution, a brief introduction to the computational methods currently available to study photoreactions of organic substrates will be presented. The main target will be to introduce a non-specialist reader into the features of computational organic photochemistry. This research field has their own methods, concepts and tools, advantages and drawbacks, usually different from those found in relatively closer topics such as ground state computational chemistry or experimental photochemistry. Thus, theoreticians trained to study thermal reactions or experimental photochemists trying to complement their research will find here a starting point. However, it is far beyond the scope of this chapter to comprehensively explore the state-of-the-art in computational photochemistry. Instead, a general overview will be presented and the reader will be invited to follow the references for an in-depth treatment of some sections. Due to the optimum balance between accuracy and computational cost for medium-sized organic compounds, the strategy based in the CASPT2//CASSCF level of theory will be especially explained. Some case studies will also be discussed in order to provide the reader with a general idea of both the importance and capacity of these tools to get a comprehensive picture of relevant processes for photochemical reactions at the molecular-level. Limitations as well as some perspectives into the near future will also be discussed.

Chapter 8 - The authors review recent studies of molecular chirality using circularly polarized light, along with the birth and evolution of life and planetary systems. Terrestrial life consists almost exclusively of one enantiomer, *left-handed* amino acids and *right-handed* sugars. This characteristic feature is called homochirality, whose origin is still unknown. The route to homogeneity of chirality would be connected with the origin and development of life on early Earth along with evolution of the solar system. Detections of enantiomeric excess in several meteorites support the possibility that the seed of life was injected from space onto Earth, considering the possible destruction and racemization in the perilous environment on early Earth. Circularly polarized light could bring the enantiomeric excess of prebiotic molecules in space. Recent experimental works on photochemistry under ultraviolet circularly polarized light are remarkable. Asymmetric photolysis by circularly polarized light can work for even amino acid leucine in the solid state. Amino acid precursors can be asymmetrically synthesized by circularly polarized light from complex organics. Astronomical observations by imaging polarimetry of star-forming regions are now revealing the distribution of circularly polarized light in space. Enantiomeric excess by photochemistry under circularly polarized light would be small. However, several mechanisms for amplification of the excess into almost pure enantiomers have been shown in experiments. When enantiomeric excess of amino acids appears in the prebiotic environment, it might initiate the homochirality of sugars as a catalyst. Astrobiological view on chirality of life would contribute to understanding of the origin and development of life, from the birth to the end of stars and planetary systems in space. Deep insights on terrestrial life, extrasolar life, and the origin of life in the universe, would be brought by consideration of both the place where life is able to live, *the habitable zone*, and the place where life is able to originate, what the authors shall call *the originable zone*.

Chapter 9 - Photochemical processes are important pathways for the transformation of biologically refractory organic compounds, including harmful pollutants, in surface waters. They include the direct photolysis of sunlight-absorbing molecules, the transformation photosensitised by dissolved organic matter, and the reaction with photochemically generated transient species (*e.g.* $^{\bullet}OH$, $CO_3^{-\bullet}$, $^1O_2$). This chapter provides first an overview of the main photoinduced processes that can take place in surface waters, leading to the transformation of the primary compounds but also to the production of harmful intermediates. The second part is devoted to the modelling of the main photochemical processes, such as direct photolysis and reaction with hydroxyl and carbonate radicals, singlet oxygen and the excited triplet states of chromophoric dissolved organic matter.

Chapter 10 - Overwintering conifer tree species differ in ability to adapt photochemical processes to light environments from other taxonomic groups of plants. Contrastingly to deciduous trees, they maintain leaves for many years and they adapted to conduct photochemical processes in highly varying irradiance. An ability to photosynthesize in different light environments depends on a species' light requirements and its successional status. Moreover, shade-intolerant species which appear in succession early are characterized by greater efficiency of photochemical processes in high irradiance compared with shade-tolerant, late-successional species with a better photochemical performance in shade. Evergreen shade tolerant conifers are adapted to winter photoinhibition, however, their response to winter stress also depends on the light environment of growth. Seedlings growing in high irradiance show a greater winter photoinhibition and shade-acclimated ones only a small, fast-recovering decline in maximum quantum yield of photosystem II photochemistry (F_v/F_m). On the other hand, high light acclimated seedlings develop more efficient photoprotective mechanisms for excess energy dissipation.

In natural conditions under the canopy of trees or in artificial shading the parameters characterising photochemical processes such as effective quantum yield of PSII photochemistry (Φ_{PSII}) and apparent electron transfer rate (ETR) are modified by light regimes. Tree seedlings of the same species acclimated to high irradiance have lower F_v/F_m indicating the PSII down-regulation compared with shade-acclimated ones. This temporarily occurring phenomenon protecting photosynthetic apparatus against permanent damage in PSII is associated with dissipation of excess energy as heat in the xanthophyll cycle. This protective mechanism can be indirectly estimated using non-photochemical quenching of fluorescence (NPQ) which is greater in leaves exposed to high irradiance compared with shade leaves. Down-regulation of PSII photochemistry decreases ETR and overall intensity of photosynthesis, but at the same time a risk of irreversible photoinhibition is diminished. However, when a reduction of F_v/F_m is significantly below 0.8 and it is long-lasting, the photoinhibition of PSII may lead to a permanent dysfunction of the photosynthetic apparatus. PSII down-regulation was observed in a daily scale at midday, but also in a seasonal scale during winter and in early spring at chilling or freezing temperatures. In these situations the reversible photoinhibition can play a photoprotective role unless a cumulative stress of high light and low temperature prolongs and permanently damages the structural proteins in PSII.

Overwintering evergreen shade-tolerant conifers have the ability to adapt the photosynthetic apparatus to changing irradiance which involves some plasticity of their photochemistry and developing of photoprotective mechanisms. Due to these photochemical traits shade-tolerant species can efficiently compete with fast-growing, shade-intolerant pioneer species. High photochemical capacity and well functioning photoprotective mechanisms may give advantage in natural selection and be of critical importance in achieving an evolutionary success.

Chapter 11 - Chlorophyll in chloroplasts is the light-harvesting pigment of plant leaves. Chlorophyll is excited by light energy to reach excitation state. This excitation energy is de-excited by several processes *in planta*. Of the two photochemical apparatus of photosynthetic electron transport (photosystem I and photosystem II), rate constants of photochemical reactions around photosystem II can be measured with chlorophyll fluorescence, because it is mainly emitted from photosystem II. Fluorescence itself is one of the de-excitation processes. Internal conversion and intersystem crossing also dissipates excitation energy. These three processes are called as ‘basal dissipation’. In addition to basal dissipation, excitation energy is consumed to drive ‘photochemistry’ (photosynthetic electron transport). Under illumination, a mechanism called as ‘non-photochemical quenching’ is induced by biochemical mechanisms and dissipates excessive energy as heat.

Based on the Stern-Volmer relationship between fluorescence intensity and de-excitation rate constants, relative sizes of rate constants for these de-excitation pathways (basal dissipation, photochemistry and non-photochemical quenching) can be calculated by the technique called as Pulse Amplitude Modulation (PAM). In the PAM analysis of chlorophyll fluorescence, yield of fluorescence is measured with weak pulses of illumination. PAM technique enables measurement of fluorescence yield under both dark and illuminated conditions. Illumination of the leaf results in decrease of photochemistry and increase of non-photochemical quenching. Such changes are measured by several fluorescence parameters.

The principle equation to calculate relative sizes of rate constants from chlorophyll fluorescence intensities was presented by Kitajima and Butler. Since Kitajima and Butler, PAM was introduced and non-photochemical quenching was discovered, thus the principle equation was updated. This updated principle equation will be referred to as 'Kitajima-Butler equation' in this review. Many fluorescence parameters have been proposed from calculation of Kitajima-Butler equation. These parameters were first represented by complex formulas, and recent finding even showed that relative sizes of rate constants of all de-excitation pathways are easily described as the comparison between inverse values of fluorescence intensities.

PAM fluorescence analysis provides high-throughput method to estimate photosynthetic rate, damage by stressful conditions (photoinhibition) and size of non-photochemical quenching in living plant leaves. PAM fluorescence measurement is routinely used to estimate and visualize the effects of plant genes or stress treatments on these processes.

In: Photochemistry
Editors: Karen J. Maes and Jaime M. Willems
ISBN: 978-1-61209-506-6

Chapter 1

THE PHOTOCHEMISTRY OF NOVEL ELECTRON-TRANSFER THREE-COMPONENT VISIBLE LIGHT PHOTOINITIATING SYSTEMS

Janina Kabatc
University of Technology and Life Sciences, Faculty of Chemical Technology and Engineering, Seminaryjna, Bydgoszcz, Poland

ABSTRACT

Three-component photoinitiator systems generally include a light-absorbing photosensitizer (PS), an electron donor, and an electron acceptor. To investigate a key factors involved with visible light activated free radical polymerizations involving three-component photoinitiators and 2-ethyl-2-(hydroxymethyl)-1,3-propanediol triacrylate, we used thermodynamic feasibility and kinetic considerations to study photopolymerizations initiated with polymethine dyes as the PS. The Rehm-Weller equation was used to verify the thermodynamic feasibility for the photo-induced electron transfer reaction. The key kinetic factors for efficient visible light activated initiation process are summarized in two ways: (i) to retard back electron transfer process and recombination reaction steps and (ii) to use a secondary reaction step for consuming dye-based radical and regenerating the original PS (dye).

The paper presents the short review on the influence of a second co-initiator structure on the free radical polymerization ability of photoredox pairs composed of polymethine dye cation acting as an electron acceptor and *n*-butyltriphenylborate anion acting as an electron donor. It is also presented a very efficient concept to raise the photoinitiators activity by introducing of a second co-initiator to the two-component photoinitiating system being the polymethine borate salts. Three-component systems, which contain a light-absorbing species (dye), an electron donor (borate salt), and a third component (N-alkoxypyridinum salt, 1,3,5-triazine derivative, N-methylpicolinium ester, heteroaromatic mercaptan or cyclic acetal), have emerged as efficient, visible light sensitive photoinitiators of free radical polymerization. It was found that three-component photoinitiating systems are more efficient than their two-component counterparts.

The photochemical reactions (primary and secondary) occurring in the three-component photoinitiating system after irradiation with visible light were proposed basing on the laser flash photolysis.

The primary photochemical reaction involves an electron transfer from *n*-butyltriphenylborate anion to an excited singlet state of the dye, followed by the reaction of the second co-initiator with the resulting dye radical to regenerate the original dye. This reaction simultaneously produces a second radical, which undergoes decomposition yielding radicals active in the initiation of free radical polymerization.

The polymethine dyes are able to start a specific chain of an electron transfer reaction involving different additives, giving as a result one photon - two-radicals photochemical response.

ABBREVIATIONS

IR	Infrared
VIS	Visible
UV	Ultraviolet
CT	Charge transfer
PI	Photoinitiator
PS	Photosensitizer
TICT	Twisted intramolecular charge transfer
ET	Electron transfer
PET	Electron transfer process
D	Electron donor
A	Electron acceptor
TMPTA	2-Ethyl-2-(hydroxymethyl)-1,3-propanediol triacrylate
MP	1-Methyl-2-pyrrolidinone
B2	*n*-Butyltriphenylborate
NO	N-Methoxy-4-phenylpyridinium tetrafluoroborate
T	1,3,5-Triazine derivatives
RBAX	Rose bengal derivative
DMF	N,N-Dimethylformamide
MeCN	Acetonitrile

INTRODUCTION

Direct photoinduced polymerization reactions concern the creation of a polymer through a chain reaction initiated by light. Since a formation of reactive species from a monomer by direct light absorption is not an efficient route, the initiation step of polymerization reaction requires the presence of a photoinitiator (PI) which under light excitation, is capable of generating reactive species that can start polymerization chain reaction (Scheme 1).

The intrinsic reactivity of PI which plays an important role on curing speed is directly connected with its molecular structure, which governs the intensity of the light absorbed, the absorption wavelength range, the electron transfer reactions ability and the efficiency of the photophysical and photochemical processes involved in the excited states deactivation (which determines the yield of cleavage reactions, the rate of electron transfer reactions, the yield of the quenching by monomer, oxygen or other additives such as, hydrogen donors, light stabilizers, etc.) [1].

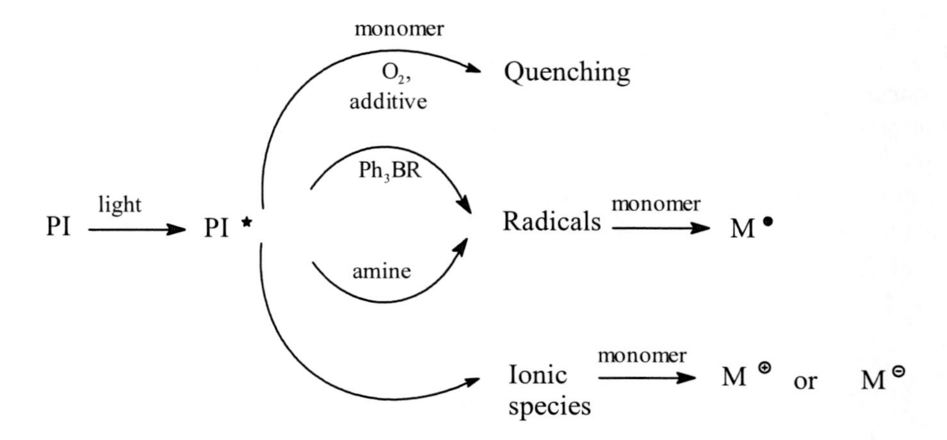

Scheme 1. Schematic presentation of reactions following after light absorption by a photoinitiator.

Talking into account the mechanism of free radical formation, the photoinitiators are classified as follows:

- The photodissociative photoinitiators (α-photodissociative, β-photodissociative, other possessing weak bonds in molecule such as: C-S, O-O, S-S, etc.).
- The ionic polymerization photoinitiators (cationic photoinitiators, anionic photoinitiators).
- The photoinitiators which generate free radicals in the hydrogen atom transfer reaction.
- The photoinitiators which generate free radical as a result of electron transfer process.

The photopolymerization is traditionally initiated by direct photolysis of a precursor to provide free radicals by bond photodecomposition. It needs the energy from the ultraviolet or blue region of a visible light. The panchromatic sensitization of vinyl polymerization requires the presence of suitable dye as a light absorber [1].

Typical photosensitive systems under visible light are classified in:

- One-component system (such as bis-acylphosphine oxides, iron arene salts, peresters, organic borates, titanocenes, iminosulfonates, oxime esters, etc.) [2].
- Two-component system (working, e.g. through electron transfer/proton transfer, energy transfer, photoinduced bond cleavage via electron transfer reaction, electron transfer).
- Three-component (where the basic idea is to try to enhance the photosensitivity by a judicious combination of several components. For example, ketone (benzophenone, thioxanthone)/amine/CBr_4 or organic bromo compound; ketone (benzophenone, thioxanthone, ketocoumarin)/amine/onium salt; dye (thioxanthene dye or eosin)/

amine/onium salt [3, 4]; dye/amine/ketone; dye/triazine/e-donor; dye/amine/ bromo derivative.

- Multi-component system (more than three partners) [3].

The initiation of free radical polymerization can occur in the presence of the dye alone (one-component photoinitiating system) or in the presence of the two-, three- or multi-component photoinitiating system. There are very often, two-component systems composed of dye molecule as a sensitizer and a second component as co-initiator (either as electron or hydrogen atom donor). There are only few dyes such as xanthene and acridine dyes which can directly photoinitiate polymerization of methyl methacrylate, styrene and acrylonitrile. The efficiency of this process is very low. Photoinitiation occurs by an electron transfer between the excited dye molecule and monomer (eqs. 1 and 2) [5, 6].

$$Dye^{\oplus}X^{\ominus} \xrightarrow{h\nu} {}^{1}(Dye^{\oplus}X^{\ominus})^{*} + R{-}CH{=}CH_2 \longrightarrow Dye^{\bullet} + [R{-}\dot{C}H{-}CH_2^{\oplus}X^{\ominus}] \quad \text{eq. (1)}$$

$$Dye^{\bullet} \xrightarrow{M} \text{Polymerization} \qquad [R{-}\dot{C}H{-}CH_2^{\oplus}X^{\ominus}] \xrightarrow{M} \text{Polymerization}$$

$$Dye^{\ominus}X^{\oplus} \xrightarrow{h\nu} {}^{3}(Dye^{\ominus}X^{\oplus})^{*} + R{-}CH{=}CH_2 \longrightarrow Dye^{\bullet} + [R{-}\dot{C}H{-}CH_2^{\ominus}X^{\oplus}] \quad \text{eq. (2)}$$

$$Dye^{\bullet} \xrightarrow{M} \text{Polymerization} \qquad [R{-}\dot{C}H{-}CH_2^{\ominus}X^{\oplus}] \xrightarrow{M} \text{Polymerization}$$

Electron deficient monomers (like acrylates) react preferentially as electron acceptors, whereas polymerization of electron rich monomers occurs by monomer oxidation [5].

Usually, photoinitiation of polymerization does not occur upon interaction of the excited singlet or triplet state of a dye [5]. Therefore, in most practical applications of photoinduced radical polymerization, the sensitizers are the UV-absorbing compounds that undergo unimolecular fragmentation in an excited state to form the initiating radicals. However, most of useful photoinitiating systems that respond to visible light are bimolecular systems. The initiation process involves a bimolecular electron transfer reaction between, for example excited dye (ketocumarin) and an electron donor (an amine) to form a radical anion and radical cation of an amine. Subsequent proton transfer results in radicals that are capable of initiating polymerization [5].

The efficient photosensitive systems for visible-laser induced polymerization reactions are of great practical use in the laser imaging, such as high speed photopolymers in computer-to-plate laser imaging systems, and the data storage area such as holographic recording systems, in high speed method of design and troubleshooting in a medicane using 3-D free radical polymerization. Low cost and long-lived of visible light sources in comparison with the UV-lamp, have made the visible light dyeing photoinitiation systems very attractive for use in commercial applications. Moreover, the visible-light induced processes are economical and environmentally friendly (no ozone emission and harmful effect on eyes).

Many attempts have been made to develop efficient photoinitiating systems that can be used upon visible light excitation. It seems rather difficult to develop initiators that generate an initiating radical by a bond cleavage upon direct excitation (type I) with a visible light. On the other hand, multi-component photoinitiating systems employing a dye as a visible-light absorber look more promising. In these systems, the photophysical energy transfer from the excited state of the dye to the second component that is supposed to yield a free radical is generally not favored in view of energetic relationship between the dye and the initiating part. Instead of an energy transfer process, a photoinduced electron transfer process plays an important role in such systems [7]. Photoinduced intermolecular electron transfer (PET), which is nonclassical, endothermic energy transfer process, uses light to initiate the electron transfer from a donor to an acceptor molecule. The process is possible because the electronically excited states are both better oxidants and reductants than their ground state equivalents [7].

Translating this to the sensitization of free radical polymerization, one can anticipate that two types of sensitization should occur:

1. Dye photoreduction leading to subsequent polymerization was first reported in 1954 when Oster identified several groups of dyes that are photoreduced in the presence of suitable reductants like ascorbic acid and triethanoloamine [8].
2. Dye photooxidation leading to the subsequent polymerization requires molecules that are strong electron acceptors in the ground state. Systems including dyes and N-alkoxypyridinium salts are excellent photoinitiators for the polymerization of acrylates.

Initiation of polymerization by photoreducible dyes in presence of different co-initiators (electron donors) is very common, whereas polymerization initiated by photooxidation of dye in presence of an electron acceptor is very rare.

The secondary reactions following electron transfer process are either proton transfer or a fragmentation reactions (cleavage). The fragmentation can occur as reductive cleavage or as oxidative cleavage, depending on whether the compound that undergoes the cleavage reaction has been initially reduced or oxidized.

The mechanism of the free radical polymerization initiated by dyeing photoinitiating systems via electron transfer process is shown in Scheme 2 [8].

$$A^* + D \underset{k_{-diff}}{\overset{k_{diff}}{\rightleftharpoons}} [A.....D]^* \underset{k_{-el}}{\overset{k_{el}}{\rightleftharpoons}} [A^{\bullet\ominus}......D^{\bullet\oplus}]$$

$$A + D \xleftarrow{k_r} \quad \xrightarrow[\text{reactions}]{\text{secondary}} \text{Polymerization initiation species}$$

$$A + D^* \overset{k_{diff}}{\rightleftharpoons} [A.....D]^* \overset{k_{el}}{\rightleftharpoons} [A^{\bullet\ominus}......D^{\bullet\oplus}]$$

Scheme 2. The mechanism of initiation of free radical polymerization by dyeing photoinitiating systems.

In Scheme 2 k_{diff} is the rate constant representing the rate of diffusion controlled encounters between reactants, k_{-diff} denotes the rate of separation of the reactants after collision, and k_{el} is the first order rate constant of electron transfer. The reverse step is designed by the rate constant k_{-el} and k_r denotes the rate of return electron transfer. The key steps of mechanism are the quenching of the chromophore excited state, either the excited singlet or triplet by electron transfer, and the various steps following the primary process (secondary reactions) [9].

It is generally accepted that the rate of electron transfer k_{el} depends not only on the free energy change but also on the distance between a donor and an acceptor. In the classical Marcus theory, the activation energy related to electron transfer reaction, $\Delta G_{el}^{\#}$, as it is shown in eq. (3) is a function of two parameters.

$$\Delta G_{el}^{\#} = \frac{\lambda}{4}\left(\frac{\lambda + \Delta G_{el}}{\lambda}\right)^2 \qquad \text{eq. (3)}$$

The first one is the free energy change for electron transfer reaction (ΔG_{el}) and second is the reorganization energy related to the entire nuclear reorganization of the electron transfer reaction (λ), involving (being the sum) both the geometry change of the reactants (λ_s). λ_s measures the energy change as the solvent dipoles change, and for spherical molecules this change can be expresses by eq. (4) first used by Marcus.

$$\lambda_s = \Delta e^2\left(\frac{1}{2r_D} + \frac{1}{2r_A} - \frac{1}{d_{DA}}\right)\left(\frac{1}{\varepsilon} - \frac{1}{n^2}\right) \qquad \text{eq. (4)}$$

where e, ε and n have the conventional meaning, r_D and r_A are the radius of donor and acceptor molecules, respectively and d_{DA} is the distance separating donor and acceptor. From both equations it is clear that the rate of electron transfer reaction depends on the distance separating donor and acceptor in a complex fashion. However, it is visible that the activation energy of ET reaction decreases as distance separating donor and acceptor is increasing, namely, the rate of the ET reaction is higher when the distance separating donor and acceptor is lower. For example, Turro had calculated the distance dependence of energy-transfer by using a simple model, and showed that the rate decreases by 10^3 when the edge-to edge distance increased from 10 Å to 15 Å [7, 10]. Oevering and co-workers demonstrated experimentally in their bridged donor-acceptor molecules that by increasing the donor-acceptor centre-to-centre separation from 11.5 Å to 13.5 Å, the rate of electron transfer decreased about 10 times in benzene [7, 1]. Hence, it is concluded that controlling a donor-acceptor distance is one of the determining factor for the rate of electron transfer reaction, and with those points in mind, it is necessary to design photopolymer system where high sensitivity is an indispensable prerequisite. There are three ways to decrease the distance between donor-acceptor as shown in Scheme 3 [7].

One of the most common and simplest way is to increase the concentration of an electron donor (co-initiator). Generally, the average distance (R_0) separating molecules in solution is proportional to the cubic root of molecules concentrations ($R_o \propto [C]^{1/3}$) [11].

The second way is to use the electrostatic interactions between a donor and an acceptor molecules. Gottschalk and co-workers reported cyanine dye-borate salt photoinitiating systems in which cationic dye sensitizer and anionic borate are close together in non-polar medium, thus performing higher photoinitiating ability [7, 12, 13]. The third way is to link a donor and an acceptor by a covalent bond. For example, photoinitiating systems composed of neutral merocyanine dye covalent linked with 2,4-bis(trichloromethyl)-1,3,5-triazine (1) described by Kawamura [7] and photoinitiating systems possessing thioxanthone linked by covalent bond with hydroxyalkyl phenyl ketone (2) [14].

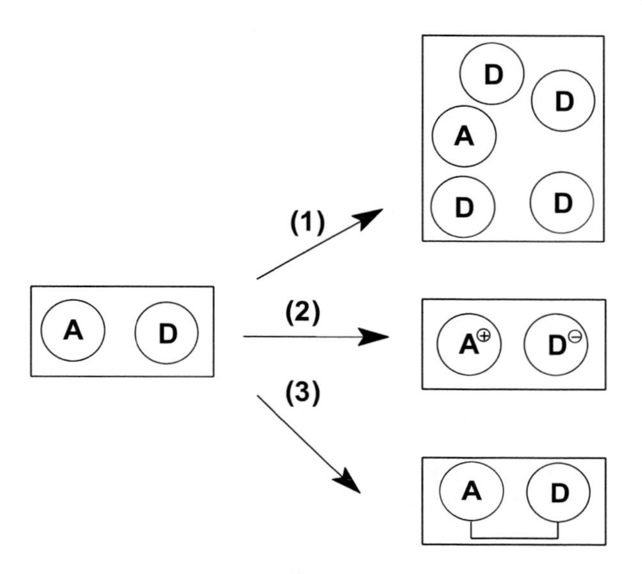

Scheme 3. Three possible ways of decreasing the distance between donor and acceptor in photoinitiating systems: (1) increase of concentration, (2) the electrostatic interaction between donor-acceptor (ion pairs), (3) linkage of donor-acceptor by a covalent bond [7].

(1) (2)

The new photoinitaiting systems composed of a benzothiazolylidene-rhodanine merocyanine (1) as a sensitizer and a substituted bis(trichloromethyl)-1,3,5-triazine (T) as a radical generator are widely employed as co-initiators in sensitized photopolymerization systems. In such photoinitiating systems, merocyanine chromophore and a bis(trichloromethyl)-1,3,5-triazine were linked by a different number of methylene chains [7, 15-17].

The dye-linked photoinitiators, in which a radical generating pair is brought close to photosensitizing dye-chromophore by a covalent bond, have been shown to be very useful as free radicals sources.

The photoreducible dyes possess maximum absorption in the visible light region. A suitable dye/co-initiator system must first exhibit a high absorption in the wavelength delivered by visible light sources (lamps, lasers) and, secondly, efficiently generate the reactive initiating radicals [18].

The cyanine dyes are an example of photoreducible sensitizers in two-component photoinitiating systems for free radical polymerization of multifunctional monomers acting via electron transfer process. The photoinitiating ability of such two-component photoinitiating systems have been mentioned in several papers [1, 12, 13, 19-34].

The cyanine dyes called also as polymethine dyes possess the conjugated double bond chain as a chromophore. Generally, the cyanine dye is defined as being built up of two nitrogen-containing ring systems in one of which the nitrogen atom is tercovalent and in the other it is tetracovalent; the two nitrogen atoms are linked by a conjugated chain of an uneven number of carbon atoms (polymethine chain) and belonging to two heterocyclic rings (Formula 3 and 4).

(3) **(4)**

There are many nitrogen-heterocyclic rings, which may be presented in cyanine dye structure. For example: quinoline, pyridine, indole, benzothiazole, benzoxazole, benzselenazole and other heterocyclic derivatives. The two ring systems may be directly linked by a bond between carbon atoms, or the union may be achieved by means of –CH= group or –CH=CH-CH= group or a longer conjugated chain containing an uneven number of carbon atoms (up to 13 methine group (–CH=)).

There are following classes of polymethine dyes:

- cyanine dyes (symmetrical), which have two nitrogen-heterocyclic rings joined by a conjugated chain of carbon atoms
- hemicyanine dyes (unsymmetrical), which have noncyclic end groups.

Structures presented in Formula 5 and 6 exemplify the cyanine and hemicyanine dyes.

where Z = O, S, $C(CH_3)_2$ **where Z = O, S, $C(CH_3)_2$**

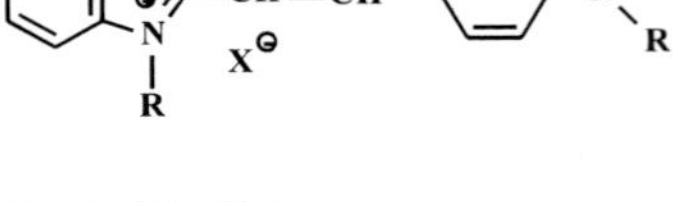

(5) **(6)**

The characteristic of polymethine dye is the positive charge of a chromophore. Therefore, these compounds are present as different salts (for example, thiocyanates, sulfides, chlorides, bromides or as difficult soluble iodides).

Because of various physical, chemical and photochemical properties, the polymethine dyes today are commonly used in numerous applications over wide areas of chemistry and modern technology [35-39]. One of the most important application of polymethine dyes are use them as saturable absorbers for lasers and fluorescence probes [40-42]. Taking into account, the possibility of formation of the stable salts with organic borate anion, the polymethine dyes are commonly used in polymer chemistry, as an effective photoinitiators of free radical polymerization [21-23].

Organoborate salts of carbocyanine dyes, described by G.B. Schuster in 1988, are the first commercial used photoinitiating systems operating in the visible light region. They are an example of the photoinitiators in which an electron transfer process occurs. In this photoinitiating system the cyanine dye acts as primary absorber of a visible light (sensitizer), but borate anion acts as an electron donor (co-initiator). The initiation of free radical polymerization by cyanine dye borate salt occurs via mechanism of photoreducible sensitization (Scheme 4).

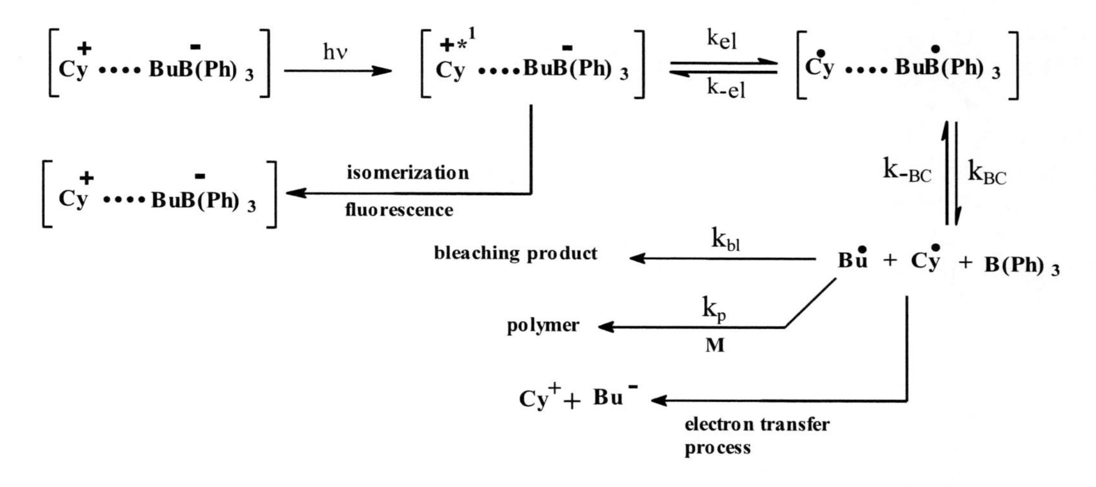

Scheme 4. The mechanism of processes occurring during the free radical polymerization initiated by cyanine borates photoredox pair.

The irradiation of cyanine borate salts (for example, *n*-butyltriphenylborate) leads to the formation of an excited singlet state of cyanine dye. Deactivation of an excited singlet state of cyanine dye may occur as a radiation process (fluorescence) or as a nonradiation process (isomerization or electron transfer process (k_{el})). In a presence of alkyltriphenylborate salts, one of the deactivation process is an electron transfer reaction from borate anion to an excited state of a day. As a result the cyanine dye radical ($Cy^{\bullet}$) and unstable boranyl radical are formed. The boranyl radical can undergo the C-B bond cleavage giving triphenylboron and alkyl radical (k_{BC}). The cyanine dye radical and alkyl radical formed can undergo recombination reaction leading to the formation of bleaching product (k_{bl}). In the radicals pair: cyanine radical/boranyl radical the return electron transfer process can occur (k_{-el}) giving dye cation and borate anion. In the case, when stable alkyl radicals are formed, the carbon-boron bond cleavage is very fast and irreversible process (250 fs).

From our studies it is known that for cyanine borate salts in some cases there is direct relationship between the rate of free radical polymerization and the rate of electron transfer process [23, 26, 28-30]. The increase of the rate of electron transfer process causes in the increase of the rate of free radical polymerization. The high values of the rate constants obtained for polymethine dye-borate ion pairs (about 10^{12} $M^{-1}s^{-1}$) are the evidence that the reaction of electron transfer occurs mostly in an intra-ion-pair assembly. The ion pair formation, therefore the small distance between cyanine cation (electron acceptor) and borate anion (electron donor) ensures an effective electron transfer process. Since the lifetime of an excited singlet state of polymethine dye is very short (about or less than 0.5 ns), therefore the formation of the tight ion pair increases the efficiency of electron transfer process [28, 34].

According to G.B. Schuster's et al. studies on symmetrical cyanine borate initiators, in non-polar or medium polarity solvents one can treat cyanine cation and borate anion as an ion pair what affects on the efficiency of electron transfer process [12, 13]. In the simplest way, the formation of contact ion pair can be enhanced by an increase of electron donor ion concentration (co-initiator) [43]. In the polymerizing mixture (acrylates, medium polarity solvents), the photoinitiating photoredox pair exists as a tight-pair and solvent separated ions. Since an efficient electron transfer between an excited singlet state of polymethine cation and borate anion is efficient only between paired components of photoredox couple, it is obvious that a dissociation of ion pair causes sharp decrease in efficiency of electron transfer. The increase of degree of dissociation of polymethine dye borate salt makes difficult the formation of an encounter complex and, hence decreases the efficiency of electron transfer process. Finally, this decreases the rate of polymerization photoinitiation.

The study on the influence of borate salt concentration (electron donor) on the rate of photoinitiated polymerization reveals a distinct increase in the rate of photopolymerization as the concentration of borate anion increases (Figure 1).

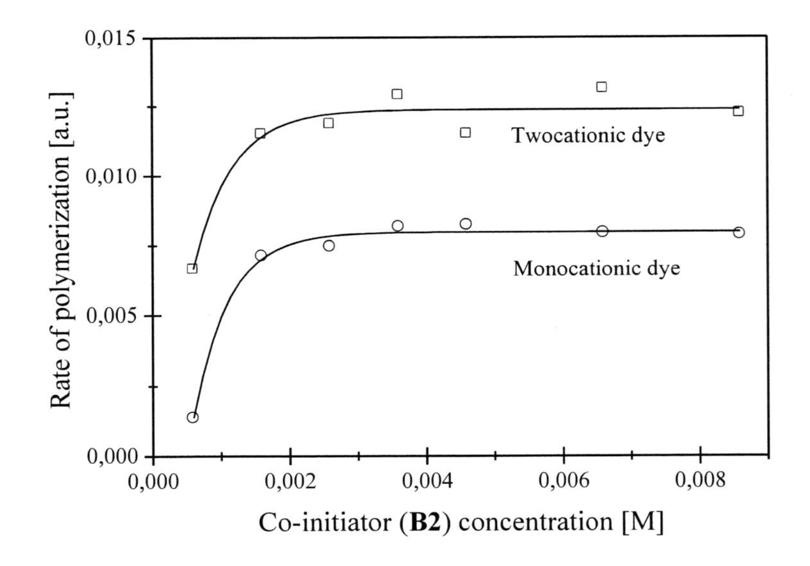

Figure 1. Dependence of the rate of photoinitiated polymerization on the concentration of the electron donor (B2). The photoinitiating system concentration was 7.5×10^{-4} M [20].

In the polymerizing mixture composed of 9 mL of 2-ethyl-2-(hydroxymethyl)-1,3-propanediol triacrylate (TMPTA), 1 mL of 1-methyl-2-pyrrolidinone (MP) and equimolar amounts of borate salt and sensitizer, only a part of photoredox couples exist as ion pairs (75 % and 20 % in the case of two- and mono-cationic dyes, respectively) [20]. In contrast to two-cationic hemicyanine dyes, mono-cationic unsymmetrical polymethine dye borate salts are almost completely dissociated in the polymerizing mixture [20]. It is obvious, that the additional amount of borate anion in the polymerizing composition shifts the equlibrium between free ions-ion pairs to a higher ion pair concentration.

The supporting experiments that can clarify the presence of both dissociated and undissociated forms of the tested initiators may come from the fluorescence quenching experiments. The absorption spectra measurements show that the exchange of iodide anion on borate anion has no effect on the electronic absorption spectra. Thus, if there is a certain contribution of a ground state equilibrium between cyanine cation (Cy^+) and borate anion ($B2^-$) leading to the formation of a non-fluorescence ion pair, then the evidence for the ground state ion pair formation may come from the fluorescence intensity quenching measurements. To estimate the contribution of the quenching deriving from the ground state ion pair (static quenching, K_S) and diffusion controlled quenching (dynamic quenching, $K_D = k_q \times \tau^0$) terms, the fluorescence quenching data for selected photoredox pair were analyzed by combining of both effects that can be described by equations (5 and 6).

$$\frac{I_0}{I_f} = (1 + K_D[Q]) \times (1 + K_S[Q]) \qquad \text{(eq. 5)}$$

where

$$K_S = \frac{[Cy^+.....B^+]}{[Cy^+] \times [B^-]} \qquad \text{(eq. 6)}$$

and:

[Q] – quencher concentration,
K_S – static quenching,
K_D – dynamic quenching.

The result of such a treatment allows to separate both type of quenching and calculate the K_S value. The stady-state fluorescence quenching data for selected polymethine dyes in the presence of *n*-butyltriphenylborate (Figures 2 and 3) leads to the conclusion that an upware curvage for the high concentration of quencher indicated the existence of both forms of polymethine borates, e.g. the form of the free ions (dissociated) and the form of the ion pairs (undissociated).

The lack of a linear relationship observed in high quencher concentrations region indicates that the quenching of fluorescence occurs by both dynamic and static mechanism. However, the fluorescence quenching of hemicyanine dye (2-methylbenzoxazole derivative) by *n*-butyltriphenylborate salt presents the classical linear Stern-Volmer relationship [19]. From the Stern-Volmer relationship obtained in Figures 2 and 3, one can calculate, based on eqs. 5 and 6, dissociation constant photoredox pairs. Assuming that for a low concentration of borate anion only dynamic quenching occurs, from the linear relationship observed in a low concentration region, one can estimate the K_D value. For tested photoredox pairs this parametr is equal 16.6 M^{-1}. The introduction of this value in eq. 6 gives K_S equal 60.58 M^{-1} and this, in turn, allows to calculate a degree of dissociation selected photoredox pair equal

92-93 % (dissociation constant K = 1.7 × 10^{-2} M). The presented above analysis clearly shows that the hemicyanine borates probably exist in a highly dissociated form, namely as solvent separated ions pair.

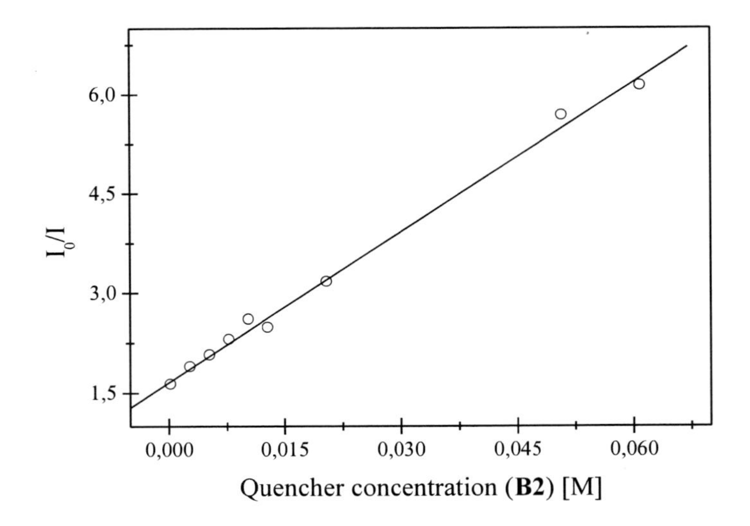

Figure 2. Stern-Volmer plot for fluorescence quenching of N-ethyl-2-(*p*-pyrrolidino)styrylbenzoxazolium iodide by tetramethylammonium *n*-butyltriphenylborate in ethyl acetate [19].

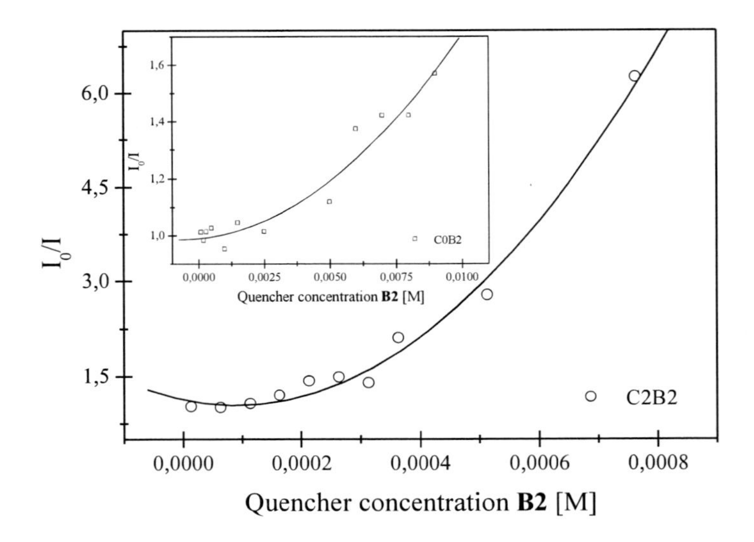

Figure 3. Stern-Volmer plot for fluorescence quenching of 1-[3-[2-(4-N,N-dimethylamino)styrylbenzothiazolo]-3-(4-methylpyridine)propane dibromide (circles) and N-ethyl-2-(4-N,N-dimethylamino)styrylbenzoxazolium iodide (squares) by tetramethylammonium *n*-butyltriphenylborate in ethyl acetate [20].

The cyanine dye borate salts are effective photoinitiating systems for free radical polymerization. However, their photoinitiating abitity is lower than the photoinitiating ability of dyeing photoinitiating systems in which after irradiation with visible light the long-lived excited triplet state of dye is formed [34].

Generally, the photoinitiation ability of photoinitiators composed of cyanine dye borate salt depends on the:

- The structure of polymethine dye (symmetrical or unsymmetrical)
- Type of heteroatom in heterocyclic ring
- Type of substituent in the "*meso*" position of polymethine chain
- Type of substituent in the phenyl ring
- The structure of an alkylamino group in the styryl moiety (unsymmetrical dyes)
- Structure of borate anion [26, 34].

The fluorescence quenching results and the relationship between the electron donor concentration and the rate of photoinitiation clearly show the need of an artifical increase of an electron donor (co-initiator) concentration in close proximity to a polymethine dye moiety. This is possible by chemical modification of sensitizer structure. Traditionally, such situation is achieved by attaching of an organic cation (pyridinium cation) that can form an ion pair with borate anion to an absorbing chromophore [20]. In cationic dyes this is possible by covalent bonding an extra organic cation that is not a part of chromophore, which can form an ion pair with an electron donor anion. The other possible way of the enhancement of photoinitiating systems efficiency is the addition to the polymerizing formulation a third component that can participate in a consecutive reaction yielding one more radical.

Taking this into account, new visible light two- and three-component photoinitiating systems are composed of mono- and multi-cationic polymethine dyes (cyanines and hemicyanines) acting as a sensitizer and alkyltriphenylborate salt as a co-initiator.

The continuous search for very fast acting and more efficient photoinitiators operating in the visible light region inclines us to continue the studies on the improvement of cyanine borate ion pairs as photoinitiators. Such improvement can be achieved by a modification of a structure of the light absorbing molecule.

Therefore, the novel conception of our studies was the description both of the kinetic and thermodynamic parameters for the electron transfer process and structure of donor-acceptor pair (dye structure and second co-initiator structure) on the efficiency of initiation of free radical polymerization. Second part was concerned on the determine a mechanism of primary and secondary reactions, which occur in the three-component photoinitiating systems after irradiation with visible light. In such three-component photoinitiating systems the polymethine dye acts as a primary absorber of visible light, borate anion and third component (for example, N-alkoxypyrdinium salt, 1,3,5-triazine derivative, N-methylpicolinium ester, cyclic acetal or heteroaromatic thiol) plays a role of an electron donor and an electron acceptor, respectively.

In present chapter the intermolecular photoinduced electron transfer phenomenon that leads to the photoinitiation of polymerization via chromophore photoreduced excited states is described. The description is focused on the important role played by a photoredox pair component, which does not absorb a light. In the literature this component is commonly called co-initiator. It is obvious that understanding the complexity of the processes of photoinitiated polymerization requires the thorough analysis of the examples illustrating the mechanistic aspects of the formation of free radicals that start the polymerization, especially the mechanism of free radicals formation describing the transformations of an electron donor.

The structures presented in Chart 1 exemplified polymethine dyes investigated.

Chart 1

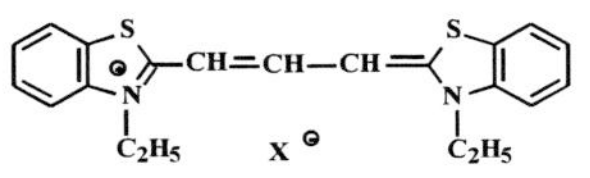

(Dye I) 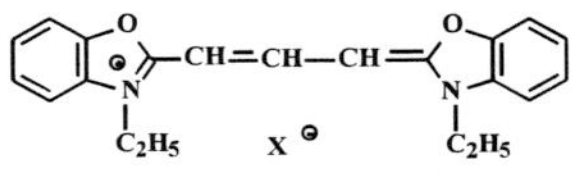 **(Dye II)**

Mono-cationic hemicyanine dyes

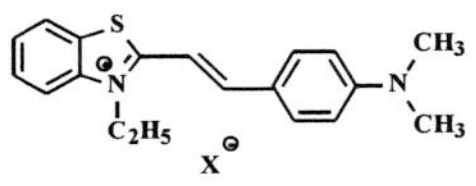

(Dye III) 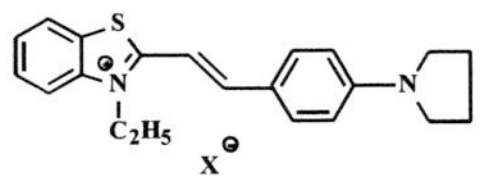 **(Dye IV)** 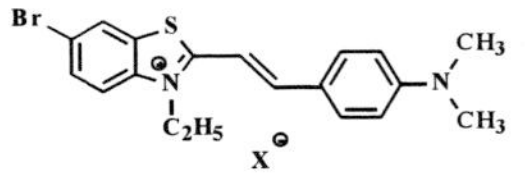 **(Dye V)** 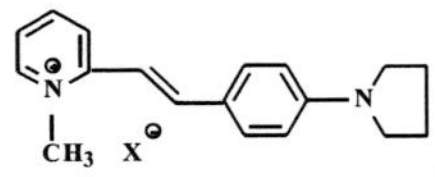 **(Dye VI)**

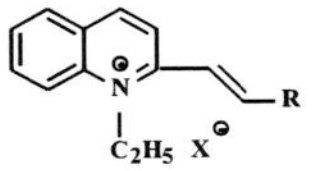

Where R:

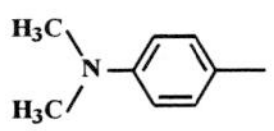

(Dyes VII)

(A) 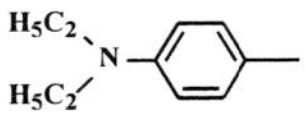 **(B)** 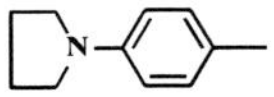 **(C)** 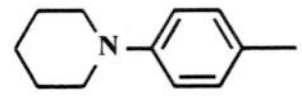 **(D)**

Modified structure cyanine dyes

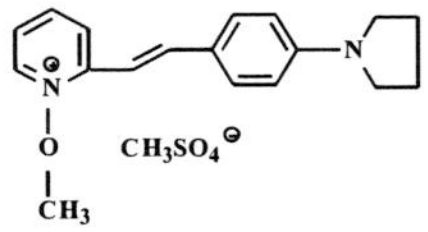

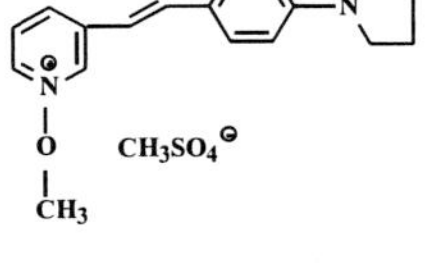

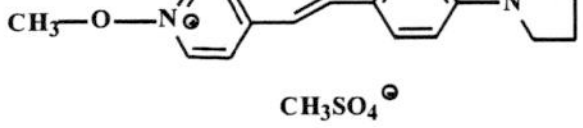

(Dyes VIII)

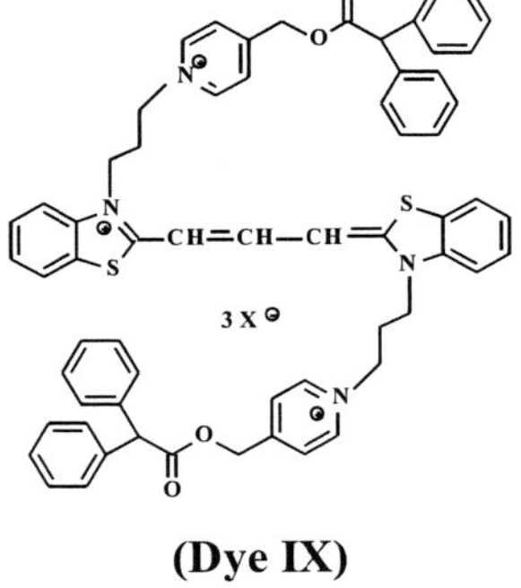

(Dye IX) 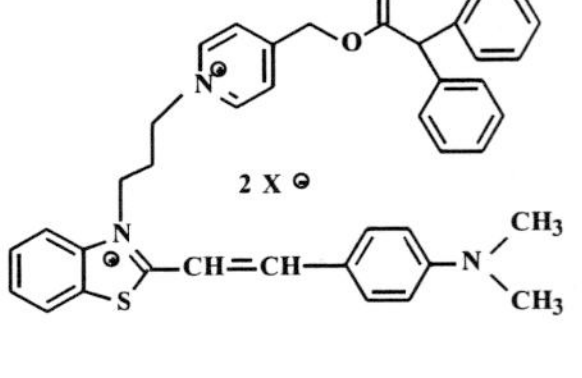 **(Dye X)**

Chart 1. (continued)

Three-cationic carbocyanine dyes

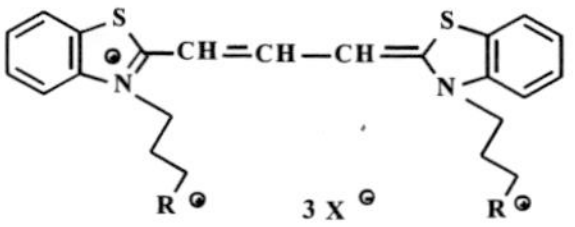

S —CH=CH—CH= S N N R⊕ 3 X⊖ R⊕

O —CH=CH—CH= O N N R⊕ 3 X⊖ R⊕

—CH=CH—CH= N N R⊕ 3 X⊖ R⊕

(Dye XI) 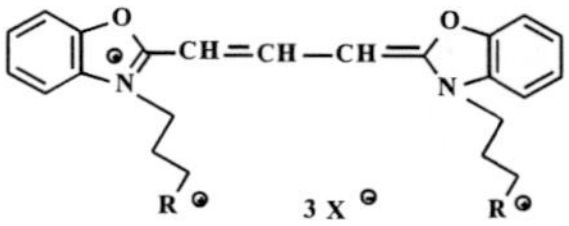 **(Dye XII)** 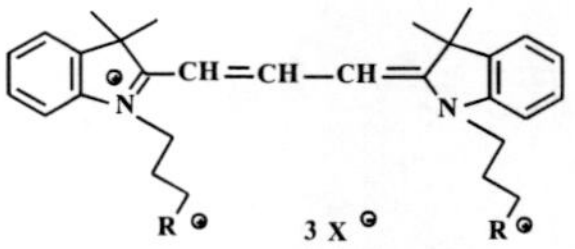 **(Dye XIII)**

Where R:

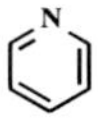

N

N CH₃

H₃C N

H₃C N O

A  B 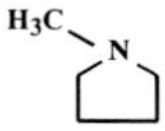 C 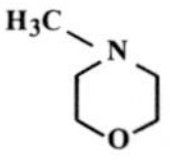 D

Monomethine dyes

Mono-cationic monochromophoric monomethine dye

O N —CH= N Br CH₃ X⊖

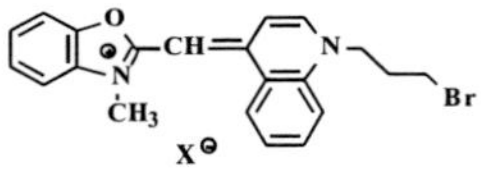

(Dye XIV)

Two-cationic monochromophoric monomethine dye

N⊕ S N N CH₃ 2X⊖

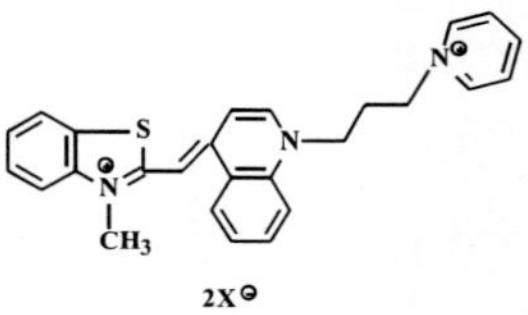

(Dye XV)

Three-cationic monochromophoric monomethine dye

S N N H₃C—N⊕ Cl H₃C—N⊕ O 3 X⊖

S N N H₃C—N⊕ O Cl N⊕ 3 X⊖

S N N CH₃ H₃C—N⊕ OH Cl H₃C—N⊕ O 3 X⊖

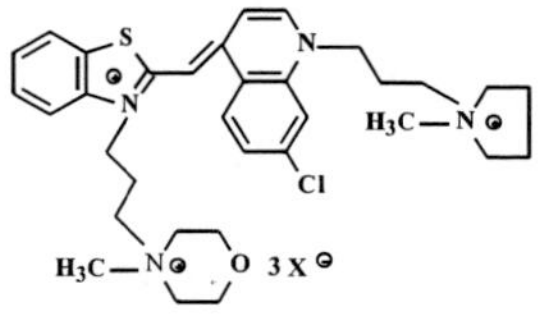

(Dye XVIA) 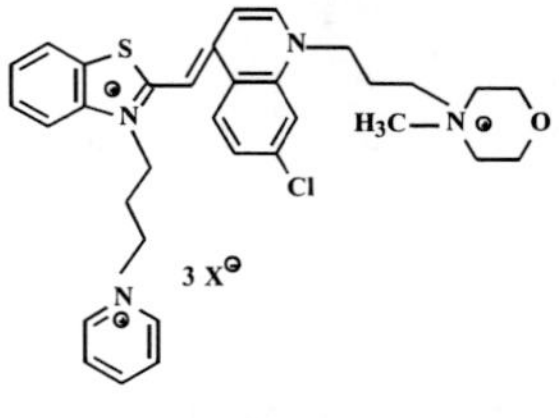 **(Dye XVIB)** 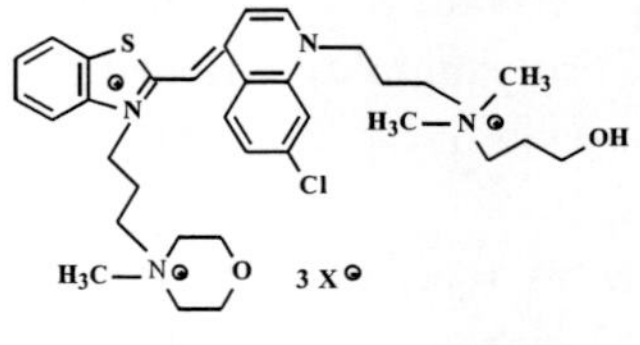 **(Dye XVIC)**

Two-chromophoric four-cationic monomethine dyes

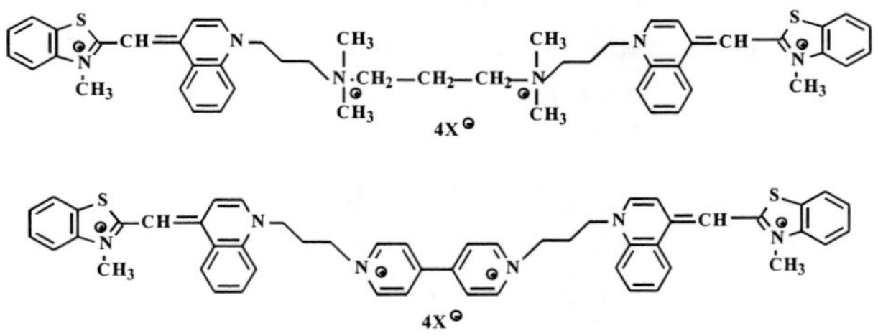

S —CH= N CH₃ CH₃ N⊕·CH₂—CH₂—CH₂—N⊕ CH₃ CH₃ CH₃ N =CH— S N CH₃ 4X⊖

S —CH= N N⊕ N⊕ N =CH— S N CH₃ CH₃ 4X⊖

(Dyes XVIIA and XVIIB)

Chart 1. (continued)

Two-cationic two-chromophoric hemicyanine dyes

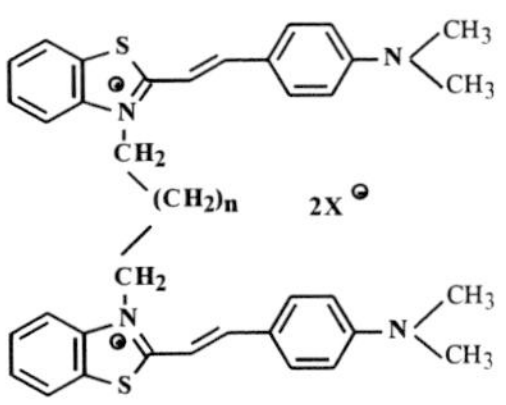

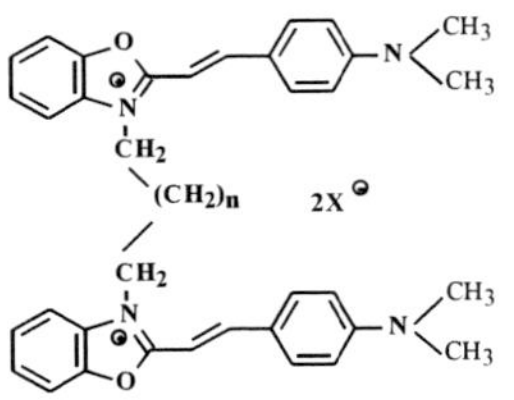

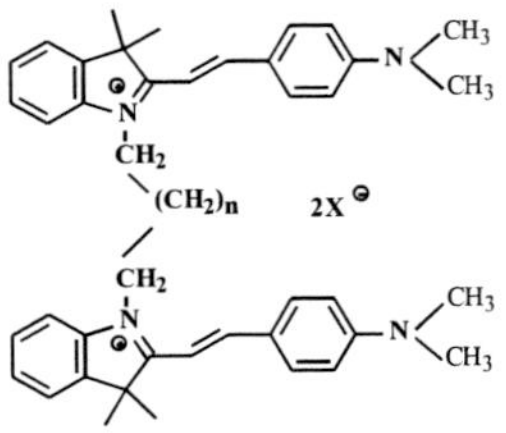

n = 3 or 5

(Dye XVIIIA) **(Dye XVIIIB)** **(Dye XVIIIC)**

Where X:

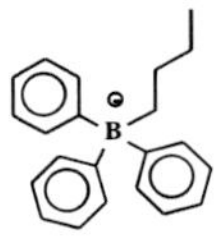

(B2)
***n*-butyltriphenylborate anion**

THREE-COMPONENT PHOTOINITIATING SYSTEMS

The visible-light activated initiators are typically two-component photoinitiating systems composed of a light absorbing PS and an electron-donating molecule. The intrinsic characteristics of two-component photoinitiating system lead to numerous kinetic limitiations [44]. For example, bimolecular quenching processes: the quenching of an excited triplet state by oxygen or by monomers can stop the initiation step and as a result in the lost of absorbed energy.

Moreover, since the back electron transfer step is invariably thermodynamically feasible, back electron transfer and radical recombination decrease the potential concentration of free radical active centers. Furthermore, an inefficient radical is often produced simultaneously in this electron transfer/proton transfer reaction step because the dye-based radical is not active for initiation but is able to terminate a growing polymer chain. These cumulative effects significantly limit polymerization kinetics of two-component photoinitiating systems and tend to make visible light photoinitiators less attractive than UV photocurring in applications where reaction rate is a primarily consideration [44].

Three-component photoinitaiting systems usually can be excited by visible light and have been found to be faster and more efficient than their two-component counterparts [45]. The three-component photoinitiating systems for free radical polymerization were rarely described in the literature. Among few, there are following examples:

- Ketone/amine/CBr_4;
- Ketone/amine/onium salt;
- Thioxanthene dye/amine/onium salt;
- Ketone/amine/ferrocenium salt;

- Ketone/bis-imidazole derivatives/thiol derivatives;
- Thioxanthene dyes/amine/1,3,5-triazine derivatives;
- Thioxanthene dyes/N-alkoxypyridinium salt/borate salt [3, 4, 46].

Like the two-component systems, the three-component photoinitiating systems include a light-absorbing moiety, which is typically a dye; an electron donor (DH) which very often is amine. A third component is usually an electron acceptor (EA) such as an iodonium or sulfonium salt or 1,3,5-triazine derivative [44]. In these systems, the third component is supposed to scavenge the chain-terminating radicals that are generated by the photoreaction between the other two components or produce an additional initiating radical (Scheme 5) [24, 47, 48]

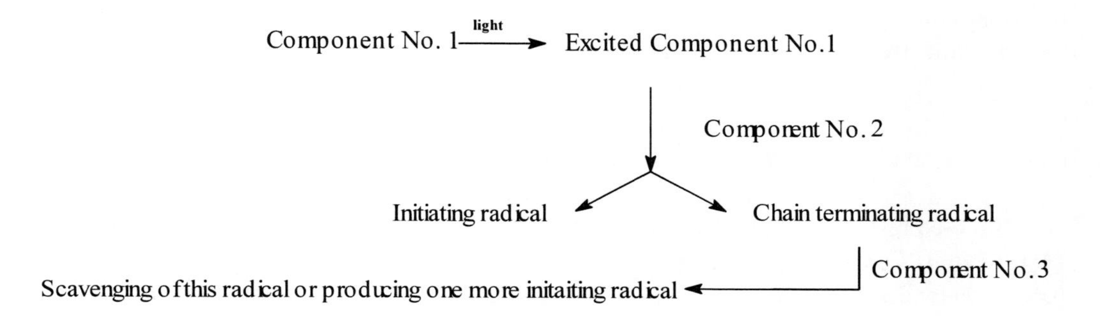

Scheme 5. The formation of the initiating radicals in the three-component photoinitiating system.

A number of hypotheses have been presented in the literature to explain the enhanced photoinitiation ability of three-component system as compared to their two-component counterparts [49].

In the basic concept developed several years ago in a three-component systems I/II/III working through electron transfer (Scheme 6) the light is absorbed by I and radicals are formed through I/II interaction. Radicals $R_i^{\bullet}$ playing a detrimental role (through a reaction with the growing macromolecular chains) are quenched by III. The use of this suitable quencher III allows to scavenge the $R_i^{\bullet}$ radicals and, if possible, to generate new initiating radicals through an electron transfer reaction. It is apparent that R_p will increase since k_t, the rate constant of termination decreases and new initiating species are generated from the deactivation process of the side radicals $R_i^{\bullet}$ (Scheme 6).

Classes of dyes that have been reported for three-component systems include ketones, xanthenes, thioxanthenes, coumarins, thiazines, cyanines, hemicyanines, merocyanines [3]. The photochemical behavior of different three-component systems such as benzophenone/amine/iodonium salt, thioxanthone/amine/CBr_4, ketocumarin/amine/iodonium salt/, eosin/amine/iodonium salt, thioxanthene dye/amine/iodonium salt or hemicyanine dye/borate salt/N-alkoxypyridinium salt have already been studied and discussed in several articles [24, 50, 51].

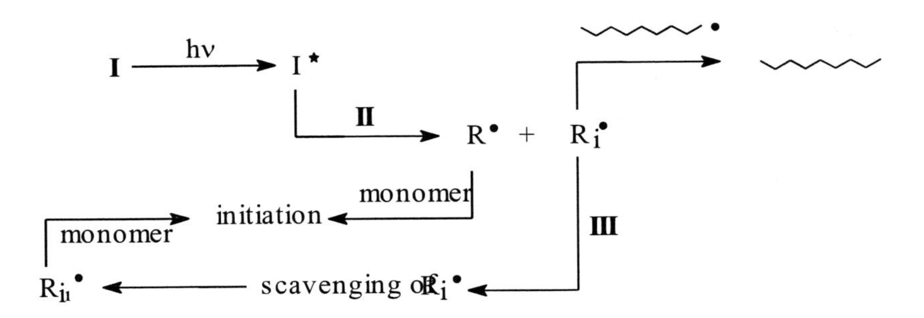

Scheme 6. The role of third component in three-component photoinitiating system.

The polymethine dyes have not been well studied by means of their three-component photoinitiating systems. The few mechanistic studies carried out have applied neutral merocyanine dyes almost exclusively [15].

Three-Component Photoinitiating Systems. Review

Three-component free-radical photoinitiating systems, first reported by Rubin [52], Monroe and Weiner [53], involve the production of radical species via a hydrogen-transfer process between an excited-state sensitizer (benzophenone) and a hydrogen donor (isopropyl alcohol) followed by an electron-proton transfer to a third component (camphorquinone) to produce a second radical which initiate free radical polymerization [54].

The electron transfer reactions in the three-component systems composed of cyanine dye as J aggregates (sensitizer), viologen acting as an electron acceptor occurring in a monolayer assembly (N,N'-dioctadecyl-4,4'-bipyridile) were described by Thomas L. Penner in 1982. In the three-component system composed of: an electron donor, J aggregates of cyanine dye and an electron acceptor, the fluorescence of sensitizer is quenched by both an electron donor and an electron acceptor. The introdution of an electron donor to the two-component systems substantially increases the concentration of viologen radicals formed as a result of electron transfer from an excited state of cyanine dye on the ground state of viologen (supersensitization). No supersensitization occurs when the dye is not J aggregated. Photooxidation and photoreduction of an excited cyanine dye were demonstrated in a monolayer assembly through fluorescence quenching when an electron acceptor or an electron donor, respectively, was incorporated into the adjacent layer in head-to-head contact. The reduced electron acceptor (viologen radical, lifetime about few minutes) has been detected by laser flash photolysis and ESR. The yield of electron transfer can been enhanced or supersensitized by the incorporation of an electron donor as a third component into the monolayer assembly containing dye and viologen. For such supersensitization to occur the dye molecules must be organized into aggregates. In such cases, the diffusion process has not effect on the electron transfer process [55].

The free radical polymerization of acrylate monomers initiated by three-component photoinitiating systems composed of ketone/amine/bromo compound was described by J.P. Fouassier and et al. in 1994. The poor stability of ketone/amine/onium salt system prompted their to change the onium salt by another substance (for example, bromo compound). The following ketones were used: benzil, 2,4-diethylthioxanthene-9-one, benzophenone, chlorothioxanthone, 2,2-dimethoxy-2-phenylacetophenone. As bromo compounds were used: tribromomethyl phenyl sulfone, 2,2-dibromoacetophenone, 2,2,2-tribromoacetophenone. 2,2'-(Methylamino)diethanol was applied as a co-initiator. The interactions between the different partners of a three-component system were studied by time-resolved laser spectroscopy and steady-state photolysis. The primary step is an electron transfer between the ketone and an amine leading to the ketyl radical and amine radical formation. The rate constants for electron transfer process are in the range from 1000 to 2000 × 10^6 $M^{-1}s^{-1}$. In polymerizing mixture, the monomer interaction is usually efficient and competes with the radical generation process (the fluorescence quenching rate constants are about 2 times lower than those observed for fluorescence quenching of ketone by amine). The next deactivation process is an electron transfer from an excited triplet state of ketone on the ground state of bromo compound (the rate constants of fluorescence quenching are in the range from 11 to 2600 × 10^6 $M^{-1}s^{-1}$). The 2,4-diethylthioxanthene-9-on radical cation formation was observed at 430 nm using the laser flash photolysis experiment (similar results were observed for the system: thioxanthone derivative/onium salt). Since ketyl radicals are known to act as terminating agents of the growing macroradicals in polymerization reactions, it is obvious that the electron transfer between ketyl radical and bromo compound has two benefits: a decrease of ketyl radicals concentration and increase of $CBr_3^{\bullet}$ radicals concentration. As a consequence, the rate of initiation should increase and the rate of termination should decrease [56].

Three-component photoinitiating systems for acrylate monomers polymerization composed of a dye, an iron arene complex, and a phenylglycine derivative were investigated and described by J.P. Fouassier in 1996. The first step of photoreaction occurs between an excited singlet or triplet state of a dye and the iron arene complex. 3-(2'-Benzothiazoryl)-7-diethylaminocoumarin and 3,3'-carbonyl-bis-7-diethylaminocoumarin were used as sensitizers. The following compounds: (η^6-chlorobenzene)(η^5-cyclopentadienyl)iron (II)-hexafluorophosphate and N-cyano-N-phenylglycine were used as co-initiators. The quenching constants of sensitizers by iron complex were estimated at 1.4 × 10^{10} $M^{-1}s^{-1}$ in ethyl acetate solution. This value is almost the same as diffusion-controlled value in ethyl acetate and greater than for fluorescence quenching by N-cyano-N-phenylglycine. Therefore, the primary photochemical process in the three-component photoinitiating system is reaction between sensitizer and iron complex. The free energy change ΔG_{el} for an electron transfer process was estimated at –28 kJ mol^{-1}. The excited singlet state of a dye or excited triplet state of a dye according to its nature reduces the iron arene complex. The reduced iron arene complex has a partially radical structure because this reduced form shows a tendency to dimerize. The reduced iron arene complex may abstract an hydrogen atom from the phenylglycine derivative that, in turn can start polymerize acrylic compounds. In the two-component (dye/amine) system, a ketyl radical of dye was generated and acts as a terminating agent for the growing polymer chains. In the system: dye/iron arene complex/phenylglycine derivative, this type of ketyl radical did not form because the most efficient process corresponded to the interaction between the dye and the iron complex [57].

In 1997, Fouassier described another example of three-component photoinitiating system consisting of a dye, an iron arene complex, and a phenylglycine derivative. The rate constant of fluorescence quenching of a dye by arene iron complex was estimated at 1.3 × 10^{10} $M^{-1}s^{-1}$ and is much higher than that of the dye/N-phenylglycine derivative system. Therefore, the first step of the reaction in three-component dye/arene iron complex/N-phenylglycine derivative system occurs between an excited singlet or triplet state of a dye and arene iron complex. From the calculated basing on the Rehm-Weller equation the values of free energy change ΔG_{el} for an electron transfer process (from –2 to 17 kJ mol^{-1}) it is obvious, that the quenching mechanism should not involve an electron transfer but an energy transfer. The laser flash photolysis experiment shown the occurrence of a physical quenching (no changes were observed in the transient absorption spectra of a dye when arene iron complex was added). In next step, the reaction between N-phenylglycine derivative and an excited arene iron complex occurs. In the kinetic studies the following compounds were applied: 3-(2'-benzothiazoryl)-7-diethylaminocoumarin and 3-2'-carbonyl-bis-7-diethylaminocoumarin were used as the dyes, (η^6-isopropylbenzene)(η^5-cyclopentadienyl) iron (II)-hexafluoro-phosphate and (η^6-hexamethylbenzene)(η^5-cyclopentadienyl) iron (II)-hexafluoro-phosphate as iron arene complex and N-(3-cyanophenyl)glycine as a phenylglycine derivative [47].

Next three-component photoinitiating system composed of Methylene blue/N-methyl-diethanoloamine/diphenyliodonium chloride was described in 2000 by K.S. Padon. This is an example of photoinitiating system, in which the dye rarely described in literature and possessing positive charge acts as a sensitizer. Therefore, electrostatic repulsion precludes direct reaction with the second co-initiator. By effectively shutting down this reaction pathway, it may be to simplify the list of possible primary and secondary reaction mechanism. Kinetic studies based upon photodifferential scanning calorimetry reveal a significant increase in polymerization rate with increasing concentration of either an amine or an iodonium salt. In such photoinitiating system all components play an important role in the photopolymerization reaction. The two-component system containing Methylene blue/N-methyl-diethanoloamine exhibits the second fastest rate, while a two-component system containing Methylene blue/diphenyliodonium chloride exhibitis a relatively slow reaction rate. Finally, it should be noted that even in the absence of amine or iodonium salt, Methylene blue causes slow polymerization to occur. The amine group present on the Methylene blue structure acts as an electron donor to a second, photoexcited Methylene blue molecule. The resulting radicals may then initiate polymerization. The primary photoreaction, which occurs in such three-component photoinitiating system is an electron transfer from an amine on an excited state of a dye. As a results of this process, the following species are formed: positively charged amine moiety (radical cation), Methylene blue-derived radical and chloride anion. In the next step, the amine radical cation transfers a proton to the chloride ion to produce hydrogen chloride and a neutral tertriary amine radical, which is active for initiation. The neutral Methylene blue radical is presumably not active for initiation, but is able to terminate the growing polymer chains. A iodonium salt reacts with dye-derived radical. This reaction involves an electron transfer from the dye radical to the iodonium salt. This reaction regenerates the original dye and produces a diphenyliodonium radical, which rapidly fragments into a molecule of phenyl iodide and a phenyl radical (initiating radical). The enhanced reaction rate observed in the presence of an iodonium salt is explained in part by the fact that the phenyl radical, unlike the Methylene blue radical is active for initiation. In this way a Methylene blue radical that is active only in termination is effectively replaced by a phenyl radical that is active in initiation. In summary, the diphenyliodonium salt plays a double role: it replaces an inactive, terminating radical with an active, initiating radical, and it regenerates the Methylene blue dye [49].

The next described in 2000 by J.P. Fouassier and et al. three-component photoinitiating systems were based on isopropylthioxanthone, amine and bifunctional benzophenone-ketosulfone photoinitiator. The addition of isopropylthioxanthone to the photoinitiating system consisting of benzophenone-ketosulfone/amine clearly enhances the efficiency of the photopolymerization of clear or pigmented coatings. This is a result of an interaction between an excited triplet state of isopropylthioxanthone and benzophenone-ketosulfone photoinitiator. The three-component photoinitiating system exhibits a high reactivity. The rate of polymerization is 1.5 times higher than for two-component system: 2-isopropyl-thioxanthone/methyl-diethanolamine. The calculated using the Rehm-Weller equation values of the free energy changes for electron transfer process shown, that the electron transfer from an excited state of 2-isopropylthioxanthone on the benzophenone-ketosulfone photoinitiator is not favorable. The excited state of sensitizer is efficiently quenched by an amine. The mechanism of primary and secondary processes for three-component photoinitiating system was proposed basing on the laser flash photolysis experiment. The addition of N-methyl-diethanolamine to two-component system: 2-isopropylthioxanthone/benzophenone-ketosulfone (a) shortens the lifetime of the 2-isopropylthioxanthone triplet state observed at 640 nm, (b) generates the ketyl radical of thioxanthone (480 nm), (c) does not produce the kethyl radical of benzophenone-ketosulfone photoinitiator, (d) forms an absorption at 750 nm, which is the superimposition of the three transients: triplet state of 2-isopropylthioxanthone, 2-isopropylthioxanthone radical anion and benzophenone-ketosulfone-derived radical cation [50].

Addition of a bis(trichloromethyl)-substituted-1,3,5-triazine to a dye/amine photoinitiating system leads clearly to an increased efficiency of polymerization under visible light irradiation. 1,3,5-Triazine derivative acts mainly as an inhibitor scavenger. The involved inhibitor is reduced by reduced dye (dye$^{\bullet}$) arising from the first photochemical reaction between the excited states of the dye and an amine. The following 1,3,5-triazine derivatives were applied by C. Grotzinger (2001) to study the kinetic of free radical polymerization: 2,4,6-tris(trichloromethyl)-1,3,5-triazine, 2-(4'-methoxyphenyl)-4,6-bis(trichloromethyl)-1,3,5-triazine, 2-(4'-methoxy-1'-naphtyl)-2,4-bis(trichloromethyl)-1,3,5-triazine. Rose bengal, Eosin Y and phenosafranine were used as the sensitizers. N-Methyl-diethanoloamine, N,N,N-triallylamine, N,N,N-triethylamine, N-phenylglycine, 1,4-diazabicyclo[2,2,2]-octane and N,N-dimethylaniline were applied as co-initiators. In the presence of two-component photoinitiating systems: dye/1,3,5-triazine derivative the polymerization does not occur. Obviously, no efficient generation of initiating radicals by direct sensitization of the triazines occurs compared to the dye/N-methyl-diethanoloamine system. However, addition of 1,3,5-triazine derivative to the dye/amine system produced a synergistic effect in the polymerization reactions. An increased efficiency of the polymerization after addition of 1,3,5-triazine derivative was observed only in the case of amines capable of readily donating a proton after oxidation, such as N-methyl-diethanoloamine and N,N,N-triethylamine. The quenching rate constants of the excited singlet state of Rose bengal, Eosin and Safranine are about 8×10^{9} $M^{-1}s^{-1}$. The calculated free enthalpies of the electron transfer reactions between the excited states of the dye and 1,3,5-triazine derivative show that an electron transfer from the singlet and triplet excited states of the dye to the triazine are both feasible, leading to oxidation of the dye and reduction of 1,3,5-triazine derivative. On the other hand, the excited singlet and triplet states of sensitizers are quenched by amines tested. The main process is an electron transfer between an excited dye and an electron donor (amine) which can be followed, depending on the amine, by a proton transfer from an electron donor radical cation ($AH^{\bullet+}$) to the dye radical anion ($dye^{\bullet-}$), thus leading to a neutral aminoalkyl radical ($A^{\bullet}$) which is a good polymerization initiator, and to the reduced dye ($dyeH^{\bullet}$). But, due to the fact that the back electron transfer reaction is expected more significant in a singlet state than in a triplet state, the main source of radicals is issued from a triplet state. For xanthenic dyes, the quenching of the excited states (singlet and triplet) of the dye by 1,3,5-triazine derivative and N-methyl-diethanoloamine has comparable interaction rates. Therefore, it seems that the improvement of the polymerization efficiency for the system dye/N-methyl-diethanoloamine/1,3,5-triazine derivative compared to dye/N-methyl-diethanoloamine is a result of the secondary reactions between 1,3,5-triazine derivative and the species arising from the first interaction between an excited state of the dye and amine. It is well known, that xanthene dye radicals can be a scavenger of the growing macromolecular chains and of the aminoalkyl radicals. The interaction between dye radicals and 1,3,5-triazine derivative probably makes it possible to reduce the scavenging effect of the dye radicals and could lead to the formation of new initiating radicals [17].

K.S. Padon in 2001 described the mechanism of initiation of free radical polymerization by three-component photoinitiating system composed of Eosin/N-methyl-diethanoloamine/diphenyliodonium chloride. Kinetic studies revealed that the fastest polymerization occurred when all three components were present (the next fastest was with the dye/amine pair, and the slowest was with the dye/iodonium pair). Eosin undergoes the photochemical reactions with both N-methyl-diethanoloamine and iodonium salt. However, an iodonium salt bleaches the dye much more rapidly than does the reaction between the Eosin and amine. It was concluded, that although a direct Eosin/amine reaction can produce active radicals in the three-component system, this reaction is largely overshadowed by the Eosin/iodonium reaction, which does not produce active radicals as effectively. In such case, the amine reduces the oxidized dye radical formed in the Eosin/iodonium reaction back to its original state as well as the simultaneous production of an active initiating amine-based radical. The primary photochemical process in the three-component photoinitiating system: dye/amine/iodonium salt is an electron transfer form an excited triplet state of Eosine on the ground state iodonium salt. The initiating radicals formed initiate free radical polymerization with low rate. The secondary photoreaction, electron transfer from an amine on the excited state of dye, is much more effective for producing radicals, but photoinitiating ability of this reaction is limited in the three-component system by competition from the Eosin/iodonium salt reaction. Because of the significant difference in reaction rates for the two pairwise reactions, it is likely that most of the active centers in the three-component system are generated via the dark reaction between the amine and the oxidized dye radical resulting from Eosin/iodonium salt photoreaction [58].

In 2001 J.P. Fouassier described a new multicomponent system consisting of a xanthenic dye, a ferrocenium salt, a hydroperoxide, and an amine. The initiating system was composed of: (η^5-2,4-cyclopentadien-1-yl)[(1,2,3,4,5,6,η)-(1-methylethyl)benzene]-iron (II) hexafluoro-phosphate, cumene hydroperoxide, xanthenic dyes: Rose bengal, thionine and N-methyl-diethanoloamine. Such four-component photoinitiating systems are industrial used in the pigmented coating to produce 200 – 400 μm thick film, contaning 20 % of inorganic pigments and 5 % of organic pigments. The mechanism of primary and secondary reactions occuring in the four-component photoinitiating system was established basing on the laser flash photolysis results. The photolysis of iron arene in acetonitrile solution leads to the formation of unstable iron complex, which undergoes decomposition to ferrocene Fe(I) ions. In the presence of air and hydroxyperoxide, the rate of formation of the photolysis products decreases in comparison with the results obtained when iron arene alone undergoes degradation. There are possible two processes: photoreduction of Fe(II) to Fe(I) or photooxidation of Fe(II) to Fe(III). Photolysis investigation of the system: ferrocenium salt, hyperperoxide, dye, amine provided evidence for following interactions: (a) in ferrocenium salt/oxygen, in the presence of air, the photolysis of $CpFe^{+}$ arene leads to Fe(III), (b) in cumene hydroperoxide/ferrocenium salt, in the absence of oxygen, the hyperperoxide as an oxidizing agent participates in the peroxidation mechanism leading to Fe(III), (c) in dye/ferrocenium salt, xanthenic, or thiazine type, dyes photosensitize the photodegradation reaction of $CpFe^{+}$ arene. Formation of ferric ions and reduction of the dye to its leuco form, in the presence or the absence of oxygen, both take place. There are three possible pathways in the formation of radicals capable of initiating the photopolymerization reaction of acrylates, under polychromatic light: (1) either through direct photolysis of ferrocenium salt; (2) or through photosensitization of the ferrocenium salt by the dye in the presence of an amine and hydroperoxide; (3) a primary interaction between a dye and an amine leading to the formation of dye-based radicals [59].

The several coumarin or ketocoumarin/additives combinations (bis-imidazole derivative, mercaptobenzoxazole, titanocene, oxime ester) studied by J.P. Fouassier (2001) are able to initiate quite efficiently a radical reaction. Basing on the laser flash photolysis experiment it was concluded that, the coumarins are able to form radicals through an electron transfer reaction with the different additives, whereas the ketocumarin leads to an energy transfer with bis-imidazole and to an hydrogen abstraction with benzoxazole derivative. The singlet state quenching of coumarin by titanocene and bis-imidazole derivative is very efficient and much more efficient that by the proton donors, such as: 2-mercaptobenzoxazole or 2-mercaptobenzothiazole (by about at least one order of magnitude). The laser flash photolysis of cumarin in presence of bis-imidazole derivative leads to the formation of lophyl radicals. The recombination of two lophyl radicals yields bis-imidazole derivative [60].

C. Grotzinger in 2003 described, the activity of three-component systems (containing a dye, an amine and a triazine derivative) for the initiation of the photopolymerization of multifunctional acrylates under visible light. As co-initiatiors were used: 2-(4'-methoxyphenyl)-4,6-bis(trichloromethyl)-1,3,5-triazine and N-methyl-diethanoloamine. The following compounds were used as sensitizers: Erythrosin B, Acridine Orange, Acriflavine, 3-butoxy-5,7-diiodo-6-fluorone and 3-hydroxy-2,4,5,7-tetraiodo-6-fluorone. The two-component dye/triazine derivative systems are not able to initiate a radical polymerization. This means that under sensitization conditions and using these dyes as sensitizers, the interaction between a dye and 1,3,5-triazine derivative does not efficiently lead to initiating species. The addition of 1,3,5-triazine derivatve to the dye/amine leads to an enhancement of the polymerization efficiency. The synergistic effect was only observed in the case of amines that were capable of readily donating a proton, after oxidation by an excited dye, through a primary electron transfer reaction. With the addition of a small amount of 1,3,5-triazine derivative (0.3 %), the Acridine Orange/N-methyl-diethanoloamine photoinitiating system becomes better than the well known two-component photoinitiating systems: Eosin/N-methyl-diethanoloamine or Rose bengal/N-methyl-diethanoloamine. After light absorption by a dye, the excited states of the dye are quenched by N-methyl-diethanoloamine and 1,3,5-triazine derivative (with different rate constants) but only the interaction with an amine leads to the generation of initiating radicals (the neutral aminoalkyl radical). This reaction also leads to the reduced dye (dyeH$^{\bullet}$) being a terminating agent of the growing macromolecular chains. The importance of the inhibition effect of dyeH$^{\bullet}$ depends on the efficiency of its interaction with 1,3,5-triazine derivative. DyeH$^{\bullet}$ being a reducing species, the reaction with 1,3,5-triazine derivative is probably a reduction of the latter, leading to the recovery of the dye. In such photoinitiating systems, the 1,3,5-triazine derivative can act as a ground state electron acceptor. The reaction between an excited state of the dye and an amine is thermodynamically favorable for all tested dyes, both in their singlet and triplet states [4].

The photoinitiated polymerization reaction of maleimide-vinyl ether and maleimide-allyl ether under visible light in the presence of four-component photoinitiating systems, with remarkable high rates of polymerization and nearly 100 % conversion of monomer was described in 2003 by D. Burget. The components of the multi-component photoinitiating system were following compounds: ferrocenium (I) salt derivative, Rose bengal, N-methyl-diethanoloamine and cumene hydroperoxide. The multi-component systems containing a ferrocenium salt with a hydroperoxide are fairly unstable at wavelengths above 500 nm when dye such as Rose bengal is used as a sensitizer. The presence of an intramolecular ion pair complex between the photosensitizer and the ferrocenium (I) salt makes the primary step of the reactions involve an efficient electron transfer (this process is obviously not diffusion controlled) which leads to the formation of the reduced iron arene salt. Reduced iron Fe (0) arene salt reacts with molecular oxygen leading to the oxygen radical anion, but the electron transfer with hydroperoxide leading to an alkoxy radical. The posssible quenching of an excited state of Rose bengal by maleimide, malenimide allyl ether or the charge transfer complex (vinyl ether-maleimide) does not compete with the intramolecular Rose bengal/Fe(I) salt interaction in the ion pair complex. However, maleimide can be involved in the next steps of the initiating mechanism. The reduction of maleimide by Fe(0) (leading to initiating radical in the presence of an hydrogen donor) can be in competition with the reduction of the hydroperoxide. All radicals formed (expect the ketyl type radical derived from the photoinitiator which is not an initiating radical for acrylate polymerization but a scavenging species of the growing macromolecular chains) can initiate the free radical polymerization of the maleimide-allyl or –vinyl ether formulations [61].

When the PI is a xanthene dye (such as Rose bengal or Eosin)/amine, the polymerization rates exhibit a clear increase when the system is irradiated with visible light in the presence of aromatic carbonyl compounds such as monooximes or O-acylmonooximes. With a mixture of Eosin as initiator, N-methyl-diethanoloamine as co-initiator and the acyloxime ethyl 1-phenyl-1-oxopropan-2-iminyl carbonate, it was found that rates were 2-3 times higher than those observed in absence of aromatic carbonyl compound [62].

The three-component photoinitiating system based on ketocumarins applied in high speed photopolymers for laser imaging was described by J.P. Fouassier in 2003. For example, the high efficiency of ketocoumarin dye in the presence of an amine (N-phenylglycine) and an onium salt (diphenyl iodonium chloride) toward the initiation of the photopolymerization of an acrylic monomer (phenoxy diethyleneglycol acrylate) has been evaluated. The initiating radicals are formed in the primary photochemical process (an electron transfer from amine on an excited triplet state of ketocumarin), and also as a result of an electron transfer from an excited triplet state of ketocumarin on the ground state of iodonium salt/N-phenylglycine complex. The ketyl radical after hydrogen abstraction becomes the terminating radical [62].

The photopolymerization of acrylamide initiated by the synthetic dye safranine T in the presence of triethanoloamine as co-initiator in aqueous solution has been investigated and described by M.L. Gómez in 2003. The initial polymerization rate increases with the amine concentration reaching a maximum at 0.03 M. Further increase of triethanoloamine concentration produces a decline of the polymerization rate. This behavior is different to that observed in the polymerization of 2-hydroxyethyl methacrylate with the same initiating system. The addition of diphenyliodonium chloride to the two-component system has a marked accelerating effect on the polymerization rate.

The effect of addition of acrylamide and iodonium salt on the excited triplet state properties of a dye was investigated by laser flash photolysis. The presence of acrylamide slightly reduces the triplet lifetime. Changes in the shape of the T-T spectrum are not observed in the presence of diphenyliodonium chloride up to 0.025 M. However, an increase in the triplet yield of more than 60 % is observed in the presence of iodonium salt 0.022 M. At the same time the triplet lifetime becomes shorter. The quenching of the monoprotonated triplet $^3SH^+$ by triethanoloamine does not follow a linear relationship with amine concentration. This is due to an excited state proton transfer process involving the monoprotonated triplet on an amine, leading to the deprotonated triplet state of the dye. A posterior slower electron transfer process between ^3S and amine would lead to the formation of semi-reduced safranine in the form of radical anion, and the radical cation of the amine. The latter undergoes a second fast proton transfer reaction leading to the active initiating radicals. When SH^+ is continuously irradiated at the maximum of its absorption band at 520 nm in the presence of triethanoloamine, a fast photobleaching process takes place. The accelerating properties of the iodonium salt must be due to secondary electron transfer reaction from the semi-reduced form of the dye formed in the primary photochemical process to the diphenyliodonium cation. In the ground state safranine will be rapidly protonated to the form SH^+. However, the phenyl radicals produced as a result of decomposition of iodonium radical can start a new polymer chains [63].

N-substituent effect of maleimides on 1,6-dihexanediol diacrylate polymerization initiated by three-component photoinitiating system was described by S.C. Clark in 2003. The photoinitiating systems were composed of various N-substituted maleimides in the presence of N-methyl-diethanoloamine and benzophenone. The introdution of N-aliphatic substituents has only small but measurable positive effect on the photoinitiation rate in comparison with the unsubstituted maleimide. In the case of N-aryl substituent, it is obvious that both the position and structure of the substituents on the phenyl ring play a critical role in determining the efficiency of photoinitiation. Upon excitation of benzophenone a chemical sensitization process involving the well-known electron-proton photoreduction of excited triplet state of benzophenone occurs to give an amine-centered radical and the semipinacol radical. A second electron-proton transfer process occurs from the semipinacol radical to the maleimide, hence resulting in the succinimidyl radical and ground state benzophenone. This chemical sensitization process results in two radicals (the amine centered radical and the succinimidyl radical) which initiate free radical polymerization [64].

Cobaltic accelerator for Methylene blue/triethanoloamine two-component photoinitiating system in aqueous ethoxylated trimethylolopropane triacrylate solution was described by J. Yang in 2004. The three-component photoinitiating systems containing cobaltic complexes as accelerator were found to be very efficient in initiating the polymerization of the water-soluble monomer, and the efficiency was higher than that initiated by Methylene blue/triethanoloamine/diphenyliodonium chloride. Twin redox cycles between Methylene blue and cobaltic complexes were proposed to explain the acceleration effect of the cobaltic complexes. The following compounds were used as accelerators: hexaamminecobalt(III) chloride and cobalt(III) acetylacetonate [65].

The polymerization of 1,6-hexanedioldiacrylate initiated by three-component photoinitiating systems composed of 2-methylmaleic anhydride or 2,3-dimethylmaleic anhydride was studied and described in 2004 by T.B. Cavitt. As the sensitizers were applied: benzophenone, 4-benzoylbiphenyl or isopropylthioxanthone with a tertiary amine (N-methyl-diethanoloamine) as a co-initiator. Concentrations of less than 0.1 weight percent of 2,3-dimethylmaleic anhydride added to 1,6-hexanedioldiacrylate containing any of the aforementioned diarylketones and N-methyl-diethanoloamine results in an increase in the polymerization rate maximum by factor of as much as three times that attained for two-component photoinitiating system diarylketone/amine. In these three-component systems, the diarylketyl radical, which is produced by the hydrogen transfer from an amine to an excited triplet diarylketone, is oxidized by the electron deficient N-substituted maleimide to give a ground state diarylketone and the corresponding succinimidyl radical. The three-component system thus produces two radicals, an alpha amino radical and the succinimidyl radical, both of which initiated free radical polymerization. The diarylketyl terminating radical is replaced by an initiating radical resulting in a substantial increase in initiation rate. For each diarylketone triplet quenching constants by N-methyl-diethanoloamine and malenic anhydrides are of the same order of magnitude ($1\text{-}6 \times 10^9\ M^{-1}s^{-1}$). For benzophenone and isopropylthioxanthone the quenching rate constants are very high, indicating that quenching of benzophenone and isopropylthioxanthone by both species is very efficient. In the three-component system consisting of a diarylketone, N-methyl-diethanoloamine and maleic anhydride, two quenching processes are possible. N-Methyl-diethanoloamine quenches excited state diarylketone by up to eleven times faster than quenching by maleic anhydride [66].

Three-component photoinitiators comprised of an N-arylphthalimide, a diarylketone, and a tertriary amine were investigated for their initiation efficiency of 1,6-hexanedioldiacrylate by T.B. Cavitt (2004). The addition of N-arylphthalimide to the two-component photoinitiating system composed of isopropylthioxanthone/N-methyl-diethanoloamine results in the increase of the observed rate of polymerization of bifunctional acrylate of about 2 times. The highest rates of polymerization were observed for N-arylphthalimide derivatives possessing a strong electron-rich substituents (for example, -CN). Each of the diarylketones is readily guenched by an N-arylphthalimide with quenching constants on the one order of magnitude lower than those for diarylketones by N-methyl-diethanoloamine. The quechning constants of semipinacol radical ($\tau = 30.6\ \mu s$) by N-arylphthalimide derivative are about $1.5 \times 10^6\ M^{-1}s^{-1}$ and are caused probably by an electron-proton transfer process to the N-arylphthalimide. The introduction of an electron-donating substituent to the N-arylphthalimide molecules causes that the quenching of semipinacol radical does not occur. For each N-arylphthalimide/diarylketone/N-methyl-diethanoloamine three-component photoinitiating system, the primary initiation mechanism entails an electron-proton transfer from N-methyl-diethanoloamine to an excited triplet state of a diarylketone forming an α-

amino radical (which initiates polymerization) and the corresponding ketyl radical. In the presence of N-arylphthalimide, the ketyl radical can be oxidized to form the radical anion of N-arylphthalimide derivative and a stabilized cation of the ketyl radical. Subsequently, a proton transfer occurs to regenerate a ground state diarylketone and the N-arylphthalimidyl ketyl radical, which presumably either initiates polymerization or rearranges to form a benzoyl-type radical that initiates polymerization [67].

The role of diphenyliodonium salt in three-component photoinitiator systems containing Methylene blue and an electron donor was described by D. Kim in 2004. It was concluded that the diphenyliodonium salt enhances the photopolymerization kinetics in two ways: (1) it consumes an inactive Methylene blue neutral radical and produces an active phenyl radical, thereby regenerating the original Methylene blue, and (2) it reduces the recombination reaction of Methylene blue neutral radical and amine radical cation. Methylene blue is a cationic dye, thus it does not undergo direct interaction with diphenyliodonium (which is also cationic) because of electrostatic repulsion. N-Phenylglycine is the most effective amine for active center generation with Methylene blue, with an initial polymerization rate four or five times higher than the other amines (N-methyl-diethanoloamine, triethanoloamine and triethylamine). The effectiveness of N-phenylglycine may be attributed to aforementioned unimolecular fragmentation reaction that prevents back electron transfer. In the three-component photoinitiating systems both competitive processes occur: the recombination of dye-based radical and an amine radical cation, and an electron transfer from an excited state of sensitizer to the ground state diphenyliodonium salt [68].

Other example of three-component photoinitiating systems consisting of 2-mercaptobenzothiazole as a co-initiator were described by S. Suzuki. The sensitization mechanism in a photoinitiating system that consist of an aminostyryl sensitizing dye, 2-[*p*-(N,N-diethylamino)styryl]naphto[1,2-d]thiazole (NAS), and a radical generator 2,2'-bis-(2-chlorophenyl)-4,4',5,5'-tetraphenyl-1,1'-bi-1*H*-imidazole (HABI). The improvement of the photosensitivity was observed by the addition of 2-mercaptobenzohiazole is suggested as an electron transfer toward aminostyryl dye that loose an electron by singlet electron transfer to imidazole derivative or hydrogen donor toward imidazolyl radicals. The mechanism of the processes occurring in three-component system was proposed basing on the laser flash photolysis experiment. The transient absorption spectrum of sensitizer in the presence of the 2-mercaptobenzothiazole was similar to that of the direct excitation of aminostyryl dye, but the initial absorption increased significantly with increasing of 2-mercaptobenzothiazole concentration. Two possibilities are considered about the transient species. First possibility is the donating an electron from 2-mercaptobenzotiazole to an excited state of aminostyryl dye, which is thermodynamically allowed, and a generated thiyl radical which has good ability to initiate radical polymerization. Another possibility is to generate of the triplet state of aminostyryl dye. To confirm the possibility of the path to generate the triplet state, Suzuki applied triplet-triplet energy transfer form Michler's ketone to aminostyryl dye. In the three-component system an electron transfer occurs via both singlet and triplet state of aminostyryl dye, resulting in the increase of imidazolyl radicals concentration and hence, in the improvement of the photoinitiating ability [69].

In 2005 J.D. Oxman described the free radical/cationic hybrid photopolymerizations of acrylates and epoxides initiated using a three-component initiator system comprised of camphorquinone as a photosensitizer, an amine as an electron donor, and a diaryliodonium salt. Thermodynamic considerations revealed that the oxidation potential of an electron donor must be less than 1.34 V for electron transfer with an excited singlet state of camphorquinone and must be less than 0.94 V for electron transfer with an excited triplet state of camphorquinone. This electron transfer leads to a production of the active centers for hybrid polymerization (two radicals and a cation). Second requirement is low basicity of an electron donor (amine) and a iodonium salt must have a non-nucleophilic counterion, such as hexafluoroantimonate or hexafluorophosphate. Because of, the protonation of the electron donor may compete with the cationic propagation reaction, therefore, the basicity of the electron donor, as characterized by the pK_b, is a good indicator of the proton-scavenging efficiency of the electron donor. The electron donor pK_b must be greather than 8. Nine different electron donors were investigated, including tetrahydrofuran, 1, 4-dimethoxybenzene, N,N-dimethylacetamide, 1,2,4-trimethoxybenzene, N,N-dimethylbenzy-lamine, ethyl 4-N,N-dimethylaminobenzoate, 4-N,N-dimethylaminobenzoic acid, dimethylphenylamine and 4-*tert*-butyl N,N-dimethylaniline. Experiments performed using a combination of electron donors revealed that the onset of the hybrid system's cationic polymerization can be advanced or delayed by controlling the concetration and composition of the electron donors [70].

Three-component photoinitiator systems containing N-substituted maleimide (photoinitiator II)/ketocumarin/tertiary amine have been used for visible light photo-polymerization of acrylate and thiol-end monomers and described in 2006 by A.F. Senyurt. The photoinitiating mechanism involves the photoreduction of ketocumarin with a tertiary amine through an electron/proton transfer reaction. Similarly, as it was previously described for other photoinitiating systems, ketyl and aminyl radicals are formed. Ketyl radical is then oxidized by a ground state maleimide via a second electron/proton transfer reaction. The resultant succinimidyl and aminyl radicals are both effective in initiating acrylate polymerization. In the case of thiol-end monomers, the addition of a maleimide to the two-component photoinitiating system: ketocumarin/tertiary amine does not improve the rate of polymerization. It is due to fact that the hydrogen abstraction of the labile hydrogen on the thiol by the ketyl radical is much more efficient than the electron/proton transfer reaction between ketyl radical and the maleimide [71].

Phenylonium salts as third component in the photoinitiator system: Safranine O/triethanoloamine were studied and described in 2007 by M.L. Gómez. The efficiency of the photoinitiator system composed of Safranine and triethanoloamine for the polymerization of acrylamine in water was improved by the incorporation of an onium salt. The phenylonium salts employed were diphenyliodonium chloride, triphenylsulfonium triflate, tetraphenylphosphonium chloride and tetraphenylarsonium chloride hydrate. The increase of rate of polymerization is observed only in the case of two onium salts: diphenyliodonium chloride and triphenylsulfonium triflate (the polymerization rate increases 2.5 and 1.5 times, respectively). The presence of an onium salt affects the photobleaching of a sensitizer. The photobleaching process in the presence of triethanoloamine is faster than in the presence of an onium salt. The bleaching quantum yield is about one order of magnitude lower for diphenyliodonium chloride. But tetraphenylphosphonium chloride and tetraphenylarsonium chloride hydrate do not effect on the quantum yields of photobleaching of Safranine [72].

Bis-acylphosphine oxide derivative/isopropylthioxanthone/amine, Rose bengal/amine/ benzyldimethyl-ketal, Rose bengal/N-methyl-diethanoloamine/vanilin and Rose bengal/N-methyl-diethanoloamine/flavone are the examples of three-component photoinitiating systems used in the industry (J.P. Fouassier, 2007). Phenols are not good radical scavengers in the absence of oxygen but become strong inhibitors (through a synergistic effect) in the presence of oxygen [73].

Three-component photoinitiating systems: ketocumarin/amine/maleimide (photoinitiator I) were described by Ch.K. Nguyen in 2007. An *N*-substituted maleimide has been used in conjunction with ketocumarins and a tertiary amine to initiate the polymerization of 1,6-hexadioldiacrylate in both the UV and visible region of the electromagnetic spectrum. The rate of polymerization initiated by ketocumarin/tertiary amine combinations is significantly increased by the addition of *N*-substituted maleimide, presumably due to oxidation of the coumarin ketyl radical formed by interaction between a triplet state of the ketocumarin and a tertiary amine. These systems are interesting since maleimides are very effective oxidants of ketyl radical requiring their use at low concentrations. In addition they readily copolymerize with acrylates and are thus incorporated into the network produced by the photopolymerization process. The efficiencies of these three-component systems are shown to approach that of a typical cleavage photoinitiator in the UV region [74].

X. Guo in 2008 described the free radical polymerization of dental resins initiated by three-component photoinitiating systems composed of camphorquinone/2-(dimethylamino)ethyl methacrylate and camphorquinone/ethyl 4-dimethylaminobenzoate in the presence of 1 % of diphenyliodonium hexafluorophosphate. All components of photoinitiating system are water soluble. As a light source a conventional dental lamp was used. Water absorption in resins has negative effect on their mechanical properties: the more water the polymers absorb, the mechanical properties decrease. The conversion of the resin decreased from 54 % to 25 % when more than 20 % water is added. In dental adhesive formulations, hydrophobic components (camphorquinone) and hydrophilic components (2-(dimethylamino)ethyl metacrylate) are employed. The hydrophobic components are usually employed to enhance the mechanical properties and compatibility with restorative resin composities. Hydrophilic components are employed to improve water-compatibility and enhance the infiltration of adhesives into wet, demineralized dentin. The presence of water and monomer phase separation affects the polymerization rate and degree of conversion. Because of the ability of diphenyliodonium chloride to enhance radical efficiency, addition of an iodonium salt significantly increases the degree of monomer conversion (2,2-bis-[4-(2-hydroxy-3-methacryloxypropoxy))phenyl]-propane/hydroxyethylmethacrylate (60/40)) up to 70 %. Addition of diphenyliodonium salt dramatically increased polymerization rates both of the two-component systems with or without water present. Although diphenyliodonium salt is generally hydrophilic due to its ionic nature, phenyl radical produced by this molecule is relatively hydrophobic. But the active aminyl radicals are hydrophilic. Besides the reactivity, the hydrophilicity/hydrophobicity of the initiator components and water played a critical role in determining the initiating polymerization behavior of adhesives. Diphenyl iodonium salt, which is an electron acceptor, has the following double roles: one is to regenerate the dye molecules by replacing inactive, terminating radicals with active, phenyl initiating radicals, and another role is generate additional active phenyl radicals. In the presence of water the monomer rich phase is hydrophobic, so the initiator systems which generate active hydrophobic radical are more tolerant to water. The optimal initiator system should be able to generate enough both hydrophilic and hydrophobic active radicals in the case of phase separation. The photoinitiating ability of the three-component photoinitiating systems increases in the presence of 0-15 % water, because of hydrophilic/hydrophobic character of initiating radicals [75].

Efficient photoinitiators have been regarded as a basic requirement for complete polymerization, especially for deep cavities. Another example of photoinitiator system used in dental resins are formulations composed of a sensitizer, an amine as a co-initiator and an iodonium salt for polymerization of 2-(hydroxyethyl)methacrylate/2,2-bis-[4-(2-hydroxy-3-methacryloxypropoxy)phenyl]propane with a mass ratio 45/55 in the of 8.3 mass.% water. The following photoinitiators were described by Q. Ye in 2009: camphorquinone as a hydrophobic photosensitizer, 3-(3,4-dimethyl-9-oxo-9H-thioxanthene-2-yloxy)-2-hydroxy-propyl]trimethylammonium chloride as a hydrophilic photosensitizer, ethyl-4-(dimethy-lamino)benzoate as a hydrophobic co-initiator, 2-(dimethylamino)ethyl methacrylate as a hydrophilic co-initiator and diphenyliodonium hexafluorophosphate as an iodonium salt (hydrophilic). Similarly as it was mentioned above, the addition of an iodonium salt to the two-component photoinitiating system produced dramatic improvements in polymerization rate and conversion up to 92.5 %, and mechanical properties (greater modulus and toughness values) when cured in the presence of water [76].

The effective photoinitiating system for visible light induced polymerization of epoxy resins was studied and described by M.A. Tehfe in 2009. This system is based on fluorinated titanocene free radical initiator bis-(cyclopentadienyl)-bis-[2,6-difluoro-(1-pyrryl)phenyl] titanium Ti, a silane (tris(trimethylsilyl)silane TTMSS) together with an onium salt (diphenyliodonium hexafluorophosphate). The efficiency of such three-component systems increases as the TTMSS concentration increases. Interesingly, a bleaching of the Ti/TTMSS/Ph_2I^+ containing film after the polymerization is noted. This can be useful for applications requiring colorless coatings, for example. In the photoinitiating system consisting of bis-(cyclopentadienyl)-bis-[2,6-difluoro-(1-pyrryl)phenyl]titanium Ti upon light irradiation, two ligands of the complex are removed leading to a titanium centered radical $Ti^•$, a cyclopentadienyl radical $Cp^•$ and a fluorinated aryl radical $Ar^•$. The recombination product (η^5-methylcyclopentadienyl pentafluorophenyl) represents the main organic photolysis compound, as revealed by steady state photolysis [77].

An efficient three-component visible light sensitive photoinitiator system for the cationic photopolymerization of epoxide monomers was described in 2009 by J.V. Crivello. The photoinitiator system consist of camphorquinone in combination with a benzyl alcohol (a hydrogen donor) to generate free radicals. The irradiation of camphorquinone in the presence of benzyl alcohol results in the formation of an α-hydroxybenzyl radical and camphor-quinone-derived radical. An α-hydroxybenzyl radical undergoes rapidly oxidation by a diaryliodonium salt giving corresponding benzaldehyde and a Brønsted acid. This latter species initiates the cationic polymerization [78].

New three-component system consisting of Rose bengal or Fluoresceine as a sensitizer for the free radical polymerization of 2-hydroxyethyl methacrylate was described in 2009 by D. Kim.

Using the thermodynamic feasibility results with experimental kinetic studies, Kim suggested three different kinetic pathways, which are (i) photo-reducible series mechanism, (ii) photo-oxidizable series mechanism, and (iii) parallel-series mechanism. These three kinetic pathways will provide useful information for selection criteria for each component as well as provide a straightforward manner for classifying the photopolymerization process. Triethylamine and N-methyl-diethanoloamine were used as electron donors. The following compounds were used as a second co-initiator: diphenyliodonium chloride and triphenylsulfonium hexafluoroarsenate. In such three-component system, an amine acts as an electron donor, but an iodonium salt is a ground state electron acceptor. On the basis on the Rehm-Weller equation, the thermodynamic feasibility from the electron donor to the photo-excited dye or from photo-excited dye to a ground state of electron acceptor was determined. The basic criterion of selection is the estimation of the value of the free energy change for electron transfer process ΔG_{el}, using the measured oxidation and reduction potentials of all components of photoinitiating system. The calculated values of the free energy changes for electron transfer process show that an electron transfer process from N-methyl-diethanoloamine to an excited state of Rose bengal or Fluoresceine and an electron transfer form an excited state of Rose bengal or Fluoresceine to the ground state of an iodonium salt are thermodynamically feasible. However, both an electron transfer form an excited state of Rose bengal or Fluoresceine on the ground state of triethylamine and an electron transfer from an excited state of Rose bengal or Fluoresceine on the ground state sulfonium salt are not thermodynamically feasible. Therefore, the free radicals formation in such three-component photoinitiating systems may results as: (i) photo-reducible series mechanism: Rose bengal/N-ethyl-diethanoloamine/sulfonium salt and Fluoresceine/N-methyl-diethanolamine/sulfonium salt, (ii) photo-oxidizable series mechanism: Rose bengal/ iodonium salt/triethylamine and Fluoresceine/iodonium salt/triethylamine, and (iii) parallel-series mechanism: Rose bengal/N-methyl-diethanoloamine/iodonium salt and Fluoresceine/ N-methyl-diethanoloamine/iodonium salt. As it is previously mentioned for the three-component photoinitiating systems containing a neutral dye serving as a photosensitizer, an amine as the electron donor, and iodonium salt as an electron acceptor, the primary photochemical process is an electron transfer form an amine to the excited singlet or triplet state of a dye and the generation of radicals followed by proton transfer from radical cation of an electron donor. The iodonium salt, consumes an inactive radical and produces an active phenyl radical, thereby regenerating the original dye in a secondary reaction step. The regenerated photosensitizer (dye) may re-enter the primary photochemical reaction. Accordingly, this kinetic mechanism is designated as a **photo-reducible series mechanism** (camphorquinone/amine, camphorquinone/amine/iodonium salt).

On the other hand, once thermodynamic feasibility only allows the photo-induced electron transfer reaction to proceed between photo-excited dye and an electron acceptor. In such a case, the initiating radicals are formed as a result of **photo-oxidizable series mechanism** (Rose bengal/iodonium salt/triethylamine, Fluoresceine/iodonium salt/triethylamine).

However, under considerations where photosensitizer (dye) has reduction potential and oxidation potential, the photo-excited dye may act as both an electron donor and an electron acceptor resulting in a **parallel-series mechanism**. In this kinetic pathway, the electron transfer between an excited dye molecule and an electron donor can compete with the corresponding electron transfer between an excited dye and an electron acceptor as the primary photochemical reaction. The Rose bengal/N-methyl-diethanoloamine/iodonium salt and Fluoresceine/N-methyl-diethanoloamine/iodonium salt initiator systems provide examples of the combined parallel-series mechanism. A comparison of the visible light activated free radical polymerization of two-component initiator systems (Fluoresceine/N-methyl-diethanoloamine and Fluoresceine/iodonium salt) with the corresponding three-component initiator system shows that the three-component initiator system produces the highest rate and final conversion. Fouassier and coworkers, reported that when the iodonium salt accepts an electron, it produces iodonium radical fragments giving an active, initiating phenyl radical along with a phenyl iodide molecule. Thus, once the iodonium salt accepts an electron from photo-excited Fluoresceine, it undergoes a rapid unimolecular fragmentation reaction that limits back electron transfer. Because of, the reduced back electron transfer between photo-excited Fluoresceine and iodonium salt, Fluoresceine/iodonium salt system leads to the generation of higher concentrations of active centers than Fluoresceine/N-methyl-diethanoloamine photoinitiating system. On the other hand, the photo-excited Fluoresceine wastes photon energy in the electron transfer step of Fluoresceine/N-methyl-diethanoloamine system because of the back electron transfer competes with separation of the radical ion pair or with proton transfer within the pair. It has generally been reported that only 10 % of the absorbed light energy may be used for photo-induced electron transfer in the bimolecular organic electron transfer reaction. Hence, the Fluoresceine/N-methyl-diethanoloamine initiator system only reached 29 % of final conversion. As expected, the Fluoresceine/N-methyl-diethanoloamine/iodonium salt three-component initiator system produced dramatically enhanced conversion (78 %) because of effective retardation of the recombination reaction step and consumption of the dye-based radical to regenerate the original photosensitizer (dye) in the secondary reaction step. The enhanced conversion relative to the two-component initiator system also arises from production of two radicals: an active initiating aminoalkyl radical and an active phenyl radical [44].

The three-component system composed of porphyrin dye/electron donor (N-methyl-diethanoloamine)/diphenyliodonium salt was studied and described by D. Kim in 2009. As the electron donors were used following compounds: N-methyl-diethanoloamine and 1,4-diazabicyclo[2,2,2]octane. These photoinitiating systems initiate free radical polymerization of 2-hydroxyethyl methacrylate with dramatically higher conversion and rate of polymerization compared with the camphorquinone/N-methyl-diethanoloamine/iodonium salt three-component photoinitiator system under the equivalent excited active photosensitizer concentration regarding incident photon flux and photosensitizer absorption. The replacement of the camphorquinone by porphyrin dye in the three-component photoinitiating system dye/N-methyl-diethanoloamine/iodonium salt produced 90 % conversion at 90 s while camphorquinone/N-methyl-diethanoloamine/iodonium salt system exhibited only 27 % conversion at 270 s. Because of the low molar absorptivity of camphorquinone, much higher initial camphorquinone concentrations can be considered without exceeding a thin film approximation (one order of magnitude higher than concentration of porphyrin dye). When the concentration of camphorquinone was increased to 1000 times, the camphorquinone/N-methyl-diethanoloamine/iodonium salt still leads to only 72 % conversion at 250 s. The photo-oxidizable series mechanism efficiently retard the back electron transfer and recombination reaction between free radicals, due to the rapid unimolecular fragmentation of iodonium salt radical in the initial step and thereby generate greater concentration of active phenyl radicals as well as the dye-based inactive radical cation. In addition, an electron donor has more chance to interact with the dye-based radical cation in a secondary reaction step, which results in the production of the initiating radical and regenerating of the original photosensitizer (dye), that re-enters the primary photochemical process (an electron transfer from an excited state of sensitizer (5, 10, 15, 20-tetraphenyl-21H,23H-porphyrin zinc) to the ground state iodonium salt. The resulted radical undergoes a rapid unimolecular irreversible fragmentation reaction that limits back electron transfer. Because of, the reduction potential of porphyrin dye is very high (-1.35 eV) relative to its low triplet state energy (1.59 eV) only an electron donor with an oxidation potential below 0.24 eV can be reacted with porphyrin dye. However, most electron donating molecules have the oxidation potential higher value than 0.24 eV (typical valus of oxidation potential of electron donors are between 0.6 an 1.35 eV). Therefore, porphyrin dye does not undergo photo-reducible transfer reaction with electron donors [44, 79] .

In summary, the improvement of the photoinitiating efficiency of the photoinitiating systems can be achieved by (1) regarding back electron transfer and recombination reaction step and (2) using a secondary reaction step for consuming dye-based radical and regenerating the original photosensitizer (dye). It was concluded from the kinetic results that the photo-oxidizable series mechanism efficiently retards back electron transfer and recombination reaction step. Both processes occur in the photo-reducible series mechanism. In addition, the photo-oxidizable series mechanism provides an efficient secondary reaction step that involves consumption of the dye-based radical and regeneration of the original photosensitizer (dye). However, back electron transfer process limites the kinetic in the photo-reducible series mechanism. The photo-oxidizable series mechanism significantly enhances conversion and rates of polymerization in comparison with the photo-reducible series mechanism [44, 80].

RESULTS AND DISCUSSION

Polymethine Dyes as Sensitizers in the Three-Component Photoinitiating Systems

The chemical modification either of the dye molecule structure or composition of photoinitiating system leads to the enhancement of the photoinitiating ability of photoinitiating system. It was achieved by the introduction of a second co-initiator to the two-component photoinitiating system composed of borate salt of polymethine dye.

The polymerization ability of such photoinitiating systems is very often more efficient than those observed for RBAX/NPG couple (typical triplet photoinitiating system).

Three-component photoinitiating systems under investigation consist of mono-cationic symmetrical carbocyanine dyes, unsymmetrical cyanine dyes (hemicyanines), three-cationic symmetrical carbocyanine dyes, two-cationic hemicyanine dyes and mono- and multi-cationic monomethine dyes as absorbing chromophore paried with *n*-butyltriphenylborate anion (B2) (electron donor) and a second co-initiator. The structures of selected two-component photoinitiating systems and second co-initiators (compounds 7 – 26) are presented in Chart 2.

In order to accelerate the rate of polymerization initiated by two-component photoinitiating systems composed of polymethine dye borate salt, the following compounds are added: N-alkoxypyridinium salt, 1,3,5-triazine derivative, N-methylpicolinium ester, cyclic acetal and heteroaromatic thiol as a second co-initiator.

The kinetic studies were carried out for photoinitiator systems composed of:

- Polymethine dye as iodide or bromide/tetramethylammonium *n*-butyltriphenylborate/ second co-initiator,
- Polymethine dye as *n*-butyltriphenylborate salt/second co-initiator.

Chart 2

Two-component photoinitiating system

where R_1 = H, Br

where X = S, O

(Dye I B2)
(Dye II B2)

(Dye VIIA B2)

(Dye III B2)
(Dye V B2)

where X = S, O, $C(CH_3)_2$

(Dye XI B2)
(Dye XII B2)
(Dye XIII B2)

Second co-initiators
N-alkoxypyridinium salts

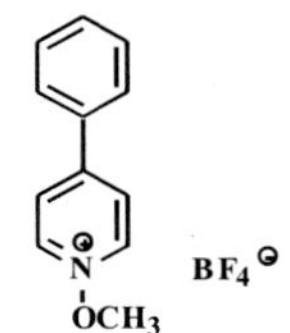

N-methoxy-4-phenylpyridinium tetrafluoroborate
NO
(7)

N,N'-dimethoxy-2,2'-bipirydilium ditetrafluoroborate
Bp1
(8)

N,N'-diethoxy-2,2'-bipirydilium ditetrafluoroborate
Bp2
(9)

Chart 2. (continued)

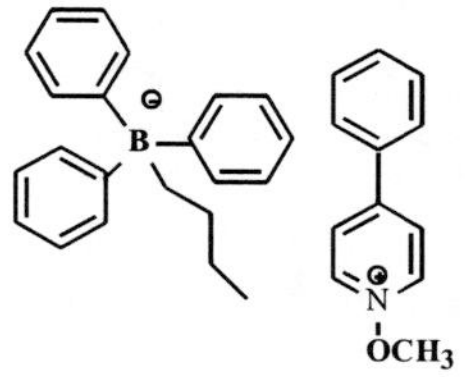

OCH_3

N-methoxy-4-phenylpyridinium *n*-butyltriphenylborate
NOB2
(10)

1,3,5-triazine derivatives

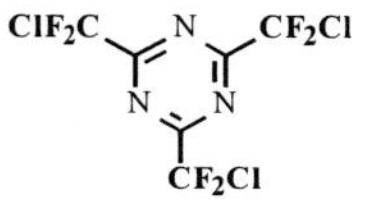

ClF_2C N CF_2Cl N N CF_2Cl

Cl_3C N CCl_3 N N OCH_3

2,4,6-Tris(chlorodifluoromethyl)-1,3,5-triazine
T1
(11)

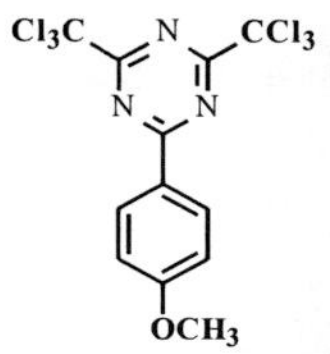

2,4-Bis(trichloromethyl)-6-(4-methoxy)phenyl-1,3,5-triazine
T2
(12)

N-methylpicolinium esters

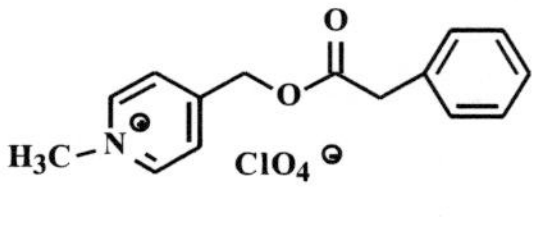

[4-(N-methyl)pyridinium]methyl phenylacetic acid perchlorate
E1A
(13)

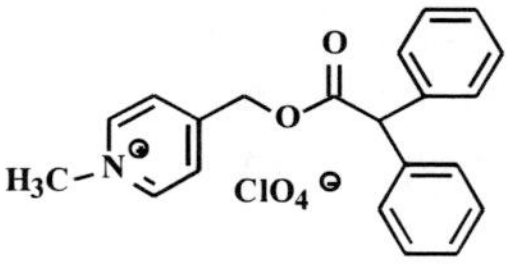

[4-(N-methyl)pyridinium]methyl diphenylacetic acid perchlorate
E1B
(14)

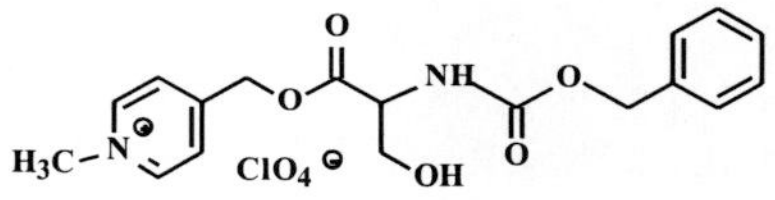

[4-(N-methyl)pyridinium]methyl L-serine acid perchlorate
E1D
(15)

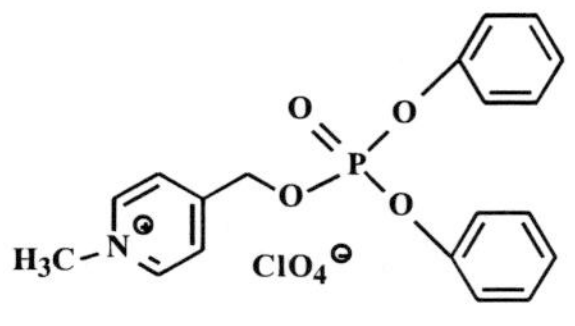

[4-(N-methyl)pyridinium]methyl diphenylphosphate acid perchlorate
E1E
(16)

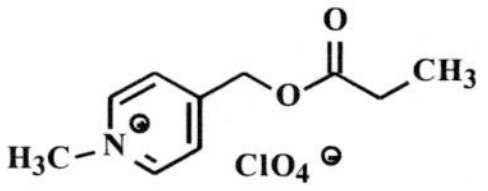

[4-(N-methyl)pyridinium]methyl proprionic acid perchlorate
E1G
(17)

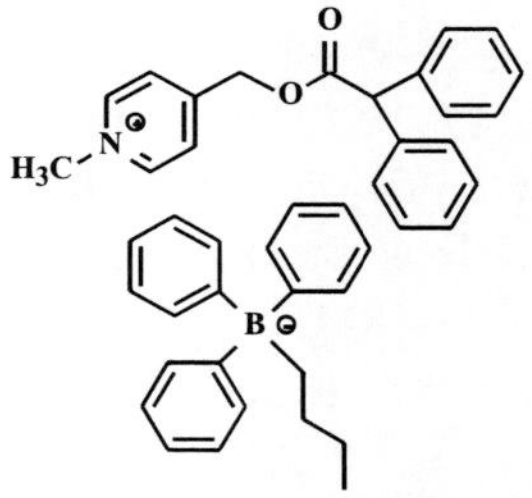

[4-(N-methyl)pyridinium]methyl diphenylacetic acid *n*-butyltriphenylborate
E1BB2
(18)

Cyclic acetals

Chart 2. (continued)

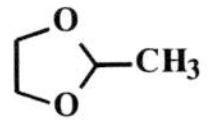

2-Methyl-1,3-dioxolan K1 (19)	2-Methoxy-1,3-dioxolan K2 (20)	1,3-Benzodioxolan K3 (21)	2-Phenyl-1,3-dioxolan K4 (22)	Glycerol formal K5 (23)

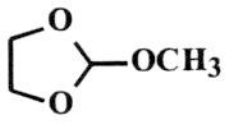

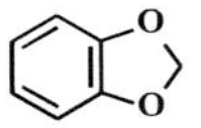

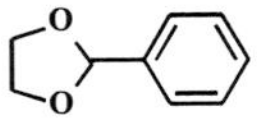

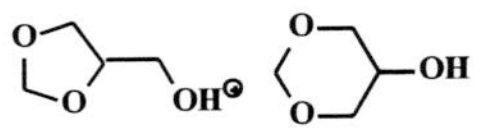

Heteroaromatic thiols

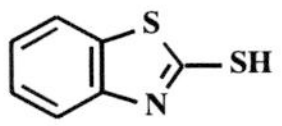

2-Mercaptobenzothiazole (MS) (24)	2-Mercaptobenzoxazole (MO) (25)	2-Mercaptobenzimidazole (MI) (26)

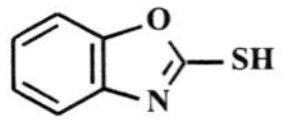

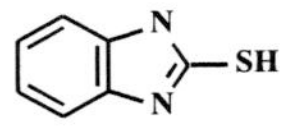

Three-Component Photoinitiating Systems Composed of Polymethine Dye/*n*-Butyltriphenylborate/N-Alkoxypyridinium Salt

Gould and Farid described the reactions of photoinitiated polymerization that apply the electron transfer occurring between an excited state of dye and N-alkoxypyridinium salts, acting as the efficient ground-state electron acceptors [81]. The three-component photoinitiating systems, possessing hemicyanine dye, borate salt and N-alkoxypyridinium salt studied by I.R. Gould are very interesting [81]. In these systems the double fragmentation occurs. For neutral dye, besides the radical formed in the sensitized cleavage of N-alkoxypyridinium salt, a radical cation of hemicyanine dye (oxidized donor $D^{\bullet+}$) is formed. Gould tried to utilize the chemical energy stored in radical cation for further enhance of the desired photopolymerization process. Thus, to the two-component photoinitiating system the third ingredient (alkyltriphenylborate salt) was added. As it was mentioned above, the oxidative cleavage of borate salt yields an alkyl radical. The radical cation of the dye ($D^{\bullet+}$) formed in the sensitized reducive cleavage of an N-alkoxypyridinium (RO-Py^{+}) could take part in a second electron transfer reaction with an alkyltriphenylborate to generate a second radical. In such photoinitiating systems, two initiating radicals could be formed for each absorbed photon, $RO^{\bullet}$ and $R^{\bullet}$ [81]. The mechanism of the reactions occurring in such compositions during initiation of free radical polymerization was given by I.R. Gould (Scheme 7) [81].

From the experiments performed, it is evident that the 1:1 combination of the three components clearly showed an approximate doubling in the rate of the free radical polymerization compared with those when initiator was used alone [81].

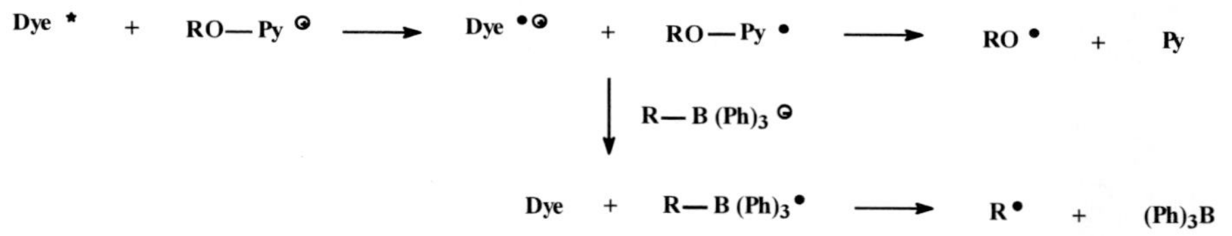

$$Dye^{*} + RO{-}Py^{\oplus} \longrightarrow Dye^{\bullet\oplus} + RO{-}Py^{\bullet} \longrightarrow RO^{\bullet} + Py$$

$$\downarrow R{-}B(Ph)_3^{\ominus}$$

$$Dye + R{-}B(Ph)_3^{\bullet} \longrightarrow R^{\bullet} + (Ph)_3B$$

Scheme 7. The primary and secondary reactions for the three-component photoinitiating systems composed of polymethine dye/alkyltriphenylborate salt/N-alkoxypyridinium salt after irradiation with a visible light.

The photochemistry of such electron transfer reactions was, in part, clarified by Schuster et al., who concluded that for the singlet state reaction the nitrogen-oxygen bond cleavage competes successfully with the back electron transfer. When reaction occurs in an overall triplet state, back electron transfer cannot occur, and solvation and nitrogen-oxygen bond cleavage to form alkoxy radical are competitive [82].

There are two significant structural differences between the main system under study in my work and that reported by Schuster [82]. First, the absorbing dyes are positively charged and this after an electron transfer allows to obtain a neutral radical. Second, we use both co-initiators paired either with different counterions or in form of an ion pair, which allows substrates of electron transfer reactions in a specific form to organize and minimalized a diffusion effect on overall efficiency of photoinitiation. Therefore, the overall rate of free radical formation reaction is not controlled either by the back-electron transfer reaction or by the solvation process. Under this condition the rate of electron transfer, the diffusion, or spatial arrangement of both, substrates and short-lived intermediates limit the rate of photoinitiated polymerization.

The reactivity of several photoinitiation three- and two-component systems composed of positively charges 3-ethyl-2-(*p*-(N,N-dimethylamino)styryl)benzothiazolium iodide (dye III), 3-ethyl-2-(*p*-pyrrolidinestyryl)benzothiazolium iodide (dye IV), N,N'-diethylthio-carbocyanine iodide (dye I), 6-bromo-N-ethyl-2-((*p*-substituted)styryl)benzothiazolium iodide (dye V) and N-ethyl-2-((*p*-substituted)styryl)quinolinium iodide (dyes VII) acting as light absorbers, *n*-butyltriphenylborate ($B2^-$) anion acting as electron donor, and alkoxypyridinium (NO^+) cation acting as ground-state electron acceptor has been compared [24, 48].

Both borate anion and N-alkoxypyridinium cation in polymerizing formulations can be present either as tetramethylammonium salt (B2) and tetrafluoroborate salt (NO), respectively, or can form an ion pair composed of borate anion and N-alkoxypyridinium cation (NO^+B2^-) [24].

Figures 4 and 5 present the kinetic traces recorded during an argon laser photoinitiated polymerization of TMPTA/MP (9:1) mixture in the presence of 3-ethyl-2-(*p*-(N,N-dimethylamino)styryl)benzothiazolium iodide and N,N'-diethylthiocarbocyanine iodide as the light absorber and functioning as co-initiators: (i) *n*-butyltriphenylborate tetramethylammonium salt, (ii) mixture of *n*-butyltriphenylborate tetramethylammonium salt and N-methoxy-4-phenylpyridinium tetrafluoroborate, and finally (iii) the ion pair composed of *n*-butyltriphenylborate anion and N-methoxypyridinium cation [24].

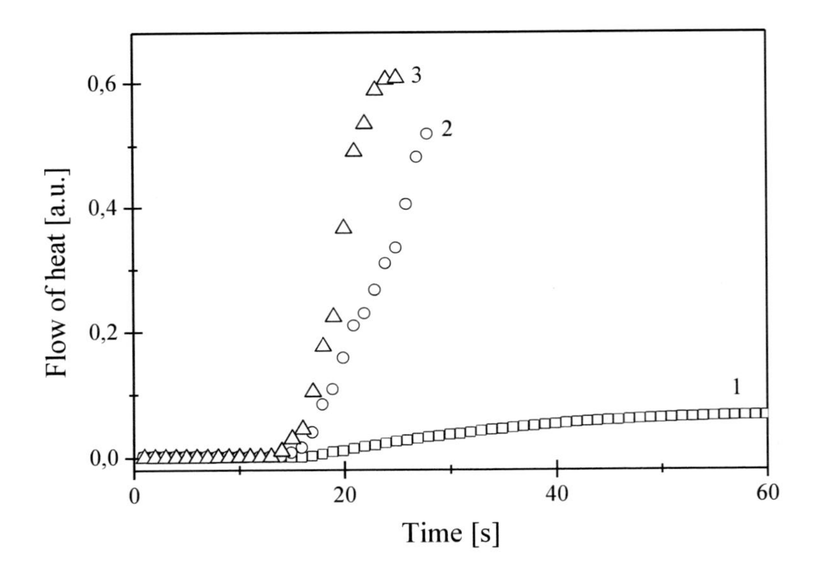

Figure 4. Family of curves recorded during the measurements of the flow of heat during the photoinitiated polymerization of the TMPTA/MP (9:1) mixture initiated by 3-ethyl-2-(*p*-N,N-dimethylaminostyryl)benzothiazolium iodide (dye III; c = 1×10^{-3} M) in the presence of (1) B2, (2) B2 and NO mixture (1:1), and (3) ion pair $NO^{+}B2^{-}$. Concentration of each co-initiator was equal 2×10^{-2} M. Light intensity was equal 64 mW/cm^2 [24].

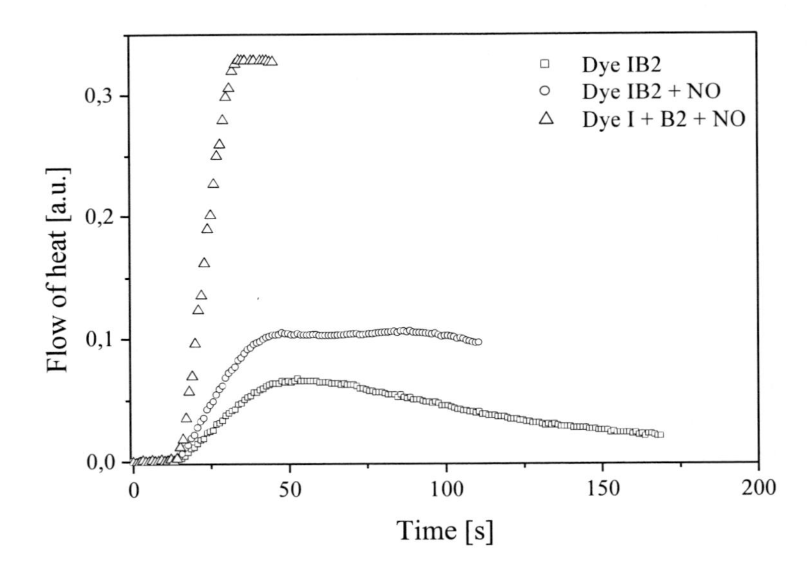

Figure 5. The effect of N-methoxy-4-phenylpyridinium tetrafluoroborate (NO) on the rate of free radical polymerization of the TMPTA/MP (9:1) mixture initiated by the two-component photoinitiating system (N,N'-diethylthiocarbocyanine iodide/*n*-butyltriphenylborate salt). The concentration of dye and co-initiators was equal 1×10^{-3} M and 3×10^{-3} M, respectively. Light intensity was equal 64 mW/cm^2.

The rate of photoinitiated polymerization is the lowest for the system containing only *n*-butyltriphenylborate salt as an electron donor. The kinetic results obtained are surprising. The addition of equimolar amount of N-methoxy-4-phenylpyridinium tetrafluoroborate to the two-component photoinitiating system composed of polymethine dye as *n*-butyltriphenylborate salt results in the increase of observed rate of free radical polymerization about 2-3 times in comparison to the two-component system. When, in similar two-component photoinitiating system composed of identical sensitizer, *n*-butyltriphenylborate anion and N-methoxy-4-phenylpyridinium cation as an ion pair, the rate of polymerization increases about 50-times as compared with two-component ones [24].

The next specific feature that differentiates the three-component photoinitiating systems composed of ion pair NO^+B2^- from a mixture of *n*-butyltriphenylborate tetramethylammonium salt and N-methoxy-4-phenylpyridinium tetrafluoroborate is their influence on the rate of free radical polymerization (Figure 6) [24].

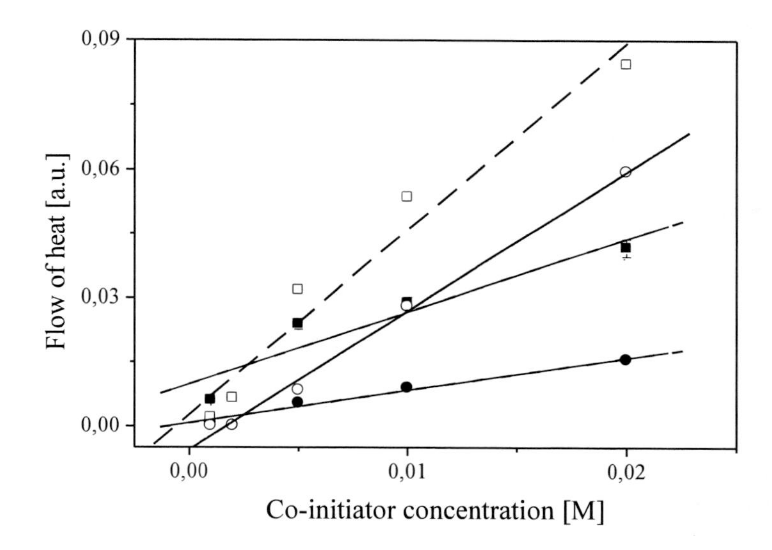

Figure 6. Dependence of the polymerization rates on co-initiator concentration. The solid points represent the system composed of B2 and NO mixture, and open points represent the NO^+B2^- pair as co-initiator. Circles represent the data for dye III and squares for dye IV [24].

On the basis of the results shown in Figure 6, it appears that the difference in co-initiation efficiency of NO^+B2^- ion pair in comparison with B2 and NO mixture is widely increasing as concentration of co-initiators increases.

We have documented earlier, that organic ion pair is partially dissociated even in medium polarity sovents [23]. A degree of dissociation depends on the concentration of solute in a complex fashion. However, in the simple approximation a degree of dissociation descreases as the concentration of solute increases (or a concentration of nondissociated salt is enhancing when concentration of solute increases). This behavior explains the relationship presented in Figure 6. For a co-initiator introduced into solution as an ion pair, its concentration increase causes higher concentration of nondissociated salt and, this in turn, starts to prefer the mechanism of polymerization described in Scheme 8 by the lower path [24]. It is, thus reasonable to conclude that the secondary reactions yielding free radicals occur in three-component encounter complex.

The highest rate of polymerization initiated by the three-component photoinitiating system composed of: hemicyanine dye/*n*-butyltriphenylborate anion/N-alkoxypyridinium cation is observed for the formulation in which *n*-butyltriphenylborate and N-methoxy-4-phenylpyridinium ion pair is acting as co-initiating pair.

In such three-component photoinitiating systems the light intensity has only insignificant influence on the rate of free radical polymerization of TMPTA. In Figure 7 it is shown that if the light intensity was decreases from 50 to 15 mW/0.785 cm^2 the efficiency of initiation of free radical polymerization by three-component photoinitiator system is still very high.

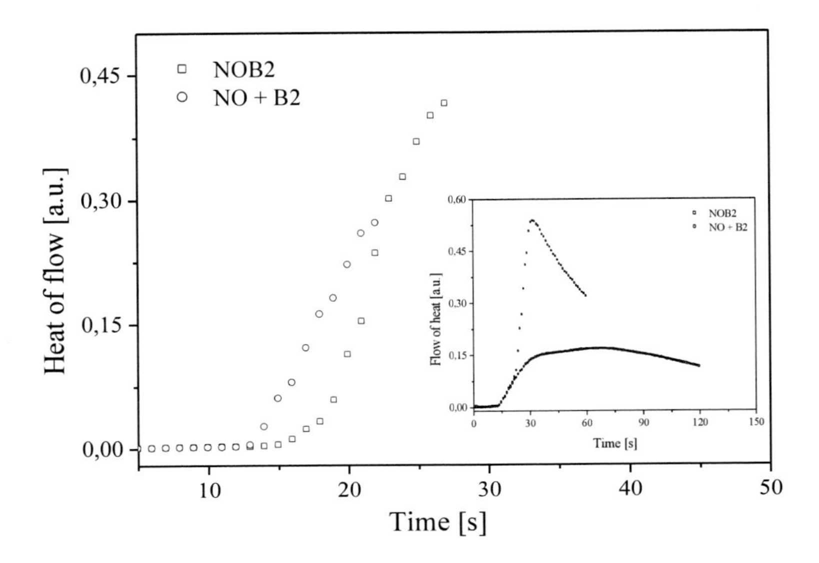

Figure 7. The kinetic curves of TMPTA/MP mixture polymerization initiated by 6-bromo-3-ethyl-2-(4-N,N-dimethylaminostyryl)benzothiazole ethyl sulfate (dye V) (c = 1 × 10^{-3} M) in the presence of B2 + NO and NOB2, respectively. Concentration of each co-initiator equal 1 × 10^{-3} M. Light intensity equal 50 mW/0.785 cm^2 (inset I_a = 15 mW/0.785 cm^2) [83].

There are some observations that are pertinent to the properties of co-initiating systems under study:

- Two-component photoinitiating systems composed of polymethine dye/N-alkoxypyridinum salt do not initiate free radical polymerization
- The photoinitiating system being the mixture of dye cation-borate ion pair and N-alkoxypyridiunium salt exhibits higher photoinitiation ability in comparison to the ability presented by dye cation-borate anion pair [24].

All this specific behaviors can be explained assuming that there is an additional factor affecting the rate of photoinitiated polymerization. It was believed that the observed difference comes from the specific spatial arrangement of all components of photoinitiating system. On the basis of the photochemistry of borate anion [12, 13] and photochemistry of N-alkoxypyridinium cation [81, 84] the mechanism of photochemical processes occurring in the three-component photoinitiating system which is consistent with laser flash photolysis was proposed (Scheme 8) [13, 24, 33, 82, 84, 85].

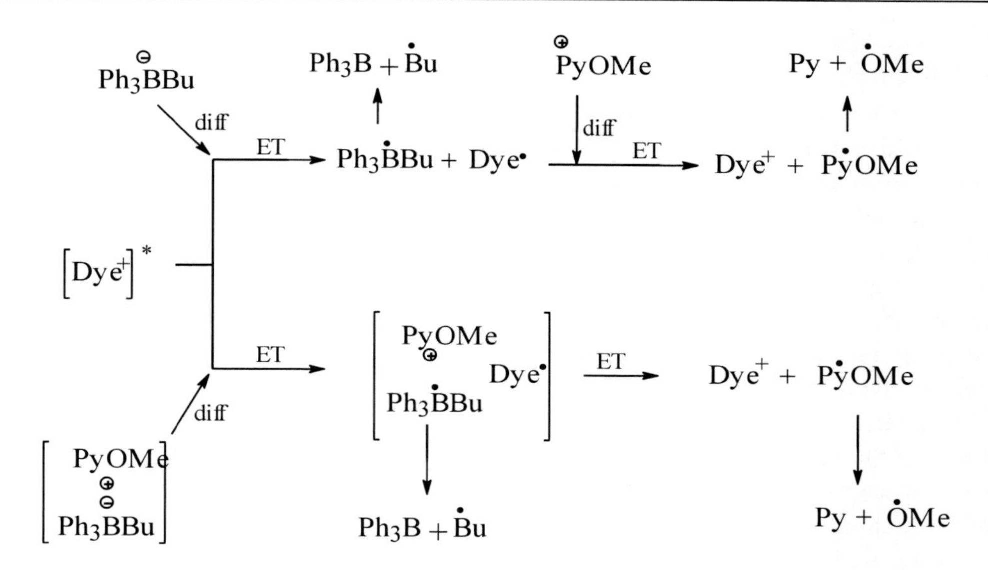

Scheme 8. The primary and secondary processes occuring in three-component photoinitiating system composed of: polymethine dye/*n*-butyltriphenylborate salt/N-alkoxypyridinium salt after irradiation with the visible light.

In this scheme, two possible mechanisms of free radical generation are considered. The upper path describes the processes that can occur when all initiating components are not organized; e.g., they are present in formulation as salts of photochemically inert counterions.

After excitation, to make an electron transfer effective, an electron donor and an electron acceptor must diffuse to each other to form encounter complex, in which electron-transfer reaction takes place. The resulting boranyl radical decomposes, yielding neutral triphenylboron and butyl radical [12, 13, 85]. The other product of electron-transfer reaction, dye radical, could participate in a second electron-transfer reaction with alkoxypyridinium cation to form alkoxypyridinium radical and in this way generate a second initiating radical as a result of N-O bond cleavage [24]. The effective rate of dye radical quenching by alkoxypyridinium cation can be expressed by the equation (7):

$$k_{obs} = \tau_T^{-1} + k_q[NO^+] \qquad \text{eq. (7)}$$

in which τ_T is the lifetime of the dye radical in the absence of an N-alkoxypyridinium cation (NO^+). It is obvious that the NO^+ concentration is uniform in an entire volume of the solution when N-alkoxypyridinium cation is uses as its tetrafluoroborate salt. Situation is different when co-initiators are present in solution as ion pair. After excitation, the dye and co-initiators ion pair diffuse and form encounter complex. An electron transfer from borate ion follows this process forming boranyl and dye radicals. However, in this case the borate ion that is reaching excited dye is accompanied by N-alkoxypyridinium cation. This artificially enhances the concentration of N-alkoxypyridinium cation in proximity to the dye radical. This, in turn, effective increases the rate of dye radical quenching. As a result, one observes an increase in speed of photoinitiated polymerization.

The mechanism of photochemical reactions that occur after electron transfer for the three-component photoinitiating systems composed of: polymethine dye/borate salt/N-alkoxypyridinium salt was proposed on the basis of the laser flash photolysis. N-(9-Methylpuryne-6-yl)pyridinium chloride was used as a model compound. The properties of an excited state of this compound are known from Skalski's and et al. studies [86, 87]. The N-(methylpurin-6-yl)pyridinum cation (Pyr^+) was selected for because (i) the chromophore is positively charged, (ii) molecule undergoes intersystem crossing and its intermediates obtained after laser pulse and electron transfer are well described and may be monitored spectroscopically on the nanosecond time scale and (iii) the reduction potential of Pyr^+ is equal –0.57 V, e.g., is close to the corresponding value measured for N-methoxy-4-phenylpyridinium tetrafluoroborate (-0.67 V) [81]. The application of this compound given us a possibility of the study of the quenching of an excited singlet state by both *n*-butyltriphenylborate anion and N-methoxypyridinium cation.

The laser flash photolysis experiment was performed for N-(methylpurin-6-yl)pyridinium cation (Pyr^+) in the presence of (i) tetramethylammonium *n*-butyltriphenylborate (B2), (ii) N-methoxy-4-phenylpyridinium tetrafluoroborate (NO), and (iii) ion pair NO^+B2^- in acetonitrile (MeCN) solution [24, 33]. The transient spectra of N-(methylpurin-6-yl)pyridinium chloride (Pyr) alone and in the presence of *n*-butyltriphenylborate tetramethylammonium salt (B2) are shown in Figure 8 [24].

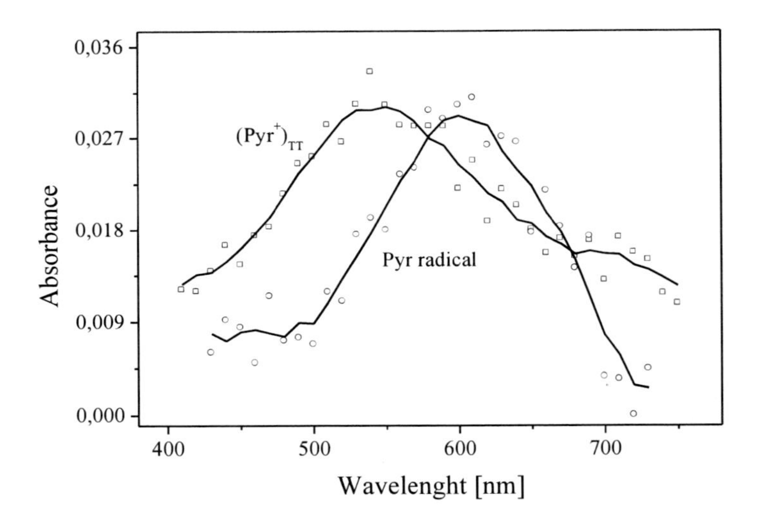

Figure 8. Transient absorption spectra recorded 100 ns after laser flash (355 nm) for N-(methylpurin-6-yl)pyridinium chloride (Pyr^+) in MeCN (squares) and 500 ns after flash for N-(methylpurin-6-yl)pyridinium chloride in the presence of *n*-butyltriphenylborate tetramethylammonium salt (B2) (c = 5 × 10^{-3} M) (circles). Concentration of Pyr^+ equal 2 × 10^{-4} M [24].

Irradiation of Pyr^+ with 5 ns laser pulse results in instantaneous appearance of its triplet state, which is characterized by absorption at 550 nm. The Pyr^+ triplet is quenched by *n*-butyltriphenylborate tetramethylammonium salt (B2), and a new transient with absorption at 610 nm is simultaneously formed. The new transient was assigned to N-(methylpurin-6-yl)pyridinium radical ($Pyr^\bullet$) [86]. The lifetime of $Pyr^\bullet$ in MeCN solution is about 10 μs and is decreasing as concentration of NO^+ increases [24]. The Stern-Volmer plot obtained from the $Pyr^\bullet$ lifetime measurements, as is shown in Figure 9, is linear over the whole range of quencher concentration.

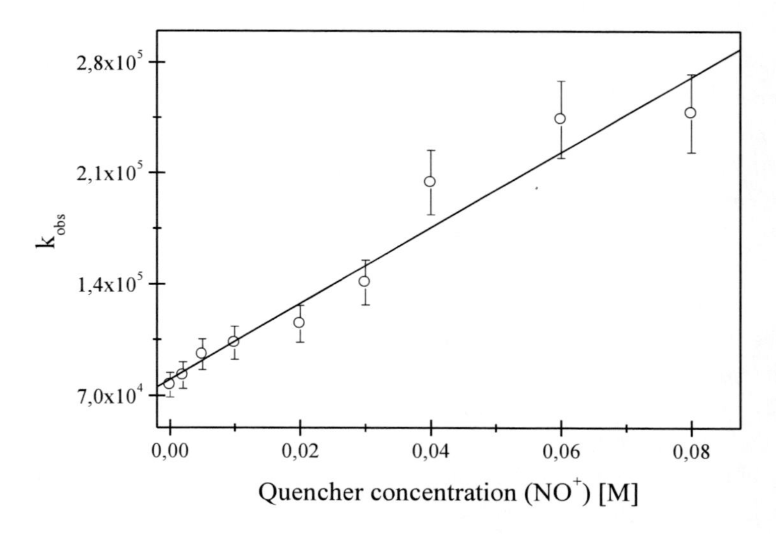

Figure 9. Stern-Volmer plot for quenching of $Pyr^{\bullet}$ by NO^+ in MeCN solution [24].

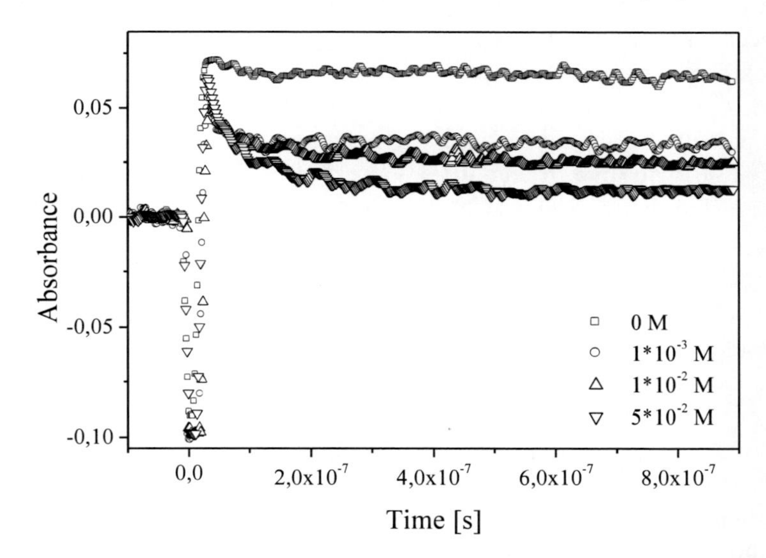

Figure 10. Kinetic traces for $Pyr^{\bullet}$ decay at 610 nm in the presence of various amounts of N-methoxy-4-phenylpyridinium cation. The concentration of NO^+ marked in Figure [24].

The established value of this rate constant is equal 3.94×10^{10} $M^{-1}s^{-1}$, e.g. reaches the diffusion-controlled limit. The electron transfer process does not occur in the two-component photoinitiating system composed of N-(9-methylpurin-6-yl)pyridinium chloride/N-methoxy-4-phenylpyridinium cation. For high concentration of *n*-butyltriphenylborate tetramethylammonium all triplets of Pyr^+ are quenched, and N-(9-methylpurin-6-yl) pyridinium radical ($Pyr^{\bullet}$) is simultaneously formed (λ_{max} = 610 nm). Bearing in mind the mechanism presented in Scheme 8, one should predict that the addition of N-methoxypyridinium cation should quench obtained radicals. Figure 10 shows the kinetic traces recorded at 610 nm for various concentrations of N-methoxy-4-phenylpyridinium cation in which triplets of Pyr^+ are quenched by *n*-butyltriphenylborate tetramethylammonium salt (B2).

The laser flash photolysis experiments confirmed that the alkoxypyridinium cation is reduced by the dye radical. This reaction yields the dye cation and alkoxypyridinium radical which undergoes N-O bond cleavage, giving stable pyridine and alkoxy radical.

The thermodynamic cycle for such free radical generation radicals is given by Gould et. al. [81]. From Gould's analysis it is evident that, when the reduction potential of donor molecule is more negative than that of N-alkoxypyridinium salt (i.e. it is harder to reduce), the photochemical electron transfer reaction will be exothermic. Thus, the energetics of electron transfer from an excited hemicyanine dye to N-alkoxypyridinium salt, and thus the efficiency of forming initiating radicals, can be a function of the difference in the reduction potential of the two reactants [81].

The dyes under study reduce at about –1.06 V for dye III and at about –0.73 V for dye IV (vs Ag-AgCl). The measured value of reduction potential for the N-methoxy-4-phenyl-pyridinium cation is about –0.67 V. Under this condition, in the presence of alkoxypyridinium cation, only dye III can weakly initiate polymerization because energetic, calculated using the Rehm-Weller equation, only for this photoredox pair is close to the region when the electron transfer becomes exothermic. A word of caution is required for the energy stored in the dye radical-N-alkoxypyridinium pair. For photoinduced electron transfer reaction between a donor and an acceptor, the energy stored in pair is define as the difference between the oxidation potential of a donor and a reduction potential of an acceptor (referred also as the redox energy). The oxidation potential of the dye radical is approximately equal to the reduction potential of the dye cation. The reduction potentials for the dyes radical under the study are –1.06 V for dye III and –0.73 V for dye IV. Thus, the driving force of electron transfer between the dye radical and N-methoxy-4-phenylpyridinium cation is –0.42 eV (-40.5 kJ mol^{-1}) for dye III and –0.06 V (-5.8 kJ mol^{-1}) for dye IV. The negative values indicate that the electron transfer between dye radicals and N-methoxy-4-phenylpyridinium cation is thermodynamically allowed [24].

The primary selection of efficient photoinitiating system should obey several major criteria:

- The interaction between the excited singlet state of the polymethine dye and *n*-butyltriphenylborate (B2) must be efficient in order to generate a large number of initiating radicals.
- The interaction between the dye radical ($dye^{\bullet}$) and N-alkoxypyridnium salt (NO) should also be efficient in order to generate a large number of the second type of initiating radical [24, 83].

Basing on the results obtained for the three-component photoinitiating systems, in the next step of the novel photoinitiating system developing should be focused on the elimination of the process that causes decrease of its photoinitiation ability. Based on Scheme 8, for the photoinitiation occurring via singlet excited state the diffusion of photoinitiator components and the back electron transfer limit an overall efficiency of photoinitiation. The simplest way of the diffusion effect elimination is the covalent bonding of a dye with either electron donor or electron acceptor. In this view three different unsymmetrical dyes, e.g. 2-((*o*-, *m*- or *p*-)-methoxypyridine)-*p*-pyrrolidine methyl sulfates (dyes VIII) were synthesized by a reaction of N-methoxymethylpyridinium sulfate salts with 4-pyrrolidinebenzaldehyde [31].

These dyes combine in one molecule the absorbing chromophore and ground-state electron acceptor. According to Gould's hypothesis [81] these dyes should be rather poor photoinitiators (E_{red} = -0.955 V). Indeed, the measurements of their photoinitiation efficiency explicitly confirmed this prediction. However, in the presence of *n*-buthyltriphenylborate tetramethylammonium salt (B2) 2-(*o*-methoxypyridine)-*p*-pyrrolidinestyrilium methyl sulfate (dye VIII) becomes extremely efficient as visible-light photoinitiator for free radical polymerization. Its photoinitiation ability is similar to that observed for initiating via triplet-state xanthene dyes described by Neckers [24, 88]. Although the mixture of dye VIII and *n*-butyltriphenylborate tetramethylammonium (B2) is less sensitive in comparison to the mixture of dye IV and ion pair (NO^+B2^-), the photoinitiating system composed of dye VIII and B2 should be considered as very effective. The difference in sensitivity between the mixture of dye IV/NO^+B2^- and the mixture of dye VIII/B2 may come from the different photoinitiation efficiency of 3-ethyl-2-(*p*-(N,N-dimethylamino)styryl)benzothiazolium dye (dye III) and 3-ethyl-2-(*p*-pyrrolidinestyryl)benzothiazolium dye (dye IV). The control measurements of the polymerization photoinitiation efficiency performed for 3-ethyl-2-(*p*-pyrrolidinestyryl)benzothiazolium iodide (dye IV) and N-methyl-2-(*p*-pyrrolidinestyryl) pyridinium) iodide (dye VI) in presence of *n*-butyltriphenylborate tetramethylammonium (B2) clearly demonstrated that the photoinitiation performance of dye IV + B2 mixture is about one order of magnitude greater in comparison to the dye VI + B2 photoredox pair (Figure 11) [24]. Therefore, one can conclude, that the type of chromophore has influence on the photoinitiation ability of photoredox pair [24].

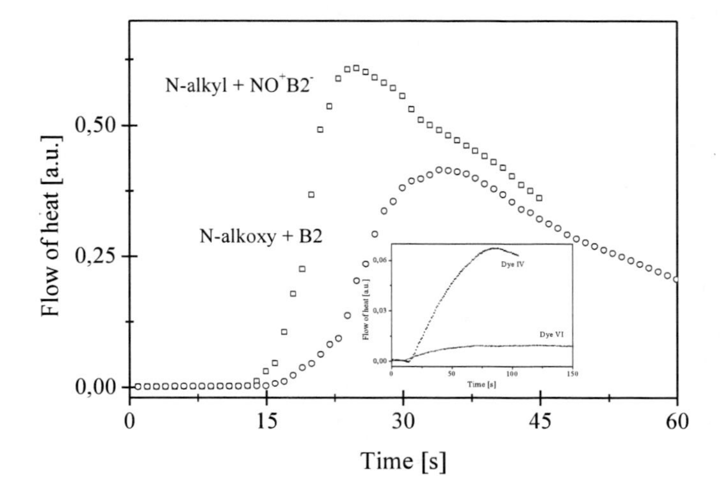

Figure 11. Kinetic curves of TMPTA/MP polymerization photoinitiated by 3-ethyl-2-(*p*-pyrrolidinestyryl)benzothiazolium iodide (dye IV) (c = 1 × 10^{-3} M) in the presence of ion pair NO^+B2^- (5 × 10^{-2} M) and 2-(*o*-methoxypyridine)-*p*-pyrrolidinestyrilium iodide (dye VIII) (c = 1 × 10^{-3} M) in the presence of *n*-butyltriphenylborate tetramethylammonium B2 (5 × 10^{-2} M). Inset: kinetic curves of TMPTA/MP mixture photopolymerization initiated by dyes IV and VI (c = 1 × 10^{-3} M) in presence of B2 (5 × 10^{-2} M) [24].

One may suppose that 2-(*o*-methoxypyridine)-*p*-pyrrolidinestyrylium methyl sulfate (dye VIII) in combination with *n*-butyltriphenylborate tetramethylammonium (B2) itself is very sensitive, and there is no one photon-two radical reaction. A simple way to verify this doubt is the comparison of the photoinitiation ability of dye VIII and dye IV (Figures 11 and 12) [24].

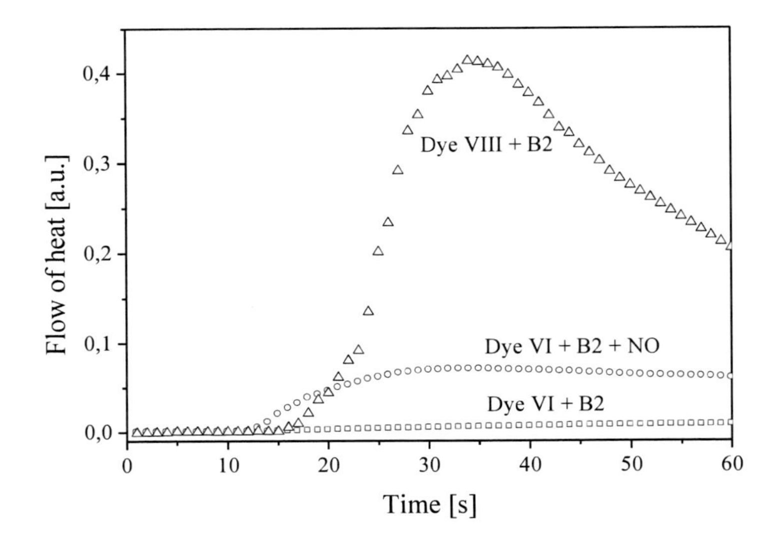

Figure 12. Kinetic traces recorded during polymerization photoinitiated by dyes VIII and VI. Photoinitiating systems marked in the figure [24].

There is the striking difference between the photoinitiating system composed of (i) dye VI acting as electron acceptor and *n*-butyltriphenylborate tetramethylammonium (B2) as electron donor, (ii) dye VI and two co-initiators B2 (electron donor) and NO (ground-state electron acceptor), and (iii) dye VIII in the presence of B2. On the basis of kinetic results presented one should recall Scheme 8 underlining the influence of dissusion of photoinitiator components on its photoinitiation ability. In the case of photoinitiating system composed of dye VI in the presence of *n*-butyltriphenylborate salt and N-alkoxypyridinium salts, after dye excitation, first, excited dye and borate salt must diffuse to each other for effective electron transfer. After the electron transfer the radical formed from the dye and N-alkoxypyridinium salt in order to form an encounter complex should diffuse to each other as well. Since 3-ethyl-2-(*p*-pyrrolidinestyryl)benzothiazolium dye (dye IV) no short-living intermediates are observed in nanosecond time scale, it is obvious that all processes occur in regime controlled by diffusion. Thus, elimination only one step that is diffusion-controlled should significantly increase the overall rate of entire reaction. The feasibility of this is demonstrated by a system composed of dye VIII (2-(*o*-methoxypyridine)-*p*-pyrrolidinestyrylium methylsulfate) and borate salt, in which after absorption of light an excited molecule and borate anion encounter (diffusion-controlled process) for effective electron-transfer reaction.

The increase of the rate of polymerization is explained as follows: after an electron transfer process, the resulting *n*-butyltriphenylboranyl radical decomposes, yielding *n*-butyl radical (first initiating radical) and triphenyl boron. The second product of electron transfer, 2-(*p*-pyrrolidinestyryl)-N-methoxypyridinium radical, being unstable, decomposes to give methoxy radical (second initiating radical) and 1-(*p*-pyrrolidenephenyl)-2-(2-pyridine)ethene. The decomposition process is not diffusion-controlled. Thus, the application of photoinitiating systems composed of the dye with N-alkoxy group attached to the quaternary nitrogen atom and alkyltriphenylborate anion eliminates one of the rate-determining steps (Scheme 9) [24].

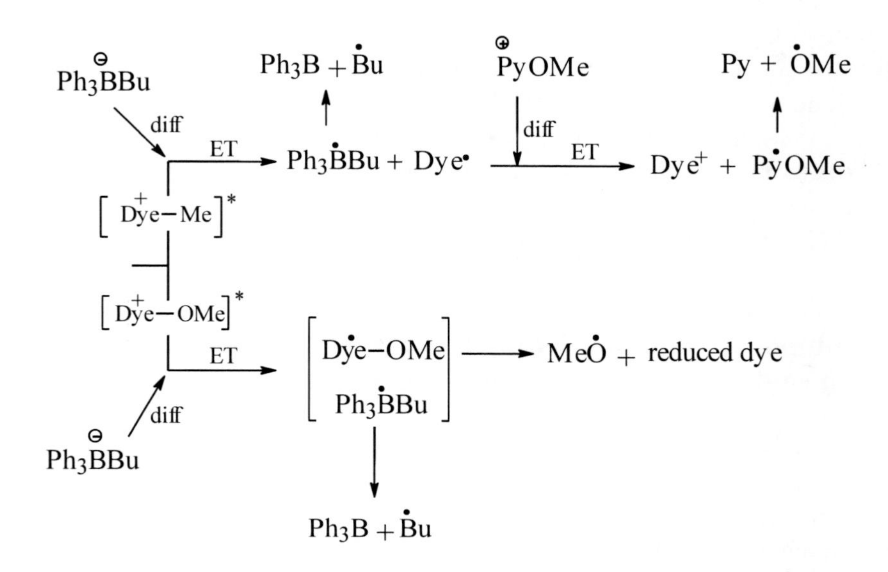

Scheme 9. The formation of radical initiating the free radical polymerization after irradiation of three-component photoinitiating system with visible light [24].

There is one more feature that can strongly increase the yield of free radicals formation. The reaction between dye VIII and *n*-butyltriphenylborate salt occurs via the singlet state. In such case a back electron transfer may be very efficient. The simplest way to avoid this energy loss is to find a competitive channel forming radicals by bond cleavage reaction. In the case of dye VIII the nitrogen-oxygen bond cleavage competes successfully with the back electron transfer [24].

Three-Component Photoinitiating Systems Composed of Polymethine Dye/*n*-Butyltriphenylborate/N,N'-Dialkoxy-2,2'-Bipyridilium Salt

Besides N-alkoxypyridinium salts as a ground-state electron acceptor other N-alkoxyheterocyclic compounds which undergo the one-electron transfer reducion and yielding the alkoxy radical initiating free radical polymerization may by used [81, 89]. Basing on the Schuster's results describing an electron transfer process between the aromatic hydrocarbons and N,N'-dialkoxy-2,2'-bipyridile it was interesting to apply N,N'-dimethoxy- and N,N'-diethoxy-2,2'-bipyridilium salts (compounds 8 and 9) as co-initiators in the photoinitiating systems.

Five different hemicyanine dyes ((6-bromo-2-ethyl-2-(*p*-alkylamino)styryl) benzo-thiazolium salts (dyes V) were used as sensitizers in photoinitiator systems. For the analysis of the photoinitiating properties were used: hemicyanine dyes as *n*-butyltriphenyl-borate salts and second co-initiator as ditetrafluoroborate salt. The irradiation of such photoinitiating systems with visible light may generate following radicals: *n*-butyl, methoxy, ethoxy that can initiate free radical polymerization of triacrylate monomers [83]. Therefore, one can conclude that on the overall efficiency of the photoinitiation of free radical polymerization can influence: (i) the rate of electron transfer process, (ii) the rate of carbon-boron bond cleavage and the rate of nitrogen-oxygen bond cleavage or (iii) the reactivity of free radicals formed. Analizing the kinetic results, one can conclude that the rate of *n*-butyltriphenylboranyl radical decomposition is greater than that of N,N'-dialkoxy-2,2'-bipyridilium radicals. The stability of radicals formed after electron transfer process is in order:

N,N'-diethoxy-2,2'-bipyridyl > N-methoxy-4-phenylpyridyl > *n*-butyltriphenylboranyl.

The two-component photoinitiating systems composed of hemicyanine dye and N,N'-dialkoxy-2,2'-bipyridilium salt do not initiated free radical polymerization [83].

Three-Component Photoinitiating Systems Composed of Polymethine Dye/*n*-Butyltriphenylborate/1,3,5-Triazine Derivatives

Among few examples of visible or near visible light photoinitiators, the substituted bis(trichloromethyl)-1,3,5-triazine derivatives are widely mentioned in the patent literature alone or in the presence of the sensitizers or/and co-initiators, such as titanocene, peroxide/amine or mercaptan [17, 90]. However, the tremendous amount of the patents dealing with triazine derivatives is in contrast with the rarity of articles relevant to photochemical characteristics of these systems. Most of them are concerned with the photochemistry of 1,3,5-triazine derivative alone or in the presence of only a sensitizer [17, 91-93]. Some authors have suggested a serial mechanism whereby the 1,3,5-triazine derivative and dye react via the known bimolecular initiating scheme [16]. Usually, in photoinitiating systems composed of sensitizer/1,3,5-triazine derivative after irradiation with a visble light an electron transfer from an excited state of sensitizer to ground state of 1,3,5-triazine occurs. The triazynyl radical (initiating polymerization) and chloride anion are formed as a result of C-Cl bond cleavage in triazynyl radical anion (Scheme 10) [94].

After electron transfer process two radicals are formed: triazynyl and sensitizer-based. The resulting 1,3,5-triazine-based radical is active for initiation, but the dye radical is a terminating radical. The third component is thought to oxidize this inactive dye-based radical, regenerating the original dye and producing an active second radical. The photochemical properties of 1,3,5-triazine derivative were recently described by Pohlers [16]. The detailed photochemical behavior of 1,3,5-triazine in the three-component photoinitiating systems composed of Rose bengal/tertiary amine/1,3,5-triazine derivative and Eozine/tertiary amine/1,3,5-triazine derivative was also recently described by C. Grotzinger [4]. In such photoinitiating systems the electron transfer form an amine to an excited state of sensitizer is primary photochemical process. The secondary process is the electron transfer from dye-based radical to the ground state of 1,3,5-triazine derivative [4]. Other photoinitiating systems were composed of merocyanine dye/1,3,5-triazine derivative and were studied by Kawamura [7, 15].

However, nothing is known about interactions in a sensitizer/borate salt/1,3,5-triazine three-component photoinitiating system.

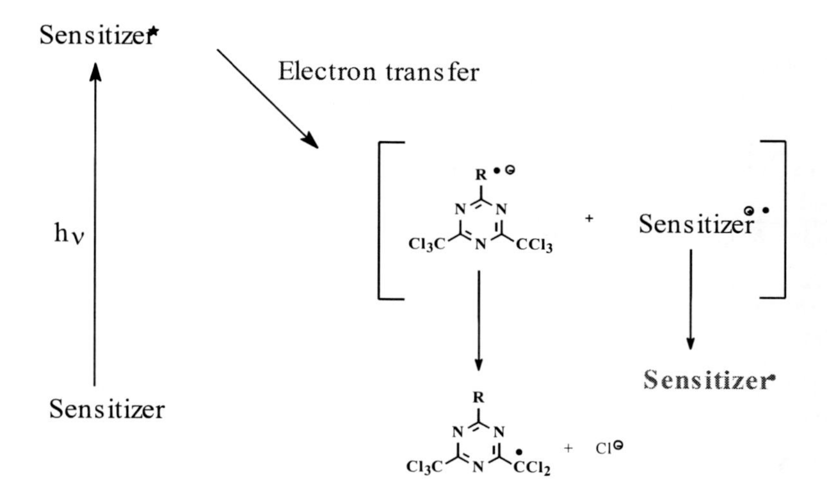

Scheme 10. The photochemical processes occurring in the two-component photoinitiating system: sensitizer/1,3,5-triazine derivative.

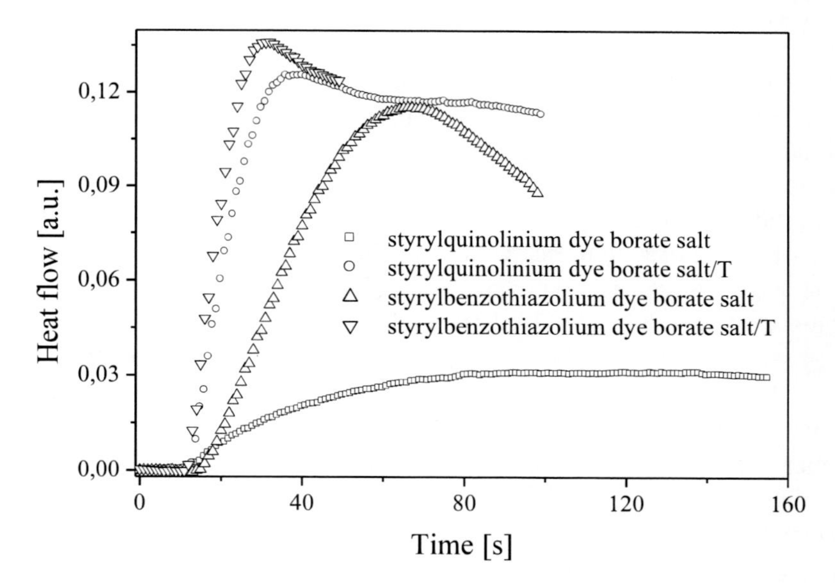

Figure 13. Family of kinetic curves recorded during the measurements of the flow of the heat during the photoinitiated polymerization of TMPTA/MP (9:1) mixture initiated by hemicyanine dyes in the presence of different co-initiators marked in Figure. The dye and co-initiators concentrations were 1×10^{-3} M and 4.5×10^{-2} M for T, respectively [94].

The photoinitiation efficiency of several combinations of polymethine dyes as iodide salts in combination with tetramethylammonium borates and 1,3,5-triazine derivatives (2,4,6-tris(chlorodifluoromethyl)-1,3,5-triazine and 2,4-bis(trichloromethyl)-6-(4-methoxy)phenyl-1,3,5-triazine) (T) (compounds 11 and 12) were used as the three-component photoinitiating systems for polymerization of TMPTA. Following polymethine dyes were used as sensitizers: N,N'-diethyloxocarbocyanine iodide (dye II), N,N'-diethylthiocarbocyanine iodide (dye I), 3-ethyl-2-(*p*-N,N-dimethylaminostyryl)quinolinium iodide (dye VII), 3-ethyl-2-(*p*-N,N-dimethylaminostyryl)benzothiazolium iodide (dye III) and three-cationic monomethine dyes (dyes XVI). The effect of 1,3,5-triazine derivative on the photoinitiating ability of three-component photoinitiating system is shown in Figure 13 and Table 1.

Table 1. The kinetic data obtained during the free radical polymerization of TMPTA

Photoinitiating system	Rate of polymerization [μmol·s^{-1}]	Relative rate of polymerization
Dye I B2	5.66	1
Dye I/NO	2.60	0.40
Dye I/Bp	2.29	0.40
Dye I/T	0.94	0.16
Dye I B2/NO	9.44	1.67
Dye I B2/Bp	8.03	1.42
Dye I B2/T	37.78	6.67
Dye I/B2/NO	42.50	7.51
Dye I/B2/Bp	113.33	20.02
Dye I/B2/T	2.36	0.42

Co-initiator concentration 3×10^{-3} M.

The two-component photoinitiating systems composed of polymethine dye/1,3,5-triazine derivative do not initiate or initiates with very low rates the free radical polymerization. Obviously, no efficient generation of initiating radicals by direct sensitization of the triazine occurs compared to the dye/borate salt system. The addition of 1,3,5-triazine to the dye/borate salt system produces a synergic effect in the polymerization reactions (Figures 13 and 14). The rates of polymerization initiated by three-component photoinitiating systems: polymethine dye/borate salt/1,3,5-triazine derivative are 3-7 times higher than those observed for two-component photoinitiating dye/borate salts initiators. The photoinitiating efficiency of three-component systems depends on the concentration of both co-initiators (Figures 14 and 15) [94].

The concentration of a second co-initiator (1,3,5-triazine derivative) affects the rate of photopolymerization up to its concentration equal 1×10^{-2} M. Further increase of 2,4,6-tris(chlorodifluoromethyl)-1,3,5-triazine (T) concentration is not changing the rate of free radical polymerization. Similarly, an additional amount of borate anion added to the photoinitiating system composed of cyanine dye and borate of equal concentration distinctly increases the rate of photoinitiation.

The effficiency of three-component photoinitiating system: polymethine dye/borate salt/1,3,5-triazine derivative is not a simple sum of the efficiences of two-component photoinitiating system: dye/borate salt and dye/1,3,5-triazine acting separately. Therefore, it seems that the improvement in photoinitiation for the system: polymethine dye/borate salt/1,3,5-triazine in comparison to the two-component photoinitiating system: polymethine dye/borate salt is a result of the secondary reactions between 1,3,5-triazine derivative and the species deriving from the first step of interaction, e.g. the electron transfer process between an excited singlet state of sensitizer and borate salt.

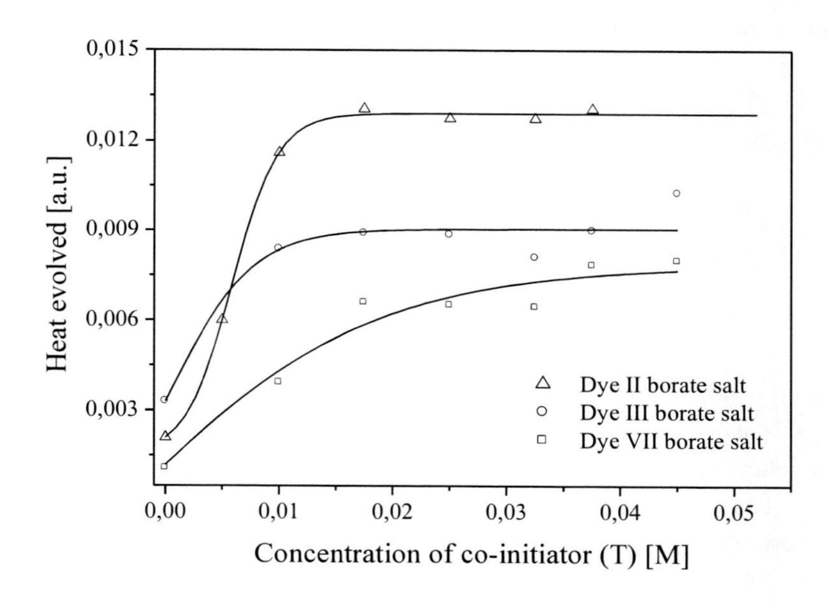

Figure 14. Effect of 2,4,6-tris(chlorodifluoromethyl)-1,3,5-triazine (T) concentration on the initial rate of free radical polymerization of TMPTA/MP (9:1) polymerizing mixture initiated by three-component photoinitiating system. Concentration of dye c = 1×10^{-3} M. Light intensity equal 50 mW/0.785 cm^2 [94].

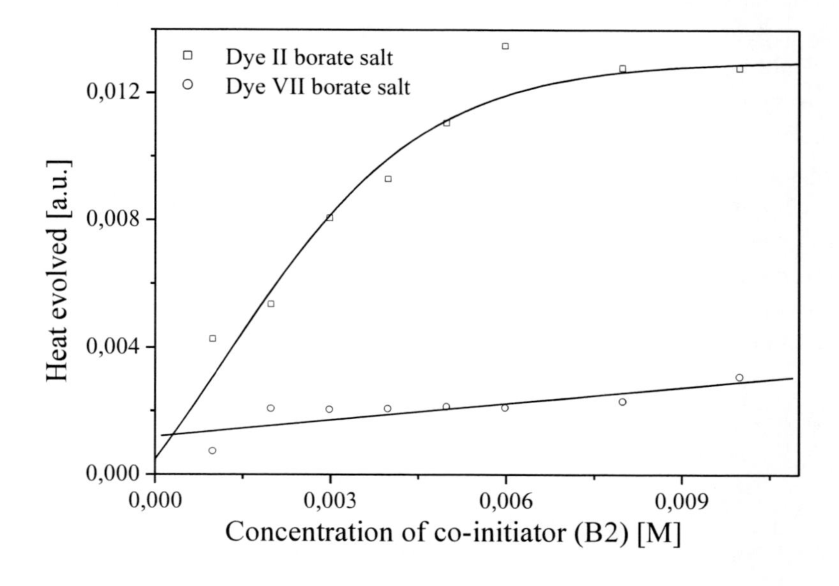

Figure 15. Effect of *n*-butyltriphenylborate (B2) concentration on the initial rate of free radical polymerization of TMPTA/MP (9:1) polymerizing mixture initiated by two-component photoinitiating system. Concentration of dye c = 1×10^{-3} M. Light intensity equal 50 mW/0.785 cm^2 [94].

For three-component systems theoretically following primary reactions are possible (eqs. 8-10):

$$Dye^{*} + B2 \longrightarrow Dye^{\bullet} + B2^{\bullet} \quad \text{eq. (8)}$$

$$Dye^{*} + T \longrightarrow Dye^{\bullet \ominus} + T^{\bullet \ominus} \quad \text{eq. (9)}$$

$$Dye^{*} + T \longrightarrow Dye^{\bullet} + T^{\bullet \ominus} \quad \text{eq. (10)}$$

where Dye* is an excited singlet state of polymethine dye, B2 is *n*-butyltriphenylborate salt, T is 1,3,5-triazine derivative. The calculated ΔG_{el} for above reactions are in the range from –0.592 eV to 0.33 eV (from 57.12 kJ·mol^{-1} to 31.84 kJ·mol^{-1}) [94]. The calculated free energies of the electron transfer reactions between the excited states of the dye and 1,3,5-triazine show that an electron transfer from the excited singlet state of the dye to the triazine is more feasible, leading to oxidation of the dye and reduction of triazine. As it was mentioned above, this reaction leads to the C-Cl bond cleavage and formation of free radicals and chloride ions [4]. Because two-component photoinitiating system dye/1,3,5-triazine derivative does not initiate free radical polymerization or initiates it with very low rates one can conclude that this process does not play a significant role in photopolymerization.

There are at least to possible explanations of observed phenomena. The first reasonable hypothesis suggests that the synergic effect of the three-component photoinitiating system behavior could be explained by an electron transfer interaction of reduced dye (dye radical) with the triazine. Cyanine dye radical is known as a weak terminator of the growing macromolecular chains. However, it reacts efficiently with an alkyl radicals, which leads to the decrease of the concentration of initiating radicals [22, 23, 26]. The linear relationship between the rate of polymerization and the square root of the light intensity absorbed confirms this postulate, suggesting that photoinitiated polymerization of the system proceeds by a conventional mechanism in which bimolecular termination occurs by the reaction between two macroradicals. This allows to conclude that the cyanine radicals do not act as terminator of polymer chain (eqs. 11 and 12) [94].

$$Dye^{\bullet} + {}^{\bullet}\text{~~~} \not\longrightarrow Dye\text{~~~} \quad \text{eq. (11)}$$

$$Dye^{\bullet} + R^{\bullet} \longrightarrow DyeR \quad \text{(Bleaching product)} \quad \text{eq. (12)}$$

Such interaction can sharply decrease an efficiency of initiation process, and this in turn, causes a decrease in observed rate of polymerization. The possible explanation of observed synergic effect for three-component system may consider possible redox reaction between cyanine dye radical and 1,3,5-triazine derivative. Similar reaction was well documented for dye radical and N-methoxypyridine cation [24]. The second explanation considers the back electron transfer reaction between the dye radical and 1,3,5-triazine radical anion. Such back electron transfer reaction can completely stop the formation of radicals from 1,3,5-triazine radical anion. This type of behavior was observed for certain cyanine dye-borate anion photoredox pairs [12, 13].

The interactions between the cyanine dye radical and 1,3,5-triazine can strongly reduce a terminating effect caused by dye radical, and additionally to form a new initiating radicals (eq. 13) [4, 24, 95].

$$\text{Dye}^{\bullet} + \text{T} \longrightarrow \text{Dye} + \text{T}^{\ominus\bullet} \rightarrow \text{Radicals}$$

eq. (13)

The following mechanism of free radical formation after irradiation of three-component photoinitiating system with visible light is proposed (Scheme 11) [94].

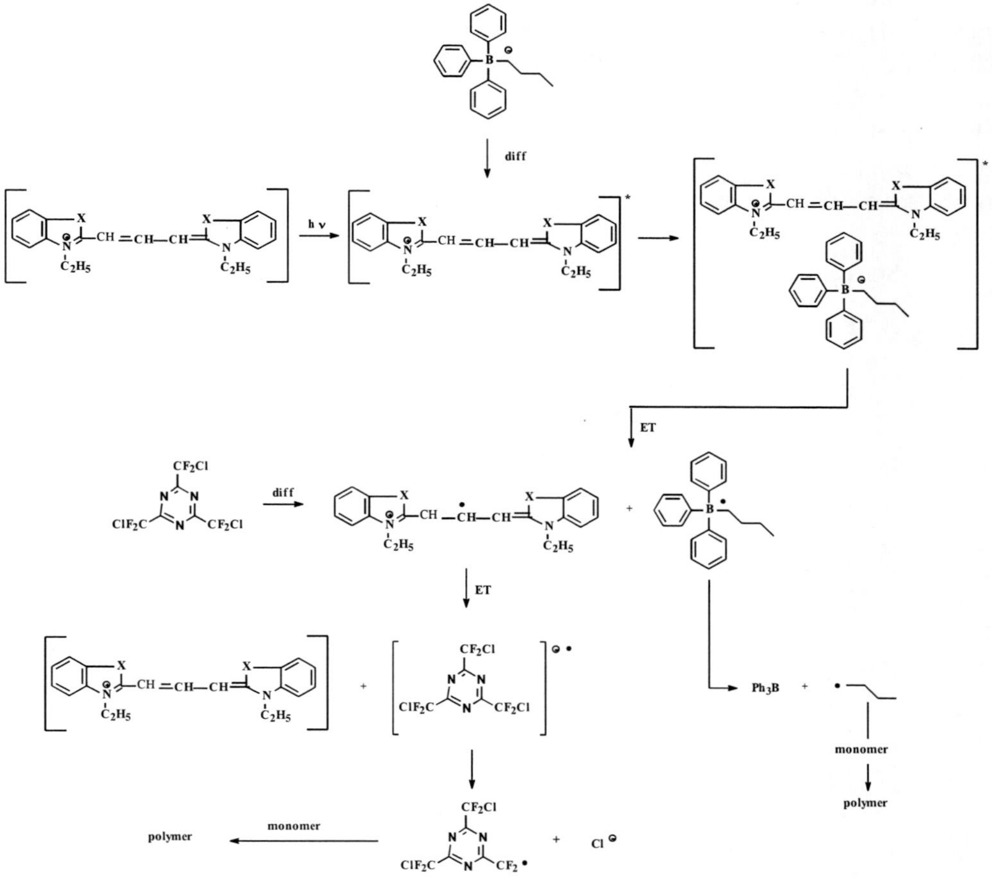

diff

X X N C2H5 CH =CH—CH= N C2H5 h ν X X N C2H5 CH =CH—CH= N C2H5 *

X X N C2H5 CH =CH—CH= N C2H5 B *

ET

CF2Cl N N ClF2C N CF2Cl diff X X N C2H5 CH —CH—CH= N C2H5 + B

ET

X X N C2H5 CH =CH—CH= N C2H5 + CF2Cl N N ClF2C N CF2Cl

Ph3B + monomer polymer

polymer monomer CF2Cl N N ClF2C N CF2 + Cl

Scheme 11. The photochemical reactions occurring between excited singlet state of polymethine dye and co-initiators [94].

The photoexcited dye molecule encounters *n*-butyltriphenylborate anion and accepts an electron from borate anion, forming boranyl radical and cyanine radical. In the next step, the electron transfer from cyanine radical to 1,3,5-triazine derivative occurs. This reaction regenerates the original dye and produces 1,3,5-triazine radical anion, which rapidly fragments into a halogene anion and triazinyl radical. The enhanced reaction rate observed in presence of the 1,3,5-triazine is explained in part by the fact that the 1,3,5-triazinyl radical, unlike cyanine dye radical is active for initiation. In addition, since the ground state of cyanine dye is regenerated in this reaction, the initiation rate will be enhanced further [94].

Three-Component Photoinitiating Systems Composed of Polymethine Dye/*n*-Butyltriphenylborate/N-Methylpicolinium Esters

Two-component photoinitiating systems composed of N-methylpicolinium esters derivatives as an electron acceptor and following photosensitizers: N,N,N',N'-tetramethylbenzydyne, 9-methylcarbazole, pyren and triphenylamine were described earlier by Ch. Sundararajan [96-98]. The laser flash photolysis experiment demonstrated the formation of unstable picolinium radical. This radical readily undergoes C-O fragmentation to yield a carboxylic acid and 4-pyridylmethyl radical [96-98]. The high quantum yields of formation of free radicals depending on the type of sensitizer used give the possibility of application of N-methylpicolinium esters as co-initiators in photoinitiating systems. These are examples of photosensitizers which undergo the C-O bond cleavage [96-98]. The mechanism of processes occurring in such photoinitiating systems is presented below (Scheme 12).

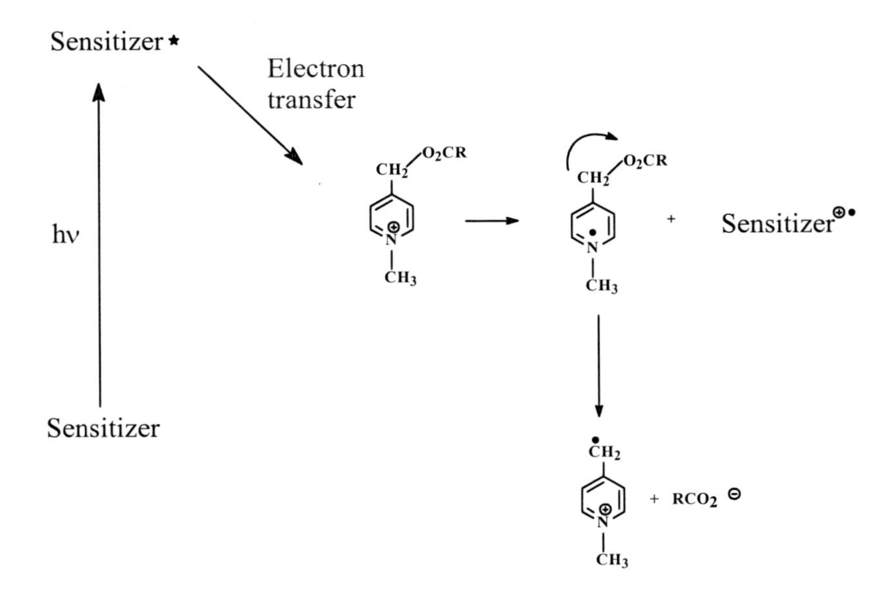

Scheme 12. Sensitized photoreduction of N-methylpicolinium ester.

Five N-methylpicolinium derivatives (compounds 13-17) were investigated to test their ability to function as second co-initiators in free radical photopolymerization [99]. N-Methylpicolinium esters derivatives were synthesized according to the method described by

Sundararajan [96-98] by the reaction of 4-pyridylcarbinol with acid chlorides and triethylamine in benzene or alternatively by a combining of the corresponding carboxylic acids with 4-pyridylcarbinol and dicyclohexylcarbodiimide in methylene chloride. A general route of the synthesis of N-methylpicolinium derivatives is shown in Scheme 13 [99].

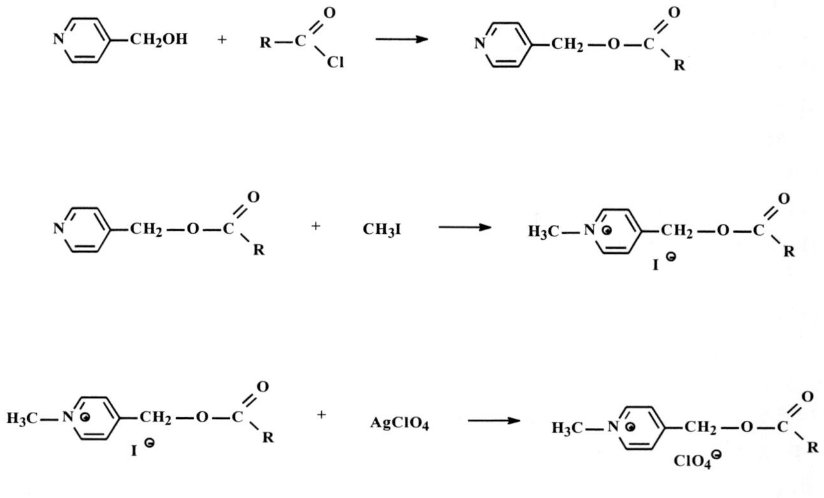

Scheme 13. Synthesis of N-methylpicolinium derivatives [99].

The synthesis of the N-methylpicolinium derivatives undergoes via three steps:

- The reaction of 4-pyridylcarbinol with appropriate organic acid or its derivative, yielding corresponding picolinium derivatives.
- The quaternization reaction of picolinium derivatives with methyl iodide that leads to the formation of corresponding N-methylpicolinium derivative iodides.
- The substitution reaction of N-methylpicolinium derivative iodides with silver perchlorate, resulting in the formation of the corresponding N-methylpicolinium perchlorates.

Pycolyl derivatives obtained in the firs step of synthesis were N-methylated with methyl iodide. The iodide salts studied display a charge-transfer absorption band around 350-400 nm, depending on the solvent. The charge-transfer absorption is attributed to the formation of picolinium/iodide ion pair. Because of the presence of CT absorption band, the iodide counterion is very often exchanged on perchlorate one to avoid any ambiguities resulting from competing absorption by the sensitizer and charge-transfer band. The perchlorate salts of N-methylpicolinium derivatives absorb below 320 nm [99].

Analizing the possible reduction-oxidation reactions between components of photoinitiating system, one should consider all processes that can occur between them. These should include interactions between all reactants and possible reactions between photoinitiating system components in their ground state and short-lived intermediates obtained after electron transfer process. Analysis of cyclovoltametric curve recorded for sensitizing dye (N,N’-diethyloxocarbocyanine iodide dye II) suggest that the dye easily reduces and does not oxidizes. This observation allows to conclude that the dye cannot be an electron donor in the two-component photoinitiating system composed of N,N'-diethyloxocarbocyanine/N-methylpicolinium derivative. Thus, adopting this observation to application of this type of photoredox pair as photoinitiating system, one can predict that this couple cannot act as effective photoinitiator.

Next possibility concerns the redox reaction between the N-methylpicolinium derivatives and borate anion. Knowing the oxidation potential of borate anion (1.16 eV), the reduction potentials of the N-methylpicolinium derivatives (from –0.60 eV to –0.94 eV), and taking into account fact that the N-methylpicolinium derivatives cannot be transferred into its excited state at 514 nm, one can easily calculate that the free energy change (ΔG_{el}) for reactions between N-methylpicolinium derivatives and borate anion oscillates from 1.760 eV to 2.10 eV (from 169.81 $kJ \cdot mol^{-1}$ to 202.62 $kJ \cdot mol^{-1}$). This observation clearly states that analyzed type of reaction is not thermodynamically allowed.

The third possibility concerns a reaction that can take place between the N,N'-diethyloxocarbocyanine radical ($E_{ox}^{\bullet}$), obtained after photoinduced electron transfer between dye cation and borate anion. Such intermediate can be treated as electron donating individuum in the reaction with N-methylpicolinium derivative ground state. The free energy change for these reactions depends on the N-methylpicolinium derivative structure and varies between –0.44 and –0.1 eV (from –42.45 $kJ \cdot mol^{-1}$ to –9.65 $kJ \cdot mol^{-1}$). The negative values of ΔG_{el}, in contrast to positive ones obtained for other possible reactions, suggest that there is a possibility of a secondary reaction between the dye radical and ground state of N-methylpicolinium derivative. From the cyclovoltammeric measurement it is known that both analyzed compounds undergo reduction. However, in the case of cyanine dye this process is fully reversible, whereas N-methylpicolinium derivative reduces irreversibly. The dye cation reduction yields dye radical that can undergo back electron transfer, namely can be easily oxidized (can be an effective reductor). On the other hand, N-methylpicolinium radical obtained does not undergo back electron transfer reaction (very fast and irreversible fragmentation leading to picolyl radical and carboxylic acid anion) [96, 97].

The photoinitiating systems composed of dyes (N,N'-diethylthiocarbocyanine iodide (dye I), N,N'-diethyloxocarbocyanine iodide (dye II) and 3-ethyl-2-(*p*-N,N-dimethylamino-styryl)benzothiazolium iodide (dye III)) as the photosensitizers with various co-initiators (borate salt and N-methylpicolinium derivatives) were used for the initiation of free radical polymerization.

Both co-initiators in polymerizing formulations were present either as tetramethylammonium salt (B2) and perchlorate, or as an ion pair. The diphenylacetic acid N-methylpyridiniummethyl ester-borate pair (compound 18) was prepared according to the method of syntesis of cyanine borate salts described by Schuster [12, 13].

Figures 16 and 17 demonstrate the effect of a second co-initiator on the rate of free radical polymerization initiated by the three-component photoinitiating system [99].

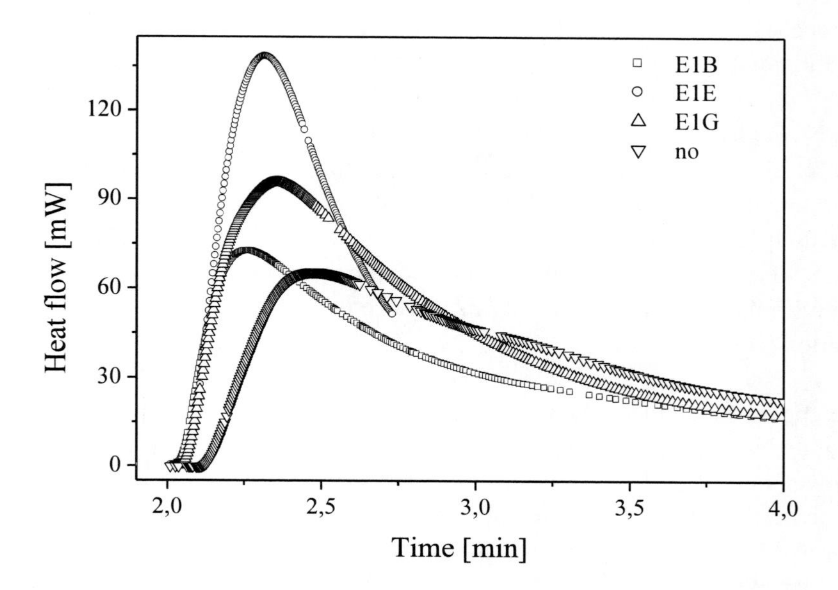

Figure 16. Family of kinetic curves recorded during the measurements of the flow of heat emitted during the photoinitiated polymerization of TMPTA/MP (9:1) mixture initiated by N,N'-diethyloxocarbocyanine *n*-butyltriphenylborate in presence of N-methylpicolinium derivative perchlorates marked in the Figure. The cyanine borate and N-methylpicolinium derivative concentrations were 5×10^{-3} M, $I_a = 20$ mW/0.196 cm^2 [99].

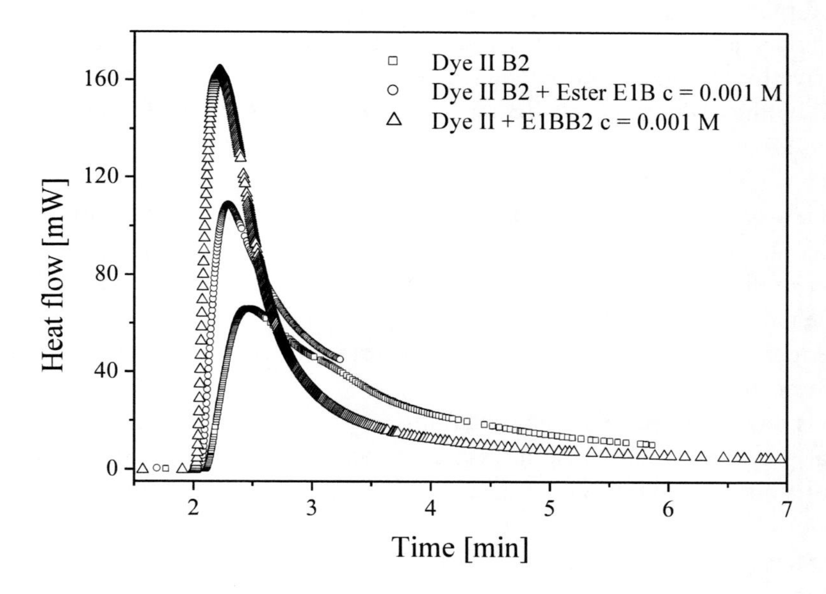

Figure 17. Family of kinetic curves recorded during the measurements of the flow of heat emitted during the photoinitiated polymerization of TMPTA/MP (9:1) mixture initiated by *n*-butyltriphenylborate N,N'-diethyloxocarbocyanine in presence of N-methylpicolinium derivative perchlorate marked in the Figure. The cyanine borate and N-methylpicolinium derivative concentrations were 5×10^{-3} M, $I_a = 20$ mW/0.196 cm^2.

Additionaly, Figure 17 presents the kinetic curves recorded during free radical polymerization of TMPTA/MP (9:1) mixture initiated by:

- N,N'-diethyloxocarbocyanine *n*-butyltriphenylborate,
- N,N'-diethyloxocarbocyanine iodide in presence of equimolar amount of tetramethylammonium *n*-butyltriphenylborate and diphenylacetic acid N-methylpyridinium methyl ester perchlorate, and finally
- N,N'-diethyloxocarbocyanine iodide in presence of the ion pair composed of *n*-butyltriphenylborate anion and diphenylacetic acid N-methylpyridinium methyl ester cation (compound 18).

The comparison of the polymerization rates observed for cyanine/borate ion pair and cyanine/borate/N-methylpicolinium ester three-component photoinitiating system indicates that the lower rate of polymerization was observed for the two-component photoinitiating system cyanine dye/borate salt. The addition of N-methylpicolinium derivative results in a significant acceleration of the polymerization process. The efficiency of the three-component photoinitiating system depends on the structure of N-methylpicolinium salt and changes in following order:

alkyl > phosphoryl > L-serine

However, the enhancement of the rate of the free radical polymerization observed (about 3-4 times) for the three-component photoinitiating system is not such significant as in the case of three-component photoinitiating systems possessing N-alkoxypyridinium salt or 1,3,5-triazine derivative as a second co-initiator. The observed lower acceleration of polymerization rate caused by the addition of picolinium derivative is probably due to much lower rates for the addition alkyl and alkoxy radicals to metacrylate double bond in comparison to similar reaction for the benzyl or benzyl-like N-methylpicolyl radicals. According to Fischer and Radom review paper, the absolute rate constant for addition alkyl radical (methyl radical) to acrylate double bond ($3.4 * 10^5$ $M^{-1}s^{-1}$) is about three orders of magnitude higher in comparison to similar reaction for benzyl radicals (430 $M^{-1}s^{-1}$) [100]. This difference is a result of the radical stabilizing effect of the phenyl group.

Furthermore, the highest rates of polymerization were observed when the both co-initiators form an ion pair.

There is one more specific feature that differentiates the efficiency of photoinitiation of free radical polymerization by the three-component photoinitiating system composed either both co-initiators as an inert salts or as an ion pair. This interesting behavior is presented in Figure 18.

The photoinitiating efficiency of co-initiators ion pair is widely increasing as concentration of co-initiators ion pair increases in comparison to both co-initiators inert salts mixture. There are also some observation and facts that are pertinent to the properties of such photoinitiating systems: (i) there is no photoinitiated polymerization when ony N-methylpicolinium ester was applied as co-initiator, (ii) the photoinitiating system being the mixture of dye cation-borate anion ion pair and N-methylpicolinium ester exhibits higher photoinitiation ability in comparison to the ability presented by dye cation-borate anion ion pair.

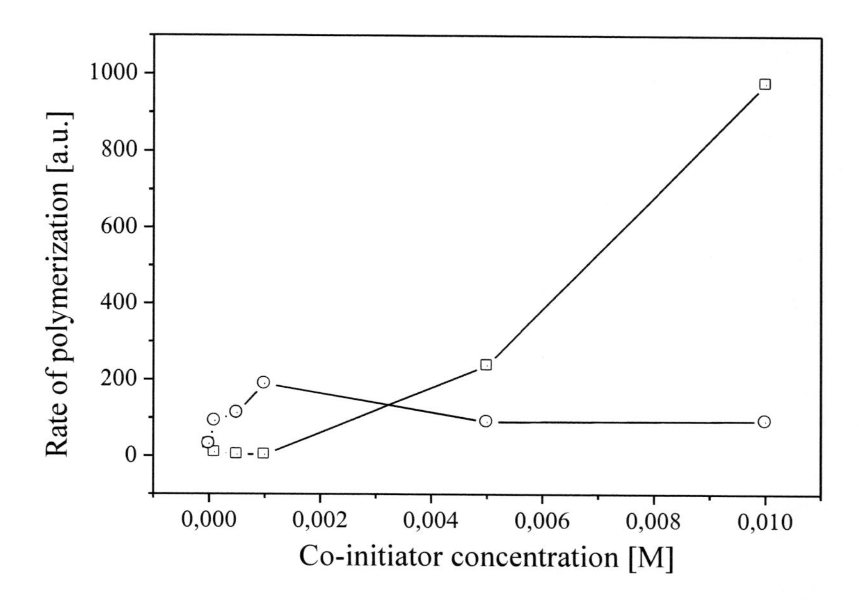

Figure 18. The dependence of the polymerization rates on a co-initiator concentration. The circles represent the data for hemicyanine dye (dye III) in presence of B2 and E1B mixture as co-initiators and squares for hemicyanine dye (dye III) in presence of ion pair of co-initiators, respectively.

Such behavior of three-component photoinitiating system is similar to that observed for the three-component photoinitiating system possessing N-alkoxypyridinium salt described above and is caused by a specific spatial organization of the reactants in the three-component encounter complex.

On the basis of the photochemical and photophysical properties of alkyltriphenylborate salts and N-methylpicolinium esters and laser flash photolysis results, a mechanism for the primary and secondary reactions yielding free radicals in the three-component photoinitiating system: cyanine dye/borate salt/N-methylpicolinium ester is proposed (Scheme 14) [99].

As it is seen in Scheme 14, two possible mechanisms of free radical generation are considered. The upper path describes the processes that can occur when all initiating components are not organized, e.g. they are present in formulation as salts of photochemically inert counterions. After excitation, in order to make an electron transfer effective the electron donor and electron acceptor must diffuse to each other to form an encounter complex, in which electron transfer reaction takes place. After irradiation of the three-component photoinitiating system with a visible light an excited singlet state of chromophore is formed. The deactivation of excited state occurs by fluorescence, photoisomerization or electron transfer process. Similarly, as it was presented for other photoinitiating system described above, in presence of alkyltriphenylborate anion, the cyanine dye undergoes one-electron reduction. The cyanine dye radical and boranyl radical are formed. The boranyl radical undergoes the C-B bond cleavage giving an alkyl radical that can start the polymerization reaction. Subsequently, the cyanine radical formed after electron transfer process in presence of N-alkylpicolinium derivative can participate in a second electron transfer process, giving cyanine cation and N-alkylpicolyl radical. This intermediate decomposes forming the pyridylmethyl radical, which can also initiate free radical polymerization.

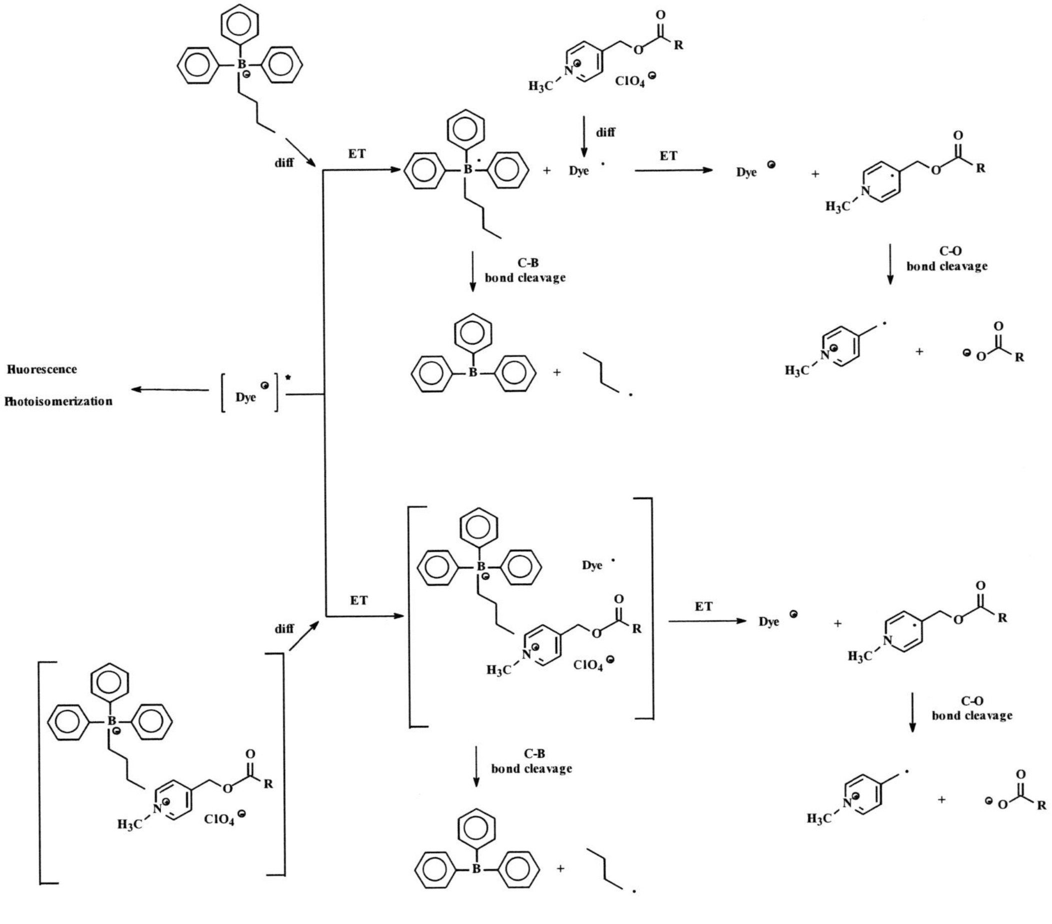

Scheme 14. The mechanism of the primary and secondary reactions for sensitized generation of free radicals after irradiation of three-component photoinitiating system composed of cyanine dye/borate salt/N-methylpicolinium ester with a visible light [99].

As it was mentioned earlier, the cyanine dye/borate salt ion pairs are partially dissociated even in medium polarity solvents. Therefore, for the co-initiators introduced into polymerizing mixture as an ion pair (Figure 18) its concentration increase causes the higher concentration of non-dissociated salt and, this in turn, starts to prefer the mechanism of polymerization described in Scheme 14 by the lower path.

The mechanism of the primary and secondary reactions for the three-component photoinitiating system was proposed on the basis of laser flash photolysis experiments.

The investigations were carried out for carbocyanine dye in presence of: (i) tetramethylammonium *n*-butyltriphenylborate, (ii) N-methylpicolinium ester perchlorate, (iii) cyanine dye *n*-butyltriphenylborate salt in presence of N-methylpicolinium perchlorate.

The transient absorption spectra of cyanine borate salt alone and in the presence of equimolar amount of N-methylpicolinium ester are presented in Figure 19 [99].

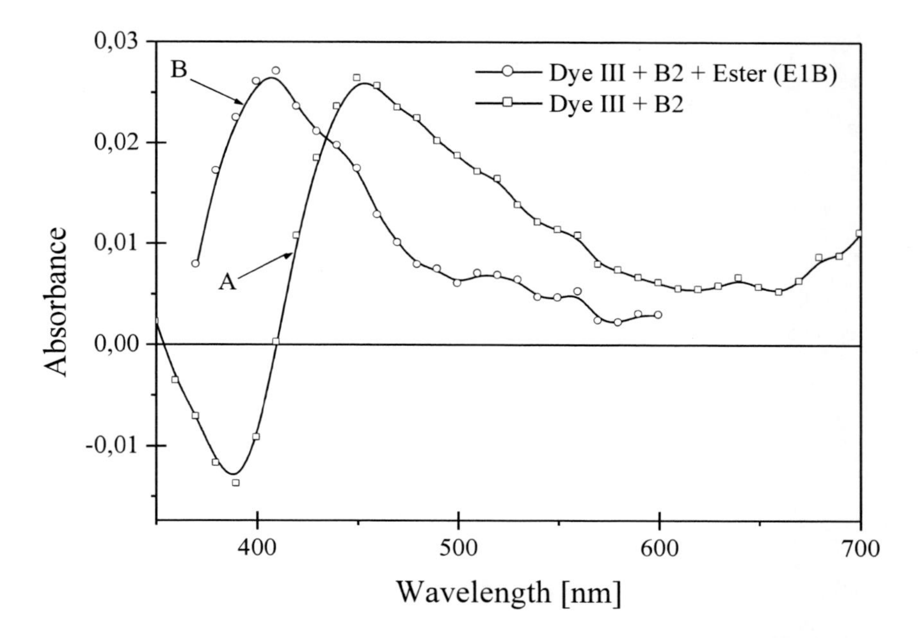

Figure 19. Transient absorption spectra of: (A) cyanine dye in presence of borate salt (B2) recorded 50 ns after laser pulse (squares) and (B) for cyanine dye in presence of equimolar ratio of tetramethylammonium *n*-butyltriphenylborate and N-methylpicolinium perchlorate (E1B) 100 ns after laser pulse (circles) (concentration of both equal 2×10^{-3} M) in acetonitrile solution. Dye concentration was 2×10^{-5} M [99].

As it was mentioned earlier, the irradiation of the carbocyanine dye with a visible light leads to an excited singlet state formation. This can be quenched in electron transfer process by *n*-butyltriphenylborate salt. The absorption band observed at 430 nm in the transient absorption spectra, after Schuster, can be assigned to the cyanine dye radical formed in electron transfer reaction [12, 13, 25]. There are no characteristic bands after irradiation of cyanine dye iodide solution in the presence of N-methylpicolinium derivative. However, laser flash photolysis recorded for three-component system composed of cyanine dye/borate salt/N-methylpicolinium derivative ($c = 2 \times 10^{-3}$ M) gives the new absorption band at 410 nm. The presence of a new band is caused in expense of the band recorded at 430 nm. This behavior is similar to that described by Sundararajan for system composed of sensitizing dye and N-alkylpicolinium ester [96-98]. This characteristic band was attributed to the presence of reduction product of N-methylpicolinium ester, for example, N-methylpicolinium radical. Taking into consideration, the results described by Sunderarajan et al. [96-98] for sensitizing dye/N-methylpicolinium ester photoredox couple, one can conclude that the addition of N-methylpicolinium derivative into two-component photoinitiating system composed of cyanine dye/borate salt causes a secondary reaction between the dye radical and N-methylpicolinium cation [99]. The N-methylpicolinium radical formation as a result of an electron transfer process is fast (τ of formation about 15 ns), but its dissappearance occurs durring the time of about 170 ns (Figure 20) [99].

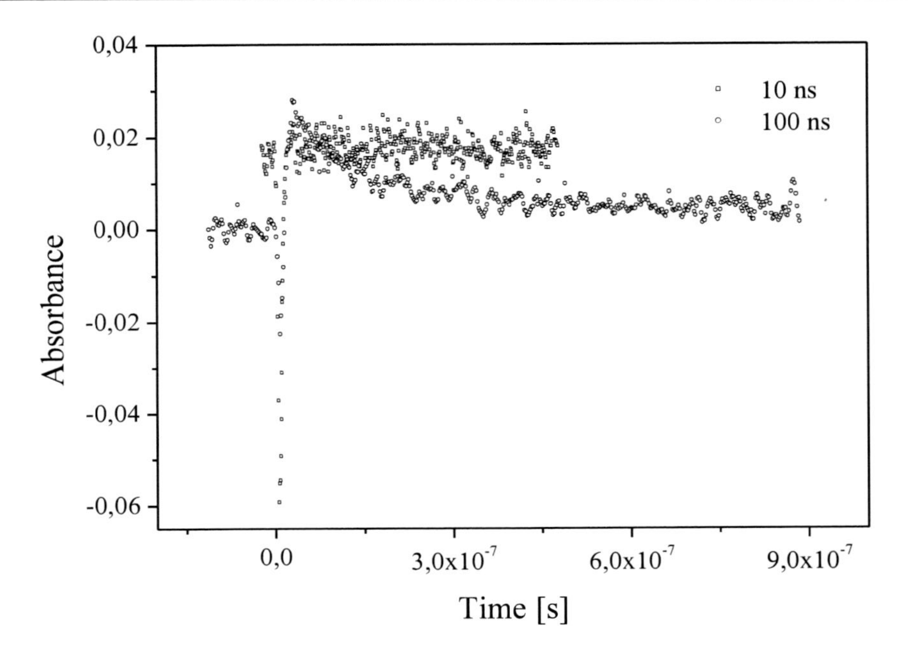

Figure 20. Transient absorption kinetic traces observed at 410 nm after different delay time (10 ns and 100 ns) for three-component system: cyanine dye/borate salt/N-methylpicolinium ester. Dye concentration equal 2×10^{-5} M, borate salt and ester concentration was 2×10^{-3} M [99].

The traces curves recorded at 410 nm and 430 nm in the case of two-component system composed of polymethine dye and borate salt are presented in Figure 21 [99].

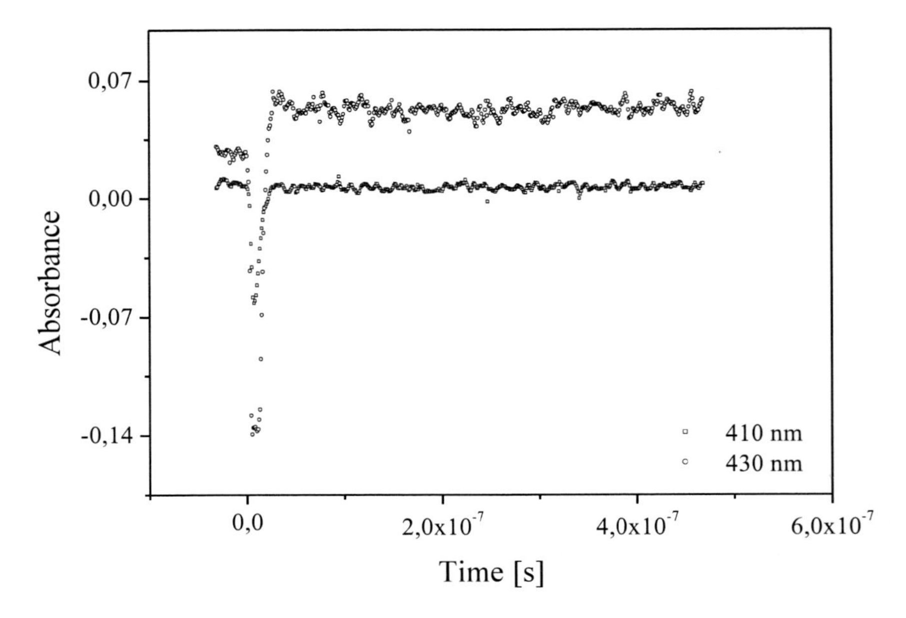

Figure 21. Transient absorption kinetic traces observed at wavelength 410 nm and 430 nm recorded for cyanine dye in presence of *n*-butyltriphenylborate salt in acetonitrile as a solvent. Borate salt concentration was 2×10^{-3} M.

In the two-component system the cyanine dye radical is formed (band at 430 nm) and no individuum at 410 nm (N-methylpicolinium radical) is observed. On the basis of the nanosecond laser flash photolysis experiments, it appears that N-methylpicolinium ester is reduced by dye radical. This reaction yields the dye cation and N-methylpicolyl radical that undergoes C-O fragmentation giving carboxylic acid and 4-(N-methyl)pyridinemethyl radical.

Further modification of photoinitiating system ability was concerned on a possibility of diffusion process elimination. This was achieved by synthesis of properly design dye possessing covalent attached diphenylacetic acid N-methylpyridiniummethyl ester group acting as an electron acceptor (dyes IX and X).

The introduction of a N-methylpicolinium moiety to the dye structure causes in significant changes in their absorption spectra (Figure 22).

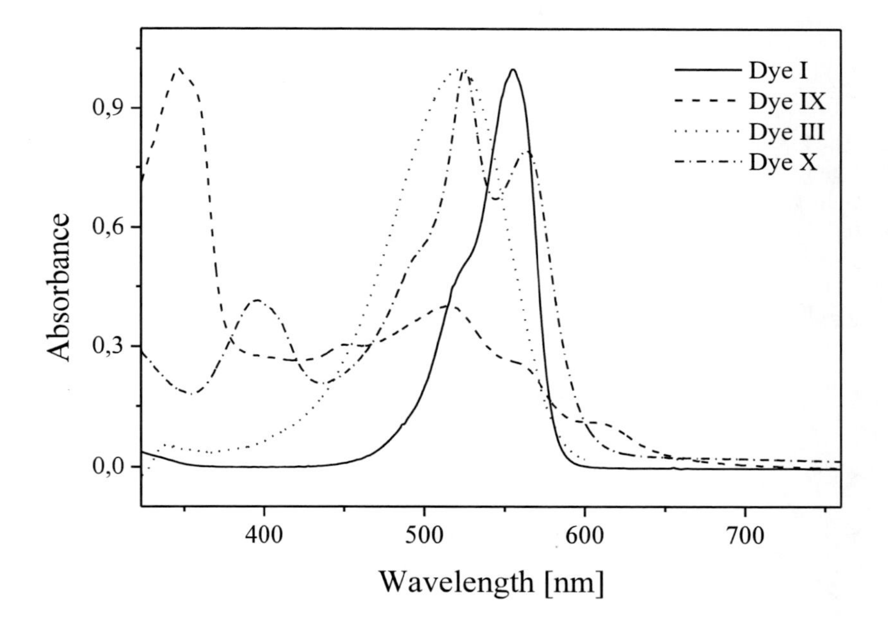

Figure 22. The illustrative electronic absorption spectra of polymethine dyes recorded in acetonitrile as a solvent.

The absorption spectra of dyes iodide salts displays a charge-transfer band about 350 nm for symmetrical and about 400 nm for unsymmetrical dyes, respectively. The position of this band depends on the solvent polarity. The change-transfer absorption band is attributed to the formation of picolinium/iodide ion pair [96, 97].

The elimination of diffussion effect by a covalent bonding of a dye with an electron acceptor causes that when it paried with *n*-butyltriphenylborate becomes extremely efficient visible light photoinitiators for free radical polymerization. Their photoinitiating ability is similar to that observed for initiating via triplet state xanthene dyes described by Neckers (Figure 23).

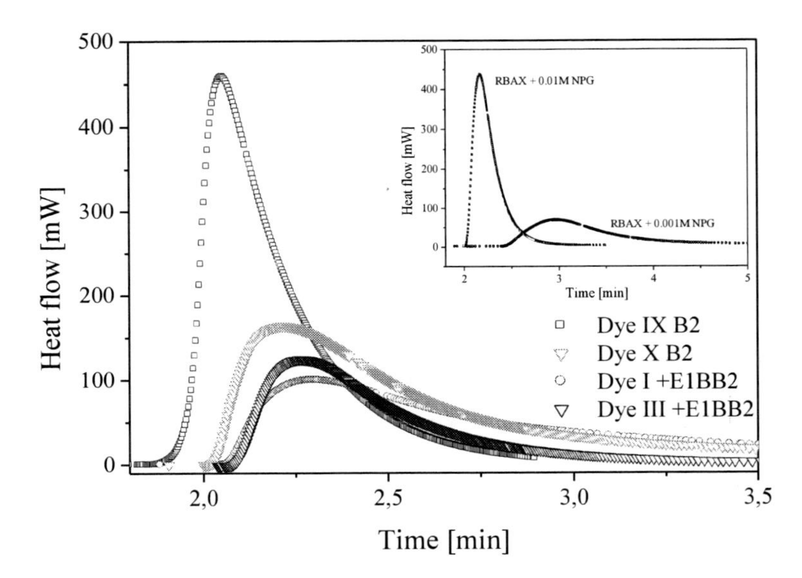

Figure 23. The kinetic curves recorded during the free radical polymerization of TMPTA/MP (9:1) polymerizing mixture initiated by N,N'-diethyltiocarbocyanine (dye I) and 3-ethyl-2-(*p*-N,N-dimethylaminostyryl)benzothiazolium iodide (dye III) (c = 1×10^{-3} M) in the presence of E1BB2 ion pair (1×10^{-2} M) and the kinetic curves of TMPTA/MP mixture polymerization initiated by dyes IX and X as *n*-butyltriphenylborate salts (c = 5×10^{-3} M). Inset: Comparison of photoinitiating ability of RBAX-NPG (N-phenylglycine concentration was 1×10^{-2} M and 1×10^{-3} M) photoinitiating system, RBAX – Rose bengal derivative (triplet state photoinitiator).

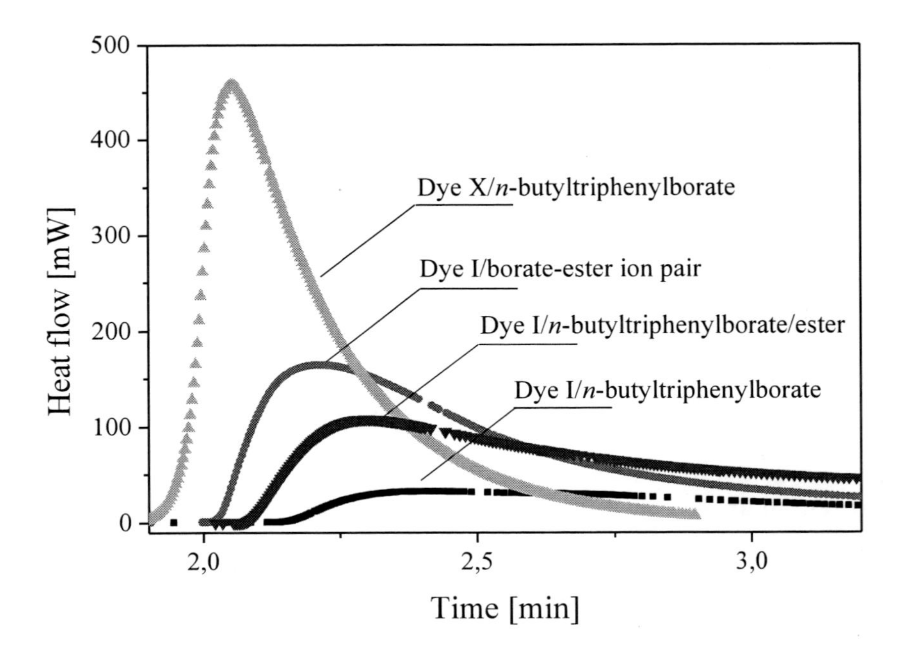

Figure 24. The kinetic curves recorded during polymerization TMPTA/MP initiated by dyes I and X. Photoinitiating systems marked in the Figure.

Although the dye IX/*n*-butyltriphenylborate ion pair is less sensitive in comparison to the mixture of dye III/E1BB2 ion pair, this photoinitiating system should be considered as very effective. The difference in sensitivity between dye III/E1BB2 and dye I/E1BB2 ion pair may come from the different photoinitiation efficiency of sensitizers tested (dyes III and I). N,N'-Diethylthiocarbocyanine *n*-butyltriphenylborate (dye I B2) initiates free radical polymerization with the rate five times higher in comparison with 3-ethyl-2-(*p*-N,N-dimethylaminostyryl)benzothiazolium *n*-butyltriphenylborate (dye III B2). Thus, probably the type of chromophore plays an important role in relative lowering of the photoinitiation ability of dye X B2 in comparison to dye IX B2 photoredox pair.

The comparison of photoinitiation efficiency of dye I in various combinations of co-initiators is presented in Figure 24.

For the photoinitiating system composed of structurally modified sensitizer the primary and secondary processes that occur after a visible light irradiation are as follows (Scheme 15):

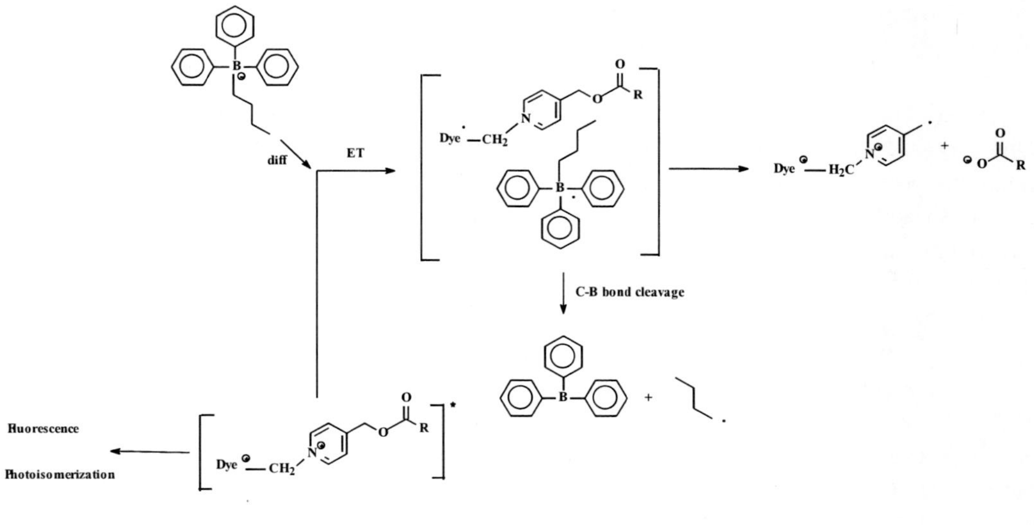

Scheme 15. Primary and secondary processes for the two-component photoinitiating system composed of dye with a covalent attached of N-methylpicolinium ester moiety as a sensitizer.

In the three-component photoinitiating system composed of N,N'-diethylthio-carbocyanine iodide, tetramethylammonium *n*-butyltriphenylborate and diphenylacetic acid N-methylpyridinemethyl ester perchlorate, after dye excitation, first, excited dye and borate salt must diffuse to each other for effective electron transfer. After electron transfer the dye radical and N-methylpicolinium ester in order to form an encounter complex should diffuse to each other as well. Since for dye I no short-living intermediates are observed in the nanosecond time-scale, it is obvious that all processes occur in regime controlled by diffusion. Thus, elimination only one step that is diffusion controlled should significantly increase the overall rate of entire reaction. The feasibility of this is demonstrated by system composed of dye IX and borate salt, in which after absorption of the light the excited molecule and borate anion encounter (diffusion controlled process) for effective electron transfer reaction. After electron transfer process resulting *n*-butyltriphenylboranyl radical decomposes yielding *n*-butyl radical (first radical) and triphenylboron. The second product of electron transfer, N,N'-[3-((4-pyrydiniummethyl)diphenylacetic acid ester propyl)]thiocarbocyanine radical, being unstable decomposes giving N,N'-[3-((4-pyridiniummethyl)propyl)]thiocarbocyanine radical (second initiating radical) and diphenylacetic acid anion. It is clear, that the decomposition process is not diffusion controlled. Thus, using the photoinitiating pair composed of dye with covalent attached diphenylacetic acid N-(methylpyridinium)methyl ester moiety (dye IX) and *n*-butyltriphenylborate anion, one eliminates one of the rate-determining step. This approach allows increase in speed of photoinitiation approximately one order in magnitude in comparison with those when either not modified sensitizer with borate salt and N-methylpicolinium ester salt was used. This dye after electron transfer besides *n*-butyl radical and triphenylboron forms N-methylpyridiniummethyl radical, organic acid anion and reduced dye.

Three-Component Photoinitiating Systems Composed of Polymethine Dye/*n*-Butyltriphenylborate/Cyclic Acetals

Cyclic acetals are examples of hydrogen donating photoinitiators [101, 102]. The monoester radical generated by the photoirradiation onto cyclic acetal compounds could initiate the polymerization of vinyl compounds and methyl methacrylate [101, 103]. Recently, photosensitized hydrogen abstraction from 2-alkyl-1,3-dioxolanes by excited triplet state of benzophenone gives corresponding 1,3-dioxolan-2-yl radicals and provides a viable alternative for synthesis of 1,4-diketones was described by Mosca [104]. More recently, Shi reported that cyclic acetals were used as hydrogen donors for bimolecular photoinitiating systems. For example, a natural component 1,3-benzodioxolane was used as a co-initiator for replacing the conventional amine for dental composite [101, 105]. Since the active hydrogen between two alkoxy groups in the cyclic acetal is abstractable and could form a radical, Elad and Yousseyfyeh proposed the photochemical rearrangement mechanism of 1,3-dioxolane compounds to give esters (Scheme 16) [106].

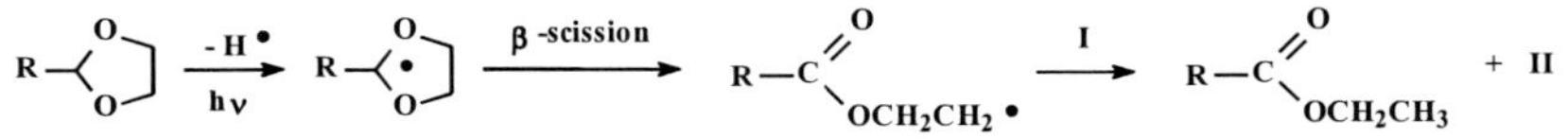

-H•, hν; β-scission; R—C(=O)OCH$_2$CH$_2$•; I; R—C(=O)OCH$_2$CH$_3$ + II

Scheme 16. The photochemical rearrangement mechanizm of 1,3-dioxolane.

Five cyclic acetals (2-methyl-1,3-dioxolane (K1), 2-methoxy-1,3-dioxolane (K2), 1,3-benzodioxolane (K3), 2-phenyl-1,3-dioxolane (K4), glycerol formal (mixtures of 40 % 4-hydroxymethyl-1,3-dioxolane and 60 % 5-hydroxy-1,3-dioxolane) (K5)) were used as a second co-initiator in the three-component photoinitiating systems for free radical polymerization (compounds 19-23). As the sensitizers following polymethine dyes were used: N,N'-diethylthiocarbocyanine idodide (dye I), N-ethyl-2-(*p*-N,N-dimethylaminostyryl) benzothiazolium iodide (dye III) and N,N'-bis-(3-pyridnio)propylthiocarbocyanine tribromide (dye XI). The example kinetic curves obtained for the polymerization of TMPTA/MP (9:1) photoinitiated by three-component photoinitiating systems composed of cyanine borates in presence of cyclic acetals, under irradiation with a visible light are shown in Figure 25.

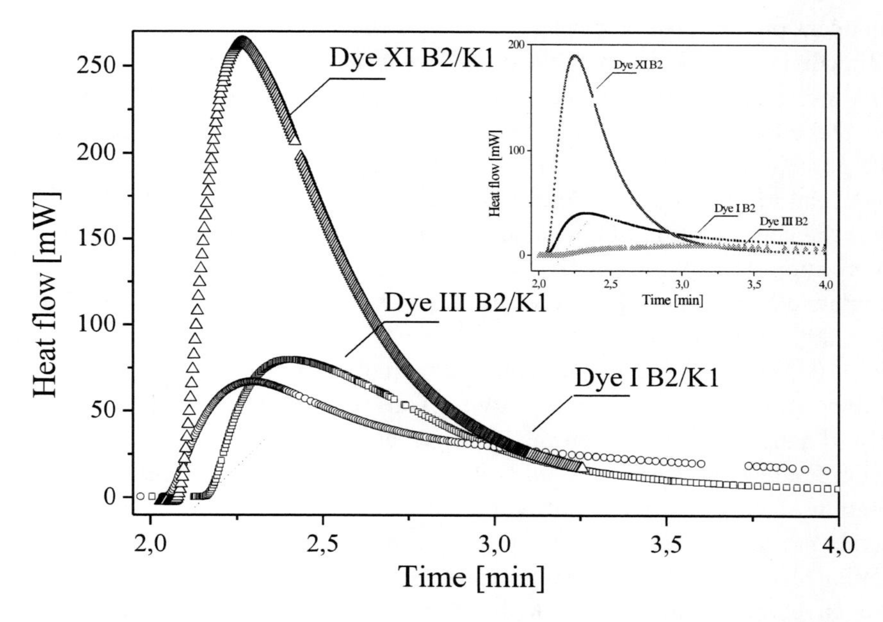

Figure 25. Family of kinetic curves recorded during the measurements of the flow of heat emitted during the photoinitiated polymerization of the TMPTA/MP (9:1) mixture initiated by three-component photoinitiating systems composed of polymethine dyes *n*-butyltriphenylborate in the presence of 2-methyl-1,3-dioxolane ($c = 1 \times 10^{-2}$ M), $I_a = 20$ mW/cm^2. Inset: Two-component photoinitiating systems: polymethine dye borate salt.

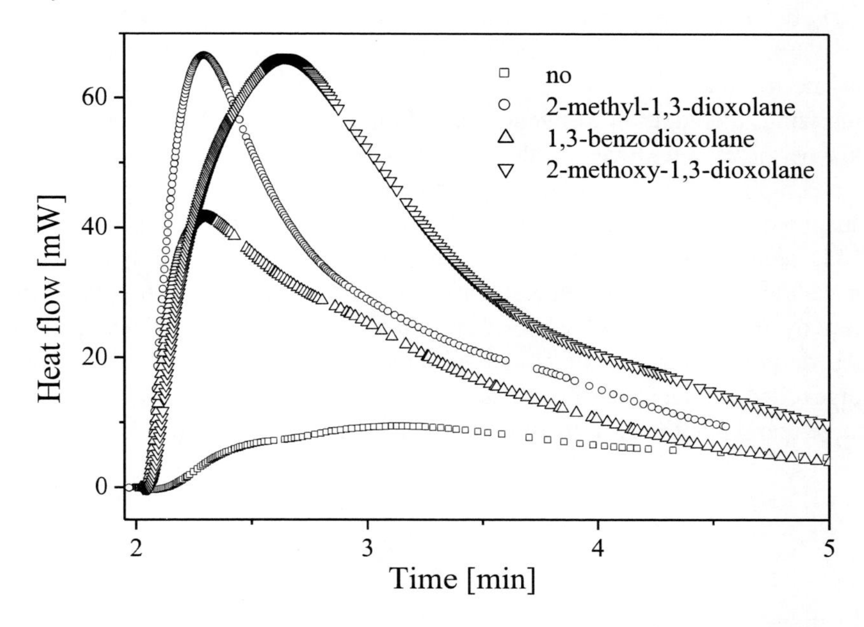

Figure 26. Effect of cyclic acetals structure on the rate of free radical polymerization of TMPTA/MP polymerizing mixture initiated by three-component photoinitiating system (dye III B2). Concentration of dye 5×10^{-2} M, $I_a = 20$ mW/0.196cm^2.

The three-component photoinitiating systems composed of cyclic acetal as a second co-initiator initiated free radical polymerization about 1.5-2 times faster in comparison to the two-component photoinitiators: cyanine dye/borate salt. The ion pair Dye III B2 (N-ethyl-2-(*p*-N,N-dimethylaminostyryl)benzothiazolium *n*-butyltriphenylborate) was the worst photoinitiating system used. However, addition of cyclic acetal into Dye III B2 considerably enhanced the efficiency of the polymerization (seven-fold incerase in the polymerization rate). The photoinitiating efficiency of the three-component photoinitiating systems depends on the concentration of both co-initiators and structure of cyclic acetal (Figure 26). The increase of the concentration of cyclic acetal from 0 to 0.1 M caused the increase of the rate of polymerization (R_p) about two times.

The relative rate of hydrogen atom abstraction by photogenerated radicals from a variety of cyclic ethers, acetals and orthoformates has been investigated using EPR spectroscopic technique [102, 107]. There was pronounced stereoelectronic effect, which produced high rates of abstraction from cyclic acetal carbon. Thus, methyl group attached to acetal carbon atom exerted a significant effect on the hydrogen abstraction. However, the activating effect of phenyl group was proved to be smaller than that of methyl group, probably because of the delocalization of the unpaired electron on to the unsaturated group came at the expense of planarisation at acetal carbon [102, 108]. In addition to the stereoelectronic factor, molecular conformation also affected the abstraction rate. For example, Malatesta et al. [107] found that the more envelope conformation of five-numbered cyclic acetals has higher hydrogen abstraction rate than that of six-numbered ones. Ouchi and Hamada [103] also reported the results that the strain of ring could affect the ability to promote the polymerization.

In our study, 2-methyl-1,3-dioxolane (K1) has relatively higher rate of polymerization than that of 2-phenyl-1,3-dioxolane (K4), which is in a good agreement with the reported rate of hydrogen abstraction from cyclic acetals. Formal glycerol (K5) with 60 % six-numbered cyclic acetal has lower reactivity. 1,3-Dioxolane (K3) is an effective co-initiator for free radical polymerization.

Summarizing, the addition of cyclic acetal to the two-component photoinitiating system induced a synergic effect. However, the efficiency of the polymethine dye/borate salt/cyclic acetal three-component photoinitiating system is not a simple sum of the efficiences of two-component photoinitaiting system: dye/borate salt and dye/acetal systems acting separately. Therefore, it seems that the improvement in photoinitiation for the system: polymethine dye/borate salt/cyclic acetal in comparison to the two-component photoinitiating system: polymethine dye/borate salt is a result of the secondary reactions between cyclic acetal and the species deriving from the first step of interaction, e.g. the electron transfer process between an excited singlet state of sensitizer and borate salt.

For three-component photoinitiating system theoretically following primary reactions are possible:

$$\mathbf{Dye}^{*} + \mathbf{B2} \longrightarrow \mathbf{Dye}^{\bullet} + \mathbf{B2}^{\bullet} \qquad \text{eq. (14)}$$

$$\mathbf{Dye}^{*} + \mathbf{K} \longrightarrow \mathbf{Dye}^{\bullet\oplus} + \mathbf{K}^{\bullet\ominus} \qquad \text{eq. (15)}$$

$$\mathbf{Dye}^{*} + \mathbf{K} \longrightarrow \mathbf{Dye}^{\bullet} + \mathbf{K}^{\bullet\oplus} \qquad \text{eq. (16)}$$

Basing on the calculated from Rehm-Weller equation values of ΔG_{el} for electron transfer process and ranging from –0.02 eV to 0.094 eV (from –1.93 $kJ{\cdot}mol^{-1}$ to 9.07 $kJ{\cdot}mol^{-1}$) it is known, that only the electron transfer from borate anion to an excited singlet state of polymethine dyes is a possible primary process. Because tested dyes do not undergo the electrochemical oxidation, an electron transfer from an excited singlet state of the dye to the ground-state cyclic acetal, leading to oxidation of the dye and reduction of acetal is not possible. From the same reason, an electron transfer process from the ground state of cyclic acetal on an excited singlet state of cyanine dye could not occur. Basing on this, one can conclude that the primary process is an electron transfer from borate anion to an excited state of cyanine dye, leading to the formation of cyanine dye radical and boranyl radical. All results, reveal that an effective interaction can take place between cyclic acetal and the products formed as a result of a primary process (cyanine dye radical, boranyl radical). If the electron transfer reaction from the cyclic acetal to an excited state of the dye occurs, leading to the formation of radicals capable of initiating polymerization of TMPTA, the initiation of free radical polymerization of TMPTA by two-component photoinitiating systems composed of polymethine dye/cyclic acetal should be observed. In the present case, no radicals capable of initiating polymerization of TMPTA are present. Thus, this process does not play a significant role in photoinitiation. One of resonable hypothesis suggests, that the synergic effect of polymethine dye/borate salt/cyclic acetal system behavior could be explained by an electron transfer interaction of the reduced dye (dye radical), boranyl radical or butyl radical (product decomposition of boranyl radical) with the cyclic acetal. As it was mentioned above, the cyanine dye radical does not act as terminator of polymer chains. In opposite case, such interaction could sharply decrease an efficiency of initiation process, and this in turn, causes a decrease in observed rate of polymerization. The possible explanation of the observed synergic effect for three-component system may consider a possible redox reaction between polymethine dye radical, boranyl radical or butyl radical and cyclic acetal. Similar reactions are well documented for dye radical and N-methoxypyridine cation or N-methylpicolinium ester. The second explanation considers the interactions between the polymethine dye radical and cyclic acetal that can strongly reduce a terminating effect caused by dye radical, or interaction between dye radical, boranyl radical or butyl radical and additionally to form a new initiating radicals as a resut of hydrogen abstraction. Similar observations were observed for three-component photoinitiating systems composed of cyanine dye/triazine/thiol [109]. The mechanism of primary and secondary reactions for three-component photoinitiating system: polymethine dye/*n*-butyltriphenylborate/cyclic acetal is presented in Scheme 17.

The photoexcited dye molecule encounters *n*-butyltriphenylborate anion and accepts an electron from borate anion, forming boranyl radical and cyanine radical. In the next step, the hydrogen abstraction from cyclic acetal by free radicals formed in the primary photochemical process occurs. This process gives free radicals, that can start the polymerization chain reaction.

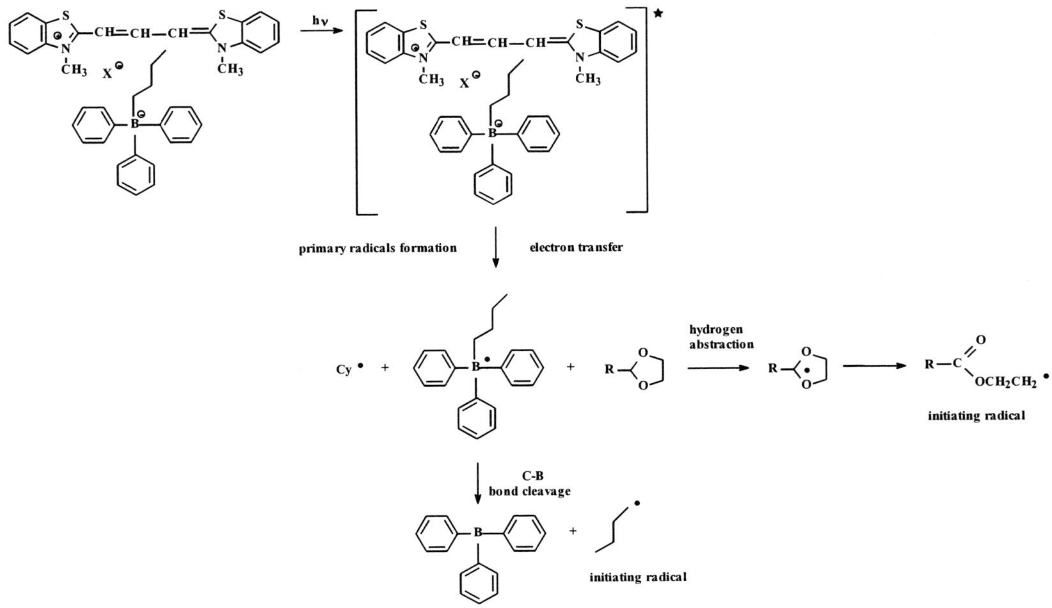

Scheme 17. The primary and secondary processes occuring in the three-component photoinitiating system composed of polymethine dye/alkyltriphenylborate salt/cyclic acetal after irradiation with a visible light.

Three-Component Photoinitiating Systems Composed of Polymethine Dye/*n*-Butyltriphenylborate/Thiol

Commonly, 2-mercaptobenzoxazole is used as a hydrogen donor (very often used also as efficient chain transfer agent). The rate of hydrogen transfer depends on the energy of H-hydrocarbon bond cleavage (hydrogen donor) and on the structure of a hydrogen acceptor [9].

In 2005 Suzuki, described the photochemical reactions occuring in the three-component photoinitiating system composed of aminostyryl dye/bis-imidazole derivative/thiol. The mechanism of free radicals formation which can start the polymerization chain reaction was proposed basing on the laser flash photolysis experiment (Scheme 18) [69].

In the case of three-component photoinitiating system, the initial absorption of imidazolyl radicals ($Im^{\bullet}$) was increased by the addition of 2-mercaptobenzothiazole. It is suggested that the quantum yield of triplet state of aminostyryl dye formation was enhanced by 2-mercaptobenzothiazole and additional sensitization, that is triplet electron transfer from aminostyryl dye to imidazole derivative might occur. Suzuki concluded, that the electron transfer occurs via both singlet and triplet state of aminostyryl dye, resulting in the improvement of the photosensitivity [69].

Dye $\xrightarrow{h\nu}$ ^{1}Dye*

(Dye ... Tiol) $\xrightarrow{h\nu}$ ^{1}(Dye ... Tiol)*

(Dye ... HABI) $\xrightarrow{h\nu}$ ^{1}(Dye ... HABI)*

^{1}(Dye ... Tiol)* $\longrightarrow$ ^{3}Dye + Tiol•

^{1}Dye* + HABI $\longrightarrow$ Dye$^{\ominus\bullet}$ + HABI$^{\ominus\bullet}$

^{1}(Dye ... HABI)* $\longrightarrow$ Dye$^{\ominus\bullet}$ + HABI$^{\ominus\bullet}$

^{3}Dye* + HABI $\longrightarrow$ Dye$^{\ominus\bullet}$ + HABI$^{\ominus\bullet}$

HABI$^{\ominus\bullet}$ $\longrightarrow$ Im$^{\ominus}$ + Im•

Scheme 18. Mechanism of processes occuring in the photoinitiating system: dye/bis-imidazole/thiol after irradiation with a visible light [69].

Other, high speed photoinitiating systems acting in the range of visible light (about 488 nm) were composed of cyanine or merocyanine dyes or 3,3'-carbonyl-bis-(coumarine) derivatives, 1,3,5-triazine derivative and a heteroaromatic mercaptan (like 2-mercaptobenzimidazole or 2-mercaptobenzothiazole) acting as an efficient chain transfer agent which greatly enhances the efficiency of photoinitiating systems [110, 111]. The excited initiator dissociates into radical fragments which abstract hydrogen from the thiol. Polymerization is subsequently mainly initiated by the sulphur-centered radical. Alternatively, the thiol can act as a co-initiator for the excited dye on its own. One of the mechanisms occuring in the three-component photoinitiating system is presented in Scheme 19 [109].

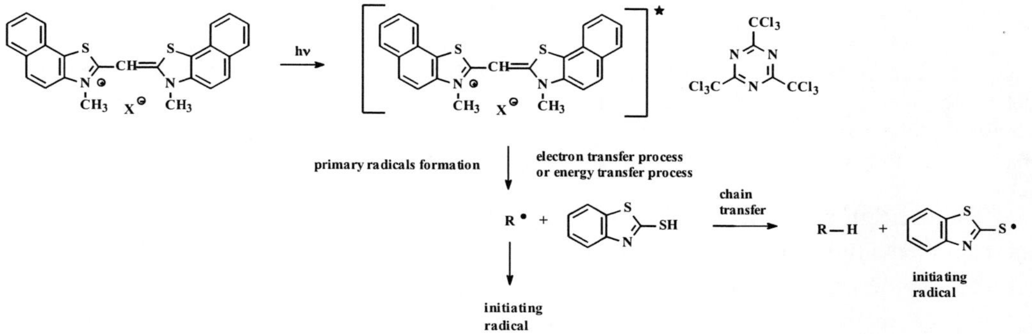

Scheme 19. Mechanism of reactions occuring in the three-component photoinitiating system composed of polymethine dye/thiol/1,3,5-triazine derivative [109].

2-Mercaptobenzoxazole (MO) and 2-mercaptobenzothiazole (MS) have also been used in two- and three-component photoinitiating systems based on coumarin derivatives [60]. Other mechanism was involved for multi-component photoinitiating system such as in the dye/bis-imidazole derivative/thiol (Scheme 20) [62].

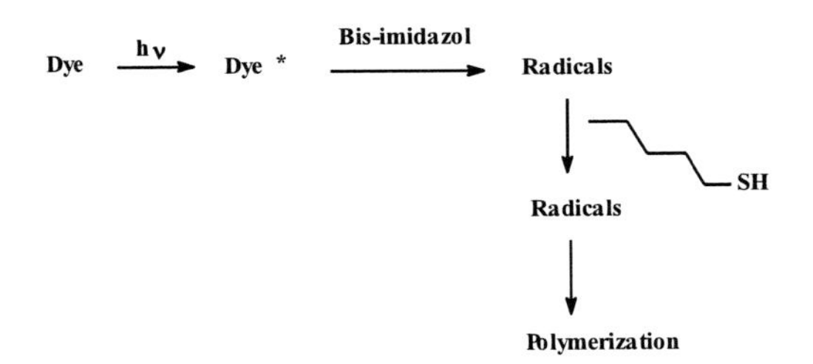

Scheme 20. Mechanism of reactions occuring in the three-component photoinitiating system: dye/bis-imidazole derivative/thiol [62].

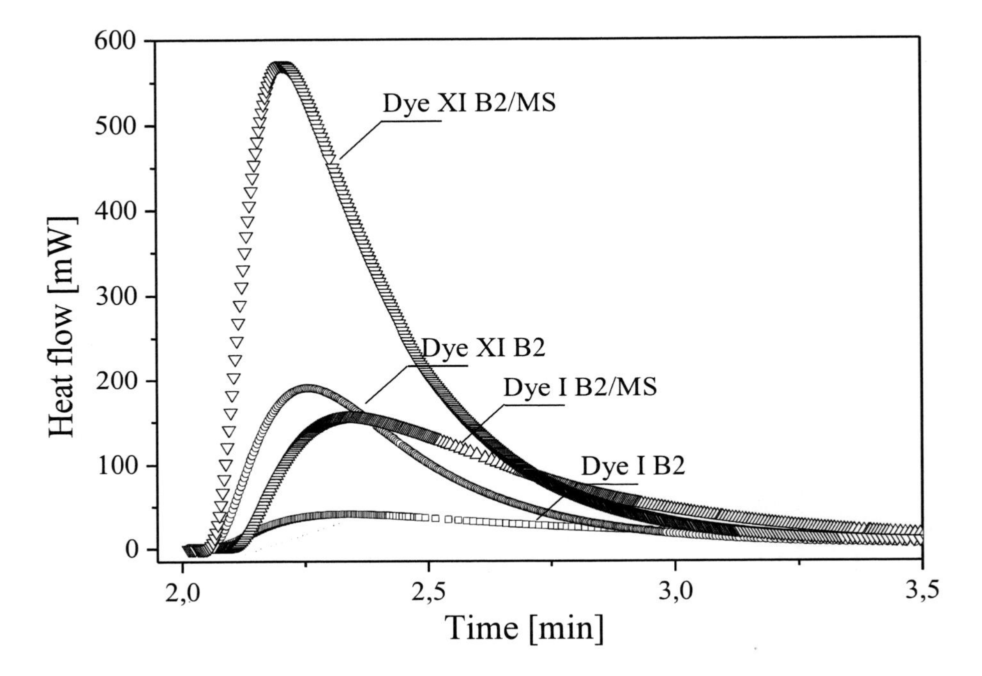

Figure 27. Family of kinetic curves recorded during the measurements of the flow of heat emitted during polymerization of the TMPTA/MP (9:1) mixture initiated by two- and three-component photoinitiating systems composed of cyanine dye *n*-butyltriphenylborate in the presence of thiols marked. The thiols concentration was 5×10^{-2} M, $I_a = 20$ mW/0.196 cm^2.

Three heteroaromatic thiols (2-mercaptobenzotiazole, 2-mercaptobenzoxazole and 2-mercaptobenzimidazole) were investigated in regard to their abilities as a co-initiator in free radical polymerization of acrylate monomers (compounds 24-26). The following polymethine dyes: N,N'-diethylthiocarbocyanine iodide (dye I), N-ethyl-2-(*p*-N,N-dimethylaminostyryl) benzothiazolium iodide (dye III) and N,N'-bis-(3-pyridinepropyl)thiocarbocyanine tribromide (dye XI) were used as sensitizers. The addition of heteroaromatic thiol to the two-component photoinitiating system composed of cyanine *n*-butyltriphenylborate increases the rate of polymerization about 2-8 times in comparison to the two-component system (Figure 27).

The photoinitiating efficiency depends on both the structure and concentration of a second co-initiator (thiol) (Figure 28).

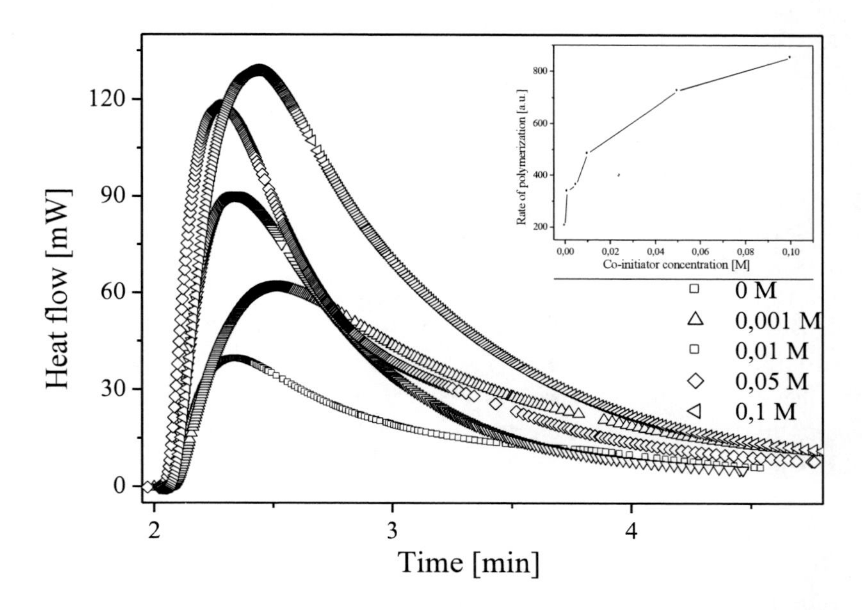

Figure 28. The kinetic curves recorded during the measurements of the flow of heat emitted during polymerization of the TMPTA/MP (9:1) mixture initiated by N,N-diethylthiocarbocyanine *n*-butyltriphenylborate in the presence of different concentrations (marked in Figure) of 2-mercaptobenzothiazole (MS). Inset: The influence of the thiol (MS) concentration on the rate of polymerization.

In distinction to the three-component photoinitiating systems described above, polymethine dyes tested in the presence of the thiols initiates the free radical polymerization of TMPTA/MP mixture but with lower photoinitiating rates (Figure 29).

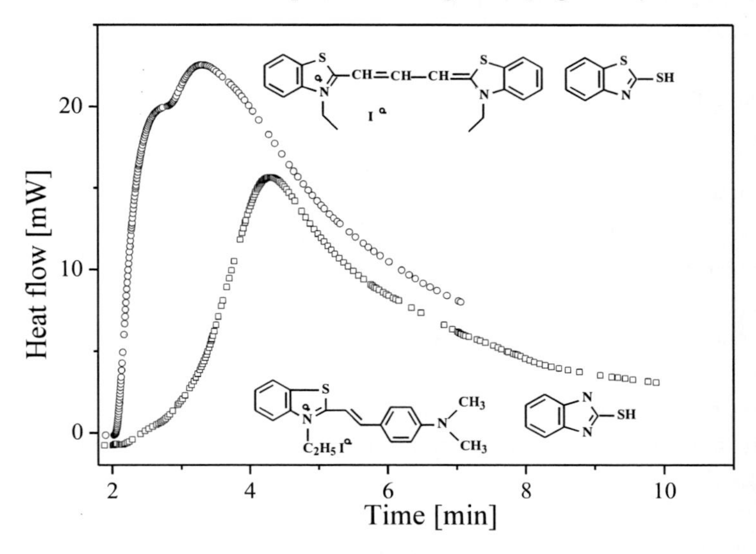

Figure 29. The kinetic curves recorded during the measurements of the flow of heat emitted during polymerization of the TMPTA/MP (9:1) mixture initiated by cyanine dye in the presence of different co-initiators (marked in Figure). The cyanine dye concentrations were 1×10^{-3} M and 5×10^{-2} M, respectively. $I_a = 20$ mW/0.196 cm^2.

For three-component photoinitiating system theoretically following primary reactions are possibile:

$$\text{Dye}^{*} + \text{B2} \longrightarrow \text{Dye}^{\bullet} + \text{B2}^{\bullet} \quad \text{eq. (17)}$$

$$\text{Dye}^{*} + \text{Thiol} \longrightarrow \text{Dye}^{\bullet} + \text{Thiol}^{\oplus\bullet} \quad \text{eq. (18)}$$

The calculated values of ΔG_{el} for above reactions are in the range from –0.454 eV to 0.094 eV (from –43.80 $kJ{\cdot}mol^{-1}$ to 9.07 $kJ{\cdot}mol^{-1}$) (Table 2). The calculated free energy changes for the electron transfer process between an excited singlet state of the dye and heteroaromatic thiols show that an electron transfer from the thiol to an excited singlet state of dye is thermodynamically more feasible, leading to the reduction of a dye and oxidation of thiol.

Table 2. Calculated free energy changes (ΔG_{el}) of the electron transfer reaction between the excited singlet state of polymethine dye and *n*-butyltriphenylborate salt or heteroaromatic thiol, respectively

Dye	Oxidation of B2 (ΔG_{el}) [eV]	Oxidation of MS (ΔG_{el}) [eV]	Oxidation of MO (ΔG_{el}) [eV]	Oxidation of MI (ΔG_{el}) [eV]
Dye I	-0.02	-0.004	-0.13	0.08
Dye III	0.094	-0.25	-0.376	-0.166
Dye XI	-0.016	-0.328	-0.454	-0.244

Calculated by $\Delta G_{el} = E_{ox}$ (Thiol lub B2) – E_{red} (dye) – E_{00}

The Coulombic stabilization term is not taken into account. It is usually negligible in polar solvents like acetonitrile.

Therefore, for two-component photoinitiating system composed of polymethine dye and thiol, the efficiency of thiyl radical formation and, hence, initiation of polymerization depends on the observed efficiency of electron transfer from a thiol to an excited singlet state of polymethine dye. The kinetic results obtained for this photoinitiating pair seems confirming this prediction. In the three-component photoinitiating systems possessing polymethine dye/*n*-butyltriphenylborate salt/thiol the effective interactions between all components can take place after light irradiation.

The following electron transfer processes can occur:

- From the borate anion to an excited singlet state of polymethine dye,
- From the ground state of heteroaromatic thiol to an excited singlet state of polymethine dye.

According to the literature the second reaction is followed by a hydrogen abstraction which leads to the formation of thiyl radicals [62, 69, 112]. These radicals initiate free radical polymerization.

The mechanism of reactions occuring after irradiation of the three-component photoinitiating system composed of polymethine dye/*n*-butyltriphenylborate salt/heteroaromatic thiol was proposed basing on the nanosecond laser flash photolysis. Irradiation of the solution of thiocarbocyanine dye in the presence of thiol in acetonitrile at 355 nm gives the transient absorption spectra with two characteristic bands at 420 i 600 nm (Figure 30).

The irradiation of carbocyanine dye with a visible light leads to an excited singlet state formation. This can be quenched by heteroaromatic thiol via electron transfer process. The absorption band observed at 420 nm in the transient absorption spectra was assigned to the cyanine dye radical formed in the electron transfer process [12, 13]. The laser flash photolysis of two-component system composed of cyanine dye/2-mercaptobenzothiazole ($c = 1 \times 10^{-2}$ M) gives the second absorption band at 600 nm. This behavior is similar to that described by Andrzejewska [112] for system composed of camphorquinone and 2-mercaptobenzothiazole, and quite different from the results obtained by Suzuki [69] for three-component photoinitiating system composed of: neutral cyanine dye/bis-imidazole derivative/thiol. This characteristic band was attributed to the presence of the oxidation product of thiol. It is ascribed to thiyl radicals on the basis on literature data [112]. The formation of thiyl radical after electron transfer process from thiol to an excited singlet state of cyanine dye and its disappearance in the time about 10 μs is shown in Figure 31.

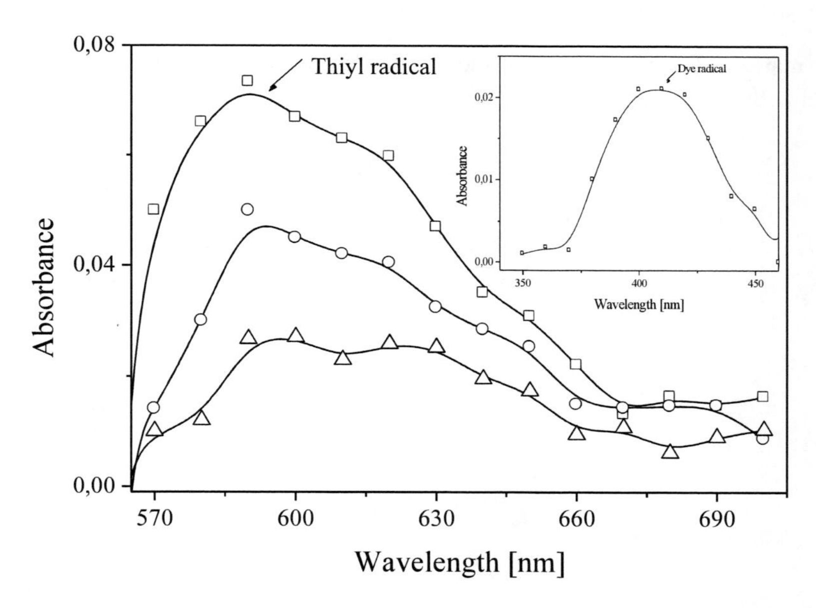

Figure 30. Transient absorption spectra of cyanine dye in a presence of 2-mercaptobenzothiazole (MS) recorded: 1 μs (squares), 4 μs (circles) and 10 μs (triangles) after laser pulse. Inset: Transient absorption spectra of cyanine dye in a presence of 2-mercaptobenzothiazole (MS) recorded 100 ns after laser pulse (circles) (concentration of 2-mercaptobenzothiazole 1×10^{-2} M) in acetonitrile solution. Dye concentration was 2×10^{-4} M.

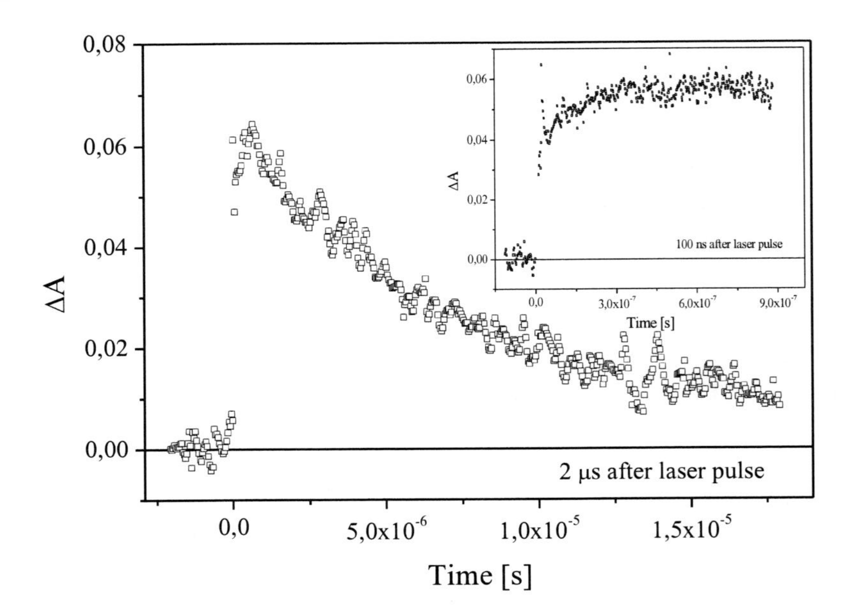

Figure 31. Transient absorption kinetic observed at 600 nm after different delay time (100 ns and 2 μs) for two-component system: N,N'-diethylthiocarbocyanine iodide/2-mercaptobenzothiazole. Dye concentration was equal 5×10^{-4} M, 2-mercaptobenzothiazole concentration was 1×10^{-2} M.

The formation of the cyanine dye radical in the two-component system: cyanine dye/2-mercaptobenzothiazole is observed at 420 nm (Figure 32).

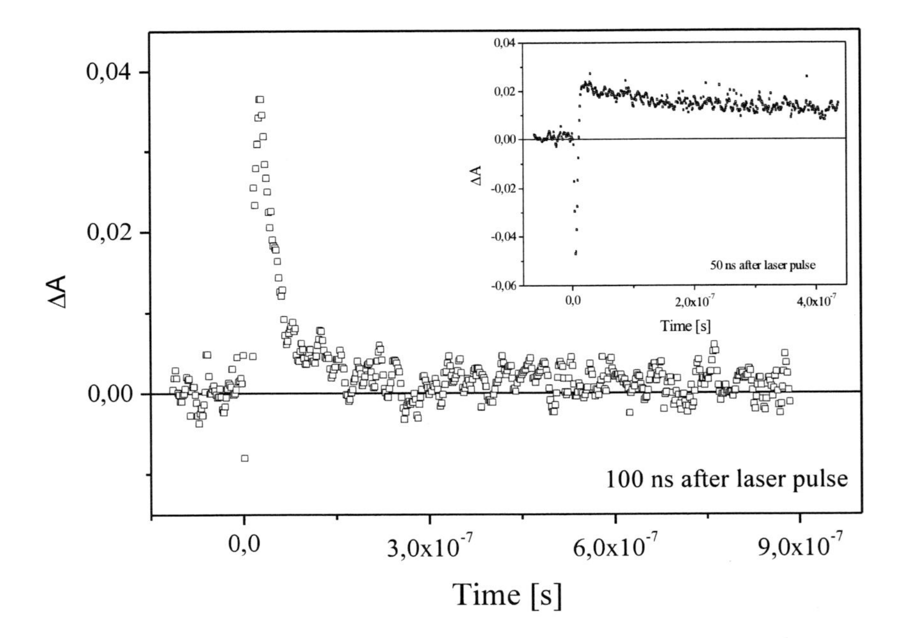

Figure 32. Transient absorption kinetic observed at 420 nm after different delay time (50 ns and 100 ns) for two-component system: N,N'-diethylthiocarbocyanine iodide/2-mercaptobenzothiazole. Dye concentration was equal 5×10^{-4} M, 2-mercaptobenzothiazole concentration was 2×10^{-2} M.

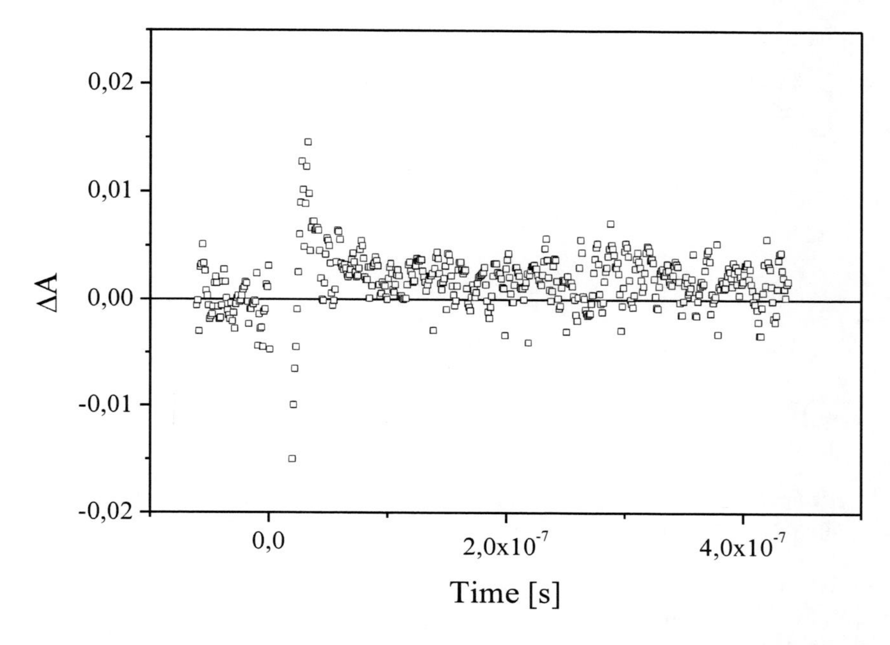

Figure 33. Transient absorption kinetic observed at 420 nm after 100 ns delay time for three-component system: cyanine dye/borate salt/thiol. Dye concentration was equal 2×10^{-5} M, borate salt and thiol concentrations were 1×10^{-2} M, respectively.

In two-component system: cyanine dye/heteroaromatic thiol, in which the sensitizer plays a role of an electron acceptor the electron transfer process occurs [112]. The kinetics of disappearance of cyanine dye radical in the three-component system: cyanine dye/*n*-butyltriphenylborate/heteroaromatic thiol is presented in Figure 33.

On the basis of the nanosecond laser flash photolysis experiment, it appears that the excited singlet state of cyanine dye can be reducted by both *n*-butyltriphenylborate anion and heteroaromatic thiol. This reactions lead to the formation of cyanine dye radical, thiyl radical and boranyl radical which undergoes fast decomposition to butyl radical and triphenylboron.

The mechanism of the primary and secondary reactions occurring in the three-component photoinitiating system composed of cyanine dye/*n*-butyltriphenylborate/heteroaromatic thiol is shown in Scheme 21.

After irradiation of the three-component photoinitiating system with a visible light, the excited singlet state of sensitizer is formed. The deactivation of the excited state occurs by fluorescence, photoisomerization or electron transfer process. In the presence of *n*-butyltriphenylborate salt the cyanine dye undergoes one-electron reduction. The cyanine dye radical and boranyl radical are formed. The boranyl radical undergoes the C-B bond cleavage, giving an alkyl radical that can start the polymerization chain reaction. On the other hand, in the presence of heteroaromatic thiol an excited singlet state of cyanine dye may undergo deactivation also via electron transfer process. The rate of this process is about two orders of magnitude lower than for cyanine borate photoredox ion pair. As a result of this process, the cyanine dye radical and thiyl radical cation are formed. The thiyl radical cation undergoes the hydrogen abstraction, giving a thiyl radical that can start the polymerization reaction.

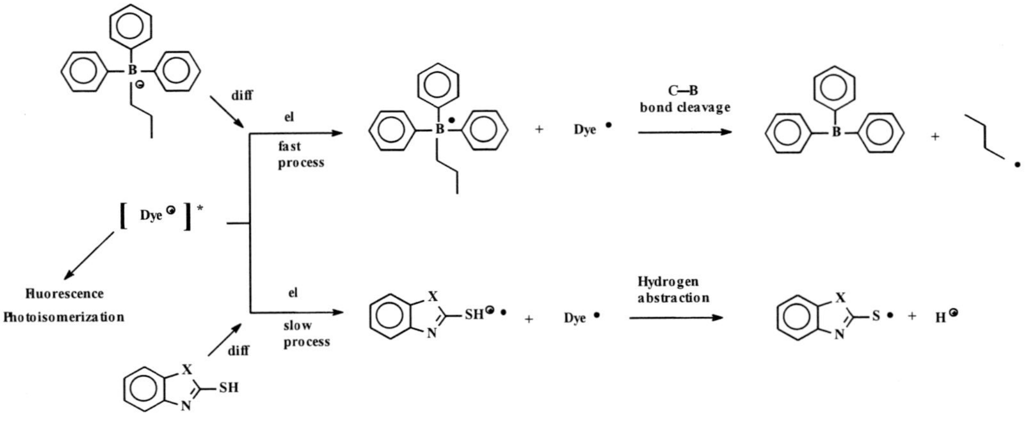

Scheme 21. Mechanism of primary and secondary reactions which occur in the three-component photoinitiating system: cyanine dye/*n*-butyltriphenylborate/thiol after irradiation with a visible light.

Generally, the addition to the two-component photoinitiating systems composed of polymethine dye/borate salt ion pair a second co-initiator is an effective method leading to the enhancement of the photoinitiating ability of photoinitiating systems. A second co-initiator may react either with the cyanine dye radical formed as a result of electron transfer process between borate anion and an excited singlet state of a sensitizer or react with the excited singlet state of a dye.

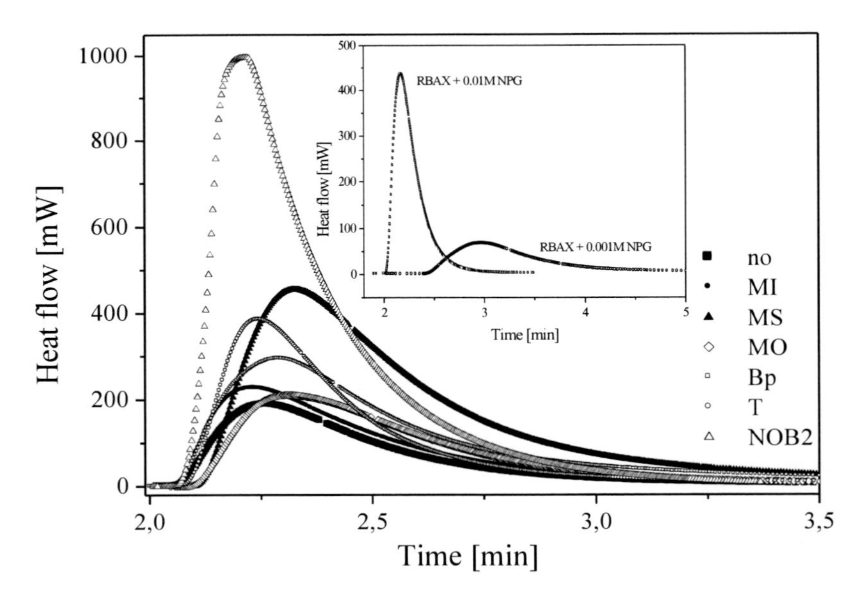

Figure 34. The family of kinetic curves recorded as a heat of flow during the free radical polymerization TMPTA/MP (9:1) polymerizing mixture initiated by three-cationic carbocyanine dye borate salt (dye XI B2) in the presence of different co-initiators (5×10^{-3} M) (marked in the figure). Inset: Comparison of the photoinitiation ability of RBAX/NPG (N-phenylglycine concentration was 1×10^{-2} M and 1×10^{-3} M), photoinitiating system, RBAX – the Rose bengal derivative (triplet excited state photoinitiator) [113].

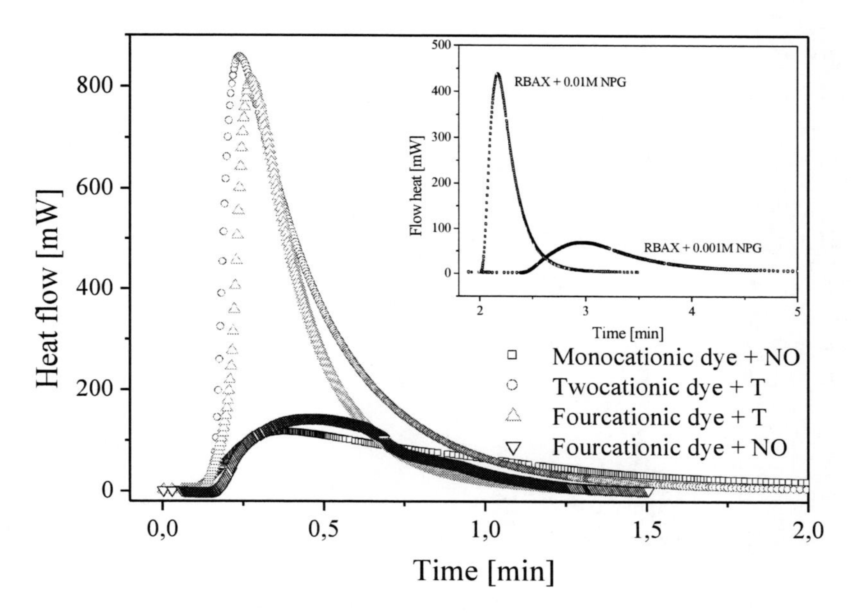

Figure 35. The family of kinetic curves recorded as a heat of flow during the free radical polymerization TMPTA/MP (9:1) polymerizing mixture initiated borate salt of mono- (Dye XIV B2), two- (Dye XV B2) and four-cationic monomethine dyes (Dye XVIIA B2 and Dye XVIIB B2) in the presence of different co-initiators (5×10^{-3} M) (marked in the figure). Inset: Comparison of the photoinitiation ability RBAX/NPG (N-phenylglycine concentration was 1×10^{-2} M and 1×10^{-3} M) photoinitiating system, RBAX – the Rose bengal derivative (triplet excited state photoinitiator).

The effect of different compounds acting as a second co-initiator on the photoinitiating efficiency of the three-component photoinitiating systems possessing tricationic carbocyanine dye or multicationic monomethine dyes as *n*-butyltriphenylborate salt is presented in Figures 34 and 35.

The rate of free radical polymerization of TMPTA/MP changes in the order:

NOB2 (12,80 μmol/s) > **MS** (5,87 μmol/s) > **T** (5,0 μmol/s) > **Bp** (3,82 μmol/s) > **MI** (2,93 μmol/s) > **MO** (2,71 μmol/s) ≈ 2,94 μmol/s (two-component photoinitiating system) [113].

The polymerization photoinitiation ability of tricationic carbocyanine dyes and multi-cationic monomethine dyes as borate salt in the presence of a second co-initiator is more efficient than this observed for RBAX-NPG couple (typical triplet photoinitiating system) (Figures 34 and 35).

In summary, the addition of an N-alkoxypyridinium salt, 1,3,5-triazine derivative, N-methylpicolinium ester or heteroaromatic thiol as a second co-initiator to the photoinitiating system: cyanine dye/borate salt causes a synergistic effect in the polymerization reactions. The rate of polymerization initiated by the three-component photoinitiating system is about 2-500 times higher than measured for dye/borate salt two-component photoinitiating system [33]. The photoinitiating ability mainly depends on the structure and concentration of a second co-initiator and also on the structure of cyanine dye (Figure 36).

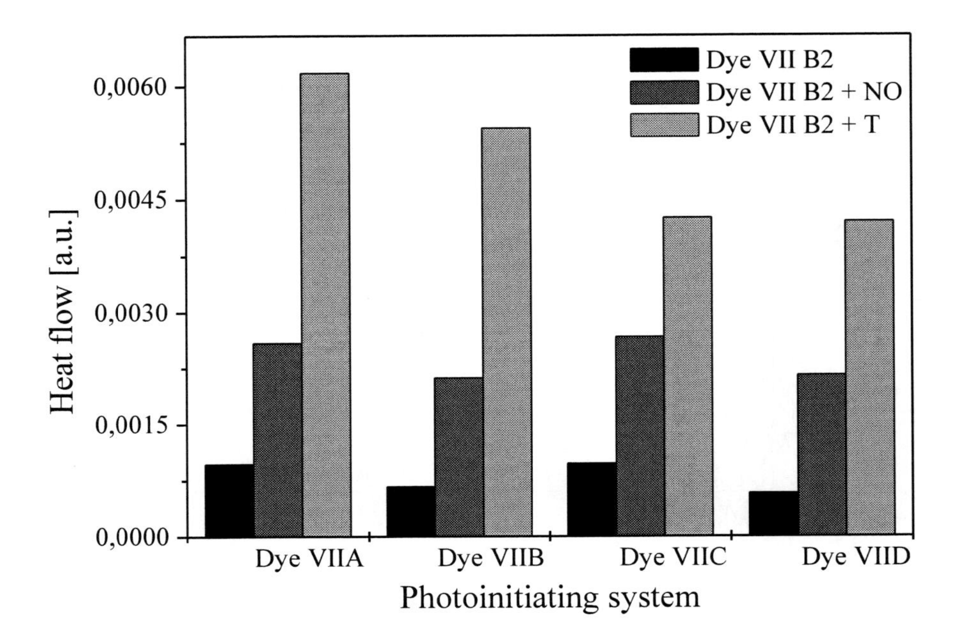

Figure 36. The comparison of the photoinitiation ability of different two- and three-component photoinitiating systems composed of the various monocationic styrylquinolinium dyes (dyes VII B2) [48].

Three-Component Photoinitiating Systems Composed of Polymethine Dye/Trichloromethyl-1,3,5-Triazine/Heteroaromatic Thiol

Trichloromethyl substituted 1,3,5-triazines, such as 2,4,6-tris(chloromethyl)-1,3,5-triazine undergo a multiple fragmentation upon irradiation with UV light. The primary photoproducts are not radicals, but dissociate into radicals in a second thermal step as shown in Scheme 22 [114].

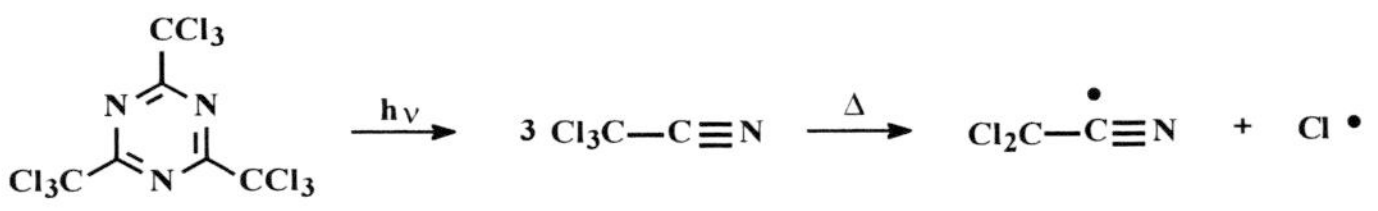

$$\text{2,4,6-}(CCl_3)_3\text{-1,3,5-triazine} \xrightarrow{h\nu} 3\ Cl_3C{-}C{\equiv}N \xrightarrow{\Delta} Cl_2C{-}\dot{C}{\equiv}N + Cl^{\bullet}$$

Scheme 22. Photolysis of 1,3,5-triazine derivative

To overcome this drawback, chromophores containing vinyl bis(trichloromethyl)-1,3,5-triazine are more frequently used as unimolecular photoinitiators. Direct fragmentation of carbon-chloride bond in a triazine derivative is the major photochemical pathway for radical formation with these compounds [115]. The most important application of trichloromethyl substituted triazines is their combination with various types of dyes as sensitizers. Both electron and energy transfer mechanisms were involved in the sensitization step. High speed photopolymer systems sensitive to light of a wavelength of 488 nm were developed on the basis of these initiating systems, using cyanine, merocyanine dyes or 3,3'-carbonyl-bis-(coumarin) derivatives [110, 111]. A key feature of these photoinitiating systems is the presence of a heteroaromatic mercaptan (like 2-mercaptobenzimidazole or 2-mercapto-benzothiazole) which acts as an efficient chain transfer agent which greatly enhances the sensitivity. The activated initiator dissociates into radical fragments which abstract hydrogen atom from the mercaptan. Polymerization is subsequently mainly initiated by the sulphur-centered radical. Alternatively, the marcaptan can act as a co-initiator for an excited dye on its own (Scheme 19) [109].

Other photoinitiating systems acting via intermolecular hydrogen abstraction from a hydrogen donor are systems composed of the long-lived triplet excited state diaryl ketones which can rapidly abstract hydrogen atom [116, 117]. In this photoinitiating system, the reaction of electronically excited ketone with a co-initiator occurs, yielding to the reduction of carbonyl group to a hydroxylic group. The reduction products can arise through two different pathways, namely: hydrogen abstraction from the hydrogen-donating compound by an excited ketone and electron transfer from an electron donor to an excited ketone and subsequent proton transfer. This type of photoinitiators is isopropylthioxanthone widely used in photocurable compositions, and camphorquinone employed in dentistry. To produce radicals efficiently camphorquinone and isopropylthioxanthone are commonly used with amine as a co-initiator, often aromatic ones. Because of many serious disadvantages of using amines (e.g., their mutagenicity and tendency to induce substrate corrosion), the investigations of new efficient co-initiators for isopropylthioxanthone and camphorquinone were important undertakings. The example, of such co-initiators were heteroaromatic thiols studied by Andrzejewska and co-workers [112]. Many thiols quench the triplet state of ketones with the subsequent formation of thiyl radicals. The resulting thiyl radicals are able to add to the monomer double bonds. The formation of thiyl radicals is facilitated by the relatively low S-H bond dissociation energies in thiols (87 kcal/mol) [118]. As it was mentioned earlier, the heteroaromatic mercaptans are used also as the co-initiators for hexaaryl-bis-imidazole derivative [69, 119]. Chemical reaction in these systems can occur either by direct hydrogen abstraction or by electron transfer giving the imidazole derivative and sulfur-centered radical. Suzuki, studied the photochemical and photophysical behavior of the three-component photoinitiating system, that consists of aminostyryl sensitizing dye, bis-imidazole derivative acting as a radical generator and 2-mercaptobenzothiazole as a co-initiator [69]. The addition of 2-mercaptobenzothiazole results in the increase of the quantum yield of the dye excited triplet state formation. The transient absorption spectra for the two-component photoinitiating system: aminostyryl dye/2-mercaptobenzothiazole was similar to that of the direct excitation of the dye to some extent and the initial absorption was increased significantly with the increasing of the thiol concentration. Two possibilities are considered about the transient species:

- The donating an electron from thiol to the aminostyryl dye, which is thermodynamically allowed, and generated thiyl radical, which can start the polymerization chain reaction.
- The generation of the triplet state of aminostyryl dye (Scheme 23) [69].

$$\text{Dye} \xrightarrow{h\nu} {}^1\text{Dye}^*$$

$$(\text{Dye} \ldots \text{Thiol}) \xrightarrow{h\nu} {}^1(\text{Dye} \ldots \text{Thiol})^*$$

$$^1(\text{Dye} \ldots \text{Thiol})^* \longrightarrow {}^3(\text{Dye} \ldots \text{Thiol})^*$$

Scheme 23. The primary and secondary processes for the photoinitiating system: dye/heteroaromatic thiol.

Suzuki concluded, that if the first mechanism occurred mainly, a radical anion of aminostyryl dye or its intermediate, the thiyl radicals should be generated more or less efficiency, and photopolymer bearing dye and 2-mercaptobenzothiazole should exhibit photosensitivity. However, this combination did not give it at all. Therefore, Suzuki considers the second mechanism presented in Scheme 23 [69].

For two-component photoinitiating system composed of aminostyryl dye (NASA) and bis-imidazole derivative (HABI), the singlet electron transfer, generating imidazolyl radical is the main reaction pathway. The improvement of the photosensitivity of two-component photoinitiating system was achieved by an addition of 2-mercaptobenzothiazole (MS). The effect of thiol is suggested as an electron donor toward aminostyryl dye that loose an electron by a singlet electron transfer to imidazole derivative or hydrogen transfer toward imidazolyl radicals [69]. The mechanism of the processes occurring in the three-component photoinitiating system was proposed based on the nanosecond laser flash photolysis (Scheme 18). In the three-component photoinitiating system composed of neutral aminostyryl dye/bis-imidazole derivative/thiol, both singlet and triplet electron transfer from dye to imidazole derivative might occur by addition of a thiol.

Takimoto shows, that the combination of 9-phenylacridine with 2-mercaptobenzimidazole, provided high photosensitivity for polymerization of pentaerythriol tetraacrylate [120]. 2-Mercaptobenzoxazole and 2-mercaptobenzothiazole were also used in two- and three-component photoinitiating systems based on coumarin derivatives [60].

The photoinitiating systems under study were composed of cyanine dye: N,N'-diethylthiocarbocyanine iodide (dye I)/2,4-bis(trichloromethyl)-6-(4-methoxy)phenyl-1,3,5-triazine/2-mercaptobenzothiazole (MS), 2-mercaptobenzoxazole (MO) or 2-mercaptobenzimidazole (MI). The kinetic studies were performed for following photoinitiating systems (Figure 37):

- Dye/1,3,5-triazine derivative
- Dye/thiol and
- Dye/1,3,5-triazine/thiol.

Both heteroaromatic thiol and 1,3,5-triazine derivative can act as a co-initiator with thiocarbocyanine dye in the two-component photoinitiating system. However, the addition of a thiol to the two-component photoinitiating system composed of thiocarbocyanine dye/1,3,5-triazine derivative leads to the increase of the polymerization rate. The observed rate of polymerization initiated by the three-component photoinitiating system composed of cyanine dye/2-mercaptobenzoxazole/1,3,5-triazine derivative is higher than those observed for two-component system possessing cyanine dye and *n*-butyltriphenylborate salt.

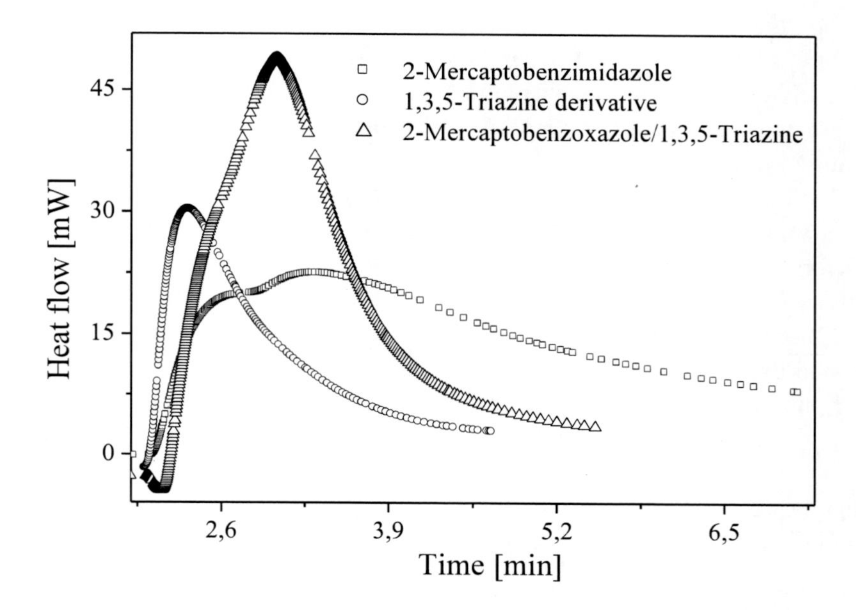

Figure 37. The family of kinetic curves recorded during the measurements of the flow of heat emitted during the photoinitiated polymerization of the TMPTA/MP (9:1) mixture initiated by N,N'-diethylthiocarbocyanine dye (dye I) in the presence of different co-initiators (marked in Figure). The cyanine dye and co-initiators concentrations were 1×10^{-3} M and 5×10^{-2} M, respectively. $I_a = 20$ mW/0.196 cm^2.

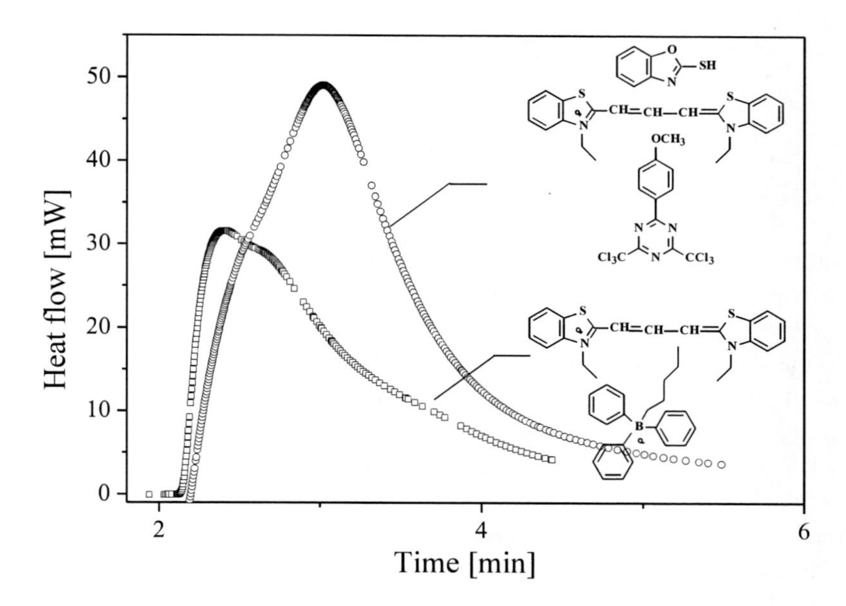

Figure 38. The family of kinetic curves recorded during the measurements of the flow of heat emitted during the photoinitiated polymerization of the TMPTA/MP (9:1) mixture initiated by *n*-butyltriphenylborate N,N'-diethylthiocarbocyanine and N,N'-diethylthiocarbocyanine iodide (dye I) in the presence of 1,3,5-triazine derivative and thiol as the co-initiators. The cyanine dye and co-initiators concentrations were 1×10^{-3} M and 5×10^{-2} M, respectively. $I_a = 20$ mW/0.196 cm^2.

Therefore, the rate of polymerization maliny depends on the fluorescence quenching rate constants for cyanine dye by co-initiators (thiol or 1,3,5-triazine derivative) and the monomer, and additionally on the quantum yields of radical formation. From the Andrzejewska's studies it is known, that the quantum yield of thiyl radical formation in the two-component photoinitiating system, composed of heteroaromatic thiol and camphorquinone was depended on the thiol structure and was in the range from 0.015 to 0.053 [121]. The high activity of 2-mercaptobenzoxazole (MO) in accelerating of the polymerization suggests the high efficiency of radical formation in the dye/1,3,5-triazine derivative/thiol photoinitiating system and/or high reactivity for 2-mercaptobenzoxazole-derived thiyl radicals.

The formation of free radicals in photoinitiating systems: dye/thiol, dye/1,3,5-triazine derivative and dye/thiol/1,3,5-triazine may occur via photoinduced electron transfer process (PET). For photoinitiating systems under study theoretically following primary reactions are possible:

- In the case of cyanine dye/thiol, the formation of radicals may occur via photoinduced electron transfer (PET) followed by proton transfer

$$Cy^{*} + R\text{—}SH \xrightarrow{ET} [Cy^{\bullet} \cdots R\text{—}SH^{\oplus\bullet}] \longrightarrow Cy^{\bullet} + R\text{—}S^{\bullet} + H^{\oplus} \quad \text{eq. (19)}$$

- In the case of cyanine dye/1,3,5-triazine derivative, the formation of free radicals may occur via photoinduced electron transfer process

$$Cy^{*} + T \xrightarrow{ET} Cy^{\oplus\bullet} + T^{\ominus\bullet} \quad \text{eq. (20)}$$

$$Cy^{*} + T \xrightarrow{ET} Cy^{\bullet} + T^{\oplus\bullet} \quad \text{eq. (21)}$$

The values of the free energy change ΔG_{el} for the electron transfer process were estimated by using the Rehm-Weller equation for following cases:

i. for the electron transfer from a thiol on an excited singlet state of the dye, using the oxidation potential of thiol, the reduction potential of sensitizer (-1.30 eV) and the energy of the excited state (2.10 eV). This process leads to the formation of thiyl radical as a result of the hydrogen abstraction.
ii. for the electron transfer process from an excited singlet state of the dye on the ground state of 1,3,5-triazine derivative, using the reduction potential of 1,3,5-triazine derivative (-0.84 eV) and the oxidation potential of N,N'-diethylthiocarbocyanine dye (1.00 eV).

The calculated values of ΔG_{el} are given in Table 3.

Tabele 3. Oxidation potentials (in V vs AgCl electrode) of the co-initiators, the values of the free energy change for the PET (ΔG_{el}) and the fluorescence quenching rate constants of N,N'-diethylthiocarbocyanine iodide (dye I) quenched by thiols and 1,3,5-triazine derivative in solution composed of ethyl acetate:1-methyl-2-pyrrolidinone (4:1)

Co-inicjator	MS	MO	MI	T
E_{ox} [V]	0.816	0.69	0.90	1.27
ΔG_{el} [$kJ \cdot mol^{-1}$]	-0.39	-12.54	7.72	-27.02
k_q [$M^{-1}s^{-1}$]	2.92×10^{10}	2.44×10^{10}	2.78×10^{10}	1.97×10^{10}

For two-component photoinitiating system composed of cyanine dye and thiol, the efficiency of thiyl radical formation and, hence initiation of polymerization depends on the observed efficiency of electron transfer from a thiol on an excited singlet state of sensitizer. For such photoredox pair the values of ΔG_{el} are in the range from –0.13 eV to 0.08 eV (e.g. –12.54 eV do 7.72 $kJ \cdot mol^{-1}$). The calculated free energy change for the electron transfer process from an excited singlet state of the dye on the ground-state 1,3,5-triazine derivative has negative value. The value of ΔG_{el} for the reverse process is very high (0.45 eV, 43.42 $kJ \cdot mol^{-1}$).

For the sensitizer employed in our studies, since the oxidation of the dye may occur, the two-component photoinitiating system composed of dye/1,3,5-triazine derivative can also be an effective photoinitiating system. The kinetic results obtained for this photoinitiating pair seem confirming this prediction. The value of ΔG_{el} is equal –0.28 eV (–27.02 $kJ \cdot mol^{-1}$). In the three-component photoinitiating system an effective interaction between all components can take place. Both electron transfer processes can occur (oxidation-reduction series mechanism):

i. from the ground state of a thiol to an excited singlet state of the thiocarbocyanine dye and
ii. from an excited singlet state of the dye to the ground-state of 1,3,5-triazine derivative.

The second reaction is followed by a C-Cl bond cleavage process which leads to the formation of radicals and chlorine anions [17]. In this case the initiating radicals are formed (Figure 37).

Observed an impact of both co-initiators presence on the rate of fluorescence decay of the sensitizer suggests that the primary photoreaction occurs between both the dye and thiol and the dye and 1,3,5-triazine derivative, respectively. Both co-initiators cause of decrease of the fluorescence lifetime, e.g. quenche an excited singlet state of cyanine dye (Figure 39).

The fluorescence quenching rate constants for cyanine dye by both co-initiators are about 10^{10} $M^{-1}s^{-1}$, e.g. are controlled by the diffusion. However, the fluorescence quenching of sensitizer by thiols occurs faster than by 1,3,5-triazine derivative. Therefore, in the case of three-component photoinitiating systems: cyanine dye/thiol/1,3,5-triazine derivative both electron transfer processes occur, e.g. from thiol molecule on an excited singlet state of sensitizer and from an excited singlet state of sensitizer to the ground-state 1,3,5-triazine derivative. The free radicals which can start the polymer chain reaction can arise from both thiol- and 1,3,5-triazine-derived radicals.

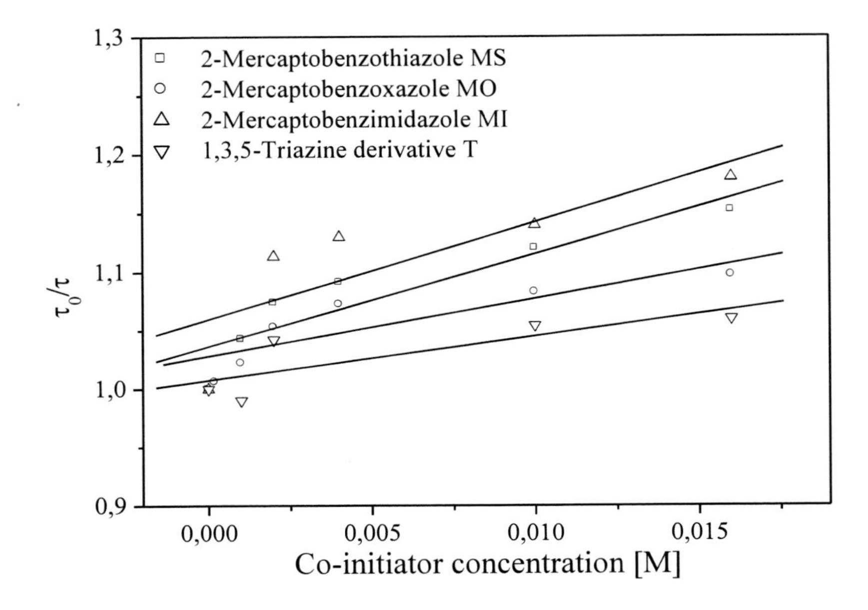

Figure 39. The Stern-Volmer relationship for the fluorescence quenching of N,N'-diethylthiocarbocyanine iodide (dye I) by the heteroaromatic thiols and 1,3,5-triazine derivative.

It should be noted, that the initiating radicals (derived from corresponding thiols) were different for all photoinitiating systems under study. Therefore, it may be expected that the activity of the thiols in the photoinitiating systems depends on the reactivity of the thiyl radical formed and the efficiency of their formation.

The mechanism of the processes occurring in the photoinitiating systems was proposed also basing on the results of nanosecond laser flash photolysis. Following laser excitation of thiocarbocyanine dye in a presence of thiol, a transient absorption spectra shows the appearance of two absorption bands at 420 and 600 nm, respectively (Figure 30). As it was metioned earlier, the absoption band at 420 nm was assigned to the cyanine dye radical formed in electron transfer reaction [12, 13]. This absorption band is not observed for the system composed of thiocarbocyanine dye in a presence of 1,3,5-triazine derivative. The second absorption band at 600 nm was attributed to the presence of a oxidation product of thiol an is ascribed to thiyl radicals [112]. The formation of thiyl radical and its disappearance is shown in the previously presented Figure 31. Whereas the curves traces observed at 420 nm for the cyanine dye radical are shown in Figure 32.

The laser flash photolysis results and calculated values of the free energy changes for electron transfer process confirmed the electron transfer process from ground-state heteroaromatic thiol to an excited singlet state of thiocarbocyanine dye. The transient absorption spectra of N,N'-diethylthiocarbocyanine iodide in a presence of 1,3,5-triazine derivative shown the presence of 1,3,5-triazine radial anion (Figure 40) and absence of the cyanine dye radical.

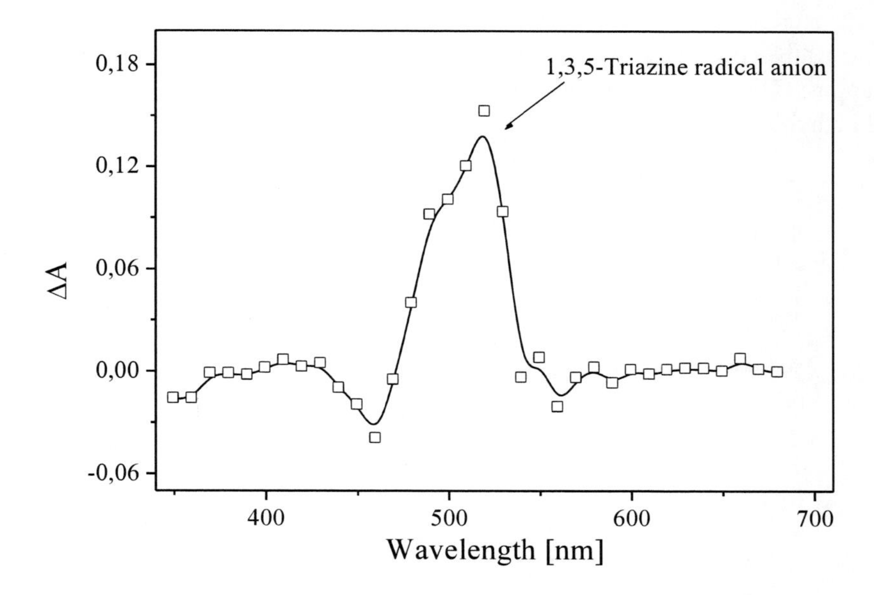

Figure 40. The transient absorption spectra of thiocarbocyanine dye (dye I) in the presence of 2,4-bis(trichloromethyl)-6-(4-methoxy)phenyl-1,3,5-triazine (T) recorded 50 ns after laser pulse. Concentration of 2,4-bis(trichloromethyl)-6-(4-methoxy)phenyl-1,3,5-triazine was equal 2×10^{-2} M in acetonitrile solution. Dye concentration was 5×10^{-4} M.

The characteristic band at 510 nm was also observed by Grotzinger in the three-component system composed of Rose bengal and 1,3,5-triazine derivative and was attributed to the product of reduction of 1,3,5-triazine derivative [17]. Basing on the obtained results, one may conclude, that an excited dye may act as an electron donor in the two-component photoinitiating system: thiocarbocyanine dye/2,4-bis(trichloromethyl)-6-(4-methoxy)phenyl-1,3,5-triazine. In the three-component photoinitiating system an excited singlet state of thiocarbocyanine dye was reduced by heteroaromathic thiol or oxidized by 1,3,5-triazine derivative. These reactions produce:

- cyanine dye radical and thiyl radical cation and
- cyanine dye radical cation and 1,3,5-triazine radical anion, which rapidly undergoes C-Cl fragmentation giving a halogene anion and triazinyl radical.

On the basis of the nanosecond laser flash photolysis and thermodynamical analysis, the mechanism for the primary and secondary reactions for sensitized generation of free radicals was proposed. After irradiation of the three-component photoinitiating system composed of thiocarbocyanine dye/heteroaromatic thiol/2,4-bis(trichloromethyl)-6-(4-methoxy)phenyl-1,3,5-triazine an excited singlet state of chromophore is formed. The deactivation of the excited state occurs by fluorescence, photoisomerization or electron transfer process. In the presence of heteroaromatic thiol the cyanine dye undergoes one-electron reduction. The cyanine dye radical and thiyl radical cation are formed. The thiyl radical cation undergoes the hydrogen abstraction, giving the thiyl radical that can start the polymerization reaction. However, in the presence of 2,4-bis(trichloromethyl)-6-(4-methoxy)phenyl-1,3,5-triazine, the electron transfer from an excited singlet state of cyanine dye to the ground-state of 1,3,5-triazine derivative occurs, giving the cyanine dye radical cation and 1,3,5-triazine-derived radical anion. The last undergoes fast fragmentation forming the chlorine anion and triazynylmethyl radical (Scheme 24).

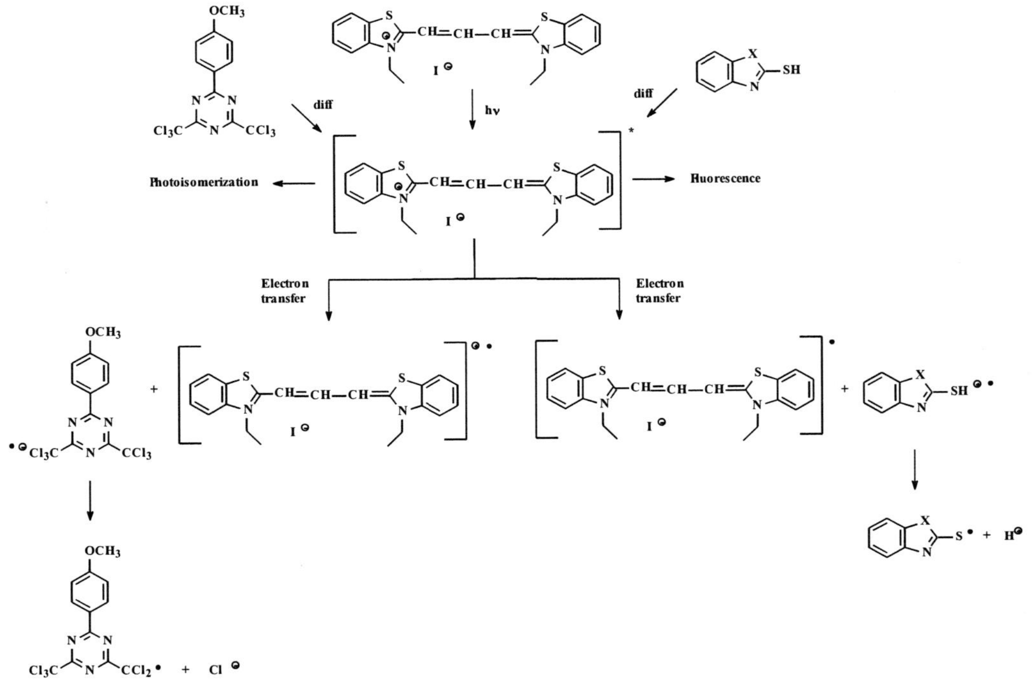

Scheme 24. The primary and secondary reactions for the three-component photoinitiating system composed of polymethine dye/1,3,5-triazine derivative/heteroaromatic thiol after a visible light irradiation.

Excited State Processes in the Excited Singlet State of Sensitizer in the Three-Component Photoinitiating Systems

In order to explain the *n*-butyltriphenylborate (B2), N-alkoxypyridinium salt, 1,3,5-triazine derivative, N-methylpicolinium ester and heteroaromatic thiol effect on the photochemical behavior of the light absorbing dye (sensitizer) in the three-component photoinitiating system the steady-state fluorescence experiments were applied to observe the fluorescence quenching as a function of a co-initiator concentration. The presence of *n*-butyltriphenylborate has significant impact on the rate of fluorescence decay of sensitizer (Figure 41). Therefore, this suggests that the primary photochemical process is an electron transfer from borate anion to the excited singlet state of sensitizer.

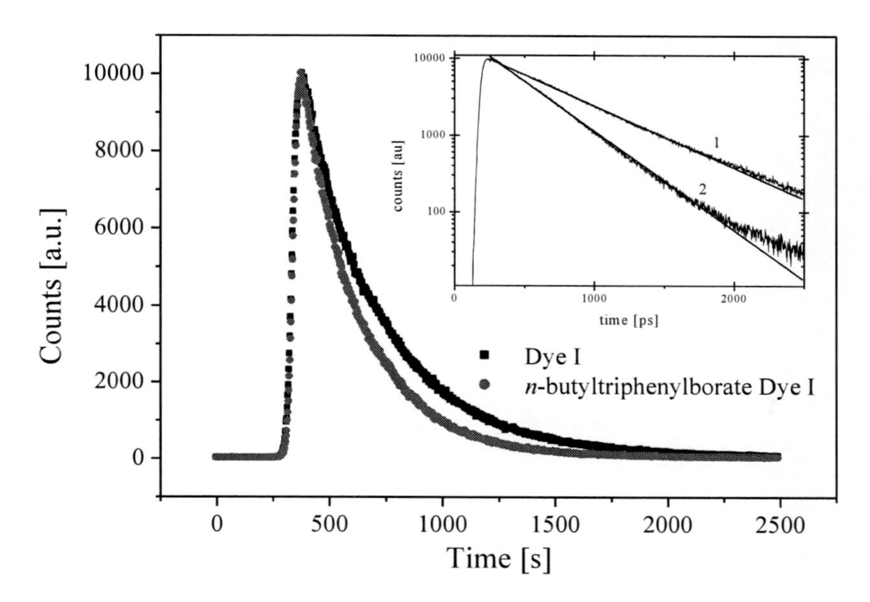

Figure 41. The effect of *n*-butyltriphenylborate on the fluorescence decay of N,N'-diethylthiocarbocyanine in a solution of 1-methyl-2-pyrrolidinone/ethyl acetate (1:9) [94].

The addition of a borate salt results in the decrease of the fluorescence lifetime of sensitizer. The fluorescence quenching of carbocyanine dyes (N,N'-diethylthiocarbocyanine iodide (dye I) and N,N'-diethyloxocarbocyanine iodide (dye II)) in the presence of borate salt (B2), 2,4,6-tris(chlorodifluoromethyl)-1,3,5-triazine (T), N-alkoxypyridinium salt (NO, Bp), N-methylpicolinium ester and heteroaromatic thiols is shown in Figures 42 and 43.

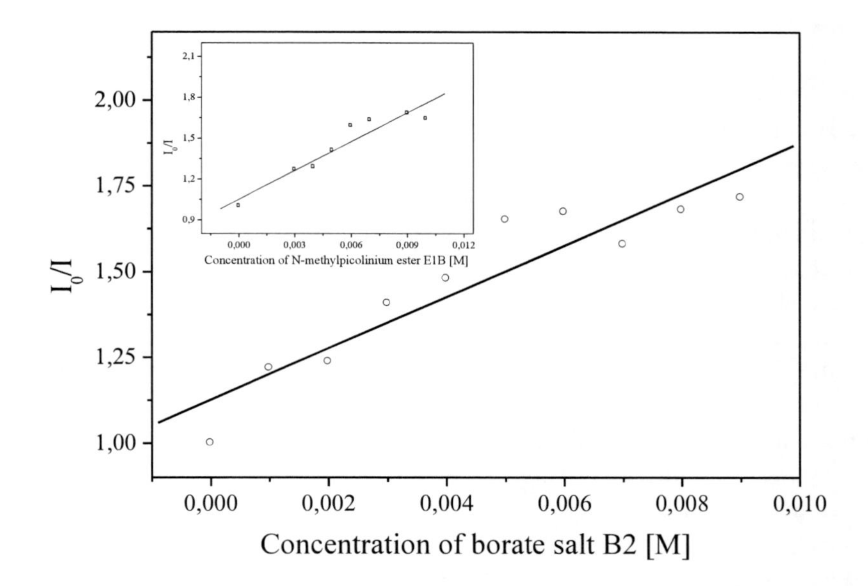

Figure 42. The Stern-Volmer relationship for the fluorescence quenching of mono-cationic dye: N,N'-diethyloxocarbocyanine iodide (dye II) by *n*-butyltriphenylborate (B2) and N-methylpicolinium ester (E1B).

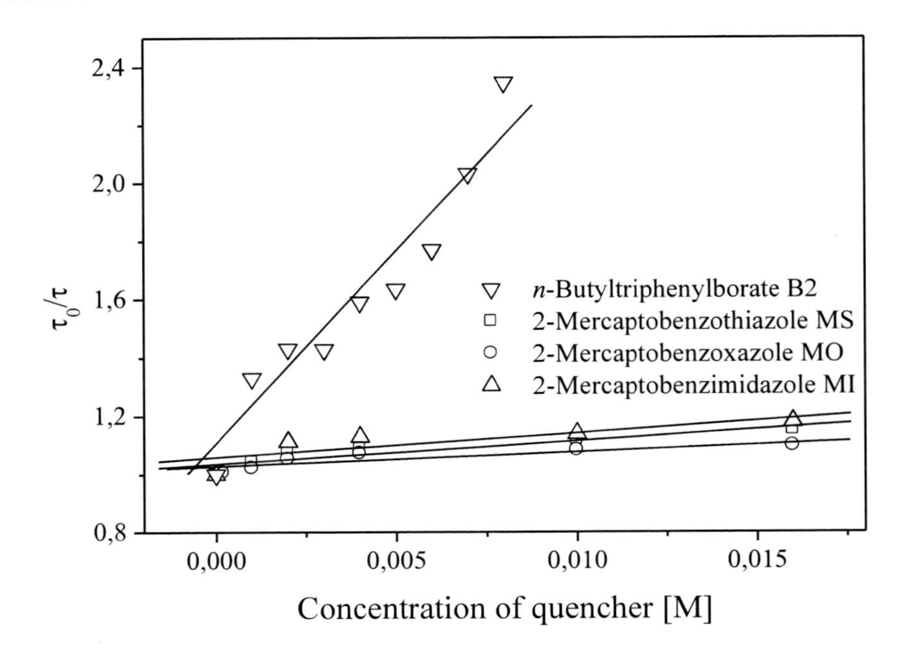

Figure 43. The Stern-Volmer relationship for the fluorescence quenching of mono-cationic dye: N,N'-diethyloxocarbocyanine iodide (dye II) by *n*-butyltriphenylborate (B2) and heteroaromatic thiols (MS, MO, MI).

The values of $k_q\tau$ (k_q the fluorescence quenching rate constant, τ the fluorescence lifetime of sensitizer) for all co-initiators are in the range from 5.01 M^{-1} for 1,3,5-triazine derivative to 11.78 M^{-1} for *n*-butyltriphenylborate salt. This values is higher for the two-component system dye/borate salt in comparison to other two-component systems. Basing on this, one should conclude, that the primary process occurring in the three-component photoinitiating system: polymethine dye/borate salt/second co-initiator is reaction between borate salt and polymethine dye in its excited state. The fluorescence lifetime of N,N'-diethyloxocarbocyanine iodide is equal 95 ps in solution of ethyl acetate:2-methyl-1-pyrrolidinone (9:1). The addition of *n*-butyltriphenylborate salt results in the decreasing of the fluorescence lifetime of the dye to 62 ps. The fluorescence quenching rate constants for all second co-initiators are about 10^{10} $M^{-1}s^{-1}$, e.g. and are controlled by the diffusion. Whereas, the fluorescence quenching rate constant for borate anion is about 10^{12} $M^{-1}s^{-1}$. The fuorescence quenching rate constant obtained for dye-borate ion pair are greater, than these commonly observed for intermolecular electron transfer reactions, since ion pairing eliminates the limitation caused by diffusion. The quenching rate constants of the cyanine excited singlet state by other co-initiators tested (1,3,5-triazine derivative, N-alkoxypyridinium salt, N-methylpicolinium ester or heteroaromatic thiol) are about one order of magnitude lower than those measured for borate anion.

In order to explain, the relationship between the observed rate of polymerization and photophysical properties of three-cationic carbocyanine dyes and multi-cationic monomethine dyes the fluorescence lifetimes were measured [122].

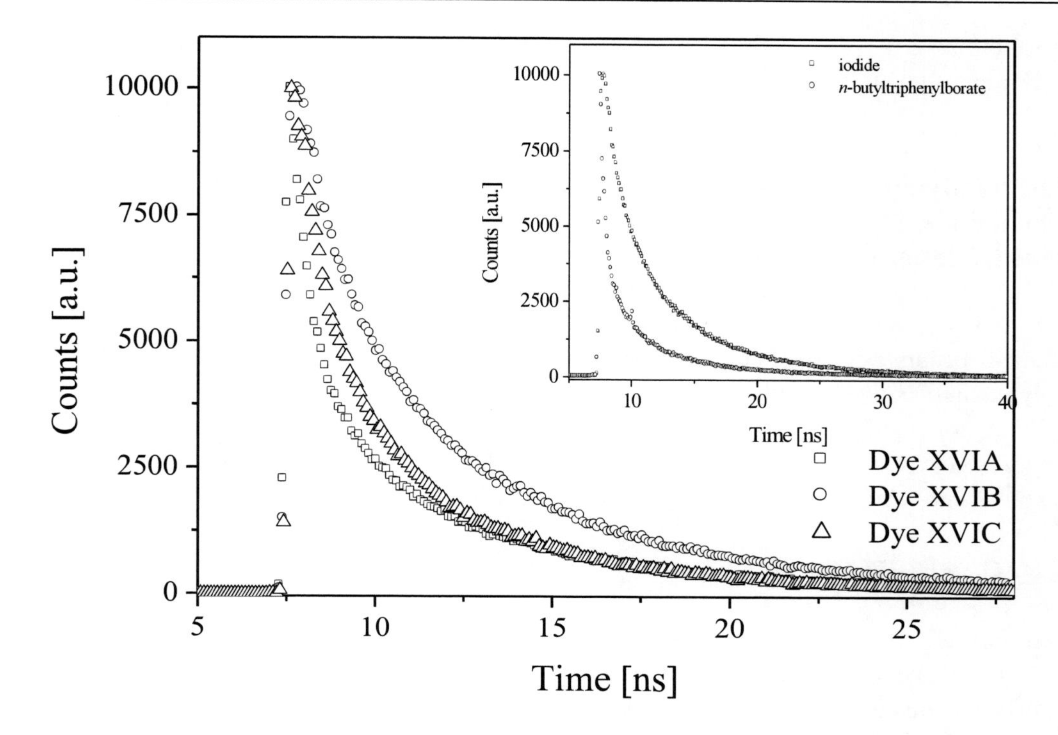

Figure 44. The effect of the monomethine dye structure on the fluorescence lifetime. The fluorescence decay was measured in 1-methyl-2-pyrrolidinone (1:4) solution. Inset: the effect of *n*-butyltriphenylborate anion on the fluorescence lifetime of three-cationic dye [122].

The fluorescence lifetime depends on the structure of sensitizer. It is in the range from 200 to 1440 ps for three-cationic thiocarbocyanine dyes. The fluorescence lifetimes of oxocarbocyanine dyes are shorter. The results obtained during fluorescence lifetime measurements for multi-cationic monomethine dyes show the multiexponential fluorescence decay. The shortest lifetime observed was about 0.5 ns and the longest lifetime oscillated from 6 to 9 ns. The percentage participation of conformer possessing the shortest lifetimes is about 5-10 % [122].

The fluorescence lifetime measurements were applied to the assignment the relationship between fluorescence lifetime of sensitizer and co-initiator concentration (Stern-Volmer relationship).

$$\frac{\tau_0}{\tau} = 1 + K_q[Q] = 1 + k_q\tau[Q] \qquad \text{eq.(22)}$$

where:

τ_0 and τ are the fluorescence lifetimes of thiocarbocyanine dye in the absence and the presence of quencher, respectively; K_q is the Stern-Volmer constant, k_q is the fluorescence quenching rate constant. The calculated from the Stern-Volmer relationship the fluorescence quenching rate constants k_q for the photoinduced electron transfer between *n*-butyltriphenylborate anion and an excited singlet state of the three-cationic carbocyanine dye are about 2×10^{11} $M^{-1}s^{-1}$, while for monomethine dyes are about 1×10^{11} $M^{-1}s^{-1}$ and are about one order of magnitude higher in comparison to the parent mono-cationic dyes [113].

Thermodynamic of Free Radicals Formation and the Application of Marcus Theory for Kinetics Study of Free Radical Polymerization of Multifunctional Acrylates

As it was mentioned above, the formation of free radicals in the photoinitiating systems studied occurs via photoinduced electron transfer process (PET). Discussion on the polymethine borates photochemistry should include the estimation of the thermodynamic driving force of photoinduced electron transfer reaction (PET). The thermodynamic requirement for a spontaneous electron transfer is that the free energy change ΔG_{el}, expressed by the Rehm-Weller equation should have negative values (eq. 23) [123].

$$\Delta G_{el} = E_{ox}(D^{\bullet +}/D) - E_{red}(A/A^{\bullet -}) - Ze^2/\varepsilon a - E_{00} \qquad \text{eq. (23)}$$

here $E_{ox}(D^{\bullet +}/D)$ is the oxidation potential of an electron donor molecule, $E_{red}(A/A^{\bullet -})$ is the reduction potential of an electron acceptor, $Ze^2/\varepsilon a$ is the Coulombic energy, normalny considered negligible in high-dielectric solvents, and E_{00} in an excited state energy of the photosensitizer. Electron transfer process competes with rapid photophysical deactivation of the sensitizer excited state. The exothermity of electron transfer process affects on the efficiency of electron transfer process. In other words, the change of the driving force of the electron transfer has the negative values of ΔG_{el}.

The values of free energy change for electron transfer process were calculated for two cases. First, for the electron transfer process from borate anion on an excited state of a dye, using the oxidation potential of borate anion (1.16 eV) and the reduction potential of sensitizer. This process leads to the formation of butyl radicals as a result of a decomposition of the boranyl radical (very fast and irreversible process). Second, the values of ΔG_{el} were calculated using the reduction potential of second co-initiator (E_{red}) and the oxidation potential of an excited singlet state of dye (E_{ox}^{*}).

For the two-component photoinitiating systems composed of borate salt of polymethine dye, the efficiency of the butyl radical formation and, hence, initiation of polymerization depends on the observed rate of electron transfer from borate anion on the excited singlet state of polymethine dye. If the electron transfer process does not depend on the rate of diffusion process, its rate determines the rate of polymerization reaction. In such a case, the polymerization rate would increase with an increase in the $-\Delta G_{el}$ values, at least up to the diffusion-controlled limit. The rate of electron transfer does not effect on the rate of free radical polymerization in the region controlled by diffusion process (does not depent on the value of ΔG_{el}). In such a case, other factors influence on the rate of free radical polymerization (for example, the reactivity of primary radicals) [112, 124]. Therefore, the efficiency of electron transfer process decreases as the viscosity of polymerizing mixture increases. Hence, the formation of free radicals via electron transfer process is effective only at the beginning of the polymerization.

Marcus determined [125] direct method which allows us to predict the kinetic of an electron transfer process on the basis on the thermodynamical and spectroscopic properties of an electron donor and an electron acceptor. Basing on the thermodynamic of electron transfer process the relationship between the rate of free radicals formation and free energy change for electron transfer process can be quantitatively predicted.

The Marcus theory [125] allows to predict the rate of the primary process e.g. the rate of photoinduced electron transfer. The use of polymethine borate salts creates a unique opportunity to study of the possibility of the application of this theory for the description of the rate of polymerization via an intermolecular electron transfer process. For these photoinitiating systems the change of the driving force of the electron transfer process has no influence on the type of the yielding free radicals. If the polymerization occurs in very viscous media, the rate of polymerization initiated via a photoinduced electron transfer can be described as follows (eq. 24) [126, 127]:

$$\ln R_p = A-(\lambda + \Delta G_{el})^2/8\lambda RT \qquad \text{(eq. 24)}$$

where A for the initial time of polymerization is the sum: $\ln k_p - 0{,}5k_t + 1{,}5\ln[M] + 0{,}5\ \ln I_a$ (here k_p, k_t denote the rate constant of polymerization and termination, respectively, [M] in the monomer concentration, I_a is the intensity of absorbed light), λ is the reorganization energy necessary to reach the transitions states both of excited molecule and solvent molecules.

Equation (24) clearly indicates that, if the primary process, e.g. the rate of electron transfer process controls the observed rate of photopolymerization, one should observe a parabolic relationship between the logarithm of polymerization rate and the free energy change ΔG_{el}.

Figures 45-47 present the normal logarithm of the TMPTA polymerization rate as a function of the ΔG_{el} value for selected initiating systems [128].

For the selected photoinitiating systems the rate of polymerization can be described by classical electron transfer theory. For the most photoinitiating systems tested the observed rate of free radical polymerization of TMPTA increase if the free energy change for electron transfer reaction increases [129].

Obtained relationships are roughly linear for three-cationic carbocyanine dyes as *n*-butyltriphenylborate salts possessing identical organic cations attached by an alkyl chain to dye heterocyclic nitrogen atom. Different slopes and different y-intercepts observed derive probably from various dissociation degrees of borate salts that are covalently attached to the dye chromophore. The variation in the dissociation degree of borate salt causes the variation of the borate anion concentration in proximity of chromophore, thus causes the changes in photoinitiating ability [99].

The values of reorganization energy (λ) is ca. –0.3 eV (ca. –29 kJ mol^{1}) for bi-chromophoric hemicyanine dyes and about –0.8 eV (ca. 80 kJ mol^{-1}) for hemicyanine dyes. These values are similar to the value reported by Schuster [12, 13] and support the postulate of an electron transfer reaction generating free radicals able to initiate free radical polymerization [26].

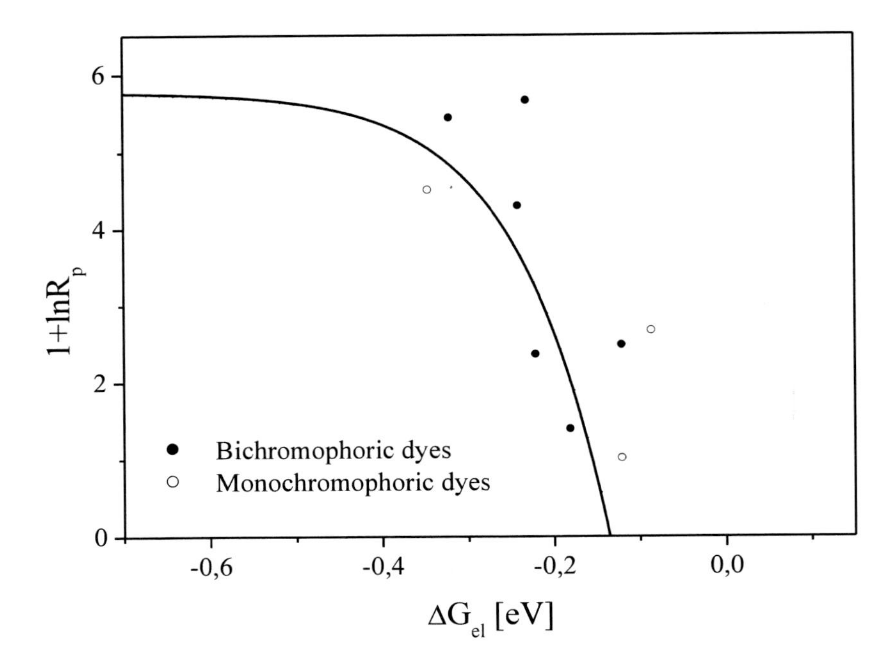

Figure 45. Rates of polymerization, R_p of TMPTA as a function of the free energy change of the electron transfer reaction from the borate anion to the excited singlet state of bichromophoric polymethine dyes [128].

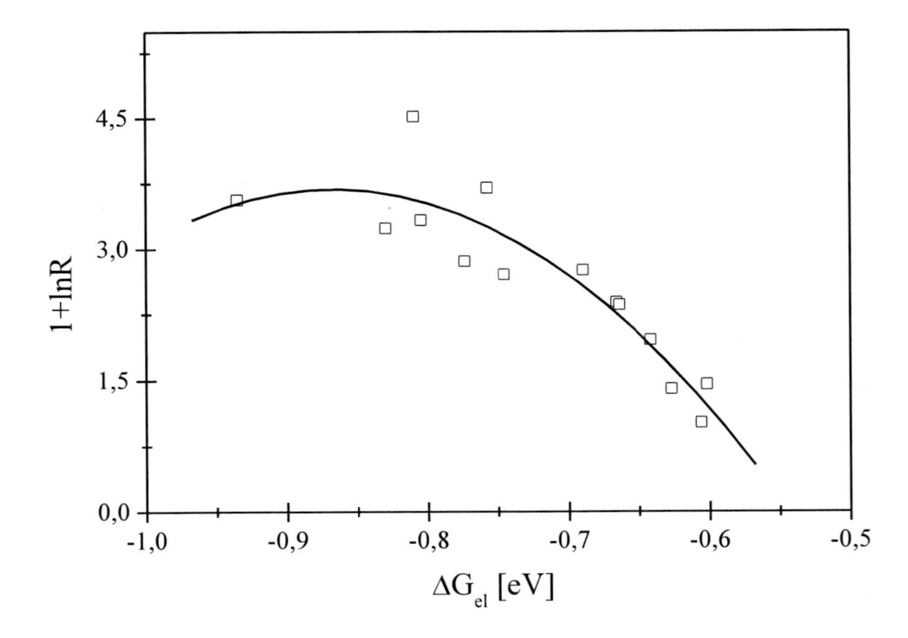

Figure 46. The Marcus relationship for *n*-butyltriphenylborate salts selected hemicyanine dyes [19, 26].

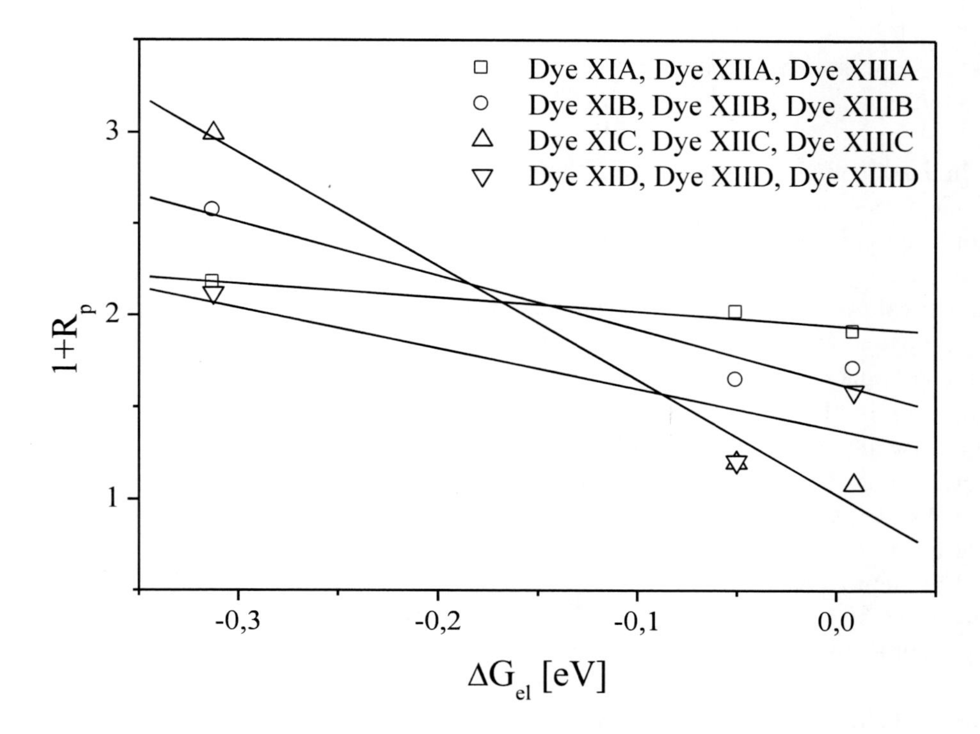

Figure 47. The relationship between the rate of free radical polymerization and the free energy change for the electron transfer from *n*-butyltriphenylborate anion to an excited singlet state of three-cationic carbocyanine dyes, possessing different chromophores and the same organic cation attached to the heterocyclic ring [113].

Conclusion

The efficiency of a visible light induced polymerization of multifunctional acrylate initiated by two- and three-component photoinitiating systems based on the polymethine dyes as sensitizer was ascertained.

Polymethine dyes with different co-initiators (alkyltriphenylborate, N-alkoxypyridinium salts, N,N'-dialkoxy-2,2'-bipyridilum salts, 1,3,5-triazine derivative, N-methylpicolinium esters, heteroaromatic thiols and cyclic acetals) can be used as the sensitizers in practical very efficient two- and multi-component visible light photoinitiating systems that are useful for photoinitiation of free radical polymerization of multifunctional acrylates.

In the two-component photoinitiating systems, in which alkyltriphenylborate salt acts as co-initiator, the initiation of free radical polymerization is a result of an electron transfer process from borate anion on an excited singlet state of polymethine dye. The alkyl radicals initiating free radical polymerization are formed as a result of the C-B bond cleavage in the product of electron transfer process. The photoinitiating ability of the photoinitiating systems composed of polymethine dye/*n*-butyltriphenylborate salt depends on:

- Structure of sensitizer
- Concentration of photoinitiator
- Concentration of co-initiator.

In the polymerizing mixture only about 20 % of *n*-butyltriphenylborate anions and mono-cationic dyes form an ion pair, but the degree of dissociacion of *n*-butyltriphenylborate multi-cationic dyes is widely lower and is equal 40 % [20].

In almost all cases the highest rates of free radical polymerization are observed for symmetrical polymethine dyes (benzothiazole derivative) without a substituent in the phenyl ring and paired with *n*-butyltriphenylborate anion. The modification of the sensitizer structure by the introduction of an electron-withdrawing atoms e.g. chloride and fluoride into the phenyl ring of benzothiazole moiety or an electron-donating group into styryl moiety causes an increase in the rate of polymerization.

The photoinitiating abilities of two-component photoinitiating systems are lower than those observed for dyeing photoinitiating systems in which after irradiation with visible light the long-lived excited triplet state of a dye is formed.

The improvement of the photoinitiating efficiency of cyanine borate photoinitiating systems can be achieved by the modification of the sensitizer structure or the composition of the photoinitiating system. There are several methods allowing to do that.

- The first, the chemical **modification of the sensitizer structure by the introduction of the group possessing second tertiary nitrogen atom**. This leads to two-times increase of the concentration of the borate anion (electron donor) in the close proximity to an excited chromophore. **The synthesis and application of the multi-cationic monochromophoric carbocyanine dyes or multi-chromophoric polymethine dyes.** The introduction of an additional positive charged group to the structure of mono-cationic dye increases of the co-initiator concentration in a close proximity to a chromophore, and hence decreases of degree of dissociation at about one order of magnitude. The free radical polymerization initiated by heterobicationic hemicyanine dyes paried with borate anion occurs with higher rates in comparison to photoinitiating systems composed of mono-cationic dyes. The rate of polymerization is comparable with the rate of polymerization initiated by the triplet state photoinitiating systems (Rose bengal/N-phenylglycine). The three-cationic carbocyanine dyes as *n*-butyltriphenylborate salts are more efficient initiating systems for free radical polymerization than parent mono-cationic carbocyanine dyes borate salts.
- The second, is **the formation of two different free radicals initiating polymerization**. It is possible **by the introduction of a second co-initiator**. Such three-component photoinitiating systems are composed of polymethine dye (sensitizer), *n*-butyltriphenylborate anion (electron donor) and a second co-initiator (N-alkoxypyridinium cation, 1,3,5-triazine derivative, N-methylpicolinium ester, cyclic acetal (ground state electron acceptors) or heteroaromatic thiol (hydrogen donor). In three-component photoinitiating systems both donor and acceptor can act separately or can be paired. The donor-acceptor pairing additionally enhances the rate of free radical formation, which is probably caused by partial elimination of the

diffusion that controls the overall efficiency of radical production. The formation of two different free radicals initiating polymerization is also possible by application of the two-component photoinitiating system composed of **cyanine dye possessing N-alkoxy or diphenylacetic acid N-(methylpyridine)methyl ester group at the tertiary nitrogen atom in the heterocyclic ring** and *n*-butyltriphenylborate salt as a co-initiator. In the first case as a result of electron transfer from borate anion on an excited singlet state of dye the pyridinium radical is formed. This radical undergoes vary fast decomposition giving alkoxy radical. From here, similar as in the three-component system there are second initiating radicals, beside *n*-butyl radicals formed after electron transfer from borate anion on the excited singlet state of cyanine dye in the polymerizing mixture. The advantage of polymethine dyes possessing N-alkoxy group attached to the tertiary nitrogen atom is that their application in the two-component photoinitiating systems eliminates the effect of the diffusion process on the rate of initiation of free radical polymerization. The application of the photoinitiating system composed of structurally modified dye for example, 2-(*o*-methoxypyridine)-*p*-pyrrolidinostyryl methylsulfate and borate salt (B2) enhances the rate of free radical polymerization in comparison with those observed for three-component photoinitiating systems composed of no modified hemicyanine dye/*n*-butyltriphenylborate salt/N-alkoxypyridinum salt. The modification of the sensitizer structure increases the rate of free radical polymerization about 300-500-times in comparison with parent hemicyanine dye borate salt.

The photoinitiating ability of the three-component photoinitiating systems depends on the value of free energy change for electron transfer process between an electron donor and an electron acceptor an also on the reactivity of free radicals formed in the secondary reactions after photoinduced electron transfer process.

The enhancement of photoinitiating efficiency of three-component photoinitiating systems composed of polymethine dye borate ion pair in the presence of a second co-initiator comes from the specific spatial arrangement of all components of photoinitiating systems. In the three-component counter complex the initiating radical are formed as a result of secondary processes. The mechanism of the primary and secondary reactions for three-component photoinitiating systems was established basing on the nanosecond laser flash photolysis.

After irradiation of the three-component photoinitiating system with a visible light an excited singlet state of chromophore is formed. The deactivation of an excited state occurs by fluorescence, photoisomerization or electron transfer process. In presence of alkyltriphenylborate anion, the cyanine dye undergoes one-electron reduction. The cyanine dye radical and boranyl radical are formed. The boranyl radical undergoes the C-B bond cleavage giving an alkyl radical that can start the polymerization reaction. Subsequently, the cyanine radical formed after electron transfer process in presence of a second co-initiator can participate in a second electron transfer process, giving cyanine cation and reduced second co-initiator. This intermediate decomposes forming free radical, which can also initiate free radical polymerization. Since the ground state of cyanine dye is regenerated in this reaction, the initiation rate will be enhanced further.

Different mechanism of primary and secondary processes was observed for the three-component photoinitiating system composed of polymethine dye/*n*-butyltriphenylborate salt/heteroaromatic thiol. In such three-component photoinitiating system two electron transfer processes occur. The electron transfer process from borate anion on an excited singlet state of cyanine dye competes with the electron transfer process from a ground state of thiol on an excited singlet state of sensitizer. The fluorescence quenching rate constants of the excited singlet state of polymethine dye by heteroaromathic thiols depend on diffusion process and are about one order of magnitude lower in comparison with those observed as *n*-butyltriphenylborate anion is used as a quencher. The electron transfer process from ground state of heteroaromathic thiol on an excited singlet state of polymethine dye causes in the enhancement of the efficiency of initiation polymerization by *n*-butyltriphenylborate salt of polymethine dye in presence of heteroaromatic thiol. A combination of heteroaromatic thiol with alkyltriphenylborate salt and suitable sensitizer enhances the photoinitiating ability of such photoinitiating systems.

In summary, from a combination of either N-alkoxypyridinium salt, 1,3,5-triazine derivative, N-methylpicolinium ester, cyclic acetal, heteroaromatic thiol and alkyltriphenylborate salt with suitable sensitizer or properly design dye, gives the dyeing photoinitiating system, in which two radicals can be generated per one absorbed photon, thus enhancing the overall polymerization efficiency. Generally, the type of applied second co-initiator, which can form the initiating radicals as a result of a secondary reaction in three-component photoinitiating system has effect on the overall efficiency of the photoinitiation.

The best photoinitiating efficiency is observed for the three-component photoinitiating systems possessing the 1,3,5-triazine derivative or N-alkoxypyridinium salt as a second co-initiator. The polymerization photoinitiation ability of three-cationic carbocyanine dyes as *n*-butyltriphenylborate salts and multi-cationic monomethine dyes as borate salts in presence of second co-initiator is more efficient than this observed for RBAX-NPG couple (typical triplet photoinitiating system).

The photoinitiating ability of the three-component photoinitiating systems depends on:

- The rate of primary electron transfer process
- The rate of carbon-boron bond cleavage
- The rate of secondary electron transfer process
- The rate of nitrogen-oxygen bond, the rate of carbon-halogen or the rate of carbon-oxygen bond cleavage
- The reactivity of free radicals formed.

In this chapter there are also examples of two- and three-component photoinitiating systems for free radical polymerization of multifunctional acrylates composed of polymethine dye/heteroaromatic thiol/1,3,5-triazine derivative. The photoinitiating ability of two-component system: polymethine dye/1,3,5-triazine derivative is comparable with those observed for well-known polymethine dye as *n*-butyltriphenylborate salt two-component photoinitiating system. The rate of polymerization increses about 2 times by the addition of the heteroaromatic thiol as a second co-initiator to the two-component photoinitiating system: dye/1,3,5-triazine derivative. The mechanism of primay and secondary processes (oxidation-reduction mechanism, e.g. parallel series mechanism) occurring in three-component photoinitiating system after absorbtion of a visible light was proposed. In such photoinitiating systems the primary photochemical reaction are both electron transfer processes: an electron transfer from the ground state of heteroaromathic thiol on an excited singlet state of polymethine dye and an electron transfer from an excited singlet state of sensitizer on the ground state of 1,3,5-triazine derivative. Thiol acts as an electron donor, reducing dye but 1,3,5-triazine derivative is an electron acceptor oxidizing the excited dye. These reactions generate free radicals, which can start the polymerization chain reaction.

The Marcus theory can be applied to the description of the kinetic of free radical polymerization. For the selected photoinitiating systems the classical parabolic relationship between the rate of polymerization and the free energy change for electron transfer process is observed.

Acknowledgment

This work was supported by The Ministry of Science and Higher Education (MNiSW) (grant No N N204 219734).

References

[1] Kabatc J., Jędrzejewska B., Pączkowski J., *Polymer Bulletin*, 2005, 54, 409.
[2] Shirai M., Suyama K., Tsunooka M., *Trends Photochem. Photobiol.*, 1999, 5, 169.
[3] Fouassier J.P., Simonin-Catilaz L., *J. Appl. Polym. Sci.*, 2001, 79, 1911.
[4] Grotzinger C., Burged D., Jacques P., Fouassier J.P.,8 *Polymer*, 2003, 44, 3671-3677.
[5] Jakubiak J., Rabek J.F., *Polimery*, 1999, 44, 447-470.
[6] Takemura F., *Bull. Chem. Soc. Jpn.,* 1962, 35, 1073.
[7] Kawamura K., *Dye-linked radical generators for visible light photoinitiating polymerization in Photochemistry and UV Curing: New Trends*, Ed. Fouassier J.P., Research Signpost Kerala, India 2006, pp. 204-207.
[8] Oster G., *Nature* 1954, 173, 300.
[9] Pączkowski J., *"Fotochemia polimerów teoria i zastosowanie"* Wydawnictwo Uniwersytetu Mikołaja Kopernika, Toruń 2003.
[10] Turro N.J., *Modern Molecular Photochemistry*, The Benjamin/Cummings Publishing, California, 316, 1978.
[11] Oevering H., Paddon-Row M.N., Heppener M., Oliver A.M., Cotsaris E., Verthoeven J.W., Hush N.S., *J. Am. Chem. Soc.*, 1987, 109, 3258.
[12] Chatterjee S., Gottschalk P., Davis P.D., Schuster G.B., *J. Am. Chem. Soc.*, 1988, 110, 2326.
[13] Chatterjee S., Davis P.D., Gottschalk P., Kurz M.E., Sauerwein B., Yang X., Schuster G.B, *J. Am. Chem. Soc.*, 1990, 112, 6329.
[14] Kohler M., Ohngemach J., European Patent, 354458, 1990, *Chem. Abstr.*, 113, 106437, 1990.
[15] Kawamura K., *J. Photochem. Photobiol. Part A Chem.*, 2004, 162, 329.
[16] Pohlers G., Sciano J.C., Sinta R., Step E., *J. Am. Chem. Soc.,* 1999, 121, 6167.
[17] Grotzinger C., Burget D., Jacques P., Fouassier J.P., *Macromol. Chem. Phys.*, 2001, 202, 3513.

[18] Pączkowski J., Kucybała Z., Ścigalski F., Wrzyszczyński A., *J. Photochem. Photobiol. Part A Chem.,* 2003, 159, 115-125.
[19] Kabatc J., Jędrzejewska B., Pączkowski J., *J. Polym. Sci., Polym. Chem*., 2003, 41, 3017-3026.
[20] Kabatc J., Jędrzejewska B., Pączkowski J., *J. Polym. Sci., Polym. Chem*., 2006, 44, 6345-6359.
[21] Kabatc J., Pietrzak M., Pączkowski J., *Macromolecules*, 1998, 31, 4651-4654.
[22] Kabatc J., Jędrzejewska B., Pączkowski J., *J. Polym. Sci., Polym. Chem*., 2000, 38, 2365-2374.
[23] Jędrzejewska B., Kabatc J., Pietrzak M., Pączkowski J., *J. Polym. Sci., Polym. Chem.,* 2002, 40, 1433-1440.
[24] Kabatc J., Pączkowski J., *Macromolecules*, 2005, 38, 9985-9992.
[25] Lan L.Y, Schuster G.B., *Tetrahedron Lett.,* 1986, 27, 4261.
[26] Kabatc J., Pietrzak M., Pączkowski J., *J. Chem. Soc., Perkin Trans*., 2002, 2, 287-295.
[27] Kabatc J., Jędrzejewska B., Bajorek A., Pączkowski J., *J. Fluoresc*., 2006, 16, 525.
[28] Kabatc J., Pączkowski J., Karolczak J., *Polimery*, 2003, 6, 425-433.
[29] Kabatc J., Pączkowski J., *Polimery*, 2003, 5, 321-328.
[30] Kabatc J., Kucybała Z., Pietrzak M., Ścigalski F., Pączkowski J., *Polymer*, 1999, 40, 735.
[31] Kabatc J., Pączkowski J., *Polymer*, 2006, 47, 2699.
[32] Kabatc J., Jędrzejewska B., Gruszewska M., Pączkowski J., *Polymer Bulletin*, 2007, 58, 691-701.
[33] Kabatc J., Two- and Three-Component Photoinitiating Systems Composed of Symmetrical Cyanine Dyes. *The Kinetic and Mechanistic Studies., in Basics and Aplications of Photopolymerization Reactions*, J.P. Fouassier and X. Allonas (Eds.), Research Signpost, Trivandrum - 695023 Kerala, India, 2010, vol.1, pp. 135-161.
[34] Pączkowski J., Kabatc J., Jędrzejewska B., Polymethine Dyes as Fluorescencje Probes and Visible-Light Photoinitiators for Free Radical Polymerization. *Heterocyclic Polymethine Dyes, Synthesis, Properties and Application, series: TOPICS in HETEROCYCLIC CHEMISTRY,* Strekowski Lucjan (Ed.), Springer-Verlag, Berlin Heidelberg, 2008, vol. 14, pp. 183-220.
[35] Ephardt H., Fromherz P.*, J. Phys. Chem*., 1989, 93, 7717.
[36] Ephardt H., Fromherz P.*, J. Phys. Chem*., 1991, 95, 6792.
[37] Fromherz P.*, J. Phys. Chem*., 1995, 99, 7188-7192.
[38] Fromherz P., Heilemann A.*, J. Phys. Chem*., 1992, 96, 6964.
[39] Strehmel B., Seifert H., Rettig W., *J. Phys. Chem*., 1997, 101, 2232.
[40] Jager W.F., Kudasheva D., Neckers D.C., *Macromolecules*, 1996, 29, 7351.
[41] Wróblewski S., Trzebiatowska K., Jędrzejewska B., Pietrzak M., Gawinecki R., Pączkowski J., *J. Chem. Soc*. Perkin Trans 2, 1999, 1909-1917.
[42] Ren K., Serguievski P., Gu H., Grinevich O., Malpert J.H., Neckers D.C., *Macromolecules*, 2002, 35, 898.
[43] Hassoon S., Sarker A., Polylarpov A.Y., Rodgers M.A.J., Neckers D.C., *J. Phys. Chem*., 1996, 100, 12386.
[44] Kim D., Stansbury J.W., *J. Polym. Sci., Part A Polym. Chem*., 2009, 47, 887-898.
[45] Yang J.Y., Neckers D.C.*, J. Polym. Sci., Part A Polym. Chem.,* 2004, 42, 3836-3841.

[46] Burged D., Grotzinger C., Fouassier J.P. w: Fouassier J.P. editor. Light Induced Polymerization Reactions. *Trends in photochemistry and photobiology*. Vol. 7., Trivandrum Trends., India, 2001, p. 71.
[47] Fouassier J.P., Morlet-Savary F., Yamashita K., Imahashi S., *Polymer*, 1997, 38, 1415-1421.
[48] Kabatc J., Pączkowski J., *J. Appl. Polym. Sci.*, 2010, 117, 2669-2675.
[49] Padon S.K., Scraton A.B., *J. Polym. Sci., Part A: Polym. Chem.*, 2000, 38, 2057-2066.
[50] Fouassier J.P., Allonas X., Lalevee J., Visconti M., *J. Polym. Sci., Part A: Polym. Chem.*, 2000, 38, 4531-4541.
[51] Burget D., Fouassier J.P., *J. Chem. Soc. Faraday Trans*, 1998, 94, 1849.
[52] Rubin M.B., *Tetrahedron Lett.*, 1969, 45, 3931.
[53] Monroe B.M., Weiner S.A., *J. Am Chem. Soc.*, 1969, 91, 450.
[54] Cavitt T.B., Phillips B., Hoyle Ch.E., Pan B., Hait S.B., Viswanathan K., Jönsson S., *J. Polym. Sci., Part A Polym. Chem.*, 2004, 42, 4009-4015.
[55] Pebber T.L., *J. Am. Chem. Soc.,* 1982, 104, 7407-7413.
[56] Fouassier J.P., Reddalane A., Morlet-Savary F., Sumiyoshi I., Harada M., Kawabata M., *Macromolecules*, 1994, 27, 3349-3356.
[57] Fouassier J.P., Morlet-Savary F., Yamashita K., Imahashi S., *J. Appl. Polym. Sci.,* 1996, 62, 1877-1885.
[58] Padon K.S., Scraton A.B., *J. Polym. Sci., Part A: Polym. Chem.*, 2001, 39, 715-723.
[59] Catilaz-Simonin L., Fouassier J.P., *J. Appl. Polym. Sci.*, 2001, 79, 1911-1923.
[60] Allonas X., Fouassier J.P., Kaji M., Miyasaka M., Hidaka T., *Polymer*, 2001, 42, 7627-7634.
[61] Burget D., Mallein C., Fouassier J.P., *Polymer*, 2003, 44, 7671-7678.
[62] Fouassier J.P., Allonas X., Burget D., *Progress in Organic Coatings*, 2003, 47, 16-36.
[63] Gómez M.L., Avila V., Montejano H. A., Previtali C. M., *Polymer*, 2003, 44, 2875-2881.
[64] Clark S.C., Hill D.J.T., Hoyle C.E., Jönsson S., Miller C.W., Shao L.Y., *Polymer International*, 2003, 52, 1701-1710.
[65] Yang J.Y., Neckers D.C., *J. Polym. Sci., Part A Polym. Chem.,* 2004, 42, 3836-3841.
[66] Cavitt T.B., Hoyle Ch.E., Kalyanaraman V., Jönsson S., *Polymer*, 2004, 45, 1119-1123.
[67] Cavitt T.B., Phillips B., Hoyle Ch.E., Pan B., Hait S.B., Viswanathan K., Jönsson S., *J. Polym. Sci., Part A Polym. Chem.*, 2004, 42, 4009-4015.
[68] Kim D., Scraton A., *J. Polym. Sci., Part A Polym. Chem.,* 2004, 42, 5863-5871.
[69] Suzuki S., Emilie P., Urano T., Takahara S., Yamaoka T., *Polymer*, 2005, 46, 2238-2243.
[70] Oxman J.D., *J. Polym. Sc., Part A Polym. Chem.,* 2005, 43, 1747-1756.
[71] Senyurt A.F., Hoyle Ch.E., *European Polym. J.,* 2006, 42, 3133-3139.
[72] Gómez M.L., Previtali C.M., Montejano H.A., *Polymer*, 2007, 48, 2355-2361.
[73] Mauguière-Guyonnet F., Burget D., Fouassier J.P., *Progress in Organic Coating*, 2007, 59, 37-45.
[74] Nguyen Ch.K., Hoyle Ch.E., Yeon Lee T., Jönsson S., *European Polym. J.*, 2007, 43, 172-177.
[75] Guo X., Wang Y., Spencer P., Ye Q., Yao X., *Dental Materials*, 2008, 24, 824-831.
[76] Ye Q., Park J., Topp E., Spancer P., *Dental Materials*, 2009, 25, 452-458.

[77] Tehfe M.A., Lalevee J., Allonas X., Fouassier J.P., *Macromolecules*, 2009, 42, 8669-8674.
[78] Crivello J.V., *J Polym Sci, Part A, Polym Chem*, 2009, 47, 866-875.
[79] Kim D., Stansbury J.W., *J. Polym. Sci., Part A Polym. Chem.,* 2009, 47, 3131-3141.
[80] Kim D., Scranton A.B., Stansbury J.W., *J. Polym. Sci., Part A Polym. Chem.*, 2009, 47, 1429-1439.
[81] Gould I.R., Shukla D., Giesen D., Farid S., *Helvetica Chimica Acta*, 2001, 84, 2796.
[82] Wölfle I., Lodaya J., Sauerwein B., Schuster G.B., *J. Am. Chem. Soc.*, 1992, 114, 9304.
[83] Kabatc J., Pączkowski J., *J. Photochem. Photobiol., Part A Chem.*, 2006, 184, 184-192.
[84] Hu S., Sarker A.M., Kaneko Y., Neckers D.C., *Macromolecules*, 1998, 31, 6476.
[85] Schuster G.B., *Adv. Electron Transfer Chem.*, Wiley, 1991, 1, 163-196.
[86] Marciniak B, Hug G.L., Rozwadowski J., Bobrowki K., *J. Am. Chem. Soc.*, 1995, 117, 127-134.
[87] Skalski B., Steer R.P., Verrall R.E., *J. Am. Chem. Soc.*, 1998, 110, 2055-2061.
[88] Tanabe T., Tores-Filho A., Neckers D.C., *J. Polym. Sci., Part A Polym. Chem.*, 1995, 33, 1691.
[89] Lorance E.D., Wolfgang H., Gould I.R., *J. Am. Chem. Soc.,* 2002, 124, 15225-15238.
[90] Kawabata M., Kimoto K., Takimoto Y., *European Patent* 211615, 1987.
[91] Buhr G., Dammel R., Lindley C.R., *Polym. Mater. Sci. Eng.*, 1989, 61, 269.
[92] Urano T., Nagasaka H., Shimizu M., Yamaoka T., *J. Imaging Sci. Technol.*, 1997, 41, 407.
[93] Humin W., Yongacai J., Qingshan L., Xiaohong Z., Shikang W., *J. Appl. Polym. Sci.,* 2002, 84, 909-915.
[94] Kabatc J., Zasada M., Pączkowski J., *J. Polym. Sci., Part A Polym. Chem.*, 2007, 45, 3626-3636.
[95] Padon K.S., Scranton A.B., *J. Polym. Sci., Part A: Polym. Chem.*, 2000, 38, 2057-2066.
[96] Sundararajan Ch., Falvey D.E., *J. Org. Chem.*, 2004, 69, 5547-5554.
[97] Sundararajan Ch., Falvey D.E., *J. Am. Chem. Soc.*, 2005, 127, 8000-8001.
[98] Sundararajan Ch., Falvey D.E., *Org. Letters,* 2005, 13, 2631-2634.
[99] Kabatc J., Pączkowski J., *J. Polym. Sci., Part A Polym. Chem.*, 2009, 47, 576-588.
[100] Fischer H., Radom L., *Angew. Chem. Int. Ed.*, 2001, 40, 1340-1343.
[101] Wang K., Nie J., *J. Photochem. Photobiol., Part A Chem.*, 2009, 204, 7-12.
[102] Shi S., Gao H., Wu G., Nie *J., Polymer*, 2007, 48, 2860-2865.
[103] Ouchi T., Hamada M., *J. Polym. Sci., Part A Chem.*, 1976, 14, 2527-2533.
[104] Mosca R., Fagnoni M., Melle M., Albini A., *Tetrahedron*, 2001, 57, 10319-10328.
[105] Shi S., Nie J., *J. Biomed. Mater. Res. Appl. Biomater.,* 2007, 23, 44-50.
[106] Elad D., Yousseyfyeh R.D., *Tetrahedron Lett.,* 1963, 30, 2189.
[107] Malatesta V., Ingold K.U., *J. Am. Chem. Soc.,* 1981, 103, 609-614.
[108] Fielding A.J., Franchi P., Roberts B.P., Smits T.M., *J. Chem. Soc. Perkin Trans 2,* 2002, 155-163.
[109] Dietliker K., *Chemistry and Technology of UV and EB Formulation for Coating, Inks and Paints. Vol. III*, Photoinitiators for Free Radical and Cationic Polymerization, SITA Technology Ltd, London, 1991.
[110] Ishikawa S.I., Iwasaki M., Tamoto K., Umehara A., *Eur. Pat. Appl.*, 109291 (Prior. 12.11.82) to Fuji Photo Film Co.

[111] Umehara A., Kondo S., Tamoto K., Matsufuji A., Nippon Kagaku Kaishi, 1984, 1, 192, 1984, *Chem. Abstr.*, 100 (20): 165332f.

[112] Andrzejewska E., Zych-Tomkowiak D., Andrzejewski M., Hug G.L., Marciniak B., *Macromolecules*, 2006, 39, 3777-3785.

[113] Kabatc J., Pączkowski J., J. Polym. Sci., *Part A Polym. Chem.*, 2009, 47, 4636-4654.

[114] Reiser A., *Photoreactive Polymers*, John Wiley and sons, New York, 1989, p. 112.

[115] Buhr G, Dammel R, Lindley CR., *Polym. Mater. Sci. Eng.* 1989, 61, 269.

[116] *Radiation Curing Science and Technology*, Pappas S.P., Ed., Plenum Press, New York, 1992.

[117] Gruber H.F., *Prog. Polym. Sci.*, 1992, 17, 953-1044.

[118] Polymer Handbook, 3rd ed., Brandrup J., Immergut E.H., *Wiley Intescience*, New York, 1989.

[119] Eaton D.F., Horgan A.G., Horgan J.P., *J. Photochem. Photobiol., Part A Chem.,* 1991, 58, 373-391.

[120] Takimoto Y., Radiation Curing in Polymer Science and Technology, Fouassier J.P., Rabek J.F., Eds. *Elsevier Applied Science Vol. III*, Chapter 8, London, 1993, pp 269-299.

[121] Andrzejewska E., Andrzejewski M., *J. Polym. Sci., Part A Polym. Chem.*, 1998, 36, 665.

[122] Kabatc J., Pączkowski J., *Dyes and Pigments* 2010, 86, 133-142.

[123] Rehm D., Weller A.*, Isr. J. Chem.,* 1970, 8, 259.

[124] Wrzyszczyński A., Filipiak P., Hug G.L., Marciniak B., Pączkowski *J., Macromolecules,* 1999, 33, 1577-1582.

[125] Marcus R.A., *J. Chem. Phys.*, 1956, 24, 966.

[126] Pączkowski J., Kucybała Z., *Macromolecules*, 1995, 28, 269.

[127] Pączkowski J., Pietrzak M., Kucybała Z., *Macromolecules*, 1996, 29, 5057.

[128] Kabatc J., Celmer A, *Polymer*, 2009, 50, 57-67.

[129] Sarker A.M., Kaneko Y., Nikolaitchik A.V., Neckers D.C., *J. Phys. Chem.,* 1998, 102, 5375.

In: Photochemistry
Editors: Karen J. Maes and Jaime M. Willems
ISBN: 978-1-61209-506-6

Chapter 2

PHOTOREACTIONS OF BENZOPHENONE IN SOLID MEDIA: RECENT DEVELOPMENTS AND APPLICATIONS

Mihaela Avadanei*
"Petru Poni" Institute of Macromolecular Chemistry, Iasi, Romania

ABSTRACT

From the very first use of photoinitiators in the crosslinking of polyethylene, benzophenone has established a reputation as an effective mediator in photoinduced modifications of polymeric materials. Most of the applications are based on the benzophenone's capability to abstract a labile hydrogen atom from the organic substrate when excited by the UVA radiation. These include the photocrosslinking and surface photografting, with the latter method aiming to modify the shallow properties of the substrates. In recent years, there has been an increasing interest in the understanding the benzophenone photophysics at the gas – solid (organic or inorganic) interface or inside the cavities of different types, which might pose some restraints on its mobility. It could be said that in benzophenone chemistry there are general rules, but there are also particular ways of interaction with various kinds of (macro)molecules. This paper summarizes some of the most interesting and successful applications of benzophenone photochemistry in materials science. The photoreactions induced by benzophenone are discussed in relation with the chemistryof the substrate.

1. INTRODUCTION

Studies regarding the benzophenone photochemistry are spread over a century and a countless number or papers have been devoted to ascertaining the main photoreactions when immersed in media with different chemical or physical properties under the action of 365 nm light. So, there is a remarkable knowledge in the field of photophysical processes and the

* "PetruPoni" Institute of MacromolecularChemistry, Iasi, ROMANIA, fax:+40-232-211299; e-mail: mavadanei@icmpp.ro

subsequent photochemistry of benzophenone, mainly in liquid systems. From a certain point of view, its photoreduction pathways are not as complicated as those undergone by aliphatic and a,b-unsaturated ketones. Their photodissociation through Norrish I or Norrish cleavage, free-radical decarbonylation, inter – and intramolecular hydrogen abstraction, photo rearrangements, or intramolecular elimination results in a larger number of photoproducts, which sometimes could not be easily identified.

The large application area, ranging from photocrosslinking, photopolymerization, and UV-curing as the original field, to employing it as an anchoring component for biomolecules on functionalized surfaces is based on its effectiveness, easiness of handling, non-toxicity, and in the fact that, having known the primary photoreaction, the secondary photoreactions and the photoproducts could be predicted to a certain extent. These are true mainly in the solid phase, where the coupling possibilities might be fewer than in liquids because the medium rigidity could impede the diffusion of the geminate radicals from the cage or the steric hindrance could affect the orientation of the radicals in respect to each other.

Because of benzophenone's low absorption extinction coefficient of 365 nm, the radiation is homogeneously absorbed inside the photosensitive system, therefore allowing a uniform distribution of photoproducts or crosslinking sites. Unlike the photoinitiators with high absorption extinction coefficients in the UVA region, benzophenone might be used for photocrosslinking of sufficiently thick samples, up to several mm, with no significant gradient of crosslinking density with depth. However, at high rates of benzophenone conversion, the photoproducts could act as inner filters that could alter the conversion kinetics.

This paper is devoted to the photochemical reactions induced by benzophenone in polymers, with emphasis on the photoreaction with the largest practical applicability that is the hydrogen atom abstraction. This process is the basis of crosslinking and UV – induced surface grafting, which in turn serve as the main methods for using benzophenone with the aim of improving the bulk and surface properties of polymeric materials. In recent years, the changing of the surface properties have found many applications in biomedical engineering, while the photoinduced crosslinking have somehow received less attention. On the other hand, benzophenone presents a growing interest as it concerns the photophysical processes involving the higher excited states or the microenvironmental influence on the emission properties. In addition, the non-specific structure of the substrate might offer to benzophenone some particular pathways to deactivation. This structure could bring forward the behavior of benzophenone in constrained media or its deactivation paths by interaction with ionic liquids. From this point of view, this paper is organized as a review, gathering a number of representative papers, some of them putting in evidence in peculiar aspects of benzophenone photochemistry. The following are briefly presented: (a) the photophysics and photochemistry of benzophenone, whose mechanisms are well established, (b) an overview of the secondary photoreactions most commonly used in polymer modification, and (c) some of the recent applications of benzophenone, not only as a photoinitiator, but also as a probe for investigating the molecular mobility in organic and inorganic systems.

2. PHOTOPHYSICS AND PHOTOCHEMISTRY OF BENZOPHENONE

2.1. Photophysical Processes of Benzophenone

The differences in the photoreduction paths of the carbonyl compounds, namely aliphatic and α,β-unsaturated vs. diaryl ketones, are connected with the character of the reactive state, from which the primary photochemical reactions occur. Aliphatic ketones are characterized by a weak absorption band around 280 nm arising from the n→π* transition. The aromatic carbonyl compounds exhibit two major electronic excitations: the allowed π→π*, due to excitation of the aromatic moiety and having an ε value greater than 10^4 $l.mol^{-1}.cm^{-1}$, and the forbidden n→π*, with an ε value of 10-100 $l.mol^{-1}.cm^{-1}$ and absorption towards higher wavelengths. In the case of excitation with light wavelength between 270 and 350 nm, the n→π*transition results from promotion in carbonyl groups of a non-bonding electron (n) into an antibonding orbital (π*) such as an electron deficiency is created at the carbonyl oxygen. The electronic singlet states are short – lived and their deactivation is achieved by radiative decay as fluorescence (F) or radiationless transition, such as internal conversion (IC, between the states with the same multiplicity) and intersystem crossing (ISC, involving the multiplicity change). Both aliphatic and unsaturated ketones undergo intersystem crossing to the triplet level by spin inversion, but the rate constant for aliphatic ketones is slower. In general, the electronic configuration of the low-lying triplet state is (n, π*). Because of the half-filled orbital localized on the oxygen atom, the (n, π*) states have got a diradicaloid nature, resembling that of the *t*-butoxy radical, but more electrophylic and selective [1]. In these circumstances, the expected primary processes from ^{1}(n, π*) and ^{3}(n, π*) states are the hydrogen atom abstraction, α-cleavage, radical addition and electron abstraction [2]. The lowest triplet state of aliphatic ketones is (n,π*) in nature and due to the fact that the ^{3}(n, π*) state is close in energy and electronic distribution with the ^{1}(n, π*) state. Photoreduction proceeds from the lowest single or triplet state or even from a mixture of them. In the aromatic ketones case, the configuration of lowest triplet state could depend on the substitution and the solvent polarity.

In close resemblance with behavior of a *t*-butoxy radical, excited aliphatic and phenyl alkyl ketones undergo α-cleavage, known as Norrish Type I process. The scission yields acyl and alkyl radicals for acyclic aliphatic ketones, and alkyl and benzoyl radicals for phenyl alkyl ketones, respectively. The energy of the C – CO bond of benzophenone is higher than the excitation energy of the ^{3}(n, π*) state, so it is not susceptible to dissociation through Norrish I reactions while excited in the lowest triplet state. Promotion in higher triplet states could be afforded by two photon absorption, from which benzophenone photodissociates in phenyl radicals and carbon monoxide [3], [4].

Photoreduction of benzophenone in 2-propanol, reported by Ciamician and Silber (1900), is one of the earliest photochemical studies [5]. .The uncertainty regarding the nature of the reactive state in hydrogen abstraction, i.e. singlet or triplet, remained unsolved almost 60 years, until Bäckstrom and Sandros [1] investigated the transfer of the triplet state energy from benzophenone to biacetyl, benzyl, and anisil with the aid of the photo-luminescence spectra. Hammond and his coworkers have used steady-state kinetics in studying photoreduction of benzophenone by benzhydrol [6]. Photocycloaddition of benzophenone to 2-methyl-2-butene has been observed by Patterni and Chieffi back in 1909 [7], and the structure of photoproducts have been confirmed in 1954 by Büchi et al. [8]. It was thus demonstrated that the reactive state of benzophenone is the triplet (n,π*) state, achieved through intersystem crossing ($k_{ISC} \cong 2 \times 10^{11}\ s^{-1}$) from the ^{1}(n, π*) to ^{3}(π, π*). The latter triplet state decays with an efficiency of almost 100% to the ^{3}(n, π*) [8], as presented in Fig. 1. Due to the very rapid ISC, no fluorescence is observed and the lifetime of the triplet state is therefore controlled by the chemical structure and the physical properties of the medium (polarity, viscosity, temperature, pressure, and rigidity etc). Generally, the triplet state lives for 80–200 ps in inert solvents such as CCl_4, fluorocarbons, acetic acid, or acetonitrile. In alcohols or alkanes, the lifetime of the triplet is much shorter, (about 100 ns). Phosphorescence of BP could be observed at liquid nitrogen temperature, but in several solvents such as water or isooctane, a weak phosphorescence emission has been detected at room temperature. Thus, in the absence of interactions with the molecular environments, the processes that contribute to the deactivation of the triplet state are the radiationless decay and phosphorescence.

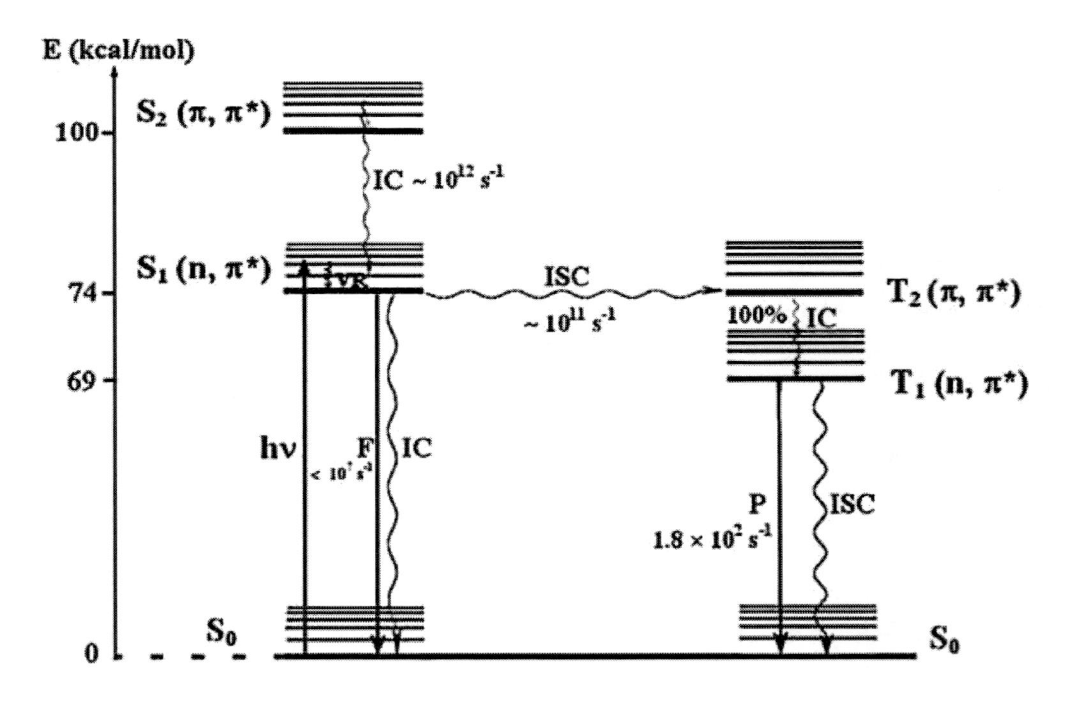

Figure 1. Jablonski diagram ofbenzophenone at 77K.

Benzophenone maintains the ^{3}(n, π*) character of its reactive state in both polar and non-polar solvents, although this state also posseses some degree of (π, π*) character [9] and the solvent polarity greatly affects its stability. The polar solvents stabilize the triplet state ^{3}(n, π*) and the non polar solvents stabilize the ^{3}(n,π*) state. By changing the polarity of the solvent from non-polar to a polar protic one, one can observe a hypsochromic shift of the (n, π*) absorption and a bathochromic shift of the (π, π*) absorption, respectively [10]. For the benzophenone and most of its derivatives, the hydrogen atom abstraction from a suitable donor environment was considered the prevailing photoreduction process. Comparison between benzophenone and 2-acetonaphtone provides a good example of how the substitution of a naphtalene for a phenyl lowers the energy level of the ^{3}(π, π*) state so that it becomes the lowest excited state. Excitation of 2-acetonaphtone is then localized in the naphtalene moiety, so the diradical character of the carbonyl group is greatly reduced. Instead, the ^{3}(π, π*) state resembles a carbon radical [2] that has insufficient energy to abstract a hydrogen atom; therefore 2-acetonaphtone is inactive towards photoreduction. In addition, unlike benzophenone, 2-acetonaphtone deactivates through fluorescence.

Substitution in the *para* position of the aromatic ring of benzophenones with strong electron donor groups, such as amines or hydroxyls, also leads to the inversion of the electronic configuration and quenching of the hydrogen abstraction [10-12]. The ^{3}(π, π*) state of *para*aminobenzophenone [10], [13-15], *para*-hidroxybenzophenone, *para*-methoxybenzophenone [16], or of 4,4'-bis-dimethylaminobenzophenone (Michler's ketone) [17–23], has a charge-transfer character (CT), highly polar, with the excitation energy localized on the heteroatom and the carbonyl oxygen becoming nucleophilic. The charge transfer complexes are less reactive towards photoreduction, but the reactivity is strongly dependent on the solvent because the interaction with the solvent must result in rising the energy level of the CT state above that of the ^{3}(n, π*) state. This is the classical case of the *para*-aminobenzophenone that is inactive with respect to H-abstraction in isopropyl alcohol [14], but reacts with cyclohexane and benzene [10], [24-26].

In consequence, the triplet state configuration can predict benzophenone's ability to abstract a hydrogen atom, having seen that the most reactive ketones have (n, π*) triplet states. Those with low reactivity are characterized by (π, π*) triplet states and the charge – transfer triplet states indicates the least photoreactive ketone [27].

2.2. The Main Photoreactions Induced by Benzophenone in Organic Media

2.2.1. Hydrogen Atom Abstraction

The common bimolecular reaction of benzophenone in alcohol and hydrocarbon media is the hydrogen atom abstraction, which generates diphenylketyl radicals and alkyl radicals. The general mechanism is depicted in Figure 2, in which the first step is represented by the excitation in the first singlet state that undergoes intersystem crossing to the lowest triplet state. The primary reaction is the abstraction of a labile hydrogen atom. By transient absorption spectroscopy, three species were detected: the benzophenone triplet, the diphenylketyl radical, and the benzophenone radical anion. The secondary reactions are the dimerization of the diphenylketyl radicals, yielding benzopinacol (reaction (3)), a combination of the two alkyl radicals and the cross-coupling products (reaction (4)). When the alkyl radicals are macromolecular chains, the crosslinking process is based on the coupling of these macroradicals (reaction (5)). The triplet state could be quenched by interaction with a benzophenone molecule in the ground state (reaction (6)).

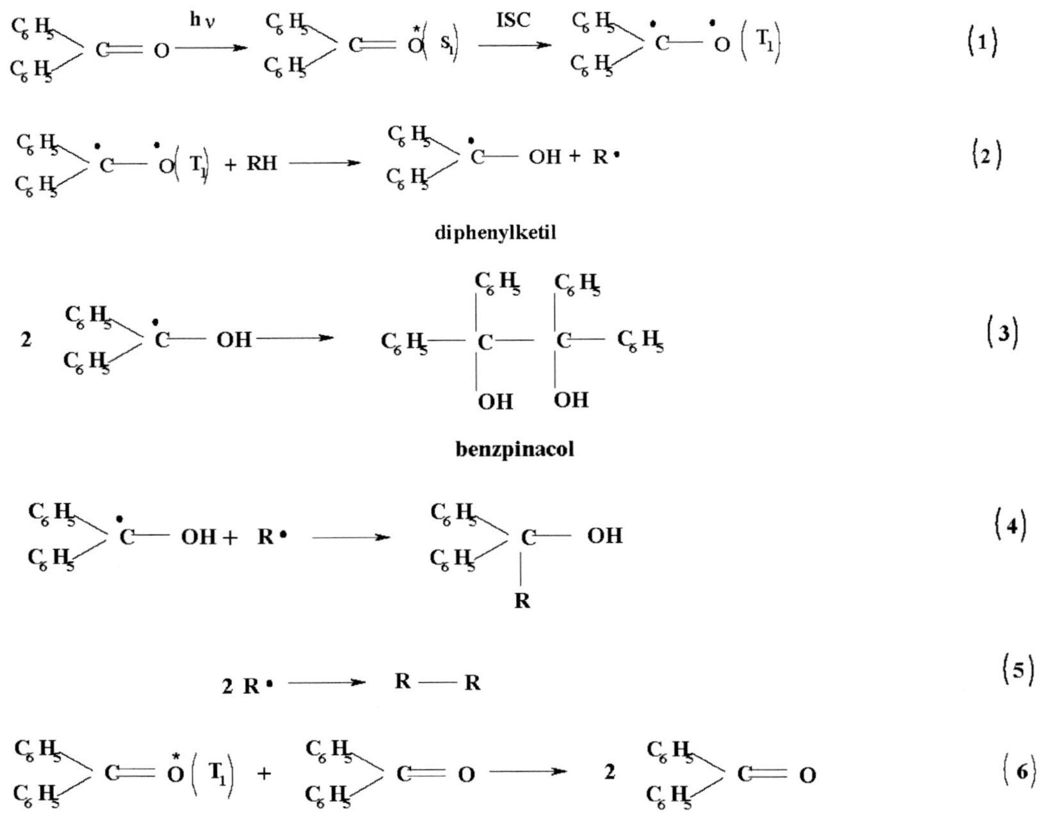

$$\begin{matrix} C_6H_5 \\ C_6H_5 \end{matrix} \!\!> C{=}O \xrightarrow{h\nu} \begin{matrix} C_6H_5 \\ C_6H_5 \end{matrix} \!\!> C{=}\overset{*}{O}(S_1) \xrightarrow{ISC} \begin{matrix} C_6H_5 \\ C_6H_5 \end{matrix} \!\!> \dot{C}-\dot{O}(T_1) \qquad (1)$$

$$\begin{matrix} C_6H_5 \\ C_6H_5 \end{matrix} \!\!> \dot{C}-\dot{O}(T_1) + RH \longrightarrow \begin{matrix} C_6H_5 \\ C_6H_5 \end{matrix} \!\!> \dot{C}-OH + R\bullet \qquad (2)$$

diphenylketil

$$2 \begin{matrix} C_6H_5 \\ C_6H_5 \end{matrix} \!\!> \dot{C}-OH \longrightarrow C_6H_5-\underset{OH}{\overset{C_6H_5}{C}}-\underset{OH}{\overset{C_6H_5}{C}}-C_6H_5 \qquad (3)$$

benzpinacol

$$\begin{matrix} C_6H_5 \\ C_6H_5 \end{matrix} \!\!> \dot{C}-OH + R\bullet \longrightarrow \begin{matrix} C_6H_5 \\ C_6H_5 \end{matrix} \!\!> \underset{R}{C}-OH \qquad (4)$$

$$2\,R\bullet \longrightarrow R-R \qquad (5)$$

$$\begin{matrix} C_6H_5 \\ C_6H_5 \end{matrix} \!\!> C{=}\overset{*}{O}(T_1) + \begin{matrix} C_6H_5 \\ C_6H_5 \end{matrix} \!\!> C{=}O \longrightarrow 2 \begin{matrix} C_6H_5 \\ C_6H_5 \end{matrix} \!\!> C{=}O \qquad (6)$$

Figure 2. Photoreactions initiated by benzophenone in alcohols and hydrocarbon media.

Photoreduction of benzophenone to benzopinacol is a common reaction in most hydrocarbon donors. The quantum yield of benzophenone disappearance is 2 in oxygen – free isopropyl alcohol [28] and the high yield of benzophenone consumption is preserved in most of the hydrogen donor media. Quantum yield of benzopinacol formation can approach unity [28 -31].

In most cases, when benzophenone consumption kinetics have been followed by UV absorption spectroscopy, it was observed the enhancement of absorption around 330 nm, in either a solid or liquid environment. Observed initially by Pitts and co-workers for photolysis of benzophenone in oxygen – free isopropyl alcohol [28], the intermediates absorbing around 330 nm are known as light absorption transients (LAT) and seem to be a regular presence in a hydrogen donor. It was observed that they act as quenchers of the benzophenone triplet state [32], [33] and are very sensitive to oxygen, the absence of which yields a lifetime that is quite long (several days). By their disappeareance from the system, the triplet absorption partially gets recovered [32]. Many studies have been carried out in an attempt to clarify the chemical structure of these intermediates, mainly in alcohol solvents [28], [32], [34 -36], but also in benzene [32] [37], n-hexane, cyclohexane [32], benzhydrol, and tetrahydrofuran [36].

The LAT structure has been attributed to some hemiketals [28] or isobenzopinacols [34], assumed to occur by dimerization of two diphenylketyl radicals [32], [34] or by a cross-coupling reaction between a diphenylketyl radical and a solvent radical created by hydrogen abstraction [33]. Demeter and Bérces reinvestigated the flash photolysis of benzophenone in ternary solutions of isopropyl alcohol or benzhydrol in acetonitrile with the purpose of understanding the LAT generation and structures [38]. By means of transient absorption spectroscopy, the authors proposed two types of photoreduction intermediates, exposed in Figure 3: one type derived from the dimerization of the diphenylketyl radicals (I) and the other being a cross-coupled product (II).

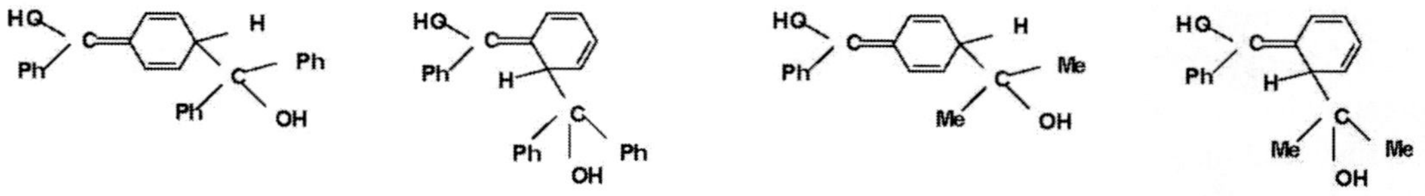

Figure 3. The chemical structures of LAT in benzophenone – isopropyl alcohol sistem (Reproduced with permission from [38]).

As it can be observed, the coupling of the diphenylketyl and dimethylketyl radicals occurs at the *orto*-or *para*-position of the aromatic ring. The intermediate I is similar to the isobenzpinacol and it was observed that its concentration accounted for less than 2% of the recombination reaction of diphenylketyl radicals. In the absence of oxygen the cross-coupling intermediates decay after a double exponential function. This behavior conducted to the identification of *para*-LAT as the major component, which decays in a longer time scale. The *orto*-LAT is in a minor amount and is quite fast removed from the system.

Lately, nanosecond time-resolved resonance Raman spectroscopy was used in order to identify the LAT structure in a study regarding the photoreduction of benzophenone from 2 propanol [39]. Du et al. have observed that hydrogen abstraction is a fast process, within 10 – 20 ns, and the LAT intermediate has been identified as the *para*-coupling cross-product, the 2-[4(hydroxylphenylmethylene)cyclohexa-2,5-dienyl]propan-2-ol.

The two types of LAT have also been detected by Scully and co-workers in photolysis of benzophenone in poly(ethylene-vinyl alcohol) films, performed at high UV lamp intensities [40]. The authors suggested that LAT decay leads to benzophenone re-formation and the rate of decay for the *para*-LAT is independent of the medium viscosity, due to the its slower disappearance. In addition, the formation and decay characteristics of LAT are not influenced by the substituents at the benzophenone phenyl ring.

2.2.2. Peculiar aspects in Photoreduction of Benzophenone in a Solid Matrix

The rigidity imposed by a solid matrix alters the rate of LAT generation and the yield of the final products. The triplet state fate is followed by phosphorescence studies and the first parameter taken into account was temperature. Many studies have been concerned with the phosphorescence decay of the benzophenone triplet at low temperatures (77K), either in glassy form [41], or dispersed in low molecular glasses, regarding the energy transfer [42-45], or when adsorbed on various substrates [46-48].

Murai and coworkers studied the hydrogen atom abstraction in ethanol in the temperature region 71 – 131 K and observed that the phosporescence lifetime decreased for temperature above 90 K [49], which was in accordance with results obtained by Godfrey et al, who detected the diphenylketyl radical at temperatures only above 100 K [50]. Murai et al.

explained this behavior by the fact the hydrogen abstraction in fact took place, but the low temperature hindered the diffusion of the geminate pair radicals and they ultimately recombined. Therefore, it seems that the diffusion plays a very significant role in photoproducts creation.

In constraining environments as the polymer matrix and for temperature below the glass transition, the restrictions on molecular mobility could make the radical pair created after the light absorption unable to diffuse from the reaction cage. In addition, if the geminate radicals did not change their orientation with respect to each other, they could recombine to re-form the original molecules. Therefore, the quantum yield of the photoreduction of benzophenone significantly decreases. From this reason, benzophenone, among other photosensitive compounds, could offer information regarding the microenvironment's rigidity and chain segment mobility of polymers [51-53].

Christoff and Atvars [52] were interested in studying the α-, β-, and γ-relaxations in polystyrene and poly(*n*-alkyl methacrylate)s (where *n* is methyl, ethyl, and butyl moieties), by monitoring the fluorescence and phosphorescence emissions of benzophenone, xanthone, and two flavones. These probes were differentiated by their shape and size, and their luminescence properties were monitored as a function of both their planar dimensions and the mobility and bulkiness of the phenyl or n-alkyl groups of the investigated polymers. The authors reported that the onset temperature of β-relaxations for short segment chains was sensed by the smaller ketones, such as benzophenone and xanthone. This fact is in contrast with the behavior of flavones, which are larger molecules and have sequentially sensed the coupled short-and main-chain motions, in a close connection with the dimension of the n-alkyl pendant groups of poly(*n*-alkyl methacrylate)s. The main conclusion was the importance of luminescent probes size in the microenvironment of the short segments and of the pendant groups. So, the authors concluded that processes with onsets at higher temperatures could be followed by using fluorescent probes only, while at lower temperatures, the relaxation processes are sensed by both fluorescent and phosphorescent molecular probes.

The effect of the glass transition on the hydrogen abstraction and on the LAT formation has been observed by Deeg and co-workers by holographic grating and absorption spectroscopy in several acrylic polymers, such as poly(methyl methacrylate) (PMMA), poly(isobutyl methacrylate) (PIBMA), poly(butyl methacrylate) (PBMA), and poly(vinyl acetate) (PVAc) [53]. The polymers were differentiated by glass transition temperature and by hydrogen donor capabilities. It was observed that the quantum yield of the hydrogen abstraction depends on the matrix rigidity, obtained at the lowest value (0.19) in the case of PMMA, which has the highest Tg, and similar values for the other hosts (0.82. – 0.89), which have lower Tg. In this experiment, the authors reported that the photoreduction of benzophenone was triggered by one photon, which contradicted their previous studies that claimed the two-photon excitation of PMMA [54]. The LAT absorption was detected in all the samples, but the conversion into the final products was very fast in the PBMA host, probably due to the softness of the matrix.

Phosphorescence studies of benzophenone dispersed in poly(methyl methacrylate) PMMA (T_g=110°C), poly(isopropyl methacrylate) (PIPMA) (80°C), and poly(methyl acrylate) (PMA) (T_g=10°C) showed that the triplet lifetime varies with increasing temperatures [55]. Below the β-transition temperature ($T_β$=-30°C for PMMA), which corresponds to the onset of the ester side-group rotation, the decay of the triplet benzophenone in the three polymers is single – exponential. In the temperature range between T_β and T_g, the triplet decays non-exponentially, attributed to the dynamic quenching of the triplet by the ester group of the host. Salmassi and Schnabel studied the phosphorescence of benzophenone in PMMA at room temperature and at high laser intensity [56]. The authors found the similar non-exponential decay of the triplet that has been attributed to an energy transfer from the triplet benzophenone to the polymer, and afterward the triplet – triplet annihilation.

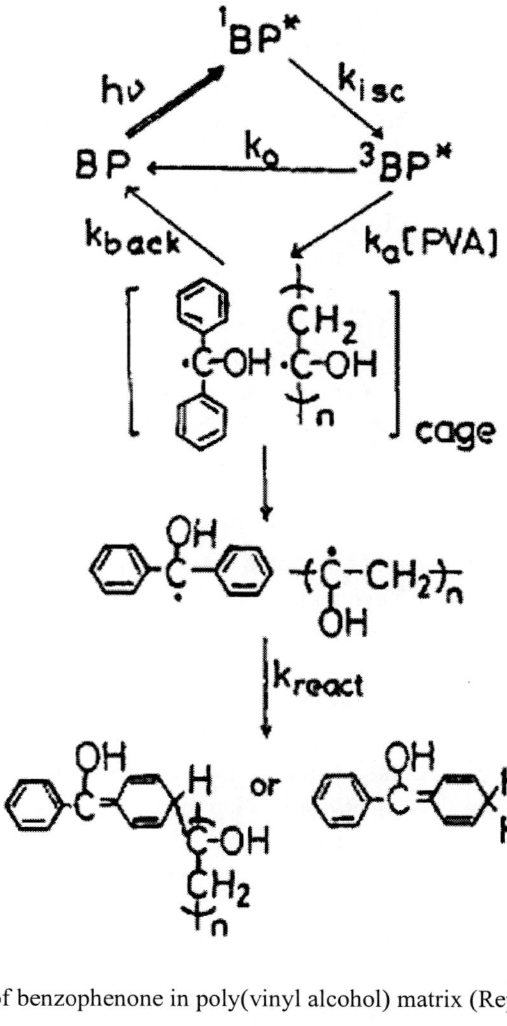

Figure 4. Photoreactions of benzophenone in poly(vinyl alcohol) matrix (Reproduced with permission from[58]).

A comparable effect due to the polymer microstructure is also seen for benzophenone dispersed in polystyrene and polycarbonate [57], or in poly(vinyl alcohol) [58].The non-exponential decay of the triplet benzophenone, commonly assigned to some interactions with the matrix, is believed to take place in polystyrene by quenching by the phenyl ring, and in polycarbonate, due to the phenylene ring, respectively. In poly(vinyl alcohol), the benzophenone triplet follows a non-exponential decay between -100°C and T_g (85°C) [58]. In this case, the deviation from the single – exponential decay, which has been observed at low temperatures, is supposed to be attributed to the onset of the benzophenone rotation so that the tertiary hydrogen abstraction from the PVA matrix became possible. Due to the high intensity of the laser irradiation, the absorption spectrum of the PVA – BP films presented a large and quite intense band center on 340 nm, assigned to LAT. The authors supposed that, for $T < T_g$, the hydrogen abstraction is diffusion – controlled and there is a significant backward disproportionation reaction inside the cage of the matrix in conditions of continuous irradiation. When temperature increases above T_g, the diphenylketyl radical escapes from the radical cage, so its transient absorption spectrum appears in milliseconds at 545 nm. On the basis of the above observations, Horie et. al proposed the following mechanism for the photoreactions of benzophenone in PVA matrix as a function of temperature (Figure 4), which also takes the creation of a *para*-LAT into account. So, Horie et. al supposed that the *para*-LAT are in the form of isobenzopinacol and as grafted moiety on the PVA chain, respectively.

Ebdon and co-workers have found a tri-exponential decay of the benzophenone phosphorescence in films of poly(acrylic acid) and acrylic acid-methyl methacrylate copolymers, when temperature has varied between 77 K and 410 K [59]. This fact was attributed by the authors to the variety of the labile sites and to the microenvironment of the benzophenone molecule experienced in each host.

Phosphorescence in films at 77 K were performed in order to investigate the influence of the matrix polarity on the triplet benzophenone decay [60]. The polymeric matrices were poly(methyl methacylate), polyethylene, isotactic, atactic polypropylene, and polystyrene, and the authors observed that benzophenone decays exponentially, irrespective the matrix structure.

All the examples presented above have dealt with the case of intermolecular hydrogen abstraction, but intramolecular abstraction could also occur. This effect has been observed in the case of ortho-substituted methylbenzophenones [61], [62], from which *cis*-and *trans*-enols result, as well as in the case of the ortho-substituted hydroxybenzophenones [63].

2.2.3. Photoreactions of Benzophenone with Olefinic or Acetylenic Compounds.

In this case, the benzophenone photochemistry involves the triplet excitation transfer from benzophenone to the compound, leading to the olefin isomerisation, or the cycloaddition of the benzophenone carbonyl group to the isolated carbon – carbon double or triple bond (Paterni – Büchi reaction) [7], [8]. Oxetane formation is efficient when the triplet energy of the carbonyl compound is not high enough for the sensitized triplet excitation of the olefin. The yield of the products formed as a consequence of photocycloaddition (oxetanes) or by hydrogen abstraction depends on the substitution degree at the double bond [8], [64], [65]. The cyclization to oxetane has been found to proceed by the formation of an intermediate as a 1,4-biradical in a triplet state, irrespective of the ketone type, i.e. aliphatic or diaryl ketone [65 (a), (b)]. This species decays to the singlet state either directly (k_{ISC}) or via a contact ion pair (CIP) (k_{IP}), as presented in Figure 5. The singlet 1,4-biradical could undergo ring closure (k_c) to the oxetane or β-scission (k_β), from which the starting compounds are re-formed.

The contact ion pair (CIP) could also be solvent separated (SSIP), or could experience a second charge transfer (k_{bet}) to the starting olefin and ketone.

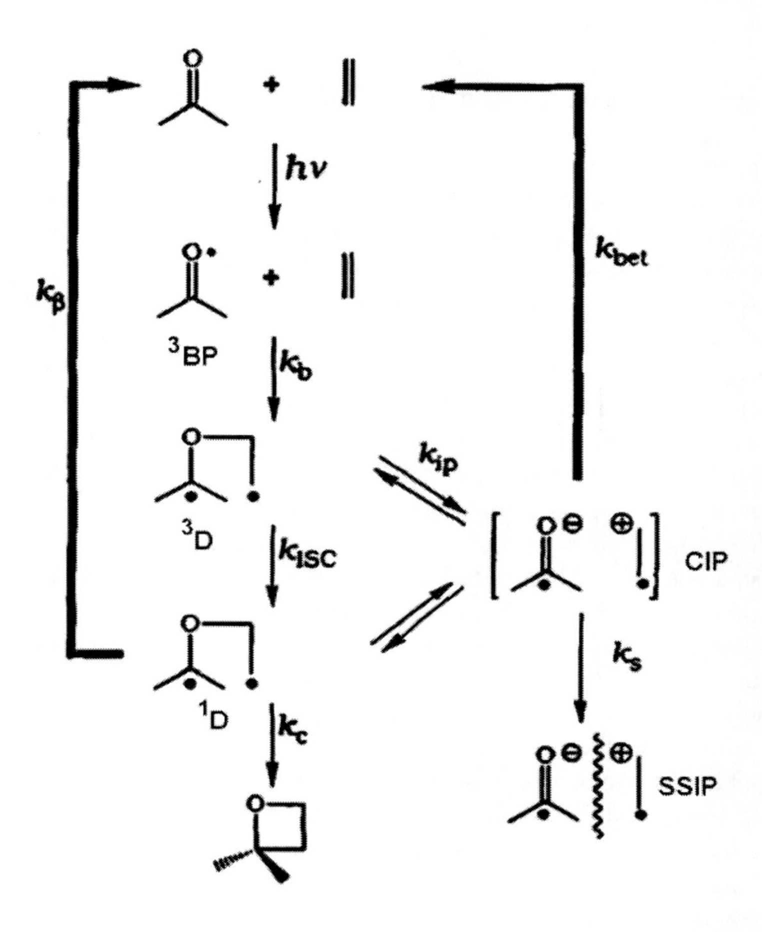

Figure 5. Mechanism of oxetane formation by photocycloaddition of a ketone to an electron – rich olefin (Reproduced by permission from [65 (c)]).

2.2.4. Photochemical Reactions of Benzophenones in the Presence of Amine Derivatives

These reactions mainly proceed through an electron transfer complex from amine to the triplet benzophenone, with formation of an excited bimolecular complex (triplet exciplex) [66 -69]. The formation of the diphenylketyl and amine radicals results from a proton transfer in the complex, with a quantum yield close to unity. It was observed that the H-abstraction took place from both – NH-and –CH-groups of the amine [70].

In polymer science, bimolecular hydrogen abstraction has found an application in photocrosslinking processes of polymers as a very efficient photoinitiator. In addition, because the lowest triplet state, $^3(n, \pi^*)$, is situated above the corresponding state of many organic compounds, benzophenone could act as a sensitizer for generating triplet states to those compounds for which the intersystem crossing is ineffective due to the alternative paths of deactivation.

3. Photocrosslinking of Polymers

3.1. Photoreactions in Saturated Hydrocarbon Polymers

The use of benzophenone and its derivatives as photocrosslinking and photo-polymerization initiators have a long and well documented history, beginning with the results obtained by Oster et al. regarding the photoinitiated crosslinking of the low density polyethylene under exposure of UVC radiation [71]. Oster considered that the 254 nm radiation triggered the benzophenone photolysis and, consequently, the chains crosslinked by macroradical recombination. Several years later, Charlesby had shown that the difference between the extinction coefficients of benzophenone at 254 and 365 nm is in the order of hundreds while the corresponding difference in the quantum yield of benzophenone conversion was only in the order of ten [72]. In addition, crosslinking is competing with chain scission and photoxidation as a result of the destructive character of the UVC radiation. Since then, benzophenone has become one of the most used photoinitiator or photosensitizer under the action of 365 nm, which has been shown to excite the carbonyl group. Its crosslinking application area is ranging from synthetic to biopolymers and from bulk (photocrosslinking) to surface modification (photografting) of polymers.

Perhaps the most comprehensive studies are concerning the photoiniated crosslinking of polyethylenes, in the form of high density, or linear and branched low density [73]. In the middle of the 1980s, Prof. Bengt Rånby reported the photocrosslinking of polyethylene in its melt, which allowed the obtaining of a high degree of crosslinking under low exposure doses [74]. In principle, the photocrosslinkable system contained polyethylene, a photoinitiator and a crosslinking agent (triallylcyanurate, in most of the cases), and a fast response at the action of the UV radiation was obtained. The main advantage of the photocrosslinking carried out near the melting point method lies in the fact that, by using cca. 1% photoinitiator and about 0.5% crosslinking agent, polyethylene samples of about 2 – 3 mm thick crosslinked rapidly after 15 seconds of exposure at 365 nm and the gel fraction reached 70 – 90%. This method has found industrial applications in manufacturing crosslinked polyethylene insulated wires and cables [75].

The photochemical reactions supposed to take place in polyethylene after BP excitation has been confirmed by photoreduction in model compounds. In consequence, comparative studies were carried out regarding the reactivity of triplet benzophenone towards the type of abstractable hydrogen, the ultimate goal being the drawing of the crosslinking mechanism photonitiated by benzophenone in polyethylene. In an elegant review, Qu summarizes the results obtained in tetracosane and dodecane as model for linear polyethylene as well as in isooctane and 3-ethylhexane that model the branched polyethylene [76]. The reactions of benzophenone triplets with unsaturated bonds have been simulated by using 1-hexene, *cis*- and *trans*-3-hexene. The presence of benzopinacol in a significant amount for most of the simulated photoreactions led to the conclusion that the main photochemical reaction undergone by benzophenone in polyethylene is hydrogen atom abstraction by the triplet, which leads to diphenylketyl and macroradicals generation. By spin – trapping electron (ESR), ^{1}H-and ^{13}C-NMR spectroscopies, Qu et. al put in evidence the existence of H – links and Y – branches appearing by abstraction of tertiary, secondary, and allylic hydrogens, respectively [73 (f)], [77]. The H links are forming by recombination of two alkyl macroradicals, while the Y – branches involve the addition of the above radicals to vinyl end groups. In several cases, an isomer of benzopinacol with quinoid structure was detected (1-phenylhydroxymethylene-4-diphenyl-hydroxymethyl-2,5-cyclohexadiene), which could give an explanation for the yellowing of the studied specimens [77]. Besides the diphenylketyl radicals' recombination, the detection of α-alkylbenzhydrols pointed out the attaching of the diphenylketyl radicals onto the chains (as pendant groups) as a side reaction. Figure 6 summarizes the photoproducts Qu and co-workers have identified in the benzophenone initiated crosslinking of polyethylene [73 (f)].

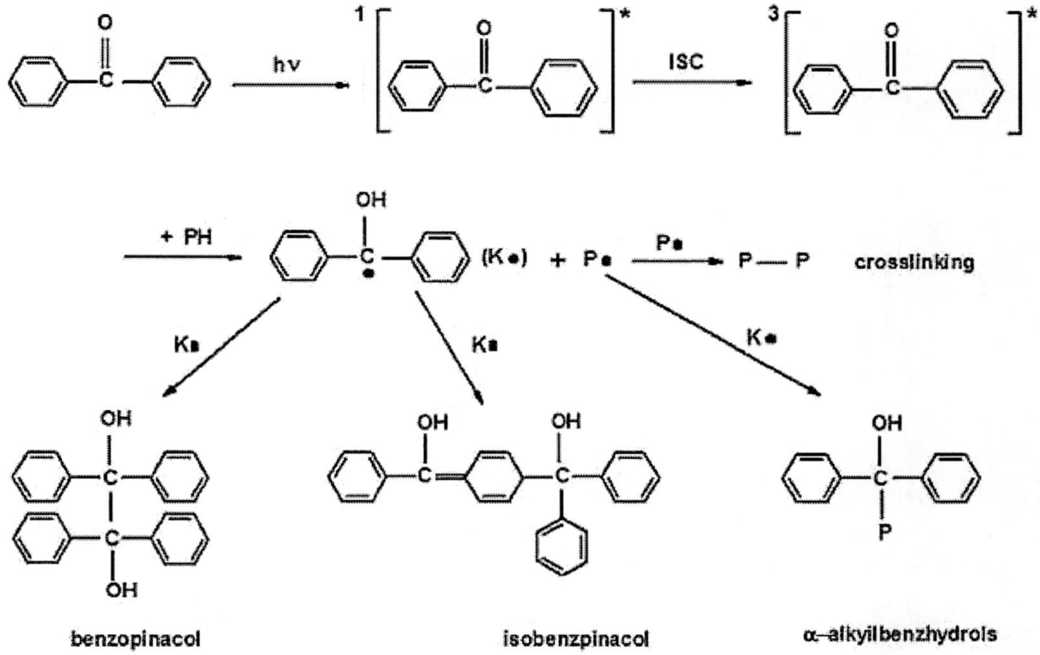

Figure 6. Mechanism of the photoinitiated crosslinking of polyethylene with the corresponding photoreductionproducts (Reproduced withpermission from [73 (f)].

The reactivity of benzophenone triplet towards the hydrogen type is obviously correlated to the dissociation bond energy and the results that Qu and co-workers had obtained are in line with the hydrogen lability series drawn by Giering et. al, that is tertiary > primary benzylic > secondary > secondary cyclohexyl > primary > benzene [78].

Chemical quenching of the benzophenone triplet by the tertiary hydrogen abstraction does not guarantee crosslinking, as has been observed when polystyrene was irradiated (365 nm), in the form of both thin films and benzene solutions in the presence of benzophenone [79], [80]. When irradiation of the films was performed *in vacuo*, Mn and Mw decreased and, consequently, the polydispersity increased, along with the presence of a high – molecular mass fraction [79], [81]. The former indicates the main chain scission, while the later is related to the crosslinking of the chains through the tertiary carbon radical. However, the breaking of the C – C bonds of the main chain dominates and this effect became more evident when photoreduction of benzophenone in polystyrene was carried out in the presence of oxygen. The effect was explained by the partial regeneration of benzophenone after hydrogen abstraction, which leads to the lowering of the its decomposition quantum yield. The increase in polydispersity shows the crosslink formation, although the hydrogen abstraction site was very sensitive to the oxygen attack. The resulting peroxy radicals could promote crosslinking, but they are also involved in hydroperoxide formation and further chain fragmentation. The formation of polyenes by chain β-scission accompanies both the irradiation *in vacuo* and in oxygen, whereas the detection of carbonyl and acetophenone end groups has been accounted for photoirradiation in air. Hence, photochemistry of benzophenone in polystyrene is summarized in Figure 7.

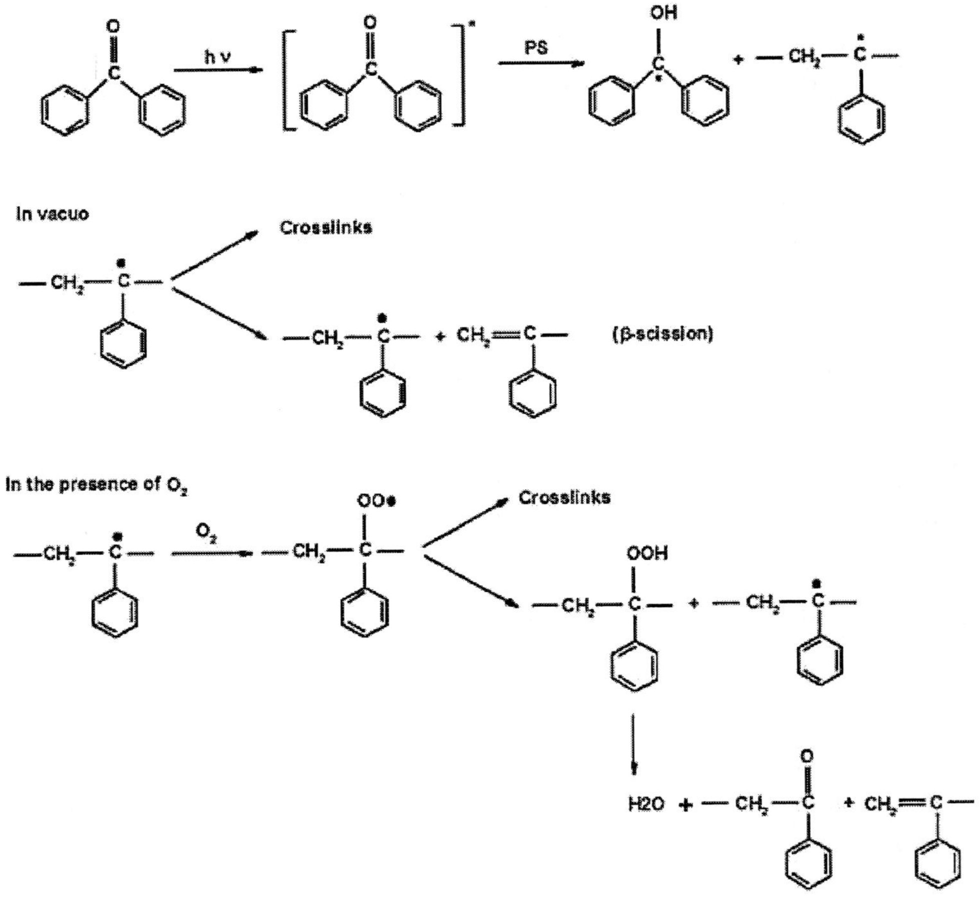

Figure 7. Photocrosslinking of PS in the presence of benzophenone (Reproduced with permission from [79]).

It's well known that the quantity of adsorbed oxygen in polymeric matrix is directly related to the length and the number of the pendant groups [82], [83] and that the oxidation is more severe in the case of polymer bearing pendant units as compared with linear chains. Hydroperoxy groups possess some interesting photophysical properties and, by absorbing energy, could produce the cleavage of the oxygen – oxygen bond [82]. In addition, hydroperoxides are decomposed by an energy transfer from the ketone groups, thus the carbonyl acts as a sensitizer of polystyrene decomposition [84].

These results have lead to the conclusion that benzophenone initiated the photodegradation of polystyrene efficiently instead of photocrosslinking. Similar results were found when solutions of benzophenone in benzene with or without polystyrene were subjected to 365 nm radiation at 60°C [80]. The decay of the reactive excited state of benzophenone occured via deactivation or hydrogen abstraction from polystyrene and benzene. Mita and coworkers [80] have observed that the disappearance of benzophenone was in inverse proportions with polystyrene concentrations and they attributed this fact to the quenching of benzophenone triplets by the later. Crosslinking and chain scission occur simultaneously, but again, the rate of crosslink formation is much lower than that of chain scission. Mita et. al concluded that, in benzene solutions, the polystyrene behaves mainly as a quencher of the triplet state rather than as hydrogen donor.

David and co-workers analyzed the ESR spectra obtained when polystyrene films containing benzophenone were irradiated at room temperature [85]. The trapping of only a small amount of isolated polystyryl radicals suggested that most of the macroradicals were involved in disproportionation reactions with diphenylketyl radicals.

The crosslinking as the main chain reaction has been reported for exposure of polystyrene films in vacuum, whilst the main chain scission dominated over polystyryl radical recombination when irradiation was performed in air [86], [87].

Accelerated photodegradation of polystyrene in the presence of benzophenone had found applications in manufacturing degradable plastics and predicting the lifetime of polymers under service conditions [88-91]. Sikkema and coworkers compared the photodegradation initiated by benzophenone and benzoin on polystyrene in the form of films and foam [92]. The authors concluded that benzophenone is the most efficient agent in the photodegradation of polystyrene foams, whereas the effects of the two photoinitiators were comparable in the film case. In order to obtain packaging materials degradable in a natural environment, Kaczmarek et. al studied the effect of benzophenone in thin films of polystyrene and poly(vinyl chloride) exposed to 254 nm radiation [93]. The benzophenone sensitized photodegradation is considerably higher in polystyrene, due to the different miscibility of the components, and thus enabling the use of this system in commercial purpose.

The high molecular mass polystyrene is insensitive to microbial growth on its surface, but as a result of photodegradation, its molar mass decreases and the microbial attack induces biodegradation [94]. The slight degradation as a result of benzophenone photolysis was observed in a study concerning the photoinduced antimicrobial properties of polymeric blends as well [95]. Here, the films of polystyrene (PS), polyethylene (PE), polypropylene (PP), and polyvinyl alcohol (PVA) with 0.5% benzophenone showed a lowering in their structural and mechanical properties after exposure to 254 nm radiation, especially in the case of PE and PP. The studied parameter, antimicrobial effectiveness, was greatest for PVA blended with BP, and lowest for the PP-BP blend. As it can be seen, the photoinitiated degradation of polymeric materials has proven its efficiency in the context of the increasing amount of global plastic waste.

In recent years, information about benzophenone employed as bulk crosslinking photoinitiator has been seen in, e.g. improving the mechanical properties of polymeric blends with inorganic fillers [96] or in designing the microfluidic devices in numerous lab-on-a-chip (LOC) applications.

It has been shown that the yield of PS crosslinking in the presence of benzophenone is quite low, and the process has been generally conducted under the glass transition temperature. In order to inhibit the thin PS films from dewetting on Si wafers and PMMA substrates, Prof. Turro's group used abi functional benzophenone derivatives (bis-3-benzoylbenzoicacidethyleneglycol (bis-BP)) as photonitiators of PS patterned photocrosslinking [97]. The crosslinking mechanism of this compound is based on the hydrogen abstraction of the benzophenone moiety. If diphenylketyl radicals are formed at both ends of the bis-BP, this molecule acts as a crosslinking agent and could link two PS macroradicals that are not in close proximity of each other by recombination of a PS radical with a ketyl radical. The immobilization onto the substrate of photocrosslinked PS is obtained when annealing above T_g or exposing it to solvent vapor.

3.2. Photoinitiated Crosslinking of Diene Polymers

The mechanism sketched for polyethylene crosslinking photoinitiated by benzophenone [73 77] seems to be followed for most of the polymers, because the breaking of the C – H bonds is the main reaction in a hydrogen donor medium. Rånby made a survey of the crosslinking process in several common diene elastomers and found that the EPDM class, with carbon – carbon double bond as a side group, is the most suitable for photocrosslinking under UV irradiation [98]. Similar observations were previously reported in the case of polybutadienes [99], [100]. On the contrary, polymers bearing the unsaturation in the main chain are mainly degraded by UV radiation.

As previoulsy presented, in the presence of benzophenone several processes could happen when the polymer contains a double or triple carbon – carbon bond: cycloaddition of the carbonyl group to the isolated carbon – carbon double or triple bond (Paterni – Büchi reaction), abstraction of a allylic hydrogen (if exists) and combination of the resulted (macro)radicals, as well as the *cis - trans* isomerization, which occurs by the transfer of the triplet excitation energy [2], [7], [8], [64], [65], [101–105]. The diphenylketyl radical that appeared by hydrogen abstraction could be further involved in disproportionation and dimerization reactions: formation of benzopinacol (by coupling of two diphenylketyls), apparition of LAT ("light absorbing transients," as *para*-coupling products of diphenylketyl and polymer, possessing a quinoid structure), or additions to the polymer chains as a pendant group. The yield of the products formed as a consequence of photocycloaddition and hydrogen abstraction depends on the substitution degree at the double bond [8], [64], [65]. When solution of *cis*-polyisoprene was irradiated in the presence of 5% benzophenone, the ^{1}H-NMR analysis revealed the formation of a copolymer of isoprene and isomeric isopren-diphenyloxetane adducts [106], where the oxetane groups were determined as being in concentrations of 25 mol%. Irradiation performed in polyisoprene film levelled out the conversion of double bonds into oxetane rings at about 70%. In polyisoprene, the cyclic structure is formed in the main chain. Polymers with the unsaturation as side groups (e. g. vinyl polymers) will present the oxetane ring as pendant moieties.

Although vinyl polymers are themselves photocrosslinking elastomers at λ less than 300 nm [107], [108] their photosensitive properties can be much improved by an addition of a photoinitiator or a photosensitizer. Photochemical crosslinking of vinyl polymers has found applications in manufacturing the photoresists in the reproducing of submicron patterns, photosensitive printing plates [109], and foamed materials obtained by radiation processes [110] etc.

A study regarding the photoinitiated crosslinking of 1,2-polybutadiene (1,2-PB) was justified by the presence of two types of reactive sites in the structure: allylic hydrogen atoms and the vinyl bond, which may compete with each other in interactions with excited benzophenone [111], [112]. Thin films of 1,2-PB with various amounts of benzophenone (1, 2.5, 5 and 10%) were irradiated in a nitrogen atmosphere with low intensity 365 nm radiation. It has been thought that the hydrogen abstraction site would be located at the allylic carbon, although it was reported that the loss of vinyl unsaturation for butadiene – styrene copolymers is from the attack of the π system of the double carbon – carbon, resulting in a macroradical [113]. The kinetics of the photoreaction followed by FTIR spectroscopy showed a very fast decay of benzophenone in the first minutes of exposure and that the vinyl group consumption takes place in two stages at least. The first step occurs quite rapidly and involves the interaction with excited benzophenone. The second stage proceeds more slowly, because most of the photoinitiator molecules have reacted. In both stages, the vinyl bonds were consumed in inter- and intra-homopolymerization reactions with subsequent formation of bridges between two chains and, most probably, 1,6-hexadiene cyclic structures between the adjacent vinyl groups of the same chain [114] or belonging to neighboring chains, respectively

It must be pointed out that the homopolymerization of vinyl bonds during UV exposure in the presence of a photoinitiator has been generally observed for copolymers containing butadiene units. Itis worth mentioning the photoinitiated crosslinking of polystyrene-*block*-polybutadiene*block*-polystyrene (SBS) [114], polyacrylonitrile-*block*-polybutadiene-*block*-polyacrylonitrile (ABA) [115], or even dicumyl peroxide– initiated crosslinking of 1,2-PB [116], where evidence of intramolecular cyclization has been observed.

No relevant information about the structure of crosslinks were obtained from the FTIR spectra of the irradiated 1,2-PB films [111], except for the apparition of a broad band assigned to linear ether entities, as well as some aromatic and hydroxyl bands, arising from semibenzopinacol moieties linked to1,2-PB chains. In addition, there was no evidence of benzopinacol formation in both FTIR and ^{1}H-NMR spectra of the extracted films. At lower photoinitiator concentrations, the degree of crosslinking slowly increased, but at higher exposure doses the additional homopolymerization of vinyl groups occurred.

Photolysis of benzophenone in 3-meythl-1-butene, used as a model compound, showed a mixture of photoproducts besides benzopinacol that was recovered as a precipitate [111]. As major components, the end groups identified with the help of HPLC/ESI-MS and ^{1}H – NMR spectroscopy are vinyl (E1), olefinic (E2), oxetane (E3, E4, and E5), and semibenzopinacol (sBzPi), as well as some ether groups, like –O-CH_2-and –O-CHR-. All of these structures are outlined in Figure 8.

The end group E1 appears due to the extraction of the allylic hydrogen atom from a 3-meythl-1-butene molecule by a BP*(T_1) excited molecule, while an E2 end group may result by the isomerization of a H_2C=CH-CMe_2· (E1) radical. The formation of E3, E4 (trace), and E5 oxetane rings can be explained by the cycloaddition of carbonyl groups in BP*(T_1) molecules to carbon-carbon double bonds belonging to 3-meythl-1-butene molecules and E2 end groups. An interesting observation of the products in discussion is the presence of ether linkages, others than those in oxetane rings.

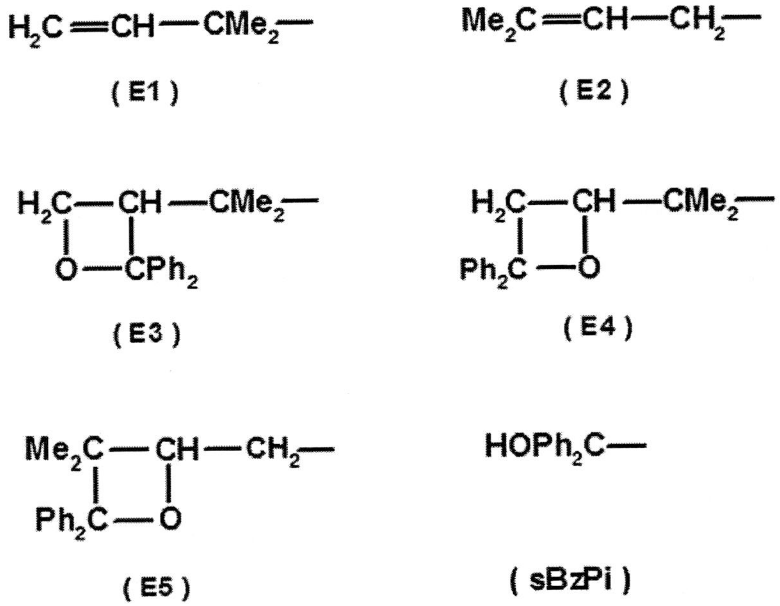

$H_2C{=}CH{-}CMe_2{-}$ (E1)

$Me_2C{=}CH{-}CH_2{-}$ (E2)

$H_2C{-}CH{-}CMe_2{-}$ / $O{-}CPh_2$ (E3)

$H_2C{-}CH{-}CMe_2{-}$ / $Ph_2C{-}O$ (E4)

$Me_2C{-}CH{-}CH_2{-}$ / $Ph_2C{-}O$ (E5)

$HOPh_2C{-}$ (sBzPi)

Figure 8. The end groups identified inthe photolysis of benzophenone in 3-methyl-1-butene.

Turning to the photochemical reactions induced by excited benzophenone in 1,2-PB, there are two pathways of its photoreduction recognized, but allylic H-abstraction prevails over [$2\pi + 2\pi$] photocycloaddition. The main difference bewtween photolysis in solid polybutadiene and in its model compound lies in the formation of benzopinacol, undetected in the former case. In thin film experiments [111], in which low intensity UV light was used, benzophenone triplets in a low minute concentration were generated and this leads to a lower rate of production of diphenylketyl radicals, which finally disappear through biradical disproportionation reactions. Benzopinacol formations by diphenylketyls recombination might also be inhibited by the slow diffusivity in the 1,2-PB matrix.

In addition, the fate of diphenylketyl radicals could be switched by partial isomerisation into enolic structures and coupled as such to a macroradical (or small radical, for photolysis performed with low molecular compounds); this intermediary is known as LAT [40], [53], [54], [58], [82]. The quinoid structure of LAT could induce the yellowing of the irradiated media [40], but this effect has not been observed in1,2-PB films.

A widely used class of elastomers with unsaturated groups as pendant units is ethylene – polypropylene – diene copolymers (EPDM), which may contain hexadiene, ethylidenenor bornene, or dicyclopentadiene comonomers. On the basis of previous experiments related to crosslinking of EPDM with 1,4 hexadiene units that was photoinitiated by benzoin derivatives, Bousquet and coworkers observed that the unsaturation played an important part in the photocrosslinking [117]. They had extended their studies by using benzophenone derivatives as photoinitiators, in comparative experiments with EPDM and ethylene–co-propylene (EPR) elastomers. Irradiations were performed in air and under N_2, respectively.

The authors reported that the main reaction step was the H-atom abstraction and, in the presence of air, the allylic sites were about 12 times more reactive than the tertiary sites. Therefore, the crosslinking could happen at the tertiary or secondary carbon atom, and the possible reaction sites are illustrated in Figure 9.

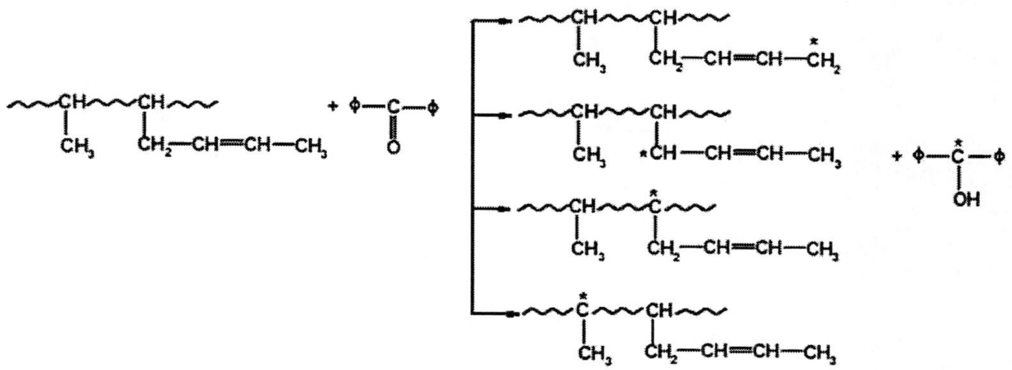

Figure 9. The main macroradicals obtained by hydrogen abstraction in the photocrosslinking of EPDM initiated by benzophenone (Reproduced with permission from [117]).

Under a N2 atmosphere, comparable abstraction rates have been obtained. In the presence of air, benzophenone sensitized the hydroperoxides decomposition and thus enhanced the photocrosslinking process.

In the photocrosslinking of EPDM, Rånby had established the reactivity order with respect to diene typse as ethylidenenorbornene > dicyclopentadiene > 1,4-hexadiene [118]. The 1,4-hexadiene units react with the photoinitiator by both addition to the double bonds and H-abstraction, but additional homopolymerization by unsaturated sidegroups was observed [118]. The allylic H-abstraction represents the major reaction in dicyclopentadiene and ethylidenenorbornene moieties [98].

Eisele and co-workers [82] studied the photoinitiated crosslinking of poly (dicyclopentenyl acrylate) (poly(DCPA)) in air and in inert atmospheres (N2) and found a faster photolysis of benzophenone under N2, due to a lack of molecular oxygen which would otherwise shorten the lifetime of the triplet state. The major photoproduct was benzopinacol and the coupling of the diphenylketyls to the allylic macroradicals was less important under N2 than in the presence of oxygen. The carbon-carbon double bonds were also consumed, in small proportions, by the addition of LAT. In addition, the removal of labile hydrogen could happen at the tertiary carbon on the main chain, therefore the crosslinking site might be located at a tertiary macroradical. The presence of a consistent insoluble fraction is then related to the linking mainly by allylic macroradicals, and the effectiveness of network formation is obviously greater under N2. Irradiation performed in the presence of oxygen revealed the great photooxidation potential due to the residual unsaturated bonds and the cyclic structures.

When polynorbornene films were exposed to radiation with l > 310 nm, the intermolecular hydrogen abstraction from the two secondary positions of the cyclopentane rings occurred [119], which could further recombine to each other (Figure 10 (a)). No photocycloaddition of benzophenone carbonyl bond was observed, nor adducts between the diphenylketyl radicals and allylic macroradicals. Instead, benzophenone sensitized the *trans – cis* isomerisation of the polymeric double carbon – carbon bonds. The polynorbornene radicals could also be generated by reactions with oxygen to produce polymer peroxy radicals, which could be converted into hydroperoxy radicals by hydrogen abstraction. The authors considered that all the available radicals participated in termination reactions, thus they were linked in a network with a non-specific structure, as presented in Figure 10 (b).

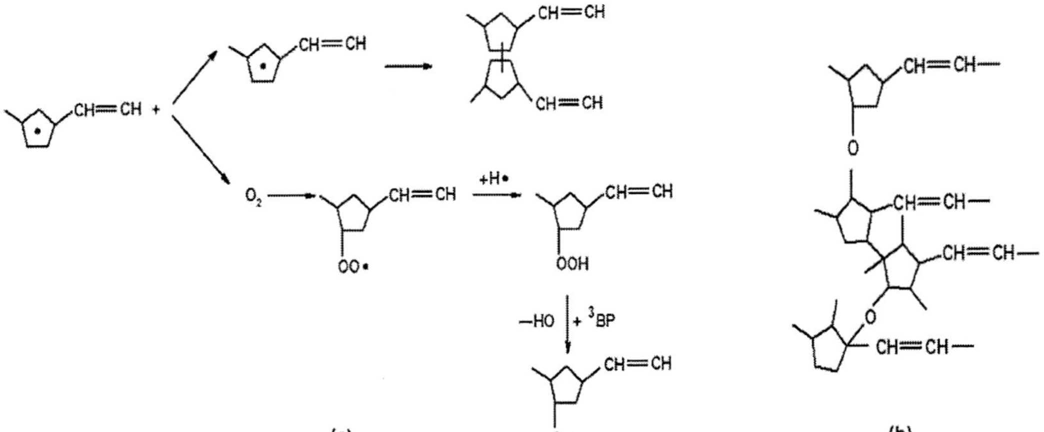

(a) (b)

3.3. Photocrosslinking of Inorganic Polymers - Polydimethylsiloxanes

In the past decade, there has been a tremendous increase in the use of polydimethylsiloxanes (PDMS) in fabricating micro-and nanometer – scale devices, for various lab-on-chip applications with biomedical purpose [120], [121]. PDMS have advantages and disadvantages regarding its properties and processing, including its transparency to radiation above 230 nm, low cost of manufacturing, chemical inertia, permeation to gases, and the fact that it can be easily molded into micron and submicron features The combination of these properties has made PDMS a favorite material in various bionanalytical applications, but, in turn, its hydrophobicity and porous structure limit its use as a biocompatible material. In PDMS – based microfluidic devices, the master is usually obtained by microfabrication process, using the traditional microelectromechanical system (MEMS) methods, and represents the negative pattern of the desired PDMS structure [120]. The need for a photosensitive PDMS was generated by the special conditions required by these technologies, i.e. equipment and clean-room facilities. In this view, two types of photo-patternable PDMS were developed. The first one is based on the negative – working class, where the PDMS contains unsaturated units (vinyl) that can be crosslinked in the presence of free – radical photoinitiators (acetophenone [122], [123] and benzophenone [124] derivatives) acting in the 300 – 400 nm domain. The positive – acting PDMS resists (also known as *photo*PDMS) consist of two – component curing mixtures (base monomer and curing agent) and a photoinitiator (usually benzophenone), therefore making them appropriate for irradiating with less than 365 nm [125], [126]. The base monomer is a vinyl – terminated

PDMS and the crosslinking agent contains silicon hydride – $OSiHCH_3$-units. It has been established that in the UV exposed areas, excited benzophenone reacts from hydrogen abstraction with both the vinyl groups of the base (reactions (1a) and (1b) in Figure 11) and the silicon hydride groups of the curing agent [126] (Figure 11, reaction (2)), resulting in a weakly crosslinked network.

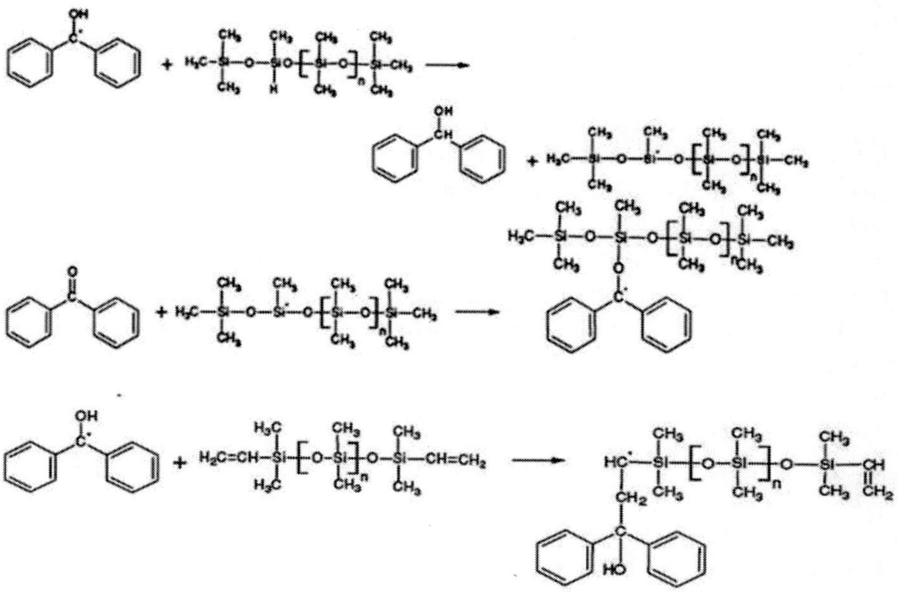

Figure 11. Photoreactions intiated by benzophenone in *photo*PDMS (Reproduced with permission from [126]).

During the thermal curing step which follows the irradiation, the crosslinking reaction between PDMS base monomer and curing agents in the unexposed regions take place by hydrosilation of the vinyl groups. As a consequence, the irradiated areas are removed in the developing solvent, while the unexposed parts are densely crosslinked. The feature sizes obtained varied from 100 mm to 2 mm. Unlike the traditional view of benzophenone as the promoter of network development, one can see that its use in the *photo*PDMS resists is resumed to a photoinhibitor. The application field of *photo*PDMS resists could be extended to the fabrication of multi-layer and multi-level structures for microfluidic devices [125], [126], substrates for the cell culture utilized for BioMEMS devices [126], or conductive materials which might be utilized as electrical connectors or physical sensors. Cong and Pan have demonstrated the use of PDMS-Ag conductive photoresists, obtained by using the same positive – tone *photo*PDMS in fabricating micro pressure sensing arrays, which afterwards have passed the antimicrobial tests [127]. The positive PDMS resists yielded a resolution of 60 microns.

4. SURFACE PHOTOGRAFTING

The surface chemical modification of polymers is usually used to change the shallow characteristics without altering the bulk structure and properties. The modified surface achieves new functionalities, whether the surface belongs to a bulk polymer or to a polymer membrane, in which case the photografting takes place inside the pores. The photografting allows the extension of the application area, from the photostabilization of substrates [128], the selective impermeability to oxygen [98], to biomedical engineering as the most promising domain. The UV-induced surface graft polymerization offers unique advantages such as no alteration of the bulk properties of the material, high speed processing made in unsophisticated equipment with a low cost of operation, and its non-toxicity.

The polymer surface modification by photografting was initiated in the mid 1950's by Oster and co-workers [71], [129] who used benzophenone as a photoinitiator under the action of radiation from the 200 – 300 nm region. In Prof. Rånby's groups, a series of photografting methods were developed and the most important results have been put together in several papers [98], [130], [131]. Very recently, the techniques and the latest advances in UV-induced surface graft polymerization have been excellently reviewed [132–134], in which benzophenone as a photoinitiator plays a significant part.

Briefly, the photochemistry of benzophenone in UV-initiated grafting is similar to the functioning in photoinitiated crosslinking. The basic principle is the hydrogen atom abstraction from the substrate by excited photonitiators, with the subsequent generation of a radical on the surface to which the monomer is added (Figure 12) [98]. The grafted chains are growing by free radical/ "living" polymerization, and they are terminated by coupling with the diphenylketyl radicals.

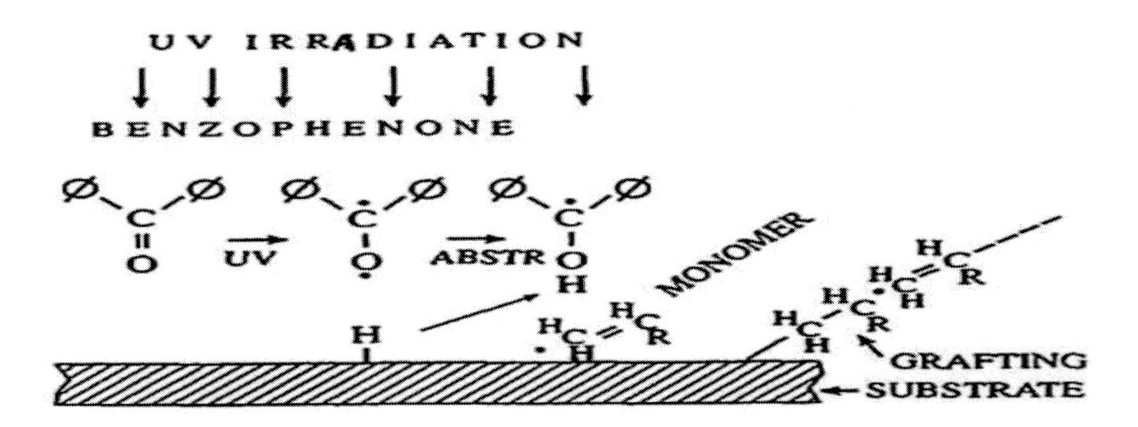

There are several photografting methods by which a surface can be chemically modified: the vapor phase [135], [136], batch phase [137], continous process [138], sandwich [139], or bulk surface polymerization [140].

The surface photografting was designed as a one- [141], [142] or two step process [143-145]. In the first method, the substrate is dipped in or covered with a solution containing the photoinitiator and the monomer, and the assembly is exposed to the UV radiation. The latter approach involves the presoaking of the substrate in a photoinitiator solution and creating the reactive surface by exposure to the UV light, and afterwards the monomer is added and grafted by photoinitiated radical polymerization. The living UV-induced graft polymerization has been designed in order to control the process in terms of reducing the homopolymer and branched/crosslinked polymer creation [143], [144]. In this approach, the semibenzopinacol-type compounds are used as dormant groups in grafting polymerization when monomers are added under appropriate conditions.

Rånby and his group investigated the graft polymerization of different monomers such as vinyl acetate and maleic anhydride on low density polyethylene [146], as well as acrylates, acrylic acid, 4-vinylpyridine, and acrylonitrile on low density polyethylene [141], [147], and poly(ethylene terephthalate), polypropylene, or polystyrene surfaces [140 (b)], [143], [144]. [148]. The experimental and monomer structural factors that influence the graft efficiency were discussed. In a special case of grafting styrene onto low density polyethylene substrates, it was found that styrene could be grafted by auto-initiation and the amount of benzophenone added has no significant importance on the grafting effectiveness [148]. By surface photografting in the vapor phase, Gao and co-workers [149] photografted maleic anhydride to the surface of low density polyethylene powder, with applications as coating material and hot – melt adhesive. Zhang and co-workers [150] had reported a novel one-step method (simultaneous) for photografting polymerization on the unable-to be-irradiated surfaces. This method uses vapors of photoinitiators (benzophenone) and monomers (acrylic acid, styrene or methyl methacrylate).

In many cases, benzophenone could be used in conjunction with another photosensitizer (e.g. isopropylthioxanthone, ITX) to increase the amount of grafted chains and surface polarity [151– 153]. By means of the "photosensitization effect", ITX transfers energy or electrons to benzophenone, leading to formation of an excited -state complex (exciplex). The efficiency of a blend of photoinitiators is improved by this synergistic effect, also observed when benzophenone was used in combination with 4,4'-bis(diethylamino)-benzophenone [154].

The use of UV-induced surface graft polymerization is now a widespread technique in changing and controlling the surface properties, which lowered the dimensions of the photografted structures to ten nanometers. In recent years, the field of obtaining micro-and nanoscale features has emerged as a distinct domain in life science by using synthetic and natural polymers, such as polyethylene, polydimethylsiloxane, polycarbonate, poly(methyl methacrylate), or the copolymer of 2-norbornene with ethylene (COC, cyclic olefin copolymer), polypropylene etc. [155–157].

The surface of the native polymers is generally hydrophobic, which may represent weak points in many biomedical applications. In bioMEMS technology, the inner polymer surfaces of the microfluidic devices or the surfaces for cell attaching must necessarily be hydrophilic, which would imply a covalently modified surface. From the wide variety of methods used to modify the polymer surfaces, the UV-induced surface graft polymerization is evidenced by the fact that it may be used in conjunction with photolithography for photopatterning in restricted geometries and on microscale.

The use of benzophenone to modify the surface of PDMS microdevices is based on the tendency of porous PDMS to adsorb analytes so that in many experiments the first step consists of adsorbing benzophenone in a shallow layer within the surface by immersing the PDMS into a benzophenone solution. This technique has been applied to graft polymerization for one of the most used monomers, namely acrylic acid [142], [158–160]. In the irradiated regions, photolysis of BP generates diphenylketyl radicals, which react with the two methyl groups of PDSM and could also initiate the polymerization of acrylic acid. The authors supposed that poly(acrylic acid) (PAA) and PDMS form an interpenetrating polymer network. In a second derivatization, the PAA chains were covalently coupled to molecules possessing positive charges, with the purpose of promoting cell or protein attachment. With the aim of fabricating particle sensors and by using the same approach, Zhao et al. have obtained amphiphilic PDMS particles of different shapes [161]. The procedure consisted in selectively photografting PAA on the one part of the particle surface, so that this side became hydrophilic and the other remains hydrophobic.

Switching the surface properties from hydrophobic to hydrophilic as a response to temperature or pH changes could be very useful in fabricating "smart" microfluidic devices acting as a "molecular trap". Temperature and pH responsive surfaces have been obtained by chemically modifying the walls of the PDMS microchannels (0.1mm deep and 0.1 or 0.5mm wide) with poly(N-isopropylacylamide) (PNIPAAm), and with poly(NIPAAm-co-AAc), respectively, by UV-induced surface graft polymerization [162]. The authors successfully demonstrated the capture and the release of the PNIPAAm-grafted nanobeads inside the microchannels as a function of temperature. Ma and coworkers have shown that the benzophenone – initiated graft polymerization of NIPAAm inside the microchannels could be successfully realized even in the case of thick substrate (greater than 1 micron) [163]. Due to the competition for reacting with excited benzophenone between the Si-H groups in the PDMS substrate and the NIPAAM molecules, two approaches have been tested. The irradiation time can be prolonged, or the residual Si-H bonds in the cured PDMS must be removed by chloroform extraction. Huck et al. have obtained surface wrinkles by patterning Au on photocrosslinked PDMS [164]. The PDMS substrate was soaked in benzophenone and exposed to UV radiation through an amplitude mask. Although the authors do not provide the crosslinking mechanism, it is supposed that benzophenone abstracts an H-atom from the methyl groups, and the chains are linked by methylene groups.

Attaching of biomolecules to polymer surfaces could also be made by a direct process, without a surface pretreatment, by using benzophenone and other photoreactive compounds as photolinkers [144], [165–167]. Dankbar and its co-worker tested PMMA, PS, and COC as substrate materials for obtaining oligonucleotide DNA microarrays by using two types of photolinkers, differentiated by their reaction mechanism [167]. The authors compared the efficiency of ketyl – reactive benzophenone and anthraquinone with nitrene/carbene – reactive nitrophenyl azide and phenyl-trifluoromethyldiazirine. According to the chemical structure of the substrates, that is the lability of the hydrogen atoms and the photochemistry of the UV-reactive compounds, it was found that phenyl-trifluoromethyldiazirine is the most efficient photolinker and the COC substrate is extremely well-suited for biomolecular attachment. Moreover, phenyltrifluoromethyldiazirine/COC seems to be the best combination in terms of reaction speed and the concentration of attached oligonucleotide DNA.

5. PHOTOPHYSICS AND PHOTOCHEMISTRY OF BENZOPHENONE ADSORBED ON VARIOUS SUBSTRATES

The electronic excited states of benzophenone are strongly influenced by the environmental characteristics; the surrounding constraints have an influence on the molecular conformation. The $^3(n, \pi^*)$ and $^3(\pi, \pi^*)$ states are very close in energy and the gap between them could vary as a result of changing the geometry around the carbonyl group [168]. Medium rigidity and/or the presence of the functional groups that act as hydrogen atom donors or are hydrogen bonded with the ketone alter the benzophenone conformation and thus its molecular motion can be severely restricted. In turn, this is directly related to the triplet state lifetime and, from a practical point of view, with its ability to abstract a hydrogen atom or to deactivate by various processes.

Due to the sensitivity to the microenvironment, benzophenone and its derivatives were intensively used as a probe for investigating the external or internal host media, with emphasis on the microenviromental influence upon the photophysics and photochemistry of these guest molecules. These properties are different when investigated at the gas – solid interface as compared to that in gas or in liquid media. The interest regarding the influence of the substrate on the interactions between the surface and adsorbate and on the processes dynamics has been directed toward exploring the behavior of benzophenone adsorbed on silica [169-171], or reversed – phase silica surfaces [172], on inert adsorbents as SiO_2, TiO_2, Al_2O_3, and Ti-Al binary oxides [173], [174], within the cavities of zeolites as silicalite [175-177], ZSM-5 or cation-exchanged ZSM-5 zeolite [177], [178], included into the cavities of cyclodextrins [179], calix[n]arene [176], [180], or within microcrystalline cellulose chains [176], [181], as well as adsorbed on the surface of the micro-[182], and mesoporous materials [183], [184]. Most of the results have been obtained by using time – resolved luminescence, steady – state, and transient absorption spectroscopy, as well as diffuse reflectance laser flash-photolysis.

By analyzing the photoluminescence spectra of benzophenone adsorbed on Al_2O_3 and Al-Ti binary oxide, Anpo et al. observed at least two types of emissive species: excited benzophenone adsorbed on the surface OH groups of the oxide by hydrogen bonding and protonated excited BP [173], [185]. The authors correlated the emission of the protonated benzophenone to the existence of the Brönsted (surface Al–OH groups) and Lewis (low-coordinated Al atoms) acid sites onto the surface, and the surface acidity caused the apparition of the two forms of benzophenone, and, consequently, the spectral changes. The detection of the diphenylketyl radical suggested the hydrogen atom abstraction from the surface OH groups.

The same emissive species, besides the triplet (n, π)* state of benzophenone, were observed by time resolved luminescence when benzophenone was included into the channels of aluminophosphate $AlPO_4$-5 [182], where the Brönsted sites are the surface hydroxyl groups (Al–OH and P–OH) and the defect sites, or in the case of cation exchanged ZSM-5zeolite [178]. In the latter case, Anpo and co-workers observed that on the surface of the ZSM-5 zeolite, benzophenone molecules exist preferentially in hydrogen bonded forms (BPH) and inside the cavities there are protonated benzophenone species (BPH^+) and that interacts with the acidic sites located on the inner surfaces of the cage. The amount of BPH^+ is closely related to the radius of the exchanged cation. Thus, for cations with smaller radius, as H^+ or Na^+, the BPH^+ species predominate, while the BPH was the main form for larger cations, as Rb^+ or Cs^+. In this case, the photolysis products of benzophenone were also associated with the size of the cations. The main photoproducts were benzopinacol and benzhydrol, whose relative yields decreased with the decreases in the size of the exchanged cations and with benzopinacol in lower concentrations than benzhydrol. The authors concluded that, inside the ZSM-5 cavities, the excited benzophenone reacts mainly through its protonated form. By transient absorption spectroscopy, the presence of the diphenylketyl radical has been demonstrated for benzophenone adsorbed on TiO2 - Al_2O_3 binary oxide surfaces [186], so Anpo et. al supposed that benzhydrol was formed from the BPH^+ species via this intermediary. When compared to the photochemistry of benzophenone on porous Vycor glass, the fact that benzopinacol was the only photoproduct identified here was correlated with the detection of BPH. This reasoning could be translated to the cation exchanged ZSM-5, where benzopinacol was identified in the case of the largest cation, Cs^+.

Vieira Ferreira et. al compared the photochemistry of benzophenone included in nanochannels of H^+ZSM5 and Na^+ZSM5, and have detected hydroxybenzophenone radicals when the inner surfaces contain only Lewis acid sites [177]. The channels of H^+ZSM5 include, additionally , the Brönsted acid sites, thus the benzophenone reactivity decreased and the diphenylketyl radicals could not be observed. The luminescence spectra has shown the presence of the benzophenone triplet state and the BPH^+.

In the case of hydrophobic medium like that of silicalite's, a de-aluminated analogue of ZSM-5 zeolite, emissions from the hydrogen bonded or the BPH^+ forms of benzophenone were not evidenced and only benzophenone phosphorescence has been detected [172], [176]. The high value of phosphorescent lifetime was explained by the fact that the narrow channels of silicalite provide a very rigid environment for the triplet benzophenone, thus decreasing the non-radiative mechanisms of deactivation. The transient absorption spectra presented two maxima that were assigned to the $T_1 \rightarrow T_2$ and $T_1 \rightarrow T_3$ transitions. All these observations indicate the high stability of benzophenone in silicalite.

Summarizing the results obtained by luminiscence spectroscopy, the emissive species of benzophenone adsorbed on surfaces were the excited triplet state (which is detected regardless the surface acidity), the protonated excited benzophenone, and the hydrogen bonded excited benzophenone, respectively, and the last two forms are leading to different photoproducts.

The reductive substrates promote hydrogen atom abstraction, clearly shown with hosts such as cellulose, calix[n]arene, cyclodextrins, or reversed-phase silicas. Cyclodextrins possess the remarkable ability to accommodate guest molecules inside their hydrophobic cavities. Monti and coworkers investigated the photoreduction of benzophenone in cyclodextrins [179] and have noticed that the triplet benzophenone abstracts a hydrogen atom from the inner walls of the cavity, but the fate of the triplet radical pair was found to depend on the cavity size. Escape of the geminate pair from the cage predominates in α- and γ-cyclodextrins, and intersystem crossing in the case of β-cyclodextrin, with regeneration of benzophenone. Similar findings were reported by Barra and Scaiano [187] and lately the photochemistry of benzophenone inside the cyclodextrin cages has been reviewed [188].

A similar dependence upon the cage dimension is seen in the case of calix[4]arene, calix[6]arene, and calix[8]arene [180], because the medium became more polar as the calixarene cavity enlarges. By inclusion in the ring, the conformation of the benzophenone molecules deviates from planarity and the reactivity is related to conformations of both benzophenone and calixarene which would favor the hydrogen atom abstraction from the hydroxyl groups of the host. The resulting radicals are diphenylketyl from benzophenone and phenoxyl from the host. Calix[4]arene behaves like a more rigid and hydrophobic environment, in a "cone" formation, thus Vieira Ferreira has not detected the creation of diphenylketyl radicals in this case. The diphenylketyls appeared in an increasing amount, going from calix[6]arene to calix[8]arene, both of them adopting a "pinch" conformation that favours the hydrogen abstraction. The interaction between excited benzophenone and calix[n]arene is summarized by Vieira Ferreira et al. in Figure 13.

$$[BZP, CLX[4]OH] \xrightarrow{h\nu} [^3BZP^*, CLX[4]OH]$$

$$[BZP, CLX[8]OH] \xrightarrow{h\nu} [^3BZP^*, CLX[8]OH] \longrightarrow [BZPH^{\bullet}, CLX[8]O^{\bullet}]$$

$$CLX[n]OH \xrightarrow{h\nu, -H^{\bullet}} CLX[n]O^{\bullet}, n = 4, 6, 8$$

Some studies on the products of benzophenone photodegradation after laser irradiation with 266 nm or lamp irradiation at 254 nm have been performed in argon atmospheres or air equilibrated conditions, on silica, reversed phase silica [189], inclusion complexes of benzophenone and calixarene [180], [189], on the MCM-41 surface [184], or inside the nanochannels of H^+ ZSM-5 and Na^+ ZSM-5 zeolites [177]. In the most cases, it was found that 2-hydroxybenzophenone (BPOH) is one of the main photodegradation products, except in the case of silicalite as a host, and appears via formation of the BPOH· radical by reaction with water traces on the surface [190]. This process competes with the α-cleavage of benzophenone, which is possible only from a higher triplet state through a π-π* transition. As a result, benzoyl and phenyl radicals were detected by diffusion reflectance laser flash photolysis, which conducted to formation of benzaldehyde, benzhydrol, biphenyl, benzoic acid, and some phenyl benzophenone isomers, identified by means of GC – MS technique. The ratio between all these photoproducts is controlled by the experimental conditions, i.e. laser vs. the lamp, and the presence of oxygen, as it was observed that the a-cleavage became important under laser irradiation as compared to the lamp.

The photochemistry of benzophenone in MCM-41 has been investigated by comparison with that in H^+-ZSM5 zeolite by J.P. Da Silva et al. [183], regarding the differences between the pore sizes (higher for MCM-41), hydrophobicity, and polarity (MCM-41 is more hydrofobic and less polar than H^+-ZSM5). While the photodegradation products were similar upon irradiation with either 254 nm (lamp) or 266 nm (laser), irradiation with 355 nm produced some discrepancy between the photoproducts. Thus, benzhydrol was identified on H^+-ZSM5 and benzopinacol on MCM-41, because of the stereochemical restrictions imposed by the pore size of H^+-ZSM5 that did not allow the coupling of two diphenylketyl radicals. In addition, the luminiscence studies showed that protonated excited benzophenone was detected in H^+-ZSM5, and only hydrogen bonded benzophenone in MCM-41, respectively. By taking into account the whole benzophenone photochemistry in the two studied substrates, J. P. Da Silva summarized the proposed photoreactions in the Figure 14.

The photochemical pathways of the benzophenone reduction in a presence of a hydrogen atom donor could be followed inside the zeolite cages, whose size promotes the "cage effect". Cizmeciyan and co-workers have studied the photochemistry and photophysics of benzophenone and cyclohexane included in supercages of the zeolite NaX [191]. The authors detected only the adduct product of the photolysis, 1-cyclohexyl-1,1-diphenylmethanol, after the complete destruction of the zeolite. This fact was explained by the entrapping of the diphenylketyl and cyclohexyl radicals inside the cage and the subsequent coupling with each other, these radicals not being able to fit through the 7.4 Å pores of the NaX zeolite.

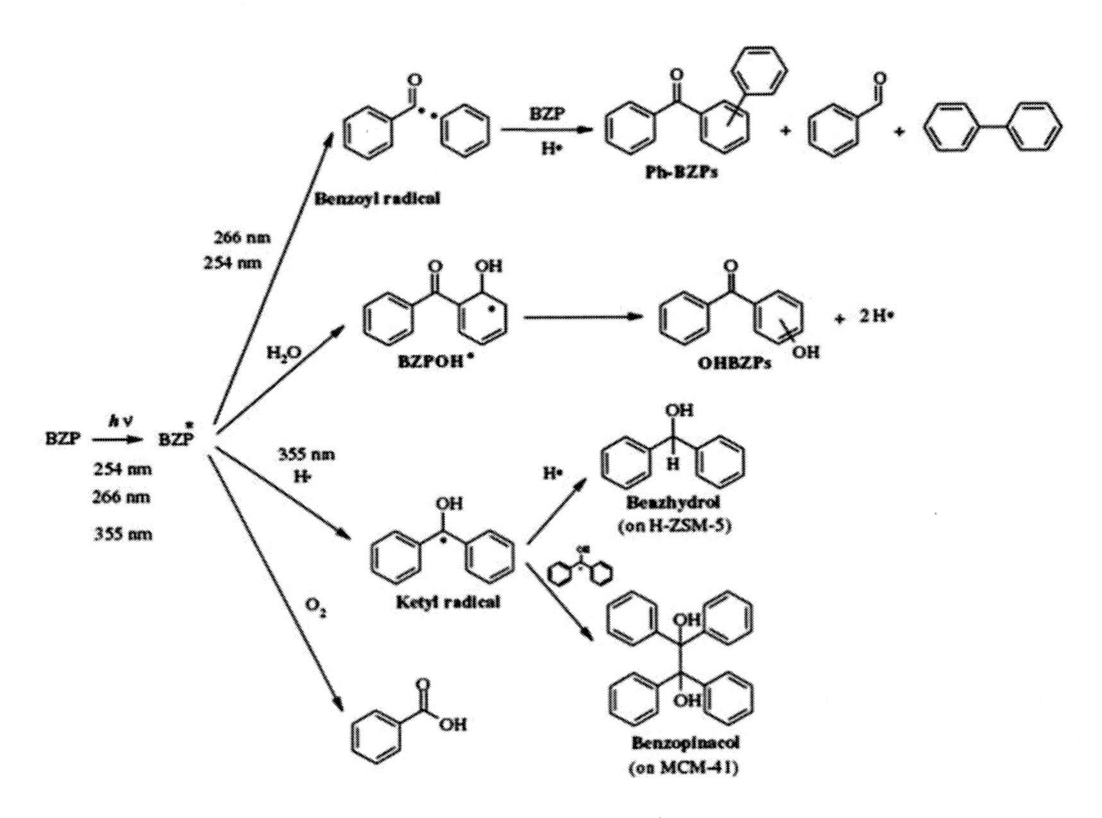

Figure 13. Proposed photoreactions of BP on MCM-41 and H^+ZSM-5 [183].

6. Conclusions

The benzophenone photochemistry is well known and is widely applied in materials science, providing a very effective and simple way to improve the polymer's characteristics. This leads to uses of benzophenone for both bulk modifications of polymers by photocrosslinking and surface functionalization in view to, e.g. tailor the response of the interface – biomolecule system. The general photoreactions supposed to appear in a system could be supplemented by specific interactions attributed to the physics and chemistry of the host. Therefore, intriguing aspects regarding its behavior in constrained environments remains partially solved.

REFERENCES

[1] Bäckstrom H. L. J., Sandros K., *Acta. Chem. Scand.*, 1960, *14*, 48–62.
[2] Turro N. J. *Modern Molecular Photochemistry*; University Science Books:Mill Valley, CA, 1991, p.229
[3] Takatori Y., Kajii Y.,Shibuya K., Obi K., *Chem. Phys.*, 1994, *180*, 99-107.
[4] Cai X., Han Z., Yao S., Lin N., *Sci. China Ser. B: Chem.*, 2001, *44* , 582-586.
[5] Ciamician G.; Silber, P. *Ber.*, 1900, *33*, 2911-2913.
[6] Moore W. M., Hammond G. S., Foss R. P., *J. Am. Chem. Soc.*, 1961, *83*, 2789-2794
[7] Paterno E., Chieffi C., *Gazz. Chim. Ital.*, 1909, *39*, 341-349.
[8] Büchi G., Inman C. G., Lipinsky E. S, *J. Am. Chem. Soc.*, 1954, *76*, 4327-4331.
[9] El – Sayed M. A., *J. Chem. Phys.*, 1962, *36*, 573-574.
[10] (a) Porter G.; Suppan P.; *Pure Appl. Chem.*, 1964, *9*, 499 – 505; (b) Porter G.; Suppan, P.; *Trans. Faraday Soc.,* 1965, *61*, 1664-1673.
[11] Pitts Jr., J.N., Johnson H.W. Jr.; Kuwana T. *J. Phys. Chem.*, 1962, *66*, 2456-2461.
[12] Piette L. H.; Sharp J.H.; Kuwana T. *J. Chem. Phys.*, 1962, *36*, 3094-3095.
[13] Porter G., Suppan P., *Trans. Faraday Soc.* 1966, *62*, 3375-3383.
[14] Cohen S. G., Saltzman M. D., Guttenplan J. B., *Tetrahed. Lett.,* 1969, 10, 4321-4324.
[15] Aspari P., Ghoneim N., Haselbach E., Von Raumer M., Suppan P., Vauthey E., *J. Chem. Soc. Faraday Trans,* 1996, *92*, 1689-1696.
[16] Bosca F.; Cosa G.; Miranda M. A.; Scaiano J.C. *Photochem. Photobiol. Sci.*, 2002, *1*, 704 708.
[17] Cohen S. G., Siddiqui M. N., *J. Am. Chem. Soc.*, 1967, *89,* 5409–5413.
[18] Brown R. G., Porter G., *J. Chem. Soc., Faraday Trans.* 1., 1977, 73, 1569-1573.
[19] Bhasikuttan A. C., Singh A. K., Palit D. K., Sapre A. V., Mittal J. P., *J. Phys. Chem. A,* 1998, *102*, 3470-3480.
[20] Singh A. K., Bhasikuttan A. C., Palit D. K., Mittal J. P., *J. Phys. Chem. A* 2000, *104*, 70027009.
[21] Singh A. K., Palit D. K., J. P. Mittal, *Res. Chem. Intermed.* 2001, *27*, 125-136.
[22] Singh A. K., Ramakrishna G., Ghosh H. N., Palit D. K., *J. Phys. Chem. A* 2004, *108*, 25832597.
[23] Pal T., Paul M., Ghosh S., *J. Molec. Struct.: THEOCHEM,* 2008, *860*, 8-12.
[24] Suppan P., J. *Molec. Spectr.*, 1969, *30,* 17-28.
[25] Ghoneim N., Monbelli A., Pilloud D., Suppan P., *J. Photochem. Photobiol. A: Chem.*, 1996, *94*, 145-148.
[26] Cheng, X.-M., Huang, Y.; Ma, J.-Y; Li, X.-Y., *Chin. J. Chem. Phys.*, 2007, *20*, 273-278.

[27] Cowan D. O., Drisko R. L., *Elements of Organic Chemistry*; Plenum Press, New York, 1977, pp.96.
[28] Pitts J. N., Letsinger R. L., Taylor R. P., Patterson J. M., Rechttenwald G., Martin R. B., *J. Am. Chem. Soc.*, 1959, *80*, 1068-1077
[29] Beckett A., Porter G., *Trans. Faraday Soc.*, 1963, *59*, 2038 – 2050.
[30] Yang N. C., Murov S. L., *J. Am. Chem. Soc.*, 1966, *88*, 2852 – 2854.
[31] Cohen S. G., Cohen J. I., *Tetrahed. Lett.*, 1968, *88*, 4823 – 4826.
[32] Chilton J., Giering L., Steel C., *J. Am. Chem. Soc.*, 1976, *98*, 1865–1870.
[33] Scaiano J. C., Abuin E. B., Stewart L. C., *J. Am. Chem. Soc.*, 1982, *104*, 5673–5679.
[34] Filipescu N., Minn F. L., J. *Am.Chem. Soc., 90* (1968) 1544-1547.
[35] Weiner S., *J. Am. Chem. Soc.*, 1971, *93*, 425 – 429.
[36] Viltres Costa C., Grela M. A., Churio M. S., *J. Photochem. Photobiol. A: Chem.*, 1996, *99*, 51-56.
[37] Dedinas J., *J. Phys. Chem.*, 1971, *75*, 181-186.
[38] Demeter A., Bérces T., *J. Photochem. Photobiol, A: Chem.*, 1989, *46*, 27–40.
[39] Du Y., Ma C., Kwok W. M., Xue J., Phillips D. L., *J. Org. Chem.,* 2007, *72*, 7148-7156.
[40] Scully A. D., Horsham M. A., Aguas P., Murphy J. K.G., *J. Photochem. Photobiol. A: Chem.*, 2008, *197*, 132–140.
[41] Mel'nik V. I., *Phys. Sol. St.* 1998, *40*, 960-963.
[42] Nakayama, T.; Sakurai, K.; Ushida, K.; Kawatsura, K.; Hamanoue, K., *Chem. Phys. Lett.*, 1989, *164*, 557-561.
[43] Nakamura T., Sohachiro H., *Jpn. J. Appl. Phys.* 1969, *8,* 85-90.
[44] Avdeev, A. V.; Erina, M. V.; Kulik ova, O. I., *J. Appl. Spectr.,* 2006, *73*, 624-626.
[45] Hayashi H., Hirota N., Nagakura S., *Chem. Lett.*, 1973, *2*, 979-983.
[46] Scharf G., Winefordner J. D., *Talanta,* 1986, *33*, 17-25.
[47] Nishiguchi H., Okamoto S., Nishimura M., Yamashita H., Anpo M., *Res. Chem. Interm.*, 1998, *24*, 849-858.
[48] Matsui K., Nozawa K., Yoshida T., *Bull. Chem. Soc. Jpn.*, 1999, *72*, 591-596.
[49] Murai H., Jinguji M., Obi K., *J. Phys. Chem.*, 1978, *82*, 38–40.
[50] Godfrey T. S., Hilpern J. W., Porter G., *Chem. Phys. Lett.*, 1967, *1*, 490-492
[51] Itagaki H., Horie K., Mita I., *Progr. Polym. Sci.*, 1990, *15*, 361-424.
[52] Christoff M., Atvars T. D. Z., *Macromolecules,* 1999, **32,** 6093-6101.
[53] Deeg F.W., Pinsl J., Bräuchle C., *J. Phys. Chem.* 1986, 90, 5715–5719.
[54] Bräuchle C., Burland D. M., Bjorklund G. C. *J. Phys. Chem.* 1981, *85*, 123-127.
[55] Horie K.; Morishita K.; Mita, I. *Macromolecules,* 1984, *17,* 1746-1750.
[56] Salmassi A., Schnabel W., *Polym. Photochem.*, 1984, *5*, 215-230.
[57] Horie K., Tsukamoto M., Morishita K., Mita, I ., *Polym. J.* , 1985, *17*, 517-524.
[58] Horie K.; Ando H.; Mita, I. *Macromolecules,* 1987, *20,* 54-58.
[59] Ebdon J. R., Lucas D. M., Soutar I., Lane A. R, Swanson L., *High Perform. Polym.,* 1999, *11*, 331-341.
[60] (a) Hrdlovič P , *Polym. Photochem.*, 1986, *7*, 359-377; (b) Hrdlovič P., Srnková K., *Eur. Polym. J.*, 1992, *28*, 1279-1287.
[61] Yang N.C.; Rivas, C. *J. Am. Chem. Soc.*, 1961, *83*, 2213-221
[62] Nakayama T., Hamanoue K., Hidaka T., Okamoto M., Teranishi H., *J. Photochem.*, 1984, 71-78.

[63] Morrison, J.; Osthoff, H.; Wan, P. *Photochem. Photobiol. Sci.*, 2002, *1*, 384-394.
[64] Arnold D. R., Hinman R. L., A. H. Glick, *Tetrahedr. Lett.,* 1964, *5*, 1425-1430.
[65] (a) Yang N. C., Nussim M., Jorgenson M. J., Murov S., *Tetrahedr. Lett.*, 1964, *5*, 3657-3664; (b) Freilich S. C., Peters K. S., *J. Am. Chem. Soc.*, 1981, *103*, 6255-6257; (c) Buschmann, H., H. -D. Scharf, Hoffmann N., Esser, P., *Angew. Chem. Int. Ed. Engl.*, 1991, *30*, 477 – 505.
[66] Parola A. H., Rose A. W., Cohen S. G., *J. Am. Chem. Soc.*, 1975, *97*, 6202 – 6209.
[67] Levin P. P., Raghavan P. R. K., *Chem. Phys. Lett.*, 1991, *182*, 663-667.
[68] Zalesskaya G. A., Yakovlev D. L., Sambor E. G., *Opt. Spektrosk.*, 2000, *88*, 782;
[69] Granchak V. M., Dilung I. I., *High Energy Chem.,* 2002, 36, 408–412.
[70] Inbar S., Linschitz H., Cohen S. G., *J. Am. Chem. Soc.*, 1980, *102*, 1419 – 1421
[71] Oster G., Oster G. K., Moroson H., *J. Polym. Sci.*, 1959, *34*, 671-684
[72] Charlesby A., Grace C. S., Pilkington F. B., *Proc. R. Soc. Lond. Ser. A: Math. Phys. Sci.*, 1962, *268*, 205-221.
[73] (a) Chen Y. L., Rånby B., *J. Polym. Sci., A: Polym. Chem. Ed.*, 1989, *27*, 4051-4075; (b) Qu B. J., Xu Z. H., Shi W. F., Rånby B, *Macromolecules*, 1992, *25*, 5220 -5224; (c) Qu B. J., Xu Z. H., Shi W. F., Rånby B., *Macromolecules*, 1992, *25*, 5215-5219; (d) Qu B. J., Rånby B., *J. Appl. Polym. Sci.*, 1993, *48*, 701 -709; (e) Qu B. J., Qu X., Xu H., Rånby B., *Macromolecules*, 1997, *30*, 1408-1413; (f) Qu B. J., Xu Y. H., Ding L. H., Rånby B., *J. Polym. Sci., A: Polym. Chem. Ed.*, 2000, *38*, 999-1005.
[74] Rånby B., Chen Y. L.., Qu B. J., Shi W. F., in „Polymers forAdvanced Technology"; M.Levin, Ed.., VCH Publishers Inc., NY, 1988, p. 162 – 181.
[75] Qu B. J., Shi W. F., Liang R. Y., Jin S., Xu Y. H., Wang Z. H., Rånby B., *Polym. Eng. Sci.*, 1995, *35*, 1005 – 1010.
[76] Qu B., *Chin. J. Polym. Sci.*, 2002, *10*, 291 – 307.
[77] (a) Xu. Y., Qu B. J., Zhang Z., *J. Photopol. Sci. Techn.*, 1996, *9*, 157 – 164; (b) Hu Z., Qu B. J. , Zhang Z., Rånby B., *J. Photopol. Sci. Techn.*, 1996, *9*, 165-172
[78] Giering L., Berger M., Steel C., *J. Am. Chem. Soc.*, 1974, *96*, 953–958.
[79] David C., Baeyens – Volant D., Delaunois G., Lu-Vinh Q., Piret W., Geuskens G., *Eur. Polym. J.*, 1978, *14*, 501-507.
[80] Mita I., Takagi T., Horie K., Shindo Y., *Macromolecules*, 1984, *17,* 2256–2260.
[81] Geuskens G., Baeyens-Volant D., Delaunois G., Lu-Vinh Q., Piret W. and C. David, *Europ. Polym. J.*, 1978, *14*, 299-303.
[82] Eisele G., Fouassier J.-P., *Macromol. Chem. Phys.,* 1996, *197*, 1731-1756.
[83] Winslow F. H., Marteyek W., Stills S. M., Polym. Prepr. (Am. Chem. SOC., Div. Polym. Chem.), 1966, 7, 390.
[84] Geuskens G., David C., *Pure Appl. Chem.*, 1979, *51*, 233—240.
[85] David C., Piret W., Sakaguchi M., Geuskens G., *Die Makrom. Chem.*, 1978, *179*, 181 – 187.
[86] Torikai A., Takeuchi T., Fueki K., *Polym. Photochem.*, 1983, *3*, 307-320.
[87] Scherzer T., Tauber A., Mehnert R., *Vib. Spectrosc.*, 2002, *29*, 125–131.
[88] Lin C. S., Liu W. L., Chiu Y. S., Ho S.-Y., *Polym. Degrad. Stab.*, 1992, *38*, 125-130.
[89] Duarte F.M., Botelho G., Machado A.V., *Polym. Test.*, 2006, *25*, 91-97.
[90] Machado A. V., Duarte F. M., Botelho G., Moura I., *J. Appl. Polym. Sci.*, 2007, *105*, 2930-2938.

[91] Kaczmarek H., Kamińska A., Świątek M., Sanya S., *Eur. Polym. J.*, 2000, *36,* 1167-1173.
[92] Sikkema K., Cross G. S., Hanner M. J., Priddy D. B., *Polym. Degrad. Stab.*, 1992, *38,* 113-118.
[93] Kaczmarek H.; Swiatek M.; Kaminska A. *Polym Degrad Stab.*, 2004, *83*, 35-45
[94] Bottino F. A.; Cinquegrani A. R.; Pasquale G.; Leonardi L.; Pollicino A. *Polym. Test.,* 2004, *23*, 405-411.
[95] Hong K. H., Sun G., *J. Appl. Polym. Sci.*, 2009, *112*, 2019–2026.
[96] (a) Yao D., Qu B., Wu Q., *Polym. Eng. Sci.*, 2007, *47*, 1761-1767; (b) Hu Y., Wu Q., Qu B., *Polym. Adv. Technol.*, 2010, *21*, 177-182.
[97] (a) Carroll G. T., Sojka M. E., Lei X., Turro N. J., Koberstein J. T, *Langmuir,* 2006, *22,* 7748-7754; (b) Carroll G. T., Turro N.J., Koberstein J. T., *J. Coll. Interf. Sci.*, 2010, *351*, 556-560.
[98] Rånby B.*, Polym. Eng. Sci.,* 1998, *38*, 1229 – 1243.
[99] Hunter, W.L.; Crabtree, P. N., *Photo. Sci. Eng* 1969, *13,* 271-280.
[100] Harita Y., Ichikawa M., Harada K., Tsunoda T., *Pol. Eng. and Sci.*, 1977, *17,* 372–376.
[101] Buchi G., Kopron J. T., Koller E., Rosenthal D., *J. Am. Chem. Soc.*, 1956, *78*, 876-877.
[102] Zimmerman H. E., Craft L., *Tetrahedr. Lett.*, 1964, 5, 2131-2136.
[103] Bryce-Smith D., Fray G. I., Gilbert A., *Tetrahedr. Lett.*, 1964, *5*, 2137-2139.
[104] Saltiel J., Coates R. M., Dauben W. G., *J. Am. Chem. Soc.*, 1966, *88*, 2745 – 2748.
[105] Encinas M. V., Scaiano J. C., *J. Am. Chem. Soc.*, 1981, *103*, 6393 – 6397.
[106] Ng H., Guillet, J. G., *Macromolecules*, 1977, *10*, 866–868.
[107] Kagiya V.T., Takemoto K., *J. Macromol. Sci., Chem.*, 1976, *A10*, 795-810.
[108] Golub M. A., *Macromolecules*, 1969, *2*, 550-552.
[109] Farber M., Worns J. R., Syndiotactic polybutadiene composition for a photosensitive printing plate, US Patent 4394435.
[110] Shikinami Y., Kimura R., Yoshikawa Y., Iida K., Hata K., Radiation process for producing 1,2-polybutadiene foamed products, US Patent 4144153.
[111] Avadanei M., PhD Thesis, *Romanian Academy*, 2008.
[112] (a) Avadanei M., Grigoriu G. E., Barboiu V., *Rev. Roum. Chim.*, 2003, *48*, 813 – 819; (b) Vranceanu N., Avadanei M., Ciovica S., Barboiu V., Grigoriu G., *Cellul. Chem. Techn.*, 2005, *39*, 423 – 426.
[113] Decker C., Nguyen Thi Viet T., *Macromol. Chem. Phys.*, 1999, *200*, 358–367; (b) Decker C., Nguyen Thi Viet T., *J. Appl. Polym. Sci.*, 2000, *77*, 1902–1912.
[114] Golub M. A., *Pure Appl. Chem.,* 1972, *30*, 105-118.
[115] Decker C., Nguyen Thi Viet T., *J. Appl. Polym. Sci.*, 2001, *82*, 2204–2216.
[116] Masaki K., Ohkawara S., Hirano T., SenoM., Sato T., J. *Polym. Sci,: Part A: Polym. Chem.*, 2004, *42*, 4437–4447.
[117] Bousquet A., Haidar B., Fouassier J P., Vidal A., *Eur. Polym. J.*, 1983, *19,* 135 – 142.
[118] Hilborn J., Rånby B., *Macromolecules,* 1989, *22*, 1159-1165.
[119] Wu S. K., Rabek J. F., *Polym. Degrad. Stab.,* 1988, *21*, 365-376.
[120] McDonald J. C., Whitesides G. M., *Acc. Chem. Res.*, 2002, *35*, 491-499.
[121] Sia S. K., Whitesides G. M., *Electrophoresis*, 2003, *24*, 3563-3576.
[122] Lotters J. C., Olthuis W., Veltink P. H., Bergveld P., *J. Micromech. Microeng.*, 1997, *7*, 145–147.
[123] Ward J. H., Bashir R., Peppas N. A., *J. Biomed. Mater. Res.*, 2001, *56*, 351–360.

[124] Tsougeni K., Tserepi A., Gogolides E., *Microelectr. Eng.*, 2007, *84,* 1104–1108.
[125] Bhagat A. A. S., Jothimuthu P., Papautsky I., *Lab Chip*, 2007, 7, 1192–1197.
[126] Jothimuthu P., Carroll A., Bhagat A. A. S., Lin G., Mark J. E., Papautsky I., *J. Micromech. Microeng.*, 2009, *19*, 1-9.
[127] Cong H., Pan T., *Adv. Funct. Mater.*, 2008, *18*, 1912–1921.
[128] He M. B., Hu X. Z., *Polym. Degrad. Stab.*, 1987, *18*, 321-328.
[129] Oster G., Shibata O., *J. Polym. Sci.*, 1957, *26*, 233–234.
[130] Rånby B., in *Current Trends in Polymer Photochemistry*, N. S. Allen, M. Edge, J. R. Bellobono and E. Selli, Eds.; E.Honvood: New York, 1995; pp.23-39.
[131] Rånby B., *Int. J. Adh. Adhes.*, 1999, *19,* 337–343.
[132] Deng J. P., Wang L. F., Liu L. Y., Yang W. T., *Progr. Polym. Sci.*, 2009, *34,* 156–193.
[133] He D. M., Susanto H., Ulbricht M., *Progr. Polym. Sci.,* 2009, *34*, 62–98.
[134] Bhattacharya A., Misra B. N., *Progr. Polym. Sci*, 2004, *29*, 767–814.
[135] Ogiwara Y.; Kanda M.; Takumi M.; Kubota H. *J. Polym. Sci. Polym. Lett. Ed.* 1981, *19*, 457-462.
[136] Allmer K.; Hult A.; Rånby B. *J. Polym. Sci. Part A: Polym. Chem.*, 1988, *26*, 2099-2111.
[137] Tazuke S.; Kimura H. *Makromol. Chem.* 1978, *179*, 2603-2612.
[138] Rånby B.; Gao Z. M.; Hult A.; Zhang P. Y., *Polym. Prepr., Am. Chem. Soc. Div. Polym. Chem.*, 1986, *27*, 38-52.
[139] YangW. T.; Rånby B., *Polym. Bull.*, 1996, *37*, 89-96.
[140] (a) Yang W. T.; Rånby B. *J .Appl. Polym. Sci.*, 1997, *62*, 533-544.; (b) Yang W. T.; Rånby B. *J. Appl. Polym. Sci.*, 1997, *62*, 545-555.
[141] (a) Deng J.P., Yang W.T., Rånby B., *J. Appl. Polym. Sci.* 2000, *77,* 1513–1521; (b) Deng J.P., Yang W.T., Rånby B., *J. Appl. Polym. Sci.* 2000, *77,* 1522–1531.
[142] Hu, S.; Ren, X.; Bachman, M.; Sims, C. E.; Li, G. P.; Allbritton, N. L. *Anal. Chem.,* 2004, *76*, 1865-1870.
[143] Yang W., Rånby B., *Macromolecules*, 1996, *29*, 3308-3310.
[144] (a) Ma H., Davis R., Bowman C.N., *Macromolecules*, 2000, *33*, 331-335; (b) Ma H., Davis R.H., Bowman C.N., *Polymer*, 2001, *42*, 8333-8338.
[145] (a) Hu S.; Ren X.; Bachman M.; Sims C. E.; Li G. P.; Allbritton N. L., *Langmuir,* 2004, *20*, 5569-5574; (b) Hu S.; Ren X.; Bachman M.; Sims C. E.; Li G. P.; Allbritton N. L., *Electrophoresis,* 2003, *24*, 3679-3688.
[146] Deng J. P., Yang W. T., Rånby B., *J. Appl. Polym. Sci.*, 2001, *80*, 1426–1433.
[147] Deng J. P., Yang W. T., Rånby B., *Polym. J.*, 2000, *32*, 834-837.
[148] Deng J. P., Yang W. T., Rånby B., *Macromol. Rapid Commun.* 2001, *22*, 535–538.
[149] Gao J., Lei J.,Su Z.,Zhang B.,Wang J., *Polym. J.*, 2001, *33*, 137 – 149.
[150] Zhang Z. D., Kong L. B., Deng J.P., Yang P., Yang W. T., *J. Appl. Polym. Sci.*, 2006, *101*, 2269–2276.
[151] Yang W.-T., Yin M.-Z., Sun Y. F., *Chin. J. Polym. Sci.*, 2000, *18*, 431–435.
[152] Cho J.-D., Kim S.-G., Hong J.-W., *J. Appl. Polym. Sci.*, 2006, *99*, 1446–1461.
[153] Yang W. T., Rånby B., *Eur. Polym. J.*, 2002, *38*, 1449-1455.
[154] Woo C.K., Schiewe B., Wegner G., *Macromol. Chem. Phys.* 2006, *207*, 148–159.
[155] Goddard J.M., Hotchkiss J.H., *Progr. Polym. Sci.*, 2007, *32,*698-725.
[156] Caldorera-Moore M., Peppas N. A., *Adv. Drug Deliv. Rev.*, 2009, *61*, 1391-1401.
[157] Wong I., Ho C.-M., *Microfluid. Nanofluid.,* 2009, *7*, 291-306.

[158] Wang Y., Lai H.-H., Bachman M., Sims C. E., Li G. P., Allbritton N. L., *Anal. Chem.* 2005, *77*, 7539-7546.

[159] Fiddes L. K., Chan H. K. C., Lau B., Kumacheva E., Wheeler A. R., *Biomat.* 2010, *31*, 315– 320.

[160] Schneider M. H., Willaime H., Tran Y., Rezgui F., Tabeling P., *Anal. Chem.* 2010, *82,* 8848–8855.

[161] Zhao L.-B., Li S.-Z., Hu H., Guo Z.-X., Guo F., Zhang N.-G., Ji X.-H., Liu W., Liu K., Guo S.-S., Zhao X.-Z., *Microfluid. Nanofluid.*, 2010, DOI: 10.1007/s10404-010-0673-5

[162] (a) Ebara M., Hoffman J. M., Hoffman A. S., Stayton P. S., *Lab Chip*, 2006, *6*, 843-848; (b) Ebara M., Hoffman J. M., Stayton P. S., Hoffman A. S., *Rad. Phys. Chem.* 2007, *76*, 1409– 1413.

[163] Ma D., Chen H. W., Shi D. Y., Li Z. M., Wang J. F., *J. Coll. Interf. Sci.* 2009, *332*, 85–90.

[164] Huck W. T. S., Bowden N., Onck P., Pardoen T., Hutchinson J. W., Whitesides G. M., *Langmuir,* 2000, *16,* 3497-3501.

[165] Dorman G., Prestwich G. D., *Biochem.*, 1994, *33*, 5661–5673.

[166] Pouliquen L., Coqueret X., *Macromol. Chem. Phys.*, 1996, *197*, 4045–4060.

[167] Dankbar D. M., Gauglitz G., *Anal. Bioanal. Chem.*, 2006, *386*, 1967–1974.

[168] Lipson M., McGarry P. F., Koptyug I. V., Staab H. A., Turro N. J., Doetschman D. C., *J. Phys. Chem.,* 1994, *98,* 7504-7512.

[169] Turro N. J., Zimmt M. B., Gould I. R., *J. Am. Chem. Soc.*, 1985, *107*, 5826 – 5827.

[170] Turro N. J., *Tetrahedron,* 1987, 43, 1589-1616.

[171] Drake J. M., Levitz P., Turro N. J., Nitsche K. St., Cassidy K. F., *J. Phys. Chem.*, 1988, *92*, 4680 – 4684.

[172] Vieira Ferreira L. F., Ferreira Machado I., Da Silva J. P., Branco T. J. F., *Photochem. Photobiol. Sci.*, 2006, *5*, 665–673.

[173] Anpo M., Nishiguchi H., Fujii T., *Res. Chem. Interm.*, 1990, *13*, 73 – 102.

[174] Kerry Thomas J., *Photochem. Photobiol. Sci.*, 2004, 3, 483–488.

[175] Casal H. L., Scaiano J. C., *Can. J. Chem.*, 1985, *63*, 1308–1314.

[176] Vieira Ferreira L.F., Vieira Ferreira M.R., Oliveira A.S., Moreira J.C., J. *Photochem. Photobiol. A: Chem.,* 2002, *153*, 11–18.

[177] Vieira Ferreira L.F., Costa A.I., I. Ferreira Machado, Da Silva J.P., *Micropor. Mesopor. Mat.,* 2009, *119*, 82-90.

[178] Okamoto S., Nishiguchi H., Anpo M., *Chem. Lett.* 1992, *6*, 1009-1012.

[179] Monti S., Flamigni L., Martelli A., Bortolus P., *J. Phys. Chem.*, 1988, *92*, 4447 – 4451.

[180] Vieira Ferreira L. F., Vieira Ferreira M. R., Oliveira A. S., Branco T. J. F., Pratab J. V., Moreira J. C., *Phys. Chem. Chem. Phys.*, 2002, *4*, 204–210.

[181] Vieira Ferreira L.F., Netto – Ferreira J. C., Khmelinskii I. V., Garcia A. R., Costa S. M. M., *Langmuir,* 1995, *11*, 231–236.

[182] Ferreira Machado I., Vieira Ferreira L. F., Branco T. J.F., Fernandes A., Ribeiro F., *J. Molec. Struct.*, 2007, *831*, 1–9.

[183] Da Silva J.P., Ferreira Machado I., Lourenc J.P., Vieira Ferreira L.F., *Micropor. Mesopor. Mater.* 2005, *84*, 1–10.

[184] Da Silva P., Ferreira Machado I., Lourenc J.P., Vieira Ferreira L.F., *Micropor. Mesopor. Mater.* 2006, *89*, 143–149.

[185] (a) Nishiguchi H., Zhang J.-L., Anpo M., Masuhara H., *J. Phys. Chem. B*, 2001, *105,* 3218– 3222; (b) Nishiguchi H., Zhang J.-L., Anpo M., *Langmuir*, 2001, *17*, 3958–3963.
[186] Ramamurthy V., Casper J. V., Eaton D. F., Erica W. K., Corbin D. R., *J. Am. Chem. Soc.*, 1992, *114*, 3882-3892.
[187] Barra M., Scaiano J. C., *Photochem. Photobiol.*, 1995, *62*, 60–64.
[188] Bortolus P., Monti S., in *Advances in Photochemistry*, D. C. Neckers, D. H. Volman, G. von Bünau, Eds.; Wiley Interscience, 2007, Vol. 21, pp. 1–133.
[189] Vieira Ferreira L. F., Vieira Ferreira M. R., Da Silva J. P., Ferreira Machado I., Oliveira A. S., Prata J. V., *Photochem. Photobiol. Sci.* 2003, *2*, 1002–1010.
[190] Baral-Tosh S., Chattopadhyay S.K., Das P.K., *J. Phys. Chem.*, 1984, *88*, 1404-1408.
[191] Cizmeciyan D., Sonnichsen L. B., Garcia-Garibay M. A., *J. Am. Chem. Soc.*, 1997, *119*, 184-188.

In: Photochemistry
Editors: Karen J. Maes and Jaime M. Willems
ISBN: 978-1-61209-506-6

Chapter 3

PHOTO-CHEMICAL METHODS AS AN ALTERNATIVE METHODOLOGY IN THE DEPOSITION OF MATERIALS WITH CHEMICAL SENSOR PROPERTIES AND PHOTO-LUMINESCENT CHARACTERISTIC

Gerardo G. Cabello*[*1], Luis A. Lillo[1], Claudia Caro[1], G.E. Buono-Core[2] and Marisol Tejos[3].

[1]Departamento de Ciencias Básicas, Facultad de Ciencias, Universidad del Bío-Bío, Chillán, Chile

[2]Instituto de Química, Facultad de Ciencias Básicas y Matemáticas, Pontificia Universidad Católica de Valparaíso, Valparaíso, Chile

[3]Facultad de Ciencias, Universidad de Valparaíso, Valparaíso, Chile

ABSTRACT

Some photochemical methods of synthesis of nanomaterials has been proposed as alternative methods in the preparation of diverse materials such as semiconductors and ceramics materials. These systems are of easy implementation and management, carried out under mild temperature conditions, utilizing simple precursor compounds showing good results in the photo-deposition of the resultant materials. Among these methods we have presented the photo-deposition in solid phase (thin films). This methodology it involves a complex precursor film is spun on to the substrate forming an optical quality film. The film is then irradiated and the precursor fragments generating metals, which remain in the film, and volatile organic by-products. When this process is carried out in the atmosphere air oxidation normally results in the formation of a metal oxide film. This methodology has been employed in the preparation of SnO and ZrO_2 thin films and their preliminary study as chemical sensor and as luminescence device respectively.

* corresponding author: [1]Departamento de Ciencias Básicas, Facultad de Ciencias, Universidad del Bío-Bío, Campus Fernando May, Chillán, Chile, gcabella@ubiobio.cl

1. Introduction

Development and innovation of alternative methods or modifications to methodologies previously probed for the deposition and manufacture of nano-structured materials with the end in nanotechnology, is subject of permanent concern mainly for two reasons:

1. The search for new materials that respond to compatibility, operating needs with other materials; functionality, with the purpose of achieving the desired properties looking for a specific application; and sophistication, devices of smaller dimensions but with large functions capacity.
2. To set in motion deposition processes that involve a low energy expense during all process and that parameters involved being of easy handling, not requiring extreme measures.

In this sense, in recent years has been reported some photochemical deposition methodologies, using UV radiation as source of energy for the preparation of thin films of metallic oxide semiconductors, and ceramic materials or metal nanoparticles deposited on substrates diverse. These systems are of easy implementation and management, carried out under mild temperature conditions, showing good results in the deposition or in the photo-deposition of these materials.

The former prepares these materials involucrate metal reduction (M^0) or their reduction partial (M^{n-1}) on a substrate originated from molecular or ionic precursors. The essential of the photochemical approach is formation of metal reduction under the conditions that prevents their deposition. The metal reduced is formed by direct photoreduction of a metal source (metal salt or complex) (Fig. 1a) or reduction of metal ions using the photochemically generated intermediates, such as excited molecules and radicals (Fig.1b). The latter method is often called photosensitization. This method uses the photoactive reagents that generate the intermediates by photo irradiation. The intermediates reduce the co-existing metal source to form the metal reduced (M^0) or partial reduction (M^{n-1}), however the source of excitation wavelength depend essentially of sensitizer more than the metal source.

In both cases the ultimate goal is the photoreduction of precursor material, however depending of photochemical method used, the photoreduction of the precursor species is only intermediate specie for subsequent reaction with other compounds and so that generate the desired material.

In this review, we overviewed of some photochemical methods for the preparation of semiconductors, ceramic materials and nanoparticle materials with a closer analysis of those based on photochemistry.

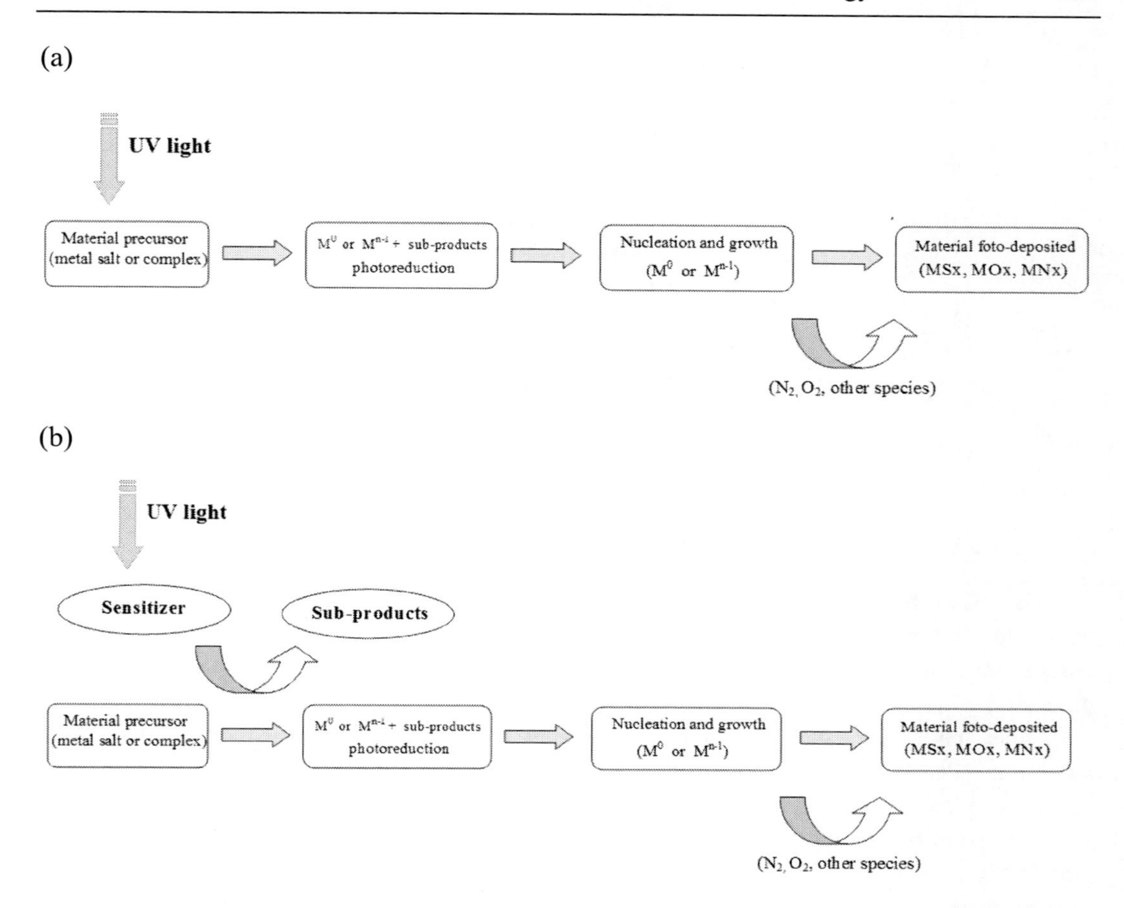

Figure 1. Systematic scheme of photochemical synthesis (a) direct photolysis (b) photosensitizacion.

1.1. Some Methods of Photodeposition

Taking the above conditions into account, some photo-deposition techniques have been proposed, among them Liquid phase photodeposition (LPPD) [1-4]. A typical LPPD procedure consists of using β-diketonate metal complexes as the precursors, which are dissolved in organic solvents and introduced in a cuvette whose interior is suspended a silicon substrate (Fig. 2). The cuvette is then irradiated at 254 and 300 nm in a photo-reactor. By means of this technique nano-particles of Pt on Al_2O_3 have been obtained to be used as catalyst in the combustion of some volatile organic compounds [1], colloidal particles and thin films of $NiCl_2$ [2], deposits of metallic halides ($SrCl_2$, SrF_2) [3], particles and nano-structured deposits of Cu on Si substrates [4], formation of colloidal metallic copper and nanometer films on quartz and silicon substrates [5], deposition of colloidal particles and nanostructured films of $CeCl_3$ [6], the preparation of photoluminescent and patterned PVC films, doped with $CeCl_3$ [7] and silver nanoparticles obtained in aqueous solutions at room temperature [8,9].

The mechanism of the photochemical reactivity this method it will be discussed further on.

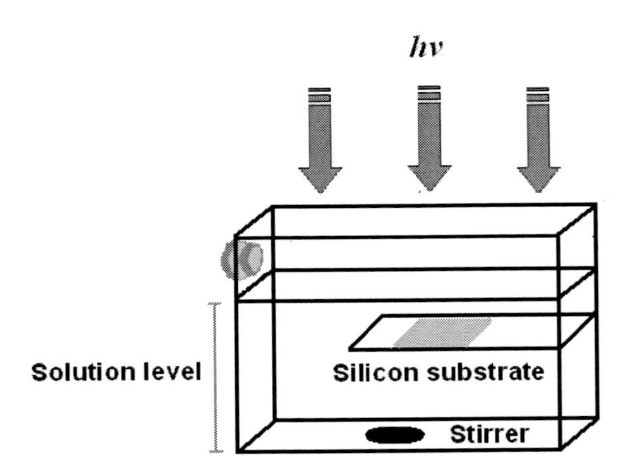

Figure 2. Cuvette for deposition on silicon substrates irradiated from above [3].

The disadvantages of this method, is in the nature of precursors whose limits hardly the deposition in the liquid middle. Only the deposition of metallic halides and particles of some metals have been reported, while there are not antecedents for the deposition of metallic oxides.

Other photo-deposition technique proposed is known as Photochemical deposition technique (PCD). This consists in the deposition from aqueous solutions of materials, a substrate is held in an aqueous solution containing a metals salt, a selenite salt, and a sulphite salt. The solution is illuminated with UV light (a high-pressure mercury lamp) (Fig. 3). The reaction is activated by the illumination and the film is deposited only on the irradiated region of the substrate. Various sulfides and selenides metallic have been deposited by this method such as Cu_xS particles and films [10,11], ZnS deposits [12,13], CdS thin films [14-17], PbS particles and films [18], InS deposits [19], $(Bi,Sb)_2S_3$ and $CuInS_2$ semiconductors films [20,21], ZnSe and CdSe thin films [22,23], oxide metallic such as SnO_2, ZnO thin films [24-26], FeS_xO_y and $ZnS_{1-x}O_x$ compounds [27,28], and metallic alloys of Se_xTe_{1-x} and $Cd_{1-x}Zn_xS$ thin films [29,30].

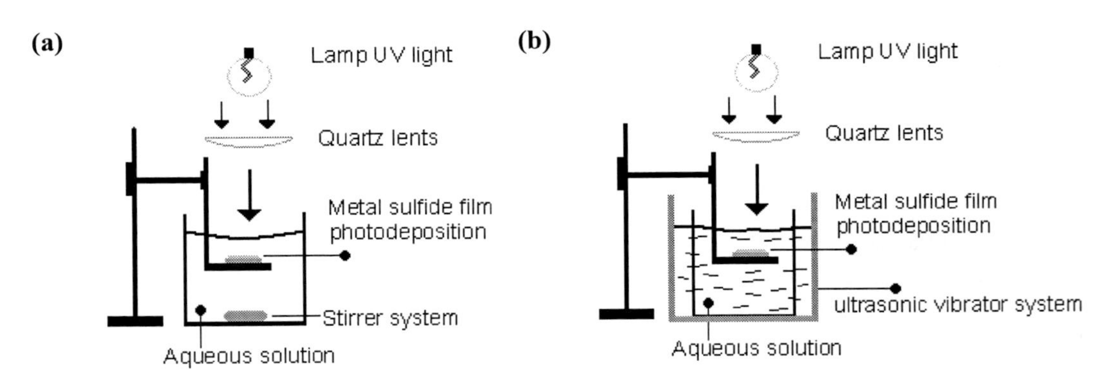

Figure 3. Schematic illustration of the Photochemical deposition (PCD) method: (a) stirrer [9] and (b) ultrasonic vibrator [13].

Thus for example has been proposed [11,13] the deposition of MS films (where M = Zn or Cu) from aqueous solutions that contain MSO_4 and $Na_2S_2O_3$ as precursors salts and H_2SO_4 for adjusting pH of the solution. The chemical reactions of MS formation are as follows. In the acidic solution, $S_2O_3^{2-}$ reacts with H^+ and releases S (Eq 1). On the other hand, the $S_2O_3^{2-}$ species absorbs the UV light with wavelength below 265 nm and is though to release S and electron (Eqs 2 and 3), and then S and electron generated produce to metallic sulfide (Eq 4).

$$2H^+ + S_2O_3^{2-} \rightarrow S^0 + H_2SO_3 \quad \text{(Eq 1)}$$

$$S_2O_3^{2-} + h\nu \rightarrow S^0 + SO_3^{2}{}_{-} \quad \text{(Eq 2)}$$

$$2S_2O_3^{2-} + \text{hv} \rightarrow S_4O_6^{2-} + 2e^- \quad \text{(Eq 3)}$$

$$M^{2+} + S + 2e^- \rightarrow MS \quad \text{(Eq 4)}$$

where M = Zn or Cu

This method has been found successful in depositing high-quality semiconductor thin films. The method seems to be very simple, inexpensive, much easier to scale up and the deposition can be carefully controlled by switching on/off the light. The applicability for the use of any substrates, either conducting or non-conducting and in some cases, the ultrasonic vibration instead of magnetic stirrer can drive chemical reactions such as oxidation, reduction and decomposition as well as in the physical state of the films, such as the surface morphology, microstructure, etc. Thus, this method has better control than the electro-deposition (ECD) and chemical bath deposition (CBD) techniques [15,16].

Nevertheless, this methodology is restricted mainly to sulfide and selenide metal deposition, due to the fact that thiosulfate ions are easily photo-excited and they can interact with the metal ions also present in the solution. A high-pressure mercury lamp (~ 500 watts) is required provide an intense light and in this way to cause the photo-excitation of the species in solution.

Another technique based on photo-excited process is known as Photodeposition from Colloid Solutions (PDCS) [31] (See Fig. 4), can be viewed as a process by which photon-material interaction results at the end in the creation of thin film structures, nanoparticles or clusters of atoms on surfaces, apart from possible bulk particles nucleation in colloid solutions. As in other photochemical deposition systems, a PDCS method incorporates three functions: light "harvesting" agents such as colloids particles, which serve as photon absorbing centers, also known as chromophores, dissolved ions which decompose or react due to the electronic photo-excited state of the chromophores, and a substrate or other adsorbing sites [31]. The chromophores may decompose, serve only as catalysts, be precipitated by photoredox reactions and create solid phases by agglomeration or direct adsorption on surfaces. Thus for example has been proposed the following scheme (Eqs 5-8) of reaction in solution [32], which specifies the interactions between the ion source (M^+), photo-electron (e^-) or photo-hole (h^+), substrate/colloid used and photon beam properties such as wavelength (λ) and fluence (F):

$$\text{Chromophore} \xrightarrow{\lambda\,(nm)\;;\;F\,(J/cm2)} e^- + h^+ \quad \text{(Eq 5)}$$

$$M^+\,(\text{solution}) + e^- \longrightarrow M^0\,(\text{metallic deposit}) \quad \text{(Eq 6)}$$

$$M^+(\text{solution}) + h^+ + H_2O \longrightarrow M_xO_y + H^+\;(\text{oxide deposit}) \quad \text{(Eq 7)}$$

$$H_2O + 2h^+ \longrightarrow 1/2O_2 + 2H \quad \text{(Eq 8)}$$

Where $M^+ = Zn^{2+}$ or Cd^{2+} from ZnS, $ZnSO_3$ and CdS, $CdSO_4$ precursors respectively.

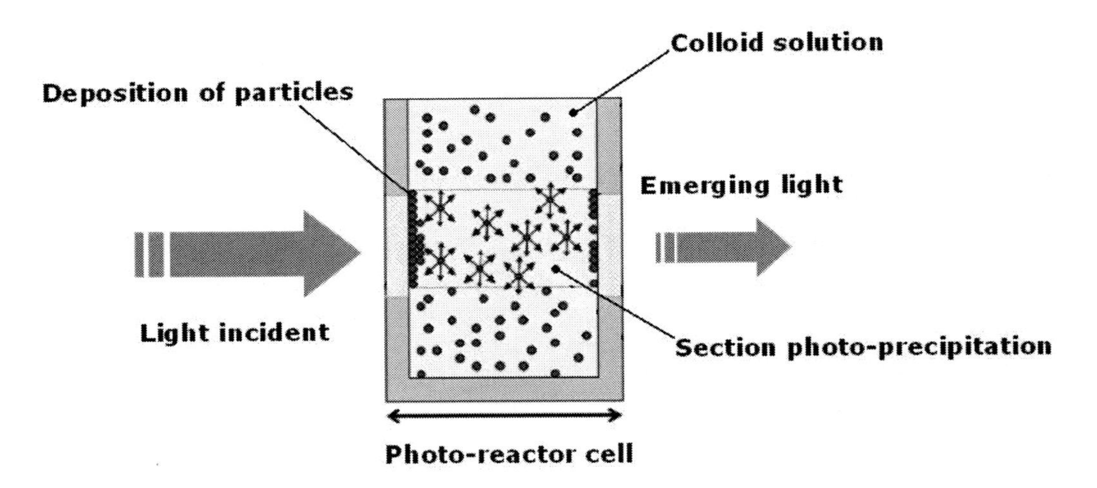

Figure 4. Schematic illustration of the Photodeposition from Colloid Solutions method (PDCS) [30].

Other photochemical methods have been reported very recently in a review which described diverse photochemical synthesis in the nanoparticles preparation [33] and they can be catalogued as a derivation of the PDCS method as example the photoreduction technique, where M^0 forms through the direct excitation of a metal source (metal salts or metal complex), the photosensitized reduction, which uses the photoactive reagents (chromophores) that generate the intermediates by photo irradiation, the intermediates reduce the co-existing metal source to form the M^0. The advantage of photosensitizacion is the fast and efficient formation of metal nanoparticles compared to photoreduction method. Flexibility of the excitation wavelength is an additional benefit because it depends on not the metal source, but the sensitizer [33]. The photocatalytic reduction of metal ions on semiconductor supports (See Fig. 5) has been found useful in designing semiconductor-metal composite catalysts, such as nanoparticles of Au, Ag and Cu has been deposited on TiO_2 as semiconductor support. The photo absorption of semiconductor supports produces positive holes and electrons. The surface-adsorbed metal ions capture the photogenerated electrons and thus get reduced at the interface [33].

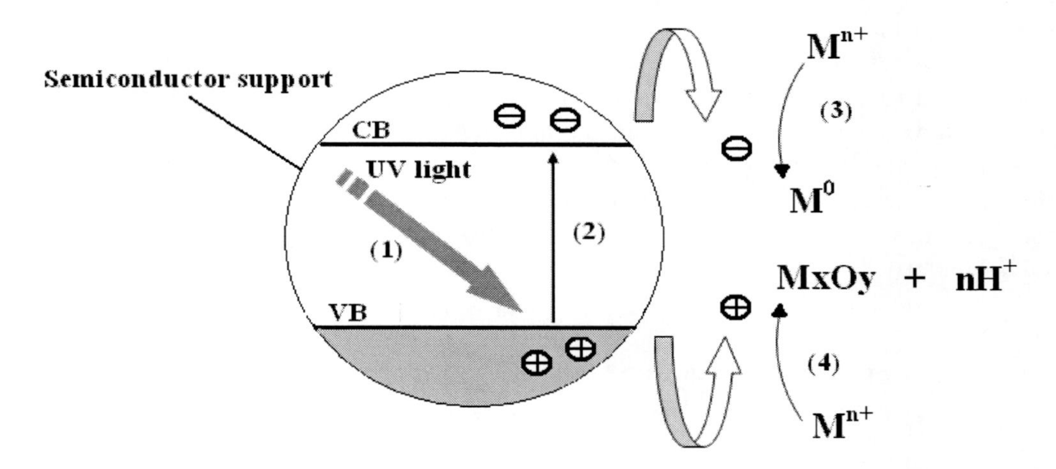

Figure 5. Schematic illustration of the Photocatalytic deposition process: (1) the photo absorption UV of semiconductor support (2) excitation of material and the generation of holes and electrons (3) reduction of metal ions and deposition of metal on semiconductor support (4) oxidation of metal ions or some other species present in the solution.

The last years has also been reported similar methods of deposition photochemistry in solution under diverse denominations: Photochemical solution deposition in the preparation of carbon nanotube into ZnO thin films [34], Photochemical deposition in the formation of silver nanoparticle onto PZT films [35] and silver nanoparticles deposited on the surface of silica spheres [36], Photo-Chemical bath deposition (PCBD) technique applied to Cu/TiO_2 composite films preparation [37], nonselective and selective deposition of Ni thin films by the photochemical reduction [38], preparation of nano-sized Pt^0 catalyst using photo-assisted deposition (PAD) method [39], light assisted chemical deposition of ZnO film [40], Photochemical synthesis of Bi_2Se_3 nanosphere and nanorods [41], Photodeposition of Pt on Colloidal CdS and CdSe/CdS [42], Preparation of Ag/ZnO nanoparticles via photoreduction [43], Photochemical reduction and deposition of metallic ions [44], Photochemical preparation of Au spherical pores [45] are only some representative examples. All These systems of materials deposition are very similar to the methods mentioned previously but with some so much differences in the parameters utilized as in the employed materials.

The development of alternative methods or modifications to already existing methodologies is the subject of permanent concern of a large range of electronic devices. Recently, the attention has been paid in the development of the photo-assisted processes useful in the chemical vapor deposition methods (CVD) [46-57] and sol-gel techniques (SG) [58-67]. Recently has been documented in an review [68] the photon-assisted oxide processing to synthesize ultra-thin oxides at relatively low temperatures, the mechanisms governing photon-assisted oxygen incorporation into growing oxide films, namely, the oxidation rate enhancement as well crystallinity improvement. In addition to these structural changes, functional properties such as electrical, magnetic and optical properties have been found to significantly improve in UV photon treated films [68].

The Photo-deposition in solid phase (thin films) method has been developed for R.H. Hill et al. [69-75]. This method is known as Photochemical metal organic deposition (PMOD). In this method, thin films of inorganic or organometallic precursors upon irradiation are converted to amorphous films of metals or oxides, depending on the reactions conditions. The development of this method requires that the precursor complexes be photo-chemically reactive and form stable amorphous thin films upon spin coating onto a suitable substrate, and that photolysis of these films results in the photo-extrusion of the ligands leaving the inorganic products on the surface (Fig.5). To date this deposition process has been applied for the deposition of a wide variety of metal and metal oxides, and irradiation of the films through a mask results in the generation of patterned films of the metal or metal oxide (photo-lithographic).

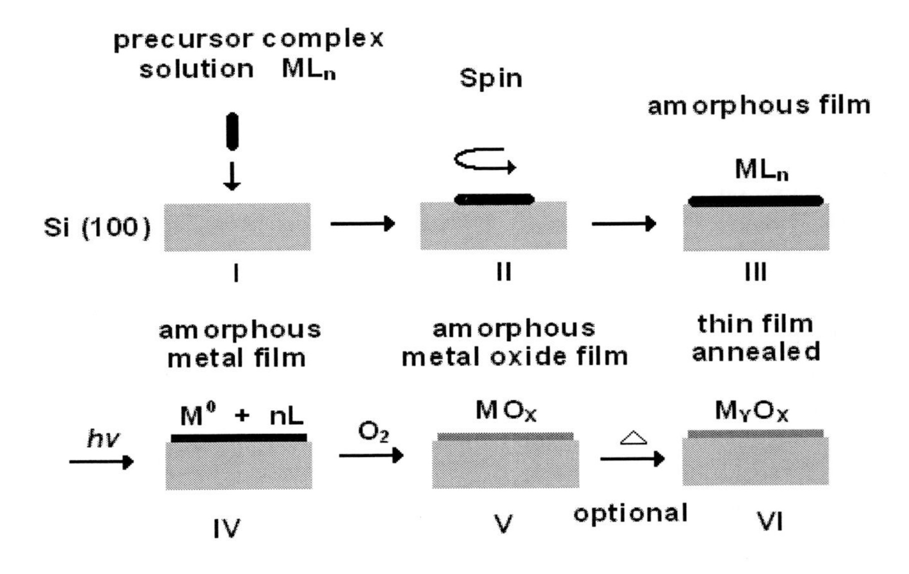

Figure 5. Schematic presentation of deposition steps of the Photochemical metal organic deposition (PMOD) method

In recent years, we have further developed this methodology for the preparation of thin films of $NiCo_X$ [76], CuO [77], SnO_X [78], In_2O_3 [79], ZnO [80], MoO_3 [81], WO_3 [82], ZnS [83] and CdS [84] some these films has been utilized in the evaluation of their gas-sensing capabilities in the presence of pollutant and toxic gases such as CO [85]. Very recently, we have used this method for the preparation of ZrO_2 and HfO_2 thin films [86] and ZrO_2 films doped with lanthanides elements such as Eu^{3+} and Er^{3+} to investigate their luminescent properties [87].

1.2. Photo-Reactivity of the Complex Precursors

As mentioned earlier, a fundamental requirement of this method of photo-deposition in solid phase, is that the precursor complex be photo-chemically reactive.

The metal β-diketonate complexes [$M(acac)_n$] have been know for many years and have been studied with potential applications in the areas of catalysis, magnetism, optical, and electronic devices and thin films. The photo-reactivity of these complexes generates a reduced metal center bound to β-diketone ligand radical.

The photo-reactivity in solution of transition metal β-diketonate complexes has been extensively documented [88,89]. In general terms, the photo-chemical reactivity, is initiated with the excitement of the precursor complex by means of the absorption of a photon that causes a ligand-to-metal charge transfer (LMCT) and subsequently the weakening and breaking of the metal-oxygen bond of the carbonyl group. This gives rise to the formation of radicals species and photo-reduction of metal, that depending on the nature of the complex and of the environmental conditions of the photolysis (nature of the solvent, inert atmosphere or oxygenated atmosphere), can unchain the formation of diverse sub-products. Thus for example has been reported [1] the photo-reactivity in solution of Pt(II) acetylacetonate complexes in ethanolic solutions, the photoreduction of $Pt(acac)_2$ to Pt(I) occurs by irradiation of the ligand-to-metal-transfer-band (LMCT). This causes the homolytic cleavage of Pt—O bond and the formation of the acetyl-acetonyl radical (Hacac) that abstracts one hydrogen atom from solvent and Pt intermediate which fast decomposes in Pt(0), then acetyl-acetonyl radical that abstracts again one hydrogen atom from the solvent giving a molecule of free ligand, Hacac. The ethanol radical can decay trough dismutation to aldehyde and alcohol [5]. The proposed photoreduction mechanism is:

$$Pt(acac)_2 + CH_3CH_2OH \xrightarrow{hv} Pt(acac) + Hacac + CH_3CHOH \quad \text{(Eq 9)}$$

$$Pt(acac) \rightarrow Pt(0) + acac \quad \text{(Eq 10)}$$

$$acac\cdot + CH_3CH_2OH \rightarrow Hacac + CH_3CHOH \quad \text{(Eq 11)}$$

$$2CH_3CHOH \rightarrow CH_3CH_2OH + CH_3CH{=}O \quad \text{(Eq 12)}$$

where: acac = 2,4 pentanedionato ligand ($CH_3COCHCOCH_3$)

Similar results have been observed in $Cu(acac)_2$ complexes [5] in similar conditions, also been studied the photolysis in the formation of halides metallic such as $NiCl_2$ [2], $SrCl_2$, SrF_2 [3] and $CeCl_3$ [6] in chlorinated solvents from β-diketonate complexes. The photochemical behavior of these complexes is typical of a solvent-initiated reaction, whose rate depends on the fraction of light absorbed by the solvent; thus it decreases with the complex concentration, in which the solvent absorbs the UV light to give hemolytic cleavage of C—Cl bond with production of chlorinated radicals leading to HCl and subsequent ligand substitution according to the following general reactions:

$$CH_2Cl_2 + hv \rightarrow CH_2Cl\cdot + Cl \quad \text{(Eq 13)}$$

$$CH_2Cl_2 + Cl\cdot \rightarrow HCl + CHCl_2 \quad \text{(Eq 14)}$$

$$M(acac)_n + nHCl \rightarrow MCl_n + nHacac + subproductos \quad \text{(Eq 15)}$$

where M = Ni^{2+}, Sr^{2+}, Ce^{3+} ions

On the other hand, the lanthanide β-diketonate complexes photo-reactivity has also been extensively studied [90-100] and currently is the subject of much research work, mainly because they present luminescent properties, when they are irradiated under certain conditions. It has been observed that the ligands efficiently absorb the UV light through the π-π*-transition, and the energy absorbed can be transferred to the lanthanide ion, which can emit visible light with more efficiency than the lanthanide ion by itself. The nature of the ligands in the sphere of coordination of these complexes, is of strong importance since they act as "antenna", absorbing and transferring energy efficiently, to the lanthanide ion and thus improving their luminescent performance [92,101]. Nevertheless, great part of these complexes are sensitive to the environment and soluble in organic solvents, and they lack the mechanical properties desired to be utilized directly as luminescent sources. This is why doping of these complexes in adequate solid matrices (host) is extensively studied [91].

It given that the degradation of most of the lanthanide-based complexes under prolonged UV irradiation, in certain cases, only a few hours, an especially the family of some lanthanide β-diketonate complexes, thus for example the complexes of the cations Ln^{3+}, which can be reduced to Ln^{2+}, are expected to display LMCT (ligand-to-metal charge transfer) absorptions at relative low energies. The LMCT states of Ln(III) compounds are apparently not luminescent, but are frequently reactive [102]. Generally, LMCT excitation of metal complexes leads to the reduction of the metal ions and the concomitant oxidation of ligands. In agreement with this notion, the photoreduction of Eu(III) and Sm(III) to Eu(II) and Sm(II) has been reported but a clear relationship to LMCT excitation has been rarely established, on the other hand the oxidation sub-products are often unknown [102]. However this degradation or fragmentation of the lanthanide β-diketonate complexes under UV radiation it has been taken advantage of to propose these complexes as precursors materials in some methods of photo-deposition for the purpose of obtain metallic deposits or metallic oxides nano-structured, depending on the conditions of reaction to that are submitted these complexes in the photo-chemical synthesis.

The photo-activity of several inorganic and organometallic complexes in solid state (thin films) has been investigated by R.H. Hill and et. [103-106]. They have developed a direct patterning method for the lithographic deposition of materials. In this process a precursor film is spun on to the substrate forming an optical quality film. This film is then exposed to light resulting in the conversion of the metalorganic into the desired material. The result of the photochemical step is the loss of the ligands from the metal center, leaving active metal. The metal may either coalesce to form a metallic film or undergo reaction with molecular oxygen resulting in the formation of metal oxide. It has been established that the nature of metallic ion and the ligand determine the reactions routes.

For example the mechanisms of photo-reactivity of In(III) [107] and Ce(III) [108] carboxylate complexes (2-ethylhexanoate) as thin films has been studied, by connecting the photolysis reactor directly to a mass spectrometer. In this way the generation of CO_2 (m/e =

44) has been determined and also two sub-products with m/e of 98 and 100. These two signals are due to a mixture of 2- and 3-heptene and heptane. Subsequently, the oxidation of the metallic ion can take place under low aired conditions. According to the sub-products generated by the irradiation they have proposed possible mechanisms:

$$M(O_2CCH(CH_2CH_3)(CH_2)_3CH_3)_3 \rightarrow M^0 + 3CO_2 + 3/2C_7H_{16} + 3/2C_7H_{14} \quad \text{(Eq 16)}$$

$$M^0 + O_2 \rightarrow M_xO_y \quad \text{(Eq 17)}$$

where: M = In(III) our Ce(III)

In another work [72] they have studied the photo-deposition of patterned aluminum oxide films from of variety of aluminum complexes (β-diketonates and alkoxides β-diketonates). The composition of the films resultant it was analyzed by Auger spectroscopy. One of its conclusions is that the use of simple β-diketonate complexes in this process resulted the contamination by significant of carbon in the composition of the films hat can be attributed to the nature of ligand of complex precursor. The replacement at least one β-diketonate ligand with an alkoxide ligand the resultant purity of the films, as also a greater % of oxygen present (greater degree of oxidation) of the films photo-deposited (Eqs 18 and 19). In this way the nature of ligand in the design of complex precursor is a determinant factor in the quality of the films obtained by this photo-chemical deposition method.

$$Al(acac)_3 \xrightarrow{h\nu} Al_2O_{3-X\,\text{(thin film)}} + \text{sub-products} \quad \text{(Eq 18)}$$

$$Al(EtOacac)_3 \xrightarrow{h\nu} Al_2O_{3\,\text{(thin film)}} + \text{sub-products} \quad \text{(Eq 19)}$$

where: acac: (β-diketonate ligand) $CH_3COCHCOCH_3$

EtOacac: (alkoxide ligand) $(CH_3CH_2O)COCHCOCH_3$

Given the photo-reactivity of these complexes in the preparation of nano-structured materials, we proposed to use of β-diketonate complexes as precursors in the photo-deposition of ZnO and HfO_2 thin films, and their evaluation as chemical sensor and photo-luminescent materials respectively.

2. Some Materials Photo-Deposited Proposed

2.1. Zno Thin Films and Their use as Gas Sensor

Since Seiyama and Taguchi used the dependence of the electrical conductivity of ZnO on the gas present on the atmosphere for gas sensing applications [109,110], many types metal oxide semiconductors such as SnO_2 [111,112], TiO_2 [113,114], WO_3 [115,116] and In_2O_3 [117,118] have been proposed as materials for gas detection.

The working temperature at which these devices are more efficient can vary depending on the gas atmosphere and on properties of the sensor material selected in very case. As these temperatures from 200 to 800 °C far from room temperature, it is necessary to implement a heating system in sensor devices. Among them, ZnO was one of the earliest sensors developed, because of its working temperature (usually in the range of 400-500 °C) and poor selectivity, ZnO sensors have not been widely used [119]. In contrast, SnO_2 sensors have found many applications in diverse fields, because of its high gas sensitivity and its morphological and chemical stability [120], but they also suffer from poor selectivity [119]. The gas sensor based on ZnO had been developed for detection and control of the gases such as CO [121,122], H_2 [121,123], NH_3 [124,125], H_2S [126,127], NO_2 [128,129], volatile organic compounds (VOCs) [130] and liquefied petroleum gas (LPG) [131]. Different additives are added/incorporated to improve their sensitivity and selectivity, and decrease the response time and the operating temperature of semiconductor sensor. All these effects arise of the increase of the catalytic activity that is produced upon adding noble metals. Most of the gas sensors developed are doped with noble metals such as Pt [132,133] and Pd [134,135].

Also, it has been demonstrated the hetero-contact with other metal oxides such as CuO [136], Fe_2O_3 [137], SnO_2 [138] and rare metal oxide into ZnO can also enhance the sensitivity and selectivity of the gas sensors.

With the advent of modern nanotechnologies, zinc oxide has been rediscovered [139,140]. In particular, a large variety of nanostructures in form of nanoparticles [141,142], nanowires [143,144], nanobelts [145,146] nanorods or nanotetrapods [147,148] have been synthesized, i.e., with morphologies ideal for gas sensing.

In this sense we have proposed us to deposit of ZnO thin films doped with some additives such as Pt and Pd and to evaluate the quality of the deposits obtained by photochemical method in solid phase and evaluate its potential use as chemical sensor in the detection of CO.

2.2. Evaluation of Characteristics Luminescence of HfO_2-Ln Thin Films (where: Ln = Er or Eu)

The application of compounds contain trivalent rare earth ions RE^{3+} is based mainly on their photoluminescence properties such as TiO_2-Eu^{3+} [149,150], TiO_2-Er^{3+} [151,152], ZrO-Eu^{3+} [153,154], ZrO_2-Er^{3+} [155,156], HfO_2-Eu^{3+} [157,158] and SiO_2-HfO_2-Er^{3+} [159,160] materials. Luminescence investigation of RE^{3+} doped in metal oxide matrices has increased in recent years due to the efficiency of energy transfer from material host to rare earth ion, thus improving their luminescent performance.

During recent years, there has been a growing interest in the study of the physical and chemical properties of hafnium dioxide (HfO_2) due to its important technological applications. Among them their rather large energy gap and low phonon frequencies the matrices are expected to be suitable hosts for RE activators and due to its hardness and transparent spectral range the infrared (10 μm) to the ultraviolet (below 300 nm) [161]. HfO_2 is one of the most commonly used high optical index (~ 2) coating materials for optical components. Although the optical properties of HfO_2 thin films have been studied at a variety of wavelengths, recent research mainly focuses on its crystalline structure. Moreover, the high density of HfO_2 more particularly when crystallized into to tetragonal phase (d ~ 10 g/cc), makes this compound attractive for host lattice activated by RE^{3+} for optical applications.

However, the efficient RE^{3+} dopant and the phase of the lattice that makes it optical properties must be determined [162].

In this context, the present work intents to apply this photochemical methodology in the preparation of HfO_2 thin films doped with Er^{3+} or Eu^{3+} to investigate its luminescent properties.

3. Experimental Details

3.1. General Procedure

Fourier Transform Infrared spectra (FT-IR) were obtained with 4 cm^{-1} resolution in a Perkin Elmer Model 1605 FT–IR spectrophotometer.

UV spectra were obtained with 1 nm resolution in a Perkin Elmer Model Lambda 25 UV–Vis spectrophotometer.

X-ray photoelectron spectra (XPS) were recorded on an XPS–Auger Perkin Elmer electron spectrometer Model PHI 1257 which included an ultra high vacuum chamber, a hemispherical electron energy analyzer and an X–ray source providing unfiltered Kα radiation from its Al anode ($h\nu$ = 1486.6 eV). The pressure of the main spectrometer chamber during data acquisition was maintained at ca 10^{-7} Pa. The binding energy (BE) scale was calibrated by using the peak of adventitious carbon, setting it to 284.6 eV. The accuracy of the BE scale was ± 0.1 eV.

The solid state photolysis was carried out at room temperature under a low–pressure Hg lamp (λ = 254 nm) equipped with two 6W tubes, in an air atmosphere. Progress of the reactions was monitored by determining the FT-IR spectra at different time intervals, following the decrease in IR absorption of the complexes.

The substrates for deposition of films were borosilicate glass microslides (Fischer, 2x2 cm) and n-type silicon(100) wafers (1x1 cm) obtained from Wafer World Inc, Florida, USA.

3.2. Preparation of Amorphous Thin Films

3.2.1. Preparation of the Precursors B-Dikentonate Complexes

The synthesis of the Bis-(1-phenyl-1,3-butanodionate)M(II), where M = Zn, Pd or Pt, it has been described and characterized in to works previous. For the case of Zn(II) complexes it has been utilized as precursor complex in the photo-deposition of ZnO thin films [80] and in turn the Pd(II) and Pt(II) complexes has been used as additive in the photochemical deposition of SnO_X-Pd and SnO_X-Pt thin films [78].

The precursors Hf(IV) and Ln(III) acetylacetonate complexes, where Ln= Er or Eu, were purchased from Aldrich Chemical Company and thin films of precursors complexes were prepared by the following procedure: A silicon chip was placed on a spin coater and rotated at a speed of 1500 RPM. A portion (0.1ml) of a solution of the precursor complex in CH_2Cl_2 was dispensed onto the silicon chip and allowed to spread. The motor was then stopped after 30 s and a thin film of the complex remained on the chip. The quality of the films was examined by optical microscopy (1000x magnification).

3.2.2. Photolysis of Complexes as Films on Si (100) Surfaces

All photolysis experiments were done following the same procedure. Here is the description of a typical experiment. A film of the complex was deposited on n-type Si(100) by spin-coating from a CH_2Cl_2 solution. This resulted in the formation of a smooth, uniform coating on the chip. The quality (uniformity, defects, etc.) of the precursor films was determined by optical microscopy (500x), while the thickness was monitored by interferometry. The FT-IR spectrum of the starting film was first obtained. The irradiation of the films, was carried out at room temperature using two low-pressure Hg lamps (6 watts, Rayonet RPR–2537 A), in air atmosphere, until the FT-IR spectrum showed no evidence of the starting material. Prior to analysis the chip was rinsed several times with dry acetone to remove any organic products remaining on the surface. In order to obtain films of a specific thickness, successive layers of the precursors were deposited by spin-coating and irradiated as above. This process was repeated several times until the desired thickness was achieved. Post-annealing was carried out under a continuous flow of synthetic air at 600 °C for 2 h. in a programmable Lindberg tube furnace.

3.3. Evaluation of Gas-Sensing Properties of ZnO-M Films (where M= Pd or Pt)

The thin films on Si(100) were located in a platform of ceramics of 24-pines to permit the electric connections. This platform was introduced in a system of flow that consists of an oven Lindberg Blue programmable, in which a tube of quartz placed was located horizontally, the one that possesses emery unions in both extremes through which they associate the exit and entrance of the gases as also the electric connections. To the entrance of this system, a controller was connected and simple channel flow meter MKS type 167-TO to permit the inlet of the gas contaminant, which this provided of a valve that permits to expose or to inhibit the film al contaminant. The electric connections originating from the device finish in the design of a circuit, which has its components in series, and that permit to measure in form simultaneous , by means of two multimeters, voltage and current that circulates through the system. One of the multimeters, this provided of a data acquisition card that permits to store the registrations and then interface to a PC, for the processing of the data (Fig 6).

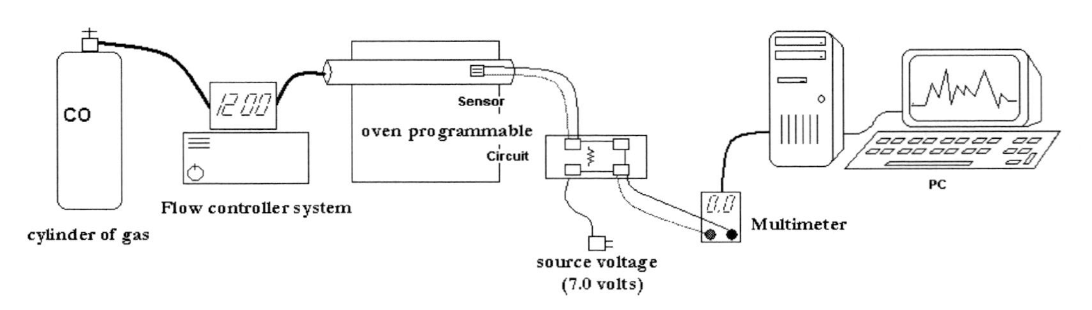

Figure 6. Schematic system experimental utilized in the sensory evaluation of the devices.

3.4. Evaluation of the Characteristics Luminescent of the HfO_2-Ln Films (where Ln= Er or Eu)

Photoluminescence (PL) emission spectra measurements were carry out in a multifrequency phase fluorometer (K2, ISS Inc., Urbana, Champaign, IL, USA) with a L type setup. Excitation was performed with a 400W UV Xe arc lamp. Excitation monochromator set at 300 nm for HfO_2-Eu and 800 nm for HfO_2-Er films which they were taken at room temperature.

4. Results and Discussion

4.1. Characterization of ZnO-Pd and ZnO-Pt Thins Films

Precursor films were prepared dissolving Bis(1-phenyl-1,3-butanodionato)Zn(II) complex with 10% mol of Bis(1-phenyl-1,3-butanodionato)M(II) complex (where M= Pt or Pd) with respect to Zn complex in CH_2Cl_2 and spin-coating at 1500 rpm on a silicon(100) substrate. After examination of the films by optical microscopy (500x magnification), they were irradiated under a UV light (254 nm) for 48 hrs under air atmosphere. Post-annealing was carried out under a continuous flow of synthetic air at 600 °C for 2 h. The post-annealing is required due have a positive effect in improving the long-term stability of the sensor.

The evaluation of the photoreactivity of the precursors of Zn(II)-Pd(II) and Zn(II)-Pt(II) β-diketonate complexes was deposited on Si(100) and were irradiated, the photolysis was monitored by FT-IR spectroscopy, (Fig 7). It was observed that the band at 1520 cm^{-1} aprox. associated with the carbonyl group of the ligand, decrease in intensity, and after 24 h of irradiation only minimal absorptions in the infrared spectrum remains. These results suggest that the diketonate groups on the precursor material are photodissociated on the surface, forming volatile products which are partially desorbed.

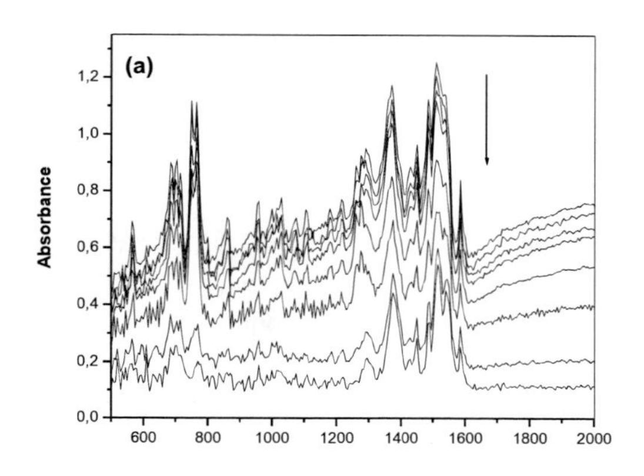

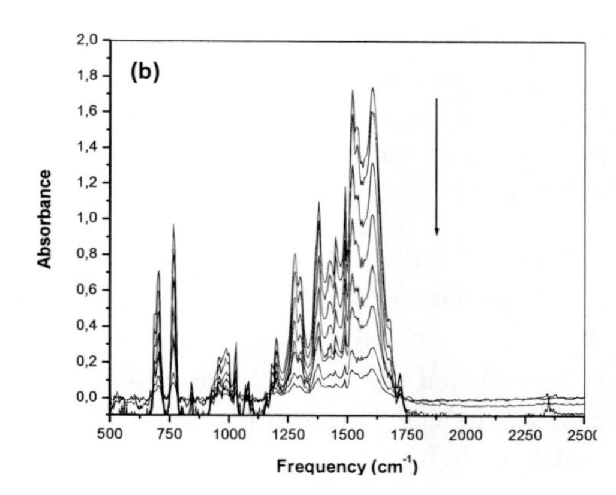

Figure 7. FT–IR spectral changes associated with photolysis (a) for 24 h of film of Bis(1-phenyl-1,3-butanodionato)Zn(II) and 10% mol of Bis(1-phenyl-1,3-butanodionato)Pd(II) (b) 24 h a film of Bis(1-phenyl-1,3-butanodionato)Zn(II) and 10% mol of Bis(1-phenyl-1,3-butanodionato)Pt(II) deposited on Si (100).

The chemical composition of the ZnO-M (M = Pd or Pt) thin films annealed was analyzed by XPS (Fig 8). The most representative signals shows the ZnO formation correspond to signals Zn $2p_{1/2}$ located at 1044.9 eV and Zn $2p_{3/2}$ located at 1022.1 eV and finally a small peak Zn 3d situated at 10.8 eV. That signal Zn $2p_{3/2}$ at 1022.1 eV it establishes the formation of ZnO [163-165].

The presence of palladium itself due to the signs Pd 3d which consists of two peaks (Fig. 8a), one of them to lower energy at 335-335.4 eV that establishes the presence of palladium metallic and another signal a higher energy at 336.3-336.8 eV corresponding to palladium (II) as PdO [166,167]. In our case one of the signals appears at 336.9 eV, associate to the formation of PdO and another located at 335.3 eV pertaining to the formation of Pd elemental state.

On the other hand, it has been reported [168-170] that the locating of the Pt $4f_{7/2}$ peak corresponding a the platinum presence, it as been determined that Pt $4f_{7/2}$ peak located at 70.7-71.2 eV corresponds to Pt metallic, that this peak displace to higher energy among 72.7-73.4 it can be assigned a Pt(II) as PtO and when this peak is located among 74.2-75 eV it can be assigned a Pt(IV) as PtO_2. In our results, the Pt $4f_{7/2}$ signal (Fig 8b) is located at 71.6 eV that can be attributed to the presence of Pt metallic only.

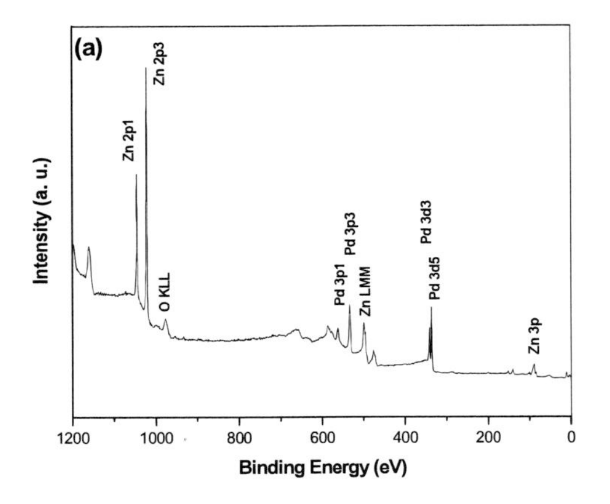

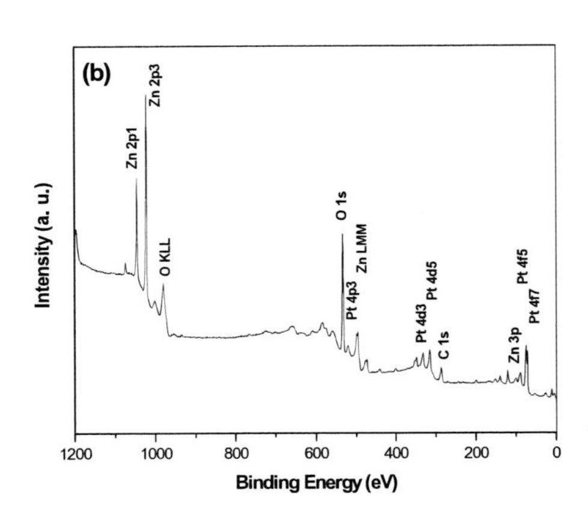

Figure 8. XPS survey scan of (a) Pd-doped ZnO thin film obtained by irradiation of a Bis(1-phenyl-1,3-butanedionato)Zn(II) and 10% mol of Bis(1-phenyl-1,3-butanedionato)Pd(II), (b) Pt-doped ZnO thin film obtained by irradiation of a Bis(1-phenyl-1,3-butanedionato)Zn(II) and 10% mol of Bis(1-phenyl-1,3-butanedionato)Pt(II) film on Si(100).

The examination of the surface morphology of the ZnO-Pd and ZnO-Pt films was reveled for AFM technique. In the Fig. 9a is exhibited the image ZnO-Pd as-deposited. It is possible to observe an amorphous deposition non-uniform, in which some random extensions are exhibited that can attributed to clusters of palladium and that contribute to roughness of the as-deposit. The values of rms specific was of 20 nm and maximum heigh calculated R_{max} was of 190 nm.

On the other hand, the ZnO-Pt films it observe a surface amorphous but with homogeneous distribution with porous characteristics and the presence of some small extensions that can be attributed to the formation of small clusters of platinum.(Fig 9b). The values of rms 17.7 nm and the maximum height R_{max} of 178 nm they are similar to the as-deposited Pd-doped ZnO films. Nevertheless the homogeneity and porous distribution the presents the as-deposited ZnO-Pt films, they do of these deposits more than adequate for their utilization as chemical sensors.

Usually, high sensitivity and quick response can be expected for a sensor made from porous materials and uniform distribution of additive on surface, because it has a comparatively large surface area. In addition, the gases will quickly diffuse into the pores and with catalyst additive and the surface reactions will take place at a higher rate [171,172].

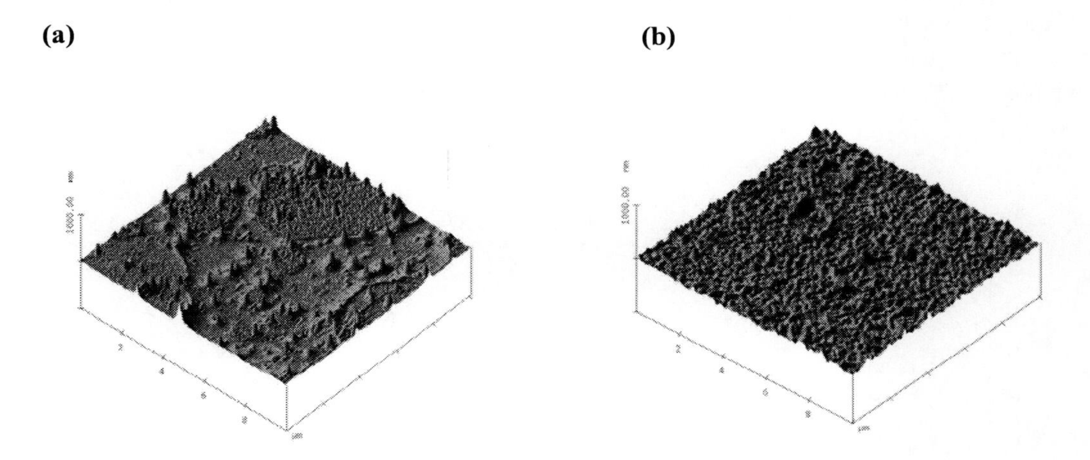

Figure 9. AFM micrography (10x10 μm) of (a) Pd doped ZnO and (b) Pt doped ZnO thin film.

4.2. Preliminary Evaluation of the Gas Sensitivity of the ZnO and ZnO-Pt Thin Films

The study of the sensitivity capacity of the ZnO and ZnO-Pt thin films it has been registered the electrical current when these films are submitted to a flow of 20 ppm CO gas, to different operation temperature in an interval of 200-400 °C. Factors such as response time, recovery time and sensitivity were considered to evaluate sensorial capacity of the films photo-deposited.

The sensitivity (S) has been defined as $S = [R_{gas}-R_N]/R_{gas} \times 100$ where R_N and R_{gas} are resistances of the films in nitrogenous atmosphere and in the target gas contaminant respectively. All the measurements were carried out under a constant voltage of 7.0 volts. Before exposing to CO gas, the ZnO and ZnO-Pt films were allowed to be stable for electrical current for 30 min which was kept in a small oven maintained at a constant temperature and the stabilized resistance was taken as R_N. CO gas diluted with high purity N_2 was passed through the test chamber at a flow rate of 1500-1700 mL/min, during 30 seconds, and to observe the variation of electrical current. During this process, gas CO was repeatedly introduced into the chamber and then evacuated. Some parameters such as response time defined as the time in which responding of device when exposed to CO gas, and time recovery defined as the time in which the device returning to its values initial of electrical current or values initial resistance after to have exposed to CO gas.

Fig. 10a shows the responses to 20 ppm of CO gas of the ZnO films, the optimum working temperature it is achieved at 300 and 350 °C. To higher or lower temperatures is not possible to obtain stable responses, for example at 250 °C not significantly changes in the electrical current registered. Probably the detection of low concentrations of gas is required to stimulate thermally the semiconductor. On the hand, the increment of the temperature to 400 °C, causes that the desorption process of the CO gas on the surface of film and consequently the changes of conductivity are minimums. The values of sensitivity calculated for ZnO films were 27% to 300 °C and 17% to 350 °C.

On the other hand, the responses test of ZnO-Pt films (see Fig. 10b), show that the responses to 20 ppm CO gas is significantly higher than that the ZnO films in similar conditions, and respond in a great interval of temperature (250-400 °C). The details of these results are show table 1.

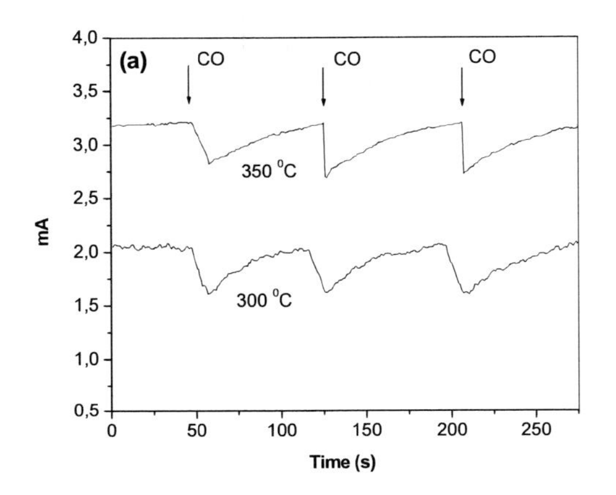

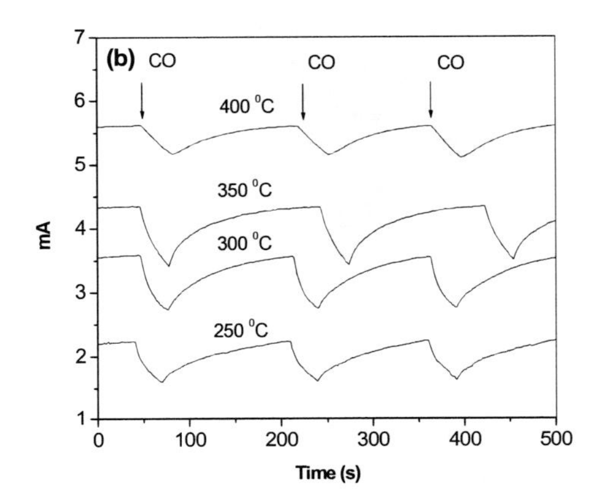

Figure 10. Sensing performance by 30 seconds period of 20 ppm CO gas (a) ZnO and (b) ZnO-Pt thin films at different working temperature.

Nevertheless it is observed that the response and the recovery times, they are greater with regard to the ZnO deposits, the major response time and time recovery this associate to that the ZnO-Pt deposits present a greater change in the conductivity when is exposed to gas contaminant, probably because now a greater interaction with the gas and with it exists a greater time in the responses test.

According to results presented, that the contribution of additive added on ZnO films, promote gas sensitivity, and has been generate a change morphologic of porous type with some extensions homogeneous distributed in the deposited, attributed the additive. This change to superficial level involves an increase in the roughness of the ZnO-Pt films compared to the ZnO deposits [80], other authors has been documented [173,174] that the increase in the roughness favors the interaction or diffusion of gas over surface of film, because it has a comparatively large surface area. Parallel this new morphology disposition is also responsible the stability of the films doped because extends to great working temperature interval.

Table 1. Sensitivity, response time and recovery time of undoped ZnO and Pt-doped ZnO thin films

Films	Temperature (^{o}C)	[CO] ppm	Time (± 1.0 s) response	Time (± 1.0 s) recovery	Sensitivity (%)
ZnO	300	20	12	62	20
ZnO	350	20	12	72	13
ZnO-Pt	250	20	34	220	22
ZnO-Pt	300	20	32	160	27
ZnO-Pt	350	20	32	146	17
ZnO-Pt	400	20	26	157	8

4.3. Characterization of HfO_2-Ln Photodeposited Thin Films (where Ln = Eu or Er)

The photoreactivity of $Hf(acac)_4$ complex has been reported previously for the preparation of HfO_2 thin films [86], as also the evaluation of the photodeposition and the study of the luminescence characteristics of the ZrO_2-Ln thin films (where Ln= Eu or Er) [87]. This time we have proposed us to conclude our luminescent studies utilizing HfO_2 thin films as material host loaded with erbium and europium.

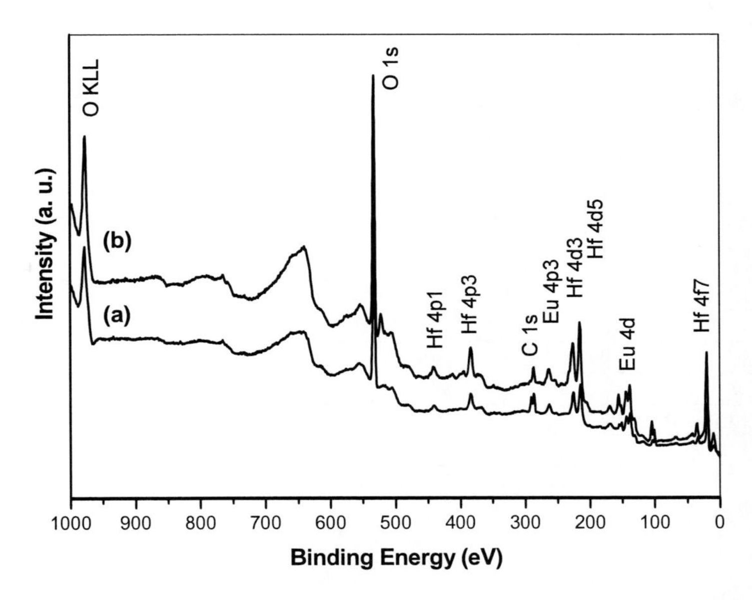

Figure 11. XPS survey spectrum of (a) an as-deposited and (b) annealed at 600 °C HfO_2-Eu thin films prepared by UV irradiation at 254 nm deposited on Si (100).

For the deposition of HfO_2-Ln thin films, solutions of $Hf(acac)_4$ with different proportions of the $Ln(acac)_3$ complexes (10, 30 and 50 mol%) were spin-coated on the Si(100) substrate and the films irradiated until no IR absorptions due to the precursors complexes were observed.

The chemical compositional changes of the HfO_2-Ln thin films, both as-deposited and annealed at 600 °C were investigated by XPS. (See Fig 11 and 12). In general for the as-deposited and annealed thin films, spectra that O, Hf and C were present. Our measured binding energy values show in table 2.

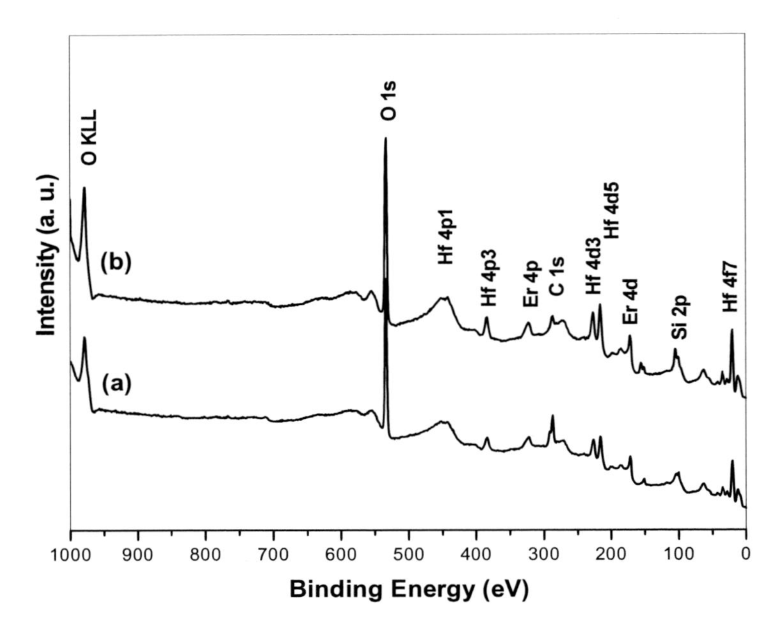

Figure 12. XPS survey spectrum of (a) an as-deposited and (b) annealed at 600 °C HfO_2-Er thin films prepared by UV irradiation at 254 nm deposited on Si (100).

Table 2. XPS results (binding energies, eV) for the HfO_2-Ln thin films. (Where Ln = Eu our Er)

Thin films	Hf $4d_{5/2}$	Hf $4f_{7/2}$	O 1s	Er $4d_{5/2}$	Eu $3d_{5/2}$
HfO_2-Eu[a]	212.6 ± 0.1	16.7 ± 0.1	529.8 ± 0.1		1135.0 ± 0.1
			531.4 ± 0.1		1124.8 ± 0.1
HfO_2-Eu[b]	213.1 ± 0.1	16.8 ± 0.1	530.2 ± 0.1		1134.9 ± 0.1
			531.8 ± 0.1		1124.0 ± 0.1
HfO_2-Er[a]	213.4 ± 0.1	16.8 ± 0.1	529.8 ± 0.1	168.8 ± 0.1	
			531.7 ± 0.1		
HfO_2-Er[b]	213.4 ± 0.1	17.0 ± 0.1	530.4 ± 0.1	168.7 ± 0.1	
			531.9 ± 0.1		

[a] as-deposited

[b] annealed at 600 °C for 2h.

As can be seen in the XPS spectra that the signal associated to Hf as an oxide corresponds to Hf $4d_{5/2}$ and Hf $4d_{3/2}$ peaks which are located at 212 eV and 223 eV. The signal located at 212 eV can be assigned to the presence of HfO_2 formation in the thin films photodeposited [175].

On the other hand, has been found the Hf 4f doublet signal at 16.8 eV (Hf $4f_{7/2}$) and 18.2 eV (Hf $4f_{5/2}$). The peak at 16.7 eV which can be attributed to Hf—O bonds corresponds HfO_2 formation [176].

The O 1s signal positions are at about 530 eV, the high resolution spectrum indicates that two subpeaks located at 530 and 531 eV. The first subpeak located at a lower energy can be assigned to lattice oxygen corresponds of Hf—O bonds in the lattice, while the second one, located at higher energy, assigned an ionization of oxygen species corresponds to oxygen weakly adsorbed. The subpeak attributed to weakly adsorbed oxygen finally weakened significantly after annealing at 600 °C.

For HfO_2-Eu thin films the Eu $3d_{5/2}$ signal (Fig. 13a) shows two well resolved peaks located at 1134 and 1124 eV binding energy. The higher binding energy peak is related to the Eu^{3+} species and lower binding energy peak corresponds to Eu^{2+} species [177]. The relative contribution of these two components to the spectrum as determined by fitting is respectively 42% associated Eu^{3+} and 58% associated Eu^{2+} for the as-deposited thin films. While for annealed at 600 °C samples the relative contribution of Eu^{3+} and Eu^{2+} states is a 91% and 9% respectively. This change in the % of contribution responds fully oxidized of Eu^{2+} to Eu^{3+} by post-deposition annealing.

On the other hand, for the HfO_2-Er thin films present a the broad signal associated Er 4 $d_{5/2}$ (Fig 13b) located at a binding energy of 168.8 and 168.7 eV for as-deposited and annealed thin films respectively, these results corresponds to Er(III) species [178,179].

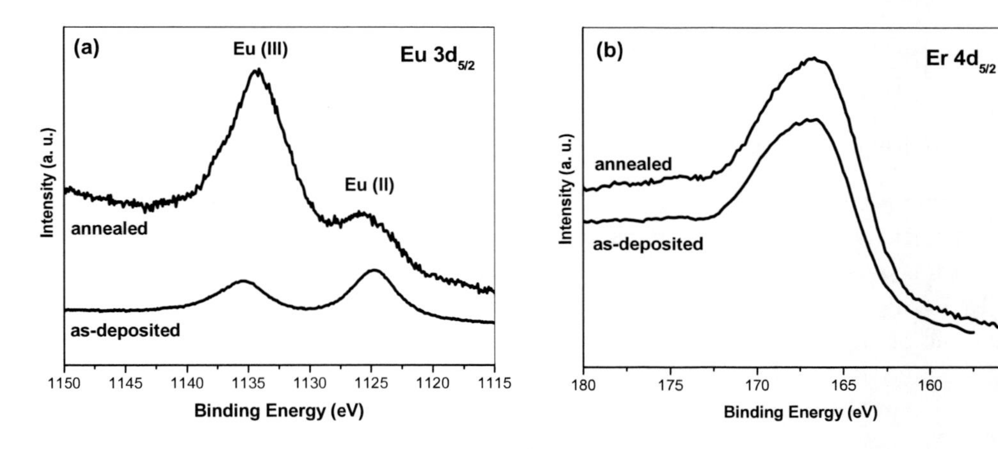

Figure 13. XPS spectra of (a) the Eu 3d and (b) the Er 4d for HfO_2-Ln thin films as-deposited and annealed at 600 °C for 2 h.

4.4. Photoluminescence Study of HfO_2-Ln Thin Films

The excitation spectra of as-deposited and annealed of HfO_2-Eu thin films are show in Fig 14. These excitation spectra were obtained by monitoring the emission at 592 nm corresponds to $^5D_0 \rightarrow {}^7F_1$ transition from Eu^{3+} [180,181]. The excitation spectra consist of a broad band principal at about 300 nm and two small signals located at 396 and 460 nm. The excitation band at 300 nm which can be assigned to charge transfer (CT) band between host lattice oxygen atoms and Eu^{3+} ions. The other bands that are located at 396 and 460 nm are assigned to the electronic transitions of $^7F_0 \rightarrow {^5L_6}$ and $^7F_0 \rightarrow {^5D_2}$ transitions respectively [182,183]. The maximum of the excitation band of sample annealed is at 296 nm, corresponding to a shift small toward lower wavelength with respect to that of as-deposited sample. The charge transfer position varies as a function of the host lattice and the variation is proportional to the Eu—O distance. With increasing bond length, the band shifts to lower energy [184].

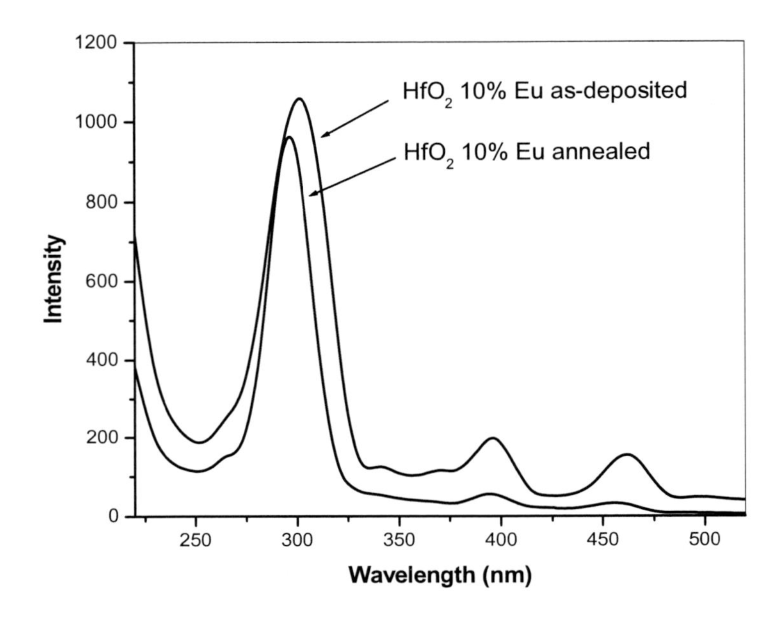

Figure 14. Excitation spectra of as-deposited and annealed of the HfO_2-10% Eu sample.

The emission spectra of Eu doped HfO_2 films excited at 300 nn are show in Fig 15. In general the characteristic emission bands of the Eu^{3+} have reported, corresponds to the following transitions: $^5D_0 \rightarrow {^7F_0}$ (550-580 nm), $^5D_0 \rightarrow {^7F_1}$ (585-600 nm), $^5D_0 \rightarrow {^7F_2}$ (605-630 nm), $^5D_0 \rightarrow {^7F_3}$ (635-670 nm) and $^5D_0 \rightarrow {^7F_4}$ (675-715 nm) aprox. [180,181,185,186]. This case we can be observed an only band between at 580-600 nm with a maximum at 592 nm (orange-red). It is possible to observe the dependence of the PL intensity spectra on Eu concentration of HfO_2 thin films reaches a maximum when the concentration is 10% mol Eu, and decrease gradually with further increase in Eu concentration, probably due to concentration quenching (self-quenching).

On the other hand, the emission intensity signal of as-deposited samples (Fig. 15a) more than higher in comparison with the signal from annealed samples (Fig. 15b). We attribute this phenomenon in the lower intensity is the fact that surface reconstruction of the annealed films may cause the formation of some clusters of Eu^{3+} ions which lead to concentration quenching of Eu^{3+} luminescence, nevertheless this aspect that deserve a subsequent study.

Recently has been reported [187] that the water molecules in layered oxides could help to improve the energy transfer from layered oxide to lanthanide ions, the water molecules in the interlayer can act "water-bridge" improve the energy transfer from Hf—O of material host to lanthanide ions. When the samples are annealed, the host band-gap excitation efficiency decrease because the "water-bridge" is destroyed, and consequently the luminescent performance diminishes.

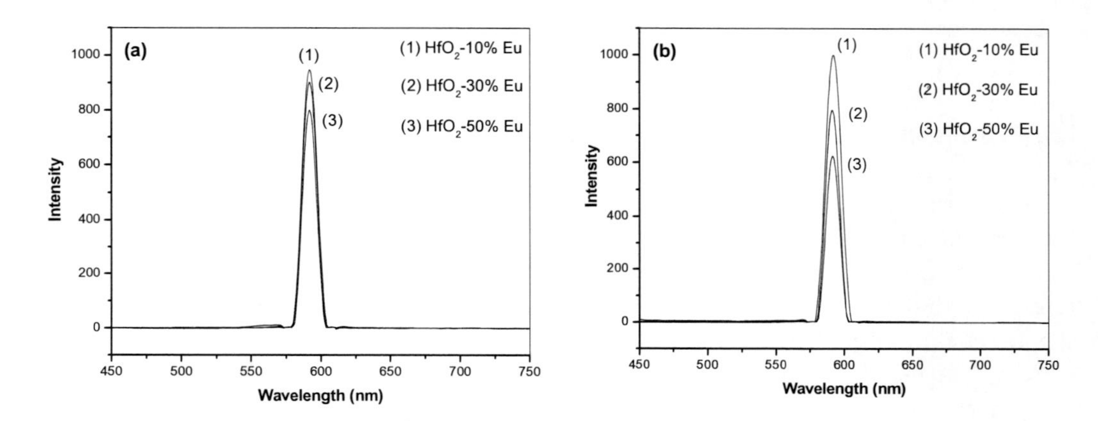

Figure 15. Emission spectra of (a) as-deposited and (b) annealed of the HfO_2 loaded with Eu thin films, excited at 300 nm.

The results of some studies of Er^{3+} ion reveal high efficiency of the emissions around 500-700 nm under 800 and 980 nm excitations [188-193]. The PL studies obtained of HfO_2-Er thin films after excitation at 800 nm show one visible emission broad band with a maximum located at 530 nm aprox, as the result upconversion process (Fig. 16). This band is assigned to $^2H_{11/2} \rightarrow {}^4I_{15/2}$ transition of the Er ion and is result of the absorption of two photons either from pumping or the energy transfer between neighboring ions [193].

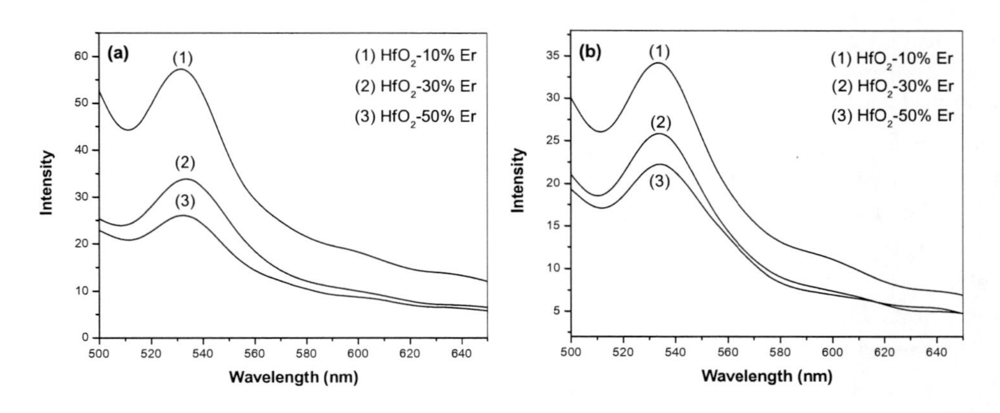

Figure 16. Emission spectra of (a) as-deposited and (b) annealed of the HfO_2 loaded with Er thin films, excited at 800 nm.

Just as in the previous case the intensity of the emissions diminishes when the concentration of erbium increases, our studies show that the intensities of the signs reach its maximum when the concentration is 10% mol of the lanthanide ion on the material host. Of this way the lanthanide ion concentrations generally are low in view of the fact that at higher concentrations the efficiency of the luminescence process usually decreases (concentration quenching).

On the other hand is possible to observe a decrease of the intensities of the signs of annealed samples in comparison with the signal from deposited samples. This lower of intensity can be associated the formation of some clusters of Er^{3+} ions, that they contribute to concentration quenching.

4.5. Proposed Photoluminescence Mechanism

This case of the HfO_2 films as material host act as an antenna which absorbs light greather than or equal in energy to its band gap. The energy is then transferred to lanthanide ion, which exhibit the characteristic color luminescence arising from f-f crystal field transitions. In the case HfO_2-Eu thin films, the λ_{max} excitation was at 300 nm (4.1 eV) in our samples at a lower energy than the band gap determined for unloaded HfO_2 thin films at ~ 5.5 eV [86]. The incorporation of lanthanides ions into HfO_2 thin films produce impurity levels between the intrinsic bands which subsequently generate new band gaps. Of this way depending on the nature of the lanthanide ions it affects of intrinsic defects of the material host, the energy levels in this case HfO_2 varied. Various types of defects have show to play an important role in energy transfer between other materials such as the semiconductors III-IV and Si and rare earth ions [193].

The energy level of these defects is critical to the energy transfer process because if it is lower in energy than the emitting state of the lanthanide ion, no sensitized luminescence can be observed. By using the well-know energies of the emitting states of the lanthanides it is possible to verify the lower limit for the energy level of the trap states in the material host that transfer energy to the ions [193].

We have maintained that the nature of the defects centers in loaded HfO_2 influence the optical properties of these films, to explain those luminescence properties, we have responded to the modified model proposed by Frindell et al. [193] and other authors [194,195], considering that these phenomena involves three factors: i) intensity of material host absorption, ii) efficiency of material host-lanthanide ion (activator) energy transfer and iii) efficiency of lanthanide ion emission.

For the case HfO_2-Eu thin films, we have proposed the following diagram (Fig 17a): (1) the UV light absorption for defects from HfO_2 host (2) followed the non-radiative energy transfer process from HfO_2 defect to $^5D_0$ excited energy level of Eu and (3) the emission process from $^5D_0$ to $^7F_1$ level of Eu ion, what is translated in an only signal which is observed.

The mechanism of the upconverted emission of Er doped in matrices various, has been well documented in diverse articles [193,196-198]. In the Fig. 17b are show a energy-level diagram for upconverted emission from HfO_2-Er samples under 800 nm excitation. The Er ion is excited between $^4I_{15/2} \rightarrow {}^4I_{9/2}$ transitions of Er, part of $^4I_{9/2}$ excited ions relaxes non-radiatively to the $^4I_{13/2}$ level, which is excited to $^2H_{11/2}$ level and from the $^2H_{11/2}$ level to $^4I_{15/2}$ the green luminescence is emitted at 530 nm. Therefore, we believe that a two-photon process involved in the upconversion mechanism is responsible for the green emission. However is necessary a study of fluorescence lifetimes for the samples for verify the proposed photoluminescence mechanism.

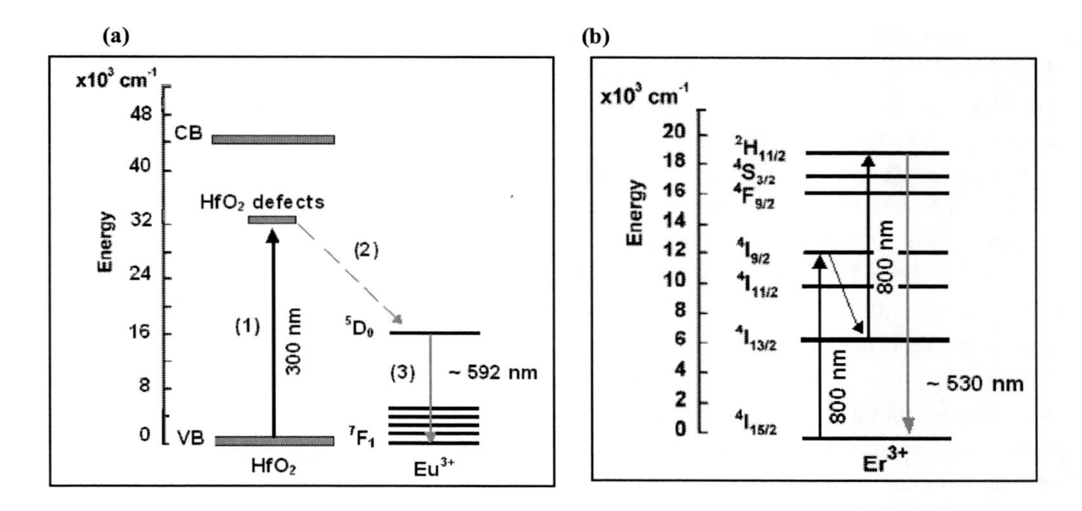

Figure 17. Proposed PL mechanism process in (a) HfO_2-Eu thin films: (1) absorption for defects material host, (2) non-radiative energy transfer process, (3) emission process from $^5D_0 \rightarrow {}^7F_1$ transition of Eu ion. (b) HfO_2-Er thin films: Energy-level diagram for the upconversion process of Er under 800 nm excitation.

Conclusions

In summary several photochemical methods for the synthesis or deposition of semiconductors and ceramic materials have been reported, as alternative methods in the manufacture of nano-structured materials. Among these methods we have presented the photo-deposition in solid phase (thin films) technique in the preparation of ZnO and HfO_2 thin films and their preliminary study as chemical sensor and as luminescence device respectively.

In the evaluate of the sensitivity capacity of the ZnO films, all showed to be sensitive in the detection to 20 ppm of CO gas, the optimum working temperatures was at 300 and 350 °C. For all the films evaluated, the introduction of additives as platinum contributed to optimize the response of these deposits in terms of sensibility and stability, mainly due to that modify surface morphology of films, facilitating the diffusion of the gas or their catalyst characteristic. It has been demonstrated that the uniform distribution of the additive in the deposit is another factor to consider since offers a greater specific surface and facilitates a greater interaction with the species gas.

On the other hand, in the study of photoluminescence properties of the HfO_2 films loaded with erbium or europium in different proportions photochemically prepared has demonstrated emissions in the visible spectrum, is as well as we observed the existence an only peak at 592 nm has been obtained from HfO_2-Eu films and upconvesion emission at 530 nm due to two photon absorption process upon excitation at 800 nm for HfO_2-Er samples.

In this sense, these photochemical methods have shown to be a valid alternative in the deposition of diverse materials with the advantage to obtain thin films of quality, the ones that can be deposited under smooth conditions (for example to room temperature) from simple precursors compounds, that contribute to diminish energy costs and sophistication in the process.

Though the first photo-chemical methods studied, they were based on the study of deposition and characterization of traditional materials recently is sought to apply this methodology in the search of new materials with some potential application (such as chemical sensors, catalyst, optical and optoelectronic devices etc.) is as well as alone a few works have shown to have some concrete application with some commercial purpose [199,200].

On the other hand, with the development of nano-science and technology and the innovation of new structures preparation as nanoparticles, nanowires, nanobelts, nanorods or nanotetrapods offers enormous potentialities that contribute to significant advances in different technological areas in this way we believe that the actual photochemical methods should be redefined ó well to harmonizing with these new advances, thus for example to design new precursors composed that be photoreact and that their deposition achieve these new nano-structures previously indicated, it involves perhaps the modification of some parameters ó well to harmonizing this method with some another different method just as occurs with the processes photo-assisted applied in the photo-CVD, photo-SG, photo-CBD process among others.

Acknowledgments

We tanks to the financial support of the FONDECYT-Chile (National Fund for Scientific and Technological Development), Grant No. 1060486 and Council of Research of the Bío-Bío University, DIUBB proyect. Grant No. 092509 3/R.

G. Cabello thanks MECESUP Chile (Project UCO 9905) for a doctoral fellowship.

References

[1] C. Crisafulli, S. Scire, S. Giuffrida, G. Ventimiglia, R. Lo Nigro, *Applied Catalisis A: General* 306 (2006) 51-57.

[2] S. Giuffrida, G. Condorelli, L. Costanzo, G. Ventimiglia, R. Lo Nigro, M. Favazza, E. Votrico, C. Bongiorno, I. Fragala, *Journal of Nanoparticle Research* 9 (2007) 611-619

[3] S. Giuffrida, L. Costanzo, G. Condorelli, G. Ventimiglia, I. Fragala, *Inorganica Chimica Acta* 358 (2005) 1873-1881.

[4] G. Conderelli, L. Costanzo, I. Fragala, S. Giufrida, G. Ventimiglia, *Journal of Materials Chemistry* 13 (2003) 2409-2411.

[5] S. Giufrida, G. Conderelli, L. Costanzo, I. Fragala, G. Ventimiglia, G. Vecchio, *Chem. Mater.* 16 (2004) 1260-1266.

[6] S. Giufrida, G. Conderelli, L. Costanzo, G. Ventimiglia, M. Favazza, S. Petralia, I. Fragala, *Inorganica Chimica Acta* 359 (2006) 4043-4052.

[7] S. Giufrida, G. Conderelli, L. Costanzo, G. Ventimiglia, A. Di Mauro, I. Fragala, *Journal of Photochemistry and Photobiology A:* 195 (2008) 215-222.

[8] S. Giufrida, G. Ventimiglia, S. Sortino, *Chemical Communications* (2009) 4055-4057.

[9] S. Scire, C. Crisafulli, S. Giuffrida, C. Mazza, P.M. Riccobene, A. Pistone, G. Ventimiglia, C. Bongiorno, C. Spinella, *Applied Catalysis A*: 367 (2009) 138-145.

[10] S.K. Mathew, N.P. Rajesh, M. Ichimura, Udayalakshmi, *Materials Letters* 62 (2008) 591-593.
[11] J. Podder, R. Kobayashi, M. Ichimura, *Thin Solid Films* 472 (2005) 71-75.
[12] M. Gunasekaran, R. Gopalakrishnan, P. Ramasamy, *Materials Letters* 58 (2003) 67-70.
[13] T. Miyawaki, M. Ichimura, *Materials Letters* 61 (2007) 4683-4686.
[14] R. Padmavathy, N.P. Rajesh, A. Arulchakkaravarthi, R. Gopalakrishnan, P. Santhanaraghavan, P. Ramasamy, *Materials Letters* 53 (2002) 321-325.
[15] S. Soundeswaran, O.S. Kumar, S.M. Babu, P. Ramasamy, R. Dhanasekaran, *Materials Letters* 59 (2005) 1795-1800.
[16] S. Soundeswaran, O.S. Kumar, S.M. Babu, P. Ramasamy, D. Kabi Raj, D.K. Avasthi, R. Dhanasekaran, *Physica B* 355 (2005) 222-230.
[17] M. Gunasekaran, M. Ichimura, *Solar Energy Materials & Solar Cells* 91 (2007) 774-778.
[18] M. Ichimura, T. Narita, K. Masui, *Materials Science and Engineering* B 96 (2002) 296-299.
[19] R. Kumaresan, M. Ichimura, N. Sato, P. Ramasamy, *Materials Science and Engineering* B 96 (2002) 37-42.
[20] H. Sasaki, K. Shibayama, M. Ichimura, K. Masui, *Journal of Cristal Growth* 237-239 (2002) 2125-2129.
[21] J. Podder, T. Miyawaki, M. Ichimura, *Journal of Cristal Growth* 275 (2005) e937-e942.
[22] R. Kumaresan, M. Ichimura, E. Arai, *Thin Solid Films* 414 (2002) 25-30.
[23] M. Ichimura, K. Takeuchi, A. Nakamura, E. Arai, *Thin Solid Films* 384 (2001) 157-159.
[24] M. Ichimura, K. Shibayama, K. Masui. *Thin Solid Films* 466 (2004) 34-36.
[25] M. Azuma, M. Ichimura, *Materials Research Bulletin* 43 (2008) 3537- 3542.
[26] M. Ichimura, A. Baoleer, T. Sueyoshi, *Physica Status Solidi C* 7 (2010) 1168-1171.
[27] H. R. Dizaji, M. Ichimira, Materials Science and Engineering B 158 (2009) 26-29
[28] M. Ichimura, K. Akita, *Physica Status Solidi C* 7 (2010) 929-932.
[29] R. Kumaresan, M. Ichimura, N. Sato, P. Ramasamy, E. Arai, *Journal of the Electrochemical Society* 149 (2002) C464-C468
[30] M. Gunasekaran, P. Ramasamy, M. Ichimura, *Journal of the Electrochemical Society* 153 (2006) G664-G668.
[31] Aaron Peled and Nina Mirchin, "Liquid Phase Photodeposition Processes from Colloid Solutions". *In Photo-excited Processes, Diagnostics and Applications;* A. Peled, editor; Kluwer Academic Publishers: Netherland, 2003, Chapter 9, p. 251-280.
[32] A. Peled, Review: State of the Art in Liquid Phase Photodeposition Processes and Applications (LPPD*), Lasers in Engineering* 6 (1997) 41-79.
[33] M. Sakamoto, M. Fujistuka, T. Majima, *Journal of Photochemistry and Photobiology C*, 10 (2009) 33-56.
[34] H. Kim, H-Ho Park, H. Jeon, H.J. Chang, Y. Chang, H.Ho Park, *Ceramics International* 35 (2009) 131-135.
[35] D. Tiwari, S. Dunn, Journal of the European Ceramic Society 29 (2009) 2799-2805.
[36] N. Luo, L. Mao, L. Jiang, J. Zhan, Z. Wu, D. Wu, *Materials Letters* 63 (2009) 154-156
[37] E. Morrison, D. Gutierrez-Tauste, C. Domingo, E. Vigil, J. A. Ayllon, *Thin Solid Films* 517 (2009) 5621-5624.

[38] N.V. Smirnova, T.B. Boistsova, V.V. Gorbunova, L.V. Alekseeva, V.P. Pronin, G.S. Kon´uhov, *Thin Solid Films* 513 (2006) 25-30.
[39] S. Shironita, K. Mori, T. Shimizu, T. Ohmichi, N. Mimura, H. Yamashita, Applied Surface Science 254 (2008) 7604-7607.
[40] Masanobu Izaki, *Chemical Communications* (2002) 476-477.
[41] S. Xu, W. Zhao, J. Hong, J. Zhu, H. Chen, *Materials Letters* 59 (2005) 319-321.
[42] G. Dukovic, M. Merkle, J. Nelson, S. Huges, A. Alivisatos, *Advanced Materials* 20 (2008) 4306-4311.
[43] T. Alammar, A. Mudring, *Journal of Materials Science* 44 (2009) 3218-3222.
[44] T. Kanki, H. Yoneda, N. Sano, A. Toyoda, C. Nagai, *Chemical Engineering Journal* 97 (2004) 77-81.
[45] F. Sun, J. C. Yu, *Angewandte Chemie International* 46 (2007) 773-777.
[46] F. Leung-Yuk Lam, X. Hu, T. M-H. Lee, K. Yu Chan, *Separation and Purification Technology* 67 (2009) 233-237
[47] Philip X. Rutkowski, Jeffrey I. Zink, *Inorganic Chemistry* 48 (2009) 1655-1660.
[48] Q. Fang, J-Y. Zhang, Z.M. Wang, J.X. Wu et al. *Thin Solid Films* 427 (2003) 391-396.
[49]] Q. Fang, J-Y. Zhang, Z.M. Wang, G. He, J. Yu, I.W. Boyd, *Microelectronic Engineering* 66 (2003) 621-630.
[50] M. Chen, X. Zhang, Q. Fang, J. Zhang, Z. Lin, I. Boyd, *Applied Surface Science* 253 (2007) 7942-7946.
[51] C.H. Liu, T.K. Lin, S.J. Chang, Y.K. Su, Y.Z. Chiou, C.K. Wang, S.P. Chang, J.J. Tang, B.R. Huang, *Surface & Coatings Technology* 200 (2006) 3250-3253.
[52] You-Lin Wu, M.H. Hsieh, H.L. Hwang, *Thin Solid Films* 483 (2005) 10-15.
[53] Q. Fang, I. Liaw, M. Modreanu, P.K. Hurley, I.W. Boyd, *Microelectronics Reliability* 45 (2005) 957-960.
[54] S. Myong, T. Kim, K. Lim, K. Kim, B. Ahn, S. Miyajima, M. Konagai, *Solar Energy Materials & Solar Cells* 81 (2004) 485-493.
[55] O. Chevaleevski, S. Myong, K. Lim, *Solid State Communications* 128 (2003) 355-358.
[56] Z.M. Wang, Q. Fang, J-Y. Zhang, J.X. Wu, Y. Di, W. Chen, M.L. Chen, I. W. Boyd, *Thin Solid Films* 453-454 (2004) 167-171
[57] M. Liu, L.Q. Zhu, G. He, Z.M. Wang, J.X. Wu, J-Y. Zhang, I. Liaw, Q. Fang, I. W. Boyd, *Applied Surface Science* 253 (2007) 7869-7873.
[58] S-Yi Chang, Yi-Chung, H-H Chu, Y-C. Hsiao, N-H. Yang, C-F. Lin, *Scripta Materialia* 59 (2008) 646-648.
[59] J.J. Yu, J-Y. Zhang, I.W. Boyd, *Applied Surface Science* 186 (2002) 190-194.
[60] J.J. Yu, Q. Fang, J-Y. Zhang, Z.M. Wang, I.W. Boyd, *Applied Surface Science* 208-209 (2003) 676-681.
[61] K.Y. Cheong, F.A. Jasni, *Microelectronics Journal* 38 (2007) 227-230.
[62] H. Liu, W. Yang, Y. Ma, J. Yao, *Applied Catalysis A: General* 299 (2006) 218-223.
[63] T. Ohishi, *Journal of Non-Crystalline Solids* 332 (2003) 87-92.
[64] B. Gao, Y. Ma, Y. Cao, J. Zhao, J. Yao, *Journal of Solid State Chemistry* 179 (2006) 41-48.
[65] I. Bretos, R. Jimenez, E. Rodriguez-Castellón, J. García-López, M.L. Calzada, *Chem. Mater.* 20 (2008) 1443-1450.
[66] C.Y. Jia, Wenxiu Que, W.G. Liu, *Thin Solid Films* (2009) 290-294.

[67] C. Millon, D. Riassetto, G. Berthomé, F. Roussel, M. Langlet, *Journal of Photochemistry and Photobiology A*: 189 (2007) 334-348.
[68] M. Tsuchiya, S. Sankaranarayanan, S. Ramanathan, *Progress in Materials Science* 54 (2009) 981-1057.
[69] J.P. Bravo-Vasquez, Ross H. Hill, *Journal of Photochemistry and Photobiology A: Chemistry* 193 (2008) 18-24.
[70] J.P. Bravo-Vasquez, Ross H. Hill, *Journal of Photochemistry and Photobiology A: Chemistry* 196 (2008) 1-9.
[71] H-Ho. Park, T-Jung Ha, H-Ho Park, T. Song Kim, Ross H. Hill. *Materials Letters* 62 (2008) 4143-4145.
[72] W.L. Law, Ross H. Hill, *Thin Solid Films* 375 (2000) 42-45.
[73] H-Ho. Park, H-Ho. Park, Ross H. Hill, *Sensors and Actuators A* 132 (2006) 429-433.
[74] S. Trudel, C. H. W. Jones, Ross H. Hill, *Journal of Material Chemistry* 17 (2007) 2206-2218.
[75] Winnie Chu, M. Jamal Deen and Ross H. Hill, *Journal of the Electrochemical Society* 145 (1998) 4219-4225.
[76] G.E. Buono-Core, G. Cabello, N. Guzman, R.H. Hill, *Materials Chemistry and Physics* 96 (2006) 98-102.
[77] G.E. Buono-Core, M. Tejos R., A.H. Klahn, G. Cabello, A. Lucero, R.H. Hill. *Journal of the Chilean Chemical Society* 52 (2007) 1126-1129.
[78] G.E. Buono-Core, G.A. Cabello, H. Espinoza, A.H. Klahn, M.Tejos, R.H. Hill, *Journal of the Chilean Chemical Society* 51(2006) 950-956.
[79] G.E. Buono-Core, G. Cabello, B. Torrejon, M.Tejos, R.H. Hill, *Materials Research Bulletin*. 40 (2005) 1765-1774.
[80] G.E. Buono-Core, G. Cabello, A.H. Klahn, R. Del Rio, R.H. Hill *Journal of Non-Crystalline Solids* 352 (2006) 4088-4092.
[81] G.E. Buono-Core, G. Cabello, A.H. Klahn, A. Lucero, M.V. Nuñez, B. Torrejón, C. Castillo, *Polyhedron* 29 (2010) 1551-1554.
[82] G.E. Buono-Core, A.H. Klahn, C. Castillo, M. J. Bustamante, E. Muñoz, G. Cabello, B. Chornik, *Polyhedron* (In Press) doi: 10.1016/j.poly.2010.10.007
[83] M. Tejos, B. Rolón, R. del Río, G. Cabello. *Journal of the Chilean Chemical Society* 52 (2007) 1257-1260.
[84] M. Tejos, B. G. Rolón, R. del Río, G. Cabello. *Materials Science in Semiconductor Processing* 11 (2008) 94-99.
[85] G. Cabello,"Foto-deposición de películas delgadas de óxidos metálicos semiconductores y su potencial aplicación como micro-sensores en la detección de gases contaminantes" *Tesis doctoral*. Instituto de Química. P. Universidad Católica de Valparaíso. Abril, 2005.
[86] G. Cabello, L. Lillo, G.E. Buono-Core, *Journal of Non-Crystalline Solids* 354 (2008) 982-988.
[87] G. Cabello, L. Lillo, C. Caro, G.E. Buono-Core, B. Chornik, M.A. Soto, *Journal of Non-Crystalline Solids* 354 (2008) 3919-3928.
[88] R.L. Lintvedt "*Concepts of Inorganic Photochemistry*" A.W. Adamson and P.D. Fleischauer editors. John Wiley & Sons, New York, 1977 Chapter 7, p. 299
[89] B. Marciniak, G. Buono-Core, *Journal of Photochemistry and Photobiology A: Chemistry* 52 (1990) 1-25.

[90] G. Buono-Core, H. LI, B. Marciniak, *Coordination Chemistry Reviews* 99 (1990) 55-87.
[91] X. Fan, X. Wu, M. Wang, J. Qiu, Y. Kawamoto, *Materials letters* 58 (2004) 2217-2221.
[92] E. Teotonio, M. Felintino, H. Brito, O. Malta, A. Trindade, R. Najjar, W. Strek, *Inorganica Chimica Acta* 357 (2004) 451-460.
[93] E. Nassar, O. Serra, P Calefi, C. Manso, C. Neri, *Materials Research* 4 (2001) 18-22.
[94] J.C. Bunzli, C. Piguet, *Chemical Society Reviews* 34 (2005) 1048-1077.
[95] G.F. de Sa, O. Malta, C. Donega, A. Simas, R. Longo, P. Santa-Cruz, E. da Silva, *Coordination Chemistry Reviews* 196 (2000) 165-195.
[96] Xi-Shi Tai, Min-Yu Tan, *Spectrochimica Acta Part A* 61 (2005) 1767-1770.
[97] H. Brito, O. Malta, M. Felinto, J. Meneses, C. Silva, C. Tomiyama, C. Carvalho, *Journal of Alloys and Compounds* 344 (2002) 293-297.
[98] X. Fan, W. Li, M. Wang, *Materials Science and Engineering* B100 (2003) 147-151.
[99] K. Binnemans "Rare-earth beta-diketonates" in *Handbook on the Physics and Chemistry of Rare Earths*. K.A. Gschneidner, Jr. J-C. G. Bunzli and V.K. Pecharsky Editors. Elsevier, Amsterdam, 2005; Vol. 35, Chapter 225, p. 107.
[100] K. Binnemans, *Chemical Reviews* 109 (2009) 4283-4374.
[101] L. Armelao, S. Quici, F. Barigelletti, G. Accorsi, G. Bottaro, M. Cavazzini, E. Tondelo, *Coordination Chemistry Reviews* 254 (2010) 487-505.
[102] A. Vogler, H. Kunkely, *Inorganica Chimica Acta* 359 (2006) 4130-4138.
[103] Ross H. Hill, *Photonics Science* News 2 (1996) 12-14.
[104] Hui Jian Zhu and Ross H. Hill, *Journal of Photochemistry and Photobiology A: Chemistry* 147 (2002) 127-133
[105] L.S. Andronic, Ross H. Hill, *Journal of Photochemistry and Photobiology A: Chemistry* 152 (2002) 259-265.
[106] H.J. Zhu, R.H. Hill, *Journal of Non-crystalline Solids* 311 (2002) 174-184.
[107] Celia L.W. Ching and Ross H. Hill, *Journal of Vacuum Science and Technology, A: Vacuum, Surfaces, and Films* 16 (1998) 897-901.
[108] W.L. Law and Ross H. Hill, *Materials Research Bulletin* 33 (1998) 69-80.
[109] T. Seiyama, A. Kato, K. Fujiisshi, M. Nagatani, *Anal. Chem.* 34 (1962) 1502-1503.
[110] N. Taguichi, *Japan Patent* 45-38200 (1962)
[111] J.D. Prades, R. Jimenez-Diaz, F. Hernandez-Ramirez, S. Barth, A. Cirera, A. Romano-Rodriguez, S. Mathur, J.R. Morante, *Sensors and Actuators B* 140 (2009) 337-341.
[112] G. Korotcenkov, B.K. Cho, *Sensors and Actuators B* 142 (2009) 321-330.
[113] Ibrahim A. Al-Homoudi, J.S. Thakur, R. Naik, G.W. Auner, G. Newaz, *Applied Surface Science* 253 (2007) 8607-8614.
[114] D. Mardare, N. Iftimie, D. Luca, *Journal of Non-Crystalline Solids* 354 (2008) 4396-4400.
[115] Vallejos, V. Khatko, J. Calderer, I. Gracia, C. Cané, E. Llobet, X. Correig, *Sensors and Actuators B* 132 (2008) 209-215.
[116] T. Siciliano, A. Tepore, G. Micocci, A. Serra, D. Manno, E. Filippo, *Sensors and Actuators B* 133 (2008) 321-326.
[117] T. Wagner, T. Sauerwald, C.-D. Kohl, T. Waitz, C. Weidmann, M. Tiemann, *Thin Solid Films* 517 (2009) 6170-6175.
[118] A. Vomiero, S. Bianchi, E. Comini, G. Faglia, M. Ferroni, N. Poli, G. Sberveglieri, *Thin Solid Films* 515 (2007) 8356-8359.

[119] H. Xu, X. Liu, D. Cui, M. Li, M. Jiang, *Sensors and Actuators B* 114 (2006) 301-307.

[120] M.C. Carotta, A. Cervi, V. di Natale, S. Gherardi, A. Giberti, V. Guidi, D. Puzzovio, B. Vendemiati, G. Martinelli, M. Sacerdoti, D. Calestani, A. Zappettini, M. Zha, L. Zanotti, *Sensors and Actuators B* 137 (2009) 164-169.

[121] Oleg Lupan, Lee Chow, Guangyu Chai, *Sensors and Actuators B* 141 (2009) 511-517.

[122] Hey-Jin Lim, Deuk Yong Lee, Young-Jei Oh, *Sensors and Actuators A* 125 (2006) 405-410.

[123] Oleg Lupan, Guangyu Chai, Lee Chow, *Microelectronic Engineering* 85 (2008) 2220-2225.

[124] G. Sarala Devi, V. Bala Subrahmanyam, S.C. Gadkari, S.K. Gupta, *Analytica Chimica Acta* 568 (2006) 41-46.

[125] M.S. Wagh, G.H. Jain, D.R. Patil, S.A. Patil, L.A. Patil, *Sensors and Actuators B* 115 (2006) 128-133.

[126] Caihong Wang, Xiangfeng Chu, Mingmei Wu, *Sensors and Actuators B* 113 (2006) 320-323.

[127] Zhaoting Liu, Tongxiang Fan, Di Zhang, Xiaolu Gong, Jiaqiang Xu, *Sensors and Actuators B* 136 (2009) 499-509.

[128] Fang-Tso Liu, Shiang-Fu Gao, Shao-Kai Pei, Shih-Cheng Tseng, Chin-Hsin J. Liu, *Journal of the Taiwan Institute of Chemical Engineers* 40 (2009) 528-532.

[129] Eugene Oh, Ho-Yun Choi, Seung-Ho Jung, Seungho Cho, Jae Chang Kim, Kun-Hong Lee, Sang-Woo Kang, Jintae Kim, Ju-Young Yun, Soo-Hwan Jeong, *Sensors and Actuators B* 141 (2009) 239-243.

[130] B. L. Zhu, C. S. Xie, W. Y. Wang, K. J. Huang, J. H. Hu, *Materials Letters* 58 (2004) 624-629.

[131] V.R. Shinde, T.P. Gujar, C.D. Lokhande, R.S. Mane, Sung-Hwan Han, *Sensors and Actuators B* 123 (2007) 882-887.

[132] N. Tamaekong, C. Liewhiran, A. Wisitsoraat, S. Phanichphant, *Sensors* 9 (2009) 6652-6669.

[133] B. Sam Kang, Hung-Ta Wang , Li- Chia Tien, Fan Ren , Brent P. Gila, David P.

[134] Norton, Cammy R. Abernathy, Jenshan Lin, and Stepehn J. Pearton, *Sensors 6* (2006) 643-666.

[135] Xiaohua Wang, Jian Zhang, Ziqiang Zhu, Jianzhong Zhu, *Colloids and Surfaces A: Physicochemical and Engineering Aspects* 276 (2006) 59-64.

[136] V.R. Shinde, T.P. Gujar, C.D. Lokhande, *Sensors and Actuators B* 123 (2007) 701-706.

[137] Seymen Aygün, David Cann, *Sensors and Actuators B* 106 (2005) 837-842.

[138] Huixiang Tang, Mi Yan, Hui Zhang, Shenzhong Li, Xingfa Ma, Mang Wang, Deren Yang, *Sensors and Actuators B* 114 (2006) 910-915.

[139] Ki-Won Kim, Pyeong-Seok Cho, Sun-Jung Kim, Jong-Heun Lee, Chong-Yun Kang, Jin-Sang Kim, Seok-Jin Yoon, *Sensors and Actuators B* 123 (2007) 318-324.

[140] Zhon Lin Wang, *Materials Science and Engineering* R 64 (2009) 33-71

[141] Xing-Jui Huang, Yang-Kiu Choi, *Sensors and Actuators B* 122 (2007) 659-671.

[142] Jin Hyung Jun, Junggwon Yun, Kyoungah Cho, In-Sung Hwang, Jong-Heun Lee, Sangsig Kim, *Sensors and Actuators B*140 (2009) 412-417.

[143] Chao Li, Zhishuo Yu, Shaoming Fang, Huanxin Wang, Yanghai Gui, Jiaqiang Xu, Rongfeng Chen, *Journal of Alloys and Compounds* 475 (2009) 718-722.

[144] N. Hongsith, C. Viriyaworasakul, P. Mangkorntong, N. Mangkorntong, S. Choopun, *Ceramics International* 34 (2008) 823-826.

[145] C. Baratto, S. Todros, G. Faglia, E. Comini, G. Sberveglieri, S. Lettieri, L. Santamaria, P. Maddalena, *Sensors and Actuators B* 140 (2009) 461-466.

[146] Y. Xi, C.G. Hu, X.Y. Han, Y.F. Xiong, P.X. Gao, G.B. Liu, *Solid State Communications* 141(2007) 506-509.

[147] Supab Choopun, Niyom Hongsith, Pongsri Mangkorntong, Nikorn Mangkorntong, *Physica E: Low-dimensional Systems and Nanostructures* 39 (2007) 53-56.

[148] Jiaqiang Xu, Yuan Zhang, Yuping Chen, Qun Xiang, Qingyi Pan, Liyi Shi, *Materials Science and Engineering B* 150 (2008) Pages 55-60.

[149] Xiaohua Wang, Jian Zhang, Ziqiang Zhu, Jianzhong Zhu, *Applied Surface Science* 253 (2007) 3168-3173.

[150] [] E.L. Prociow, J Domaradzki, A. Podhorodecki, A. Borkowska, D. Kaczmarek, J. Misiewicz, *Thin Solid Films* 515 (2007) 6344.

[151] Z. Liu, J. Zhang, B. Han, J. Du, T. Mu, Y. Wang, Z. Sun, *Microporous and Mesoporous Materials* 81 (2005) 169.

[152] C.W. Jia, J.G. Zhao, H.G. Duan, E.Q. Xie, *Materials Letters* 61 (2007) 4389.

[153] J. Zhang, X. Wang, W-T. Zheng, X-G. Kong, Y-J. Sun, X. Wang, *Materials Letters* 61 (2007) 1658.

[154] K. Kuratani, M. Mizuhata, A. Kajinami, S. Deki, *J. of Alloys and Compounds* 408-412 (2006) 711.

[155] L. Chen, Y. Liu, Y. Li, *J. of Alloys and Compounds* 381 (2004) 266.

[156] A. Patra, *Chemical Physics Letters* 387 (2004) 35.

[157] N. Maeda, N. Wada, H. Onada, A. Maegawa, K. Kojima, *Thin Solid Films* 445 (2003) 382.

[158] J. Chen, Y. Shi, T. Feng, J. Shi, *J. of Alloys and Compounds* 391 (2005) 181.

[159] S. Lange, V. Kiisk, V. Reedo, M. Kirm, J. Aarik, I. Sildos, *Optical Materials* 28 (2006) 1238

[160] R. Goncalves, G. Carturan, M. Montagna, M. Ferrari, L. Zampedri, S. Pelli, G. Righini, S. Ribeiro, Y. Messaddeq, *Optical Materials* 25 (2004) 131.

[161] Y. Jestin, C. Armellini, A. Chiappini, A. Chiasera, M. Ferrari, C. Goyes, M. Montagna, E. Moser, G. Nunzi Conti, S. Pelli, R. Retoux, G.C. Righini, G. Speranza, *J. of Non-Crystalline Solids* 353 (2007) 494.

[162] M. Villanueva-Ibañez, C. Le Luyer, O. Marty, J. Mugnier, *Optical Materials* 24 (2003) 51

[163] M. Villanueva-Ibañez, C. Le Luyer, C. Dujardin, J. Mugnier, *Materials Science and Engineering* B105 (2003) 12.

[164] H. Gong, Y. Wang, Z. Yan, Y. Yang. *Materials Science in Semiconductor Processing* 5 (2002) 31-34.

[165] X.Q. Wei, B.Y. Man, M. Liu, C.S. Xue, H.Z. Zhuang, C. Yang, *Physica B: Condensed Matter* 388 (2007) 145-152

[166] Li Li, Liang Fang, Xian Ju Zhou, Zi Yi Liu, Liang Zhao, Sha Jiang, *Journal of Electron Spectroscopy and Related Phenomena* 173 (2009) 7-11.

[167] M. Brun, A. Berthet, J.C. Bertolini. *J. of Electron Spectroscopy and Related Phenomena* 104 (1999) 55-60.

[168] K.S. Kim, A.F. Gossmann, N. Winograd. *Analytical Chemistry* 46 (1974) 197-200.

[169] R. Diaz, J. Arbiol, F. Sanz, A. Cornet, J. Morante. *Chemistry of Materials* 14 (2002) 3277-3283.
[170] F. Morazzoni, C. Canevali, N. Chiodini, C. Mari, R. Ruffo, R. Scotti, L. Armelao, E. Tondello, L. Depero, E. Bontempi. *Chemistry of Materials* 13 (2002) 4355-4361.
[171] K.S. Kim, N. Winograd, R.E. Davis. *J. Am. Chem. Soc.* 17 (1971).
[172] Hyun-Wook Ryu, Bo-Seok Park, Sheikh A. Akbar, Woo-Sun Lee, Kwang-Jun Hong, Youn-Jin Seo, Dong-Charn Shin, Jin-Seong Park, Gwang-Pyo Choi, *Sensors and Actuators B* 96 (2003) 717-722.
[173] Takeo Hyodo, Norihiro Nishida, Yasuhiro Shimizu, Makoto Egashira, Sensors and Actuators B 83 (2002) 209-215.
[174] D.H. Kim, S.H. Lee, K-H. Kim. *Sensors and Actuators B* 77 (2001) 427-431.
[175] C-H Shim, D-S Lee, S-I Hwang, M-B Lee, J-S Huch, D-D Lee. *Sensors and Actuators B* 81 (2002) 176-181.
[176] X-ray Photoelectron Spectroscopy Database 20, Version 3.5 National Institute of Standards and Technology (NIST). *http://srdata.nist.gov/xps/Default.aspx*
[177] Li-ping Feng, Z-T Liu, Y. Shen, *Vacuum* 83 (2009) 902-905.
[178] S.M.M. Ramos, B. Canut, P. Moretti, P. Thevenard, D. Poker, *Thin Solid Films* 259 483 (1995) 113.
[179] C.R. Tewell, S.H. King, *Applied Surface Science* 253 (2006) 2597.
[180] M.F. Al-Kuhaili, S.M.A. Durrani, *Thin Solid Films* 515 (2007) 2887.
[181] X. Fan, W. Li, F. Wang, M. Wang, *Materials Science and Engineering* B100 (2003) 147-151.
[182] E.O. Oh, Y.H. Kim, C.M. Whang, *J. Electroceram* 17 (2006) 335-338.
[183] K. Kuratani, M. Mizuhata, A. Kajinami, S. Deki, *J. of Alloys and Compounds* 408-412 (2006) 711.
[184] G. Bhaskar Kumar, S. Buddhudu, *Ceramics International* 35 (2009) 521-525
[185] [L. Chen, Y. Liu, Y. Li, *J. of Alloys and Compounds* 381 (2004) 266.
[186] G. Bhaskar Kumar, S. Buddhudu, *Ceramics International* 35 (2009) 521-525.
[187] G. Ehrhart, M. Bouazaoui, B. Capoen, V. Ferreiro, R. Mahiou, O. Robbe, S. Turrell, *Optical Materials* 29 (2007) 1723-1730.
[188] J. Yin, X. Zhao, *Materials Chemistry and Physics* 114 (2009) 561-568.
[189] J. Zhang, X. Wang, W-T. Zheng, X-G. Kong, Y-J. Sun, X. Wang, *Materials Letters* 61 (2007) 1658.
[190] R. Jia, W. Yang, Y. Bai, T. Li, *Optical Materials* 28 (2006) 246-249.
[191] [A. Camargo, J. Possatto. L. O. Nunes, E. Botero, E. Andreeta, D. Garcia, J.A. Eiras, *Solid State Communications* 137 (2006) 1-5.
[192] P. Salas, C. Angeles-Chávez, J. Montoya, E. De la Rosa, L. Diaz-Torres, H. Desirena, A. Martinez, M. Romero-Romo, J. Morales *Optical Materials* 27 (2005) 1295-1300.
[193] D. Solís, T. López-Luke, E. De la Rosa, C. Angeles-Chávez, *Journal of Luminescence* 129 (2009) 449-455.
[194] K. Frindell, M. Bartl, M. Robinson, G. Bazan, A. Popitsch, G. Stuky, *Journal of Solid State Chemistry* 172 (2003) 81-88
[195] C. Gao, H. Song, L. Hu, G. Pan, R. Qin, F. Wang, Q. Dai, L. Fan, L. Liu, H. Liu, *Journal of Luminescence* 128 (2008) 559-564.
[196] L. Hu, H. Song, G. Pan, B. Yan, R. Qin, Q. Dai, L. Fan, S. Li, X. Bai, *Journal of Luminescence* 127 (2007) 371-376.

[197] J. Zhang, X. Wang, W-T. Zheng, X-G. Kong, Y-J. Sun, X. Wang, *Materials Letters* 61 (2007) 1658.
[198] N. Maeda, N. Wada, H. Onada, A. Maegawa, K. Kojima, *Thin Solid Films* 445 (2003) 382.
[199] W. Que, Y. Zhou, Y.L. Lam, K. Pita, Y.C. Chan, C.H. Kam, *Applied Physics* A 73 (2001) 209-213.
[200] Ross H. Hill, Paul Roman, Seigi Suh, Xing Zhang, USA patent 20030190820 (2003).
[201] J.P. Bravo-Vasquez, Ross H. Hill, *USA patent 20030059544* (2003).

In: Photochemistry
Editors: Karen J. Maes and Jaime M. Willems

ISBN: 978-1-61209-506-6

Chapter 4

PHOTOLABILE MOLECULES AS LIGHT-ACTIVATED SWITCHES TO CONTROL BIOMOLECULAR AND BIOMATERIAL PROPERTIES

Catherine A. Goubko, Nan Cheng, and Xudong Cao*
Department of Chemical and Biological Engineering,
University of Ottawa, Ottawa, Ontario, Canada

ABSTRACT

It is of great interest to develop technologies to dynamically control the properties of biomolecules and materials in order to conduct advanced studies in cell biology and to create new medical devices. Light has been considered an ideal external stimulus to exert such control since it can be readily manipulated spatially and temporally with high precision. To this end, photolabile molecules are often employed as light switches. These photolabile molecules or photocages are light-sensitive compounds which can be covalently bound to biomolecules or materials using a variety of functional groups and can be removed upon exposure to light. A variety of biomolecules such as peptides, proteins, and nucleic acids have been synthesized to incorporate photocages that initially render them inactive. Light exposure can subsequently re-activate these biomolecules at a particular time and area in space. The number of the caged biomolecules and their applications in the life sciences is ever-growing.

More recently, a number of studies have attempted to make creative use of these photolabile molecules by designing novel materials for biological applications that can respond to light in different manners. For example, materials have been created that can degrade upon light exposure in specific locations or patterns. Alternatively, solutions that form gels upon light exposure have been reported. Biomaterials have often been designed to release biologically important molecules or drugs upon implantation with the timing and location of release being of the utmost importance; the design of materials that can release their cargo upon light exposure is now being explored. In another avenue of material development, photolabile molecules have been used to form patterns of

* To whom correspondence should be addressed: Department of Chemical and Biological Engineering, University of Ottawa, Ottawa, Ontario, K1N 6N5 (Canada), Tel: (+1) 613-562-5800 ext. 2097, Fax: (+1) 613-562-5172, e-mail: xcao@eng.uottawa.ca

biomolecules on solid surfaces towards the invention of new biochips. Yet others have employed photocaged molecules to form patterns of live cells for biological study and tissue engineering applications. The ability to dynamically alter material properties to dictate gel formation, material degradation or the location of proteins and cells will ultimately bring about a new era of highly tunable materials for improved performance in a number of biological applications.

This review will first briefly overview some commonly used photolabile molecules and then examine some of the applications they have found to date in caging biomolecules. This will set the stage for a detailed discussion on how photolabile molecules are incorporated into materials for biological applications and the future potential of these novel strategies in biomedical engineering research and biotechnology.

1. INTRODUCTION

Photolabile protecting groups, or caging groups, offer researchers involved in the biological sciences a unique tool – a light switch – which can alter material properties at will or even result in the activation of whole biomolecules. These light-responsive caging groups can be covalently bound to a functional group on a molecule of interest. Upon irradiation with light of appropriate energy, the caging group is removed while freeing the functional group in the process. Thanks to the imagination of chemists, biologists, and material scientists alike, this seemingly simple process has brought about a new generation of biological molecules and materials which can be controlled by light in both a spatial and temporal manner. The photolabile protecting group can render a biomolecule unrecognizable to its corresponding receptor, enzyme, or target due to steric hindrance or changes in charge. Biomolecules in a particular location can be re-activated any time during a process or experiment with light. In materials, the photolabile group can mask a functional group which can then be revealed with light in desired regions to allow for further chemical modifications or building of a material. Caging groups can be incorporated into crosslinkers, holding a material together, in order to allow for controlled degradation with light. Removal of caging groups can also result in charge alterations in regions of a material. Caged biomolecules can even be bound to a biomaterial and selectively activated upon irradiation. In these ways, materials can be fundamentally altered with time or patterned at or even below the micron scale for the fine-tuning of material properties.

Chemists have provided us with a variety of photolabile protecting groups which can be introduced onto biological molecules and materials. The most popular of these groups in the biological sciences are the o-nitrobenzyl derivatives [1]. Figure 1 depicts the structure of some commonly used o-nitrobenzyl derivates as well as the reaction they undergo with light. These derivatives can be covalently bound to a range of functional groups including carboxylates, amines, amides, alcohols, phenols, phosphates and more with relative ease. Despite this, there are several disadvantages associated with their use including the formation of potentially toxic strongly absorbing reaction by-products and relatively slow rates of cage release following excitation [2]. Furthermore, the light required to uncage these molecules is of relatively high energy – in the near-UV range.

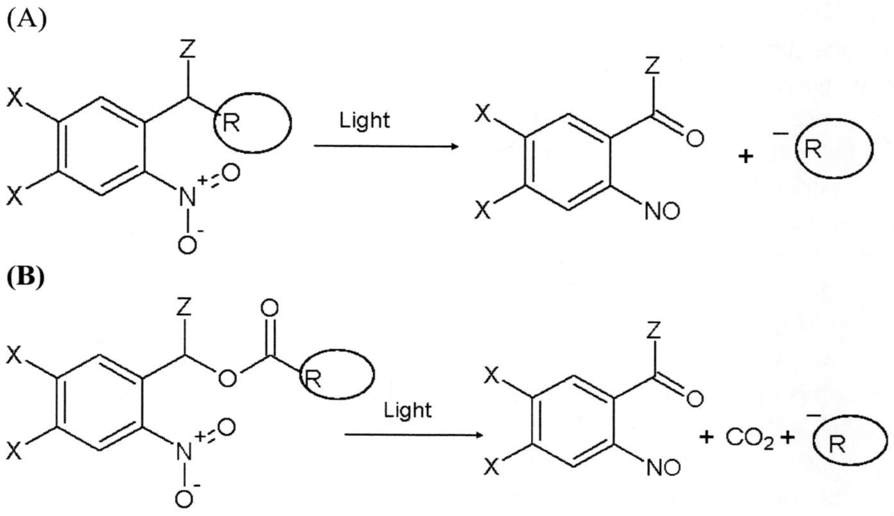

Figure 1. General structure of o-nitrobenzyl derivatives and associated light reactions. Some common substituents include: (A) X=H, Z=H o-nitrobenzyl (NB); X=H, Z= CH_3 o-nitrophenethyl (NPE), X=H, Z=COOH α-carboxy-2-nitrobenzyl (CNB); X=OCH_3, Z=H 4,5-dimethoxy-2-nitrobenzyl (DMNB); X=OCH_3, Z=CH_3 4,5-dimethoxy-2-nitrophenethyl (DMNPE) (B) X=OCH_3, Z=H 6-nitroveratryloxycarbonyl (NVOC) [2].

A group of newer photolabile cages becoming increasingly more popular are coumarin-4-ylmethyl groups (coumarins). Figure 2 depicts the general structure of some coumarins used in the literature as well as the reaction they undergo upon irradiation. These groups have been used to cage phosphates, carboxylates, amines, alcohols, phenols, and carbonyl compounds. Coumarins absorb strongly into the visible light region and demonstrate relatively quick photolysis rates. They have also been shown as well-suited for two-photon photolysis. However, these compounds often suffer from solubility isssues [3].

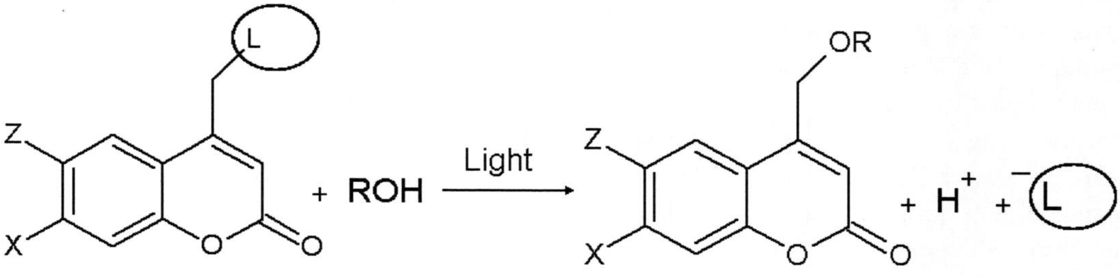

Figure 2. General structure of photolabile coumarin-4-ylmethyl groups and reaction resulting from irradiation. Some substituents found in the literature include [3]: X=OX' (X'=CH_3, H, CH_3CO, CH_3CH_2CO, or CH_2CO_2H), Z=H 7-alkoxy group; X=Z=OX' (X'=CH_3, CH_2CO_2H, or CH_2CO_2Et) 6,7-dialkoxy group; X=OX' (X'=H or CH_3CO), Z=Br 6-bromo-7-alkoxy group; X=NX'_2 (X'=CH_3CH_2, CH_3), Z=H 7-dialkylamino group.

A number of other more minor photolabile caging molecules exist and have been used in biological applications, but are beyond the scope of this review. For more information, please refer to recent comprehensive review articles [4, 5].

A number of issues must be considered when attempting to cage a molecule or material for biological applications. The first is the efficiency of photolysis, which can be considered as the percentage of molecules from which the photolabile cage is removed upon irradiation. This is dependant on the quantity of light absorbed at the wavelength of irradiation, which is related to the molecule's extinction coefficient, as well as the fraction of caged molecules that will react after absorbing a photon, described by the quantum yield [3]. The coumarins absorb strongly, but have low quantum yields [2]. As caged biomolecules often need to be dissolved in aqueous environments, solubility can also be an issue for some cages. Absorption properties and solubility can be altered and improved by carefully choosing the substituents on the ring structure of these cages. The rate of photolysis should also be considered; coumarin groups show relatively high photolysis rates upon irradiation. Another important consideration is the irradiation wavelength which should be greater than 300 nm to minimize potential damage caused by UV light. At lower wavelengths significant damage to proteins and the nucleic acids that make up cell DNA can be detected [1]. Furthermore, to be useful in biological enviroments, caged products must not hydrolyze in water to a large degree. Finally, one of the greatest considerations must be to the placement of the cage on the biomolecule or material of interest. There must be a functional group to which the caging group can bind and binding in that particular location must have a significant effect on the molecule. For example, in a large protein, placing the caging group away from the active site may not have any effect on the ability of the protein to function, rendering the light switch useless.

Once a photocaged biomolecule or material is created, the next decision is the nature of the light source for photolysis. UV lamps are inexpensive, readily accessible, and can be attachced to filters for selection of specific wavelengths. Photomasks can be placed on surfaces underneath a lamp to allow uncaging only in patterned regions allowing for spatial selectivity. Lasers can be used to deliver light of one specific wavelength corresponding to the maximal absorbance of a caged molecule. Lasers can also be focused onto small areas below the nano scale for spatial selectivity. Two photon lasers are now in use for photolysis of caged biomolecules or materials [6]. Photolabile molecules exposed to such sources absorb two photons at the same wavelength practically simultaneously to undergo photolysis. These two photons together deliver the same quantity of energy to the caged product as a single photon would in another laser. Therefore, the wavelength of light delivered by these lasers is approximately double that used in a single photon laser so infrared light is normally employed for uncaging with two-photon lasers. This is highly beneficial since cells and tissues do not strongly absorb IR light, allowing for greater depth of penetration with less scattering and less damage. Furthermore, two-photon lasers can target a small focal volume within a 3D environment, whereas single-photon lasers lack this 3D selectivity. Coumarins have been found to be much more effecient for two-photon uncaging as compared to o-nitrobenzyl derivatives, which require higher laser powers for uncaging [3].

The combined efforts of chemists in designing new caging groups with improved properties for use in the biological environment and physicists in designing improved lasers for targeted removal of photolabile caging groups have lead to the recent increase in the use of these caging groups in life sciences research and the development of materials for biological applications. To date, much effort has been devoted to developing a variety of caged biomolecules and their applications are ever-growing. Use of these molecules in biomaterials has been gaining in popularity recently and new highly-creative, light-responsive materials have been born and their applications in sensor-design, tissue engineering, and life sciences research are just beginning.

2. Caging of Biological Molecules

Over the past thirty years, great effort has been devoted to caging a variety of biological molecules. Initial work involved caging smaller molecules that nonetheless play large roles in biochemistry and moved towards complex macromolecules such as nucleic acids and proteins. Techniques to cage these molecules are well developed in the literature but still require expertise in synthetic chemistry. Nevertheless, there are relatively fewer works dealing with the application of these molecules. Here, we will provide an overview of some key biomolecules that have been bound with photolabile groups, and demonstrate some of the applications they have found.

2.1. Small Biologically Relevant Molecules

To date, hundreds of experiments have been conducted employing small caged biomolecules of which a number are currently commercially available. Strong interest in caging biomolecules began with a landmark study by Kaplan et al [7]. In 1978, they synthesized a caged ATP by binding o-nitrobenzyl onto a phosphate group. Caged ATP was introduced into red blood cells where they were used as an energy source upon irradiation to activate cellular Na:K pumps. Caged ATP cannot be hydrolyzed to form ADP - which normally generates energy for cellular processes - until it is activated with light. Caging this molecule thus allows for a number of studies requiring control over energy reserves making caged ATP an excellent research tool still in use today [8].

Calcium plays a strong role in cellular biochemistry; changes in intracellular calcium levels are linked to muscle contraction, mitosis, and neurotransmitter release among others. Since calcium cannot be covalently bound to a caging group, photo-responsive high affinity calcium chelators were developed to have a high affinity for calcium in the caged state. Upon irradiation, affinity for calcium is decreased, releasing the inorganic molecule and therefore increasing calcium levels in the cellular environment. A good review of caged calcium by Ellis-Davies exists for interested readers [9]. By employing caged calcium to spatially and temporally control available calcium levels, Gomez and Spitzer were able to determine that transient calcium level elevations control axon growth in the developing spinal cord [10]. Growth cones located at the extending tips of neuronal cells showed transient elevations of calcium as these extensions migrated. Releasing caged calcium chelators accelerated the migration, while reproducing the transient elevations with the photorelease of calcium slowed the growth.

Photolabile molecules which release nitric oxide have also been created. Nitric oxide plays a role in a large number of biological events such as certain neural physiological processes like long term potentiation and depression, constriction of smooth muscle and blood flow control, and NO has been found, under different conditions, to both enhance tumor growth and destroy it [11]. Another important small molecule that has been caged is cAMP which acts as a secondary messenger in eukaryotic cells and as such is involved in the regulation of a vast number of cellular processes [5]. Caging these molecules allows for in-depth investigations into their roles in these widely varied cell processes.

Over the years, there has existed a strong interest in caging neurotransmitters. Caged glutamate is one of the most well-known neurotransmitters thanks in part to a 1994 study by Dalva and Katz who investigated the patterns of synaptic connections in a developing visual cortex. Brain slices were irradiated to selectively uncage glutamate and generate action potentials in presynaptic neurons located at the laser focal point [12]. Since then, new caged derivatives of glutamate have been developed with improved sensitivity to two-photon photolysis allowing for better spatial selectivity during light activation. Some other reported caged neurotransmitters include GABA, glycine, and anandamides [13]. Such caged species have played important roles in elucidating the processes of neurotransmission and signal transduction [3].

In an excellent example of an application of a small caged biomolecule, Cambridge et al. designed a gene expression system responsive to light [14]. Their method was based on the "tetracycline-controlled transcriptional activation" system, which can control the transcription of transgenes in an organism or cell culture by the presence of tetracyline antibiotics or doxycycline. The group caged doxycycline so that irradiation would free the antibiotic and induce transcription. Sauers et al. also employed a photolabile doxycycline to control cell localization in a co-culture [15]. A culture of 3T3s were created that could be induced with light to express Ephrin A5. When expressed in a cell, this molecule causes repulsion towards other cells expressing the receptor, EphA7, and attraction to cells expressing a variant, EphA7-T1. After forming a pattern of Ephrin A5 expression in a monolayer of 3T3s with light, they found that EphA7-T1-expressing cells preferentially adhered to the patterned area. This method of controlling attractive/repulsive cell cues with light is an exciting and creative way to control the arrangement of cells *in vitro*.

2.2. Peptides and Proteins

Proteins and peptides play many different roles in our body; enzymes act as chemical catalysts; transporters regulate the cell environment; signaling molecules and hormones send messages across the body for coordinated action and in development; and the list goes on. These biomolecules are key to the basic processes of life and disease. However, we do not yet fully understand the roles played by many proteins. Now that we have sequenced the human genome, we must come to a better understanding of how these key gene products act in the body. Caging proteins is an often under-used tool in these important studies with the capability to temporally and spatially control protein activity. Much effort has been spent on the technique of inserting a caging group into these biomolecules, but comparatively less effort has been extended to their use in advanced biological studies. We will therefore begin with an overview of some of the key methods employed in caging proteins.

Several methodologies have been developed to synthesize photocaged proteins. Traditionally, caging groups were introduced non-specifically into whole, intact proteins. Various o-nitrobenzyl derivatives were designed to specifically react with functional groups on certain amino acids such as sulfhydryl groups, found on the amino acid cysteine, and amino groups found on lysine [16-18]. These techniques allow for the caging of larger molecules, but suffer certain drawbacks: (a) it is difficult to target a specific residue for caging, (b) the technique is limited to specific types of residues, and (c) typically only surface-exposed residues can be reached for caging. Furthermore, cysteine residues are relatively rare in proteins and may need to be introduced via mutagenesis near a protein active site for this strategy to work [3]. In addition, too many residues caged can result in difficulties reactivating the protein upon irradiation.

In comparison, when a small peptide is involved, a more direct approach to introducing photolabile molecules can be taken. Researchers have been successful in introducing a caging group onto a single amino acid and then introducing this caged unit into a larger peptide via automated synthesis. One study has even attempted to add the caging group to the peptide directly during solid phase synthesis [19]. Many different caged amino acids have been designed including Lys, Tyr, Glu, Asp, Arg, Ser, Gln, and Gly (placing the cage on the peptide backbone) [20-23]. This process allows for much improved control over the placement of the caging group. Often, several different caged amino acids are synthesized to produce peptides caged in different locations. In this manner, the caged peptide which most effectively inhibits activity prior to light exposure and allows for activation after irradiation can be selected for use [20, 22, 23]. However, this method is limited to smaller proteins and synthesis can be very time consuming.

In another synthesis strategy, caged amino acids were introduced directly into proteins via *in vitro* translation. For example, Wu et al. made use of a unique *E. Coli* tRNA/aminoacyl tRNA synthetase pair to incorporate a caged cysteine into proteins synthesized in yeast in response to a nonsense codon, TAG [24]. Others have used a similar methodology to introduce caged lysine intro proteins in mammalian cells [25, 26]. This system of caging allows for direct incorporation of a caging group into a large protein, but is fairly complex and requires highly specialized knowledge and techniques to accomplish.

Another approach involves the protein phosphorylation process. Phosphorylation is a common way for cells to regulate protein activity after translation. In one study, the caging group was placed on the phosphate groups needed to activate Smad2, a molecule known to play a role in cancer. The caged Smad2 acted like the unphosphorylated protein, while UV irradiation resulted in activity similar to the phosphorylated Smad2. This strategy could theoretically be used to cage any protein activated through phosphorylation [27]. Another interesting strategy involves ligating a synthetic moiety to the C-terminus of a peptide such that the bond is photocleavable due to the presence of an o-nitrobenzyl derivative. In one example, a photolabile lipid was ligated to a protein resulting in its re-localization mostly within cell membranes. Upon photocleavage, the protein was able to move towards the cytoplasm and nucleus. Therefore, proteins can be deactivated through relocation strategies [28].

Using the strategies discussed above, a wide variety of peptides and proteins can potentially be caged such that they can be switched on by light at will. Proteins involved in gene expression, from transcription to translation, have even been caged such that they can control the expression of a wide variety of other proteins and gene products. These caged proteins can also offer us further insight into the roles their uncaged versions play in gene expression. To this end, Chou et al. were able to cage an RNA polymerase to render it inactive until near-UV exposure. With this caged enzyme, they were able to control gene function with light in both bacteria and mammalian cells [29]. A variety of caged biomolecules involved in protein translation from RNA have been designed such as caged anisomycin which inhibits protein synthesis by binding to cell ribosomes, caged 4E-BP which is involved in translation of mRNAs with a cap structure on their end, and rapamycin which inhibits mTORC1, a complex involved in translation initiation [30, 31]. Growth factors or signaling molecules stimulate certain cell behaviors and can be caged to either study their function, or to control cell behavior experimentally. Miller et al. caged a synthetic epidermal growth factor sequence and demonstrated its use in controlling cell migration and proliferation upon activation with UV light [32].

Despite a growing library of caged proteins in the literature, relatively few studies make true use of the advantages of this strategy to control protein expression – precise temporal and spatial control. Using light as a switch to turn on proteins can allow us to target single cells in a culture or multi-cellular organism or even organelles within a cell. The state of a cell is constantly changing with time; cells divide, migrate, differentiate and undergo apoptosis. Different proteins may vary in activity or function during these different times. With light as a switch, we can turn on a particular protein during any point of a cell's life cycle to examine temporal effects.

One excellent example of the spatial control afforded by protein caging is seen in a study by Priestman and Lawrence where they investigated the role played by cofilin, an intracellular protein involved in cell motility. A caged cofilin was injected it into cells. After exposing whole cells to irradiation, they found an increase in lamellipod (a cell projection) size and formation time. They then irradiated 3 um spots on single cells and noted that cell protrusions formed near these spots in 80% of the cells. Therefore, they were able to demonstrate that by using caged cofilin proteins, cells could be induced to move in the direction of illumination. In comparison, cells without the caged protein moved randomly [33].

In a good example of the experimental potential of caged proteins for temporal control of processes, Sinha et al. designed a system for introducing a caged protein into zebrafish embryos. These creatures are ideal for experimentation due to their small size and transparency [34]. Sinha et al. genetically programmed a protein with a small site that is activated by a lipophilic caged molecule. The developing embryo was incubated in a solution of this caged inducer molecule that binds to form a caged protein complex *in vivo*. Caging was accomplished with both o-nitrobenzyl and coumarin derivatives to allow for single photon or two-photon uncaging. This protein could thus be activated in a select group of cells within the zebrafish and at any time, offering an exciting method to study the activity of a protein at specific time points in embryonic development.

2.3. Nucleotides and Nucleic Acids

Another way to photo-control gene expression is the direct caging of nucleic acids – DNA or RNA. To this end, photocages have been introduced on nucleotide bases, backbone phosphates, and hydroxyl groups on ribose sugar rings [35]. Ando et al. developed a process to cage the phosphate backbone of mRNA with a coumarin such that approximately 30 caged sites were generated per kb of RNA [36]. Eng2a mRNA was caged, which codes for a transcription factor, Engrailed2a. Embryos with the caged mRNA showed normal eye development, while those irradiated in the head region at a certain time during development to uncage the mRNA developed an eyeless phenotype. In this way, caged mRNA can be used to investigate the impact of certain genes on development. DNA has also been caged. In the past, plasmid DNA has been caged non-specifically on the phosphate backbone. Yamaguchi et al. have introduced a site-specific caging group into a plasmid attached to biotin. Streptavidin selectively binds to the biotin to provide additional steric hindrance during caging which assures the inability of transcription factors to bind to the plasmid. Irradiation cleaves the biotin thus freeing the plasmid for gene expression and subsequent transcription [37]. Caging groups have also been used to exert spatial and temporal control over the activation of small interfering RNAs (siRNAs). A little over a decade ago, it was discovered that double stranded RNA, namely siRNAs, could silence the expression of a particular gene and this process was called RNA interference. Caging groups have been introduced onto siRNAs to disrupt their interaction with a protein complex, the RISC, necessary for gene silencing. Irradiation of caged siRNAs at a time point during an experiment leads to the silencing of a particular gene product. For more information on this topic, please refer to an excellent review paper [38]. Caging of nucleic acids has occurred more recently as compared to proteins and as such fewer studies are available. It was only a little over ten years ago, when in 1999, gene expression was first controlled with a caged nucleic acid - a plasmid coding for luciferase [39]. As a result, more applications are expected as knowledge of these techniques become more widespread.

2.4. Drugs

Caging drug molecules could lead to effective targeting strategies; theoretically, free drug could be localized in irradiated areas of the body allowing for minimal side effects elsewhere. This could be particularly exciting for tumor targeting. Therapeutic antibodies have been developed to target tumors, but unfortunately, are often not specific enough leading to dangerous side-effects. Caging such antibodies to render them inactive and irradiating the tumor for local release of active antibody could lead to another level of specificity increasing their safety. It is theorized that visible areas of the body such as the skin, eye, mouth and those accessible with endoscopes or areas open during surgeries could be treated with such caged drugs [40]. Skwarczynski et al. developed a prodrug of paclitaxel bound to a coumarin derivative. This drug is commonly used in the treatment of a variety of cancers [41]. In another study, Reinhard and Schmidt investigated o-nitrobenzyl photocaged derivatives of phosphoramide mustards, which are also used in cancer therapy. These compounds have an alkylating activity that allows them to attack proliferating cells [42]. Insulin, for the treatment of diabetes, has also been caged for improved temporal control over its release; researchers envisioned its use in a glucose sensor armed with a small UV lamp to activate and release specific doses in response to blood glucose levels [43]. While the caging of drug molecules is very promising for improved targeting through controlled irradiation, as well as temporal control in response to need, many issues still remain to be addressed. New light sources that can access other areas of the body are needed, and the laser light must be able to significantly penetrate tissues without damage. We must move away from the more damaging near-UV light required by the popular o-nitrobenzyl derivatives and towards safer lower-energy wavelengths. Hopefully, multiphoton laser technologies will eventually satisfy these needs.

3. Photolabile Molecules in Biological Devices and Biomaterials

The caged biomolecules discussed in the previous section can be incorporated into biomaterials to create powerful new biotechnologies. In addition to the direct caging of biomolecules, traditional photolabile caging groups have been more directly implemented in materials, in creative ways, for the development of biological devices on solid surfaces and biomaterials that can respond dynamically to light.

3.1. Patterning Biomolecules and Cells on Solid Surfaces

Solid surfaces modified with caged compounds hold great promise for various biotechnological and biomedical applications. They can be used as a platform to make chips for drug and biomarker discovery, mapping protein-protein interactions, DNA screening and analysis of cellular processes [44, 45]. One strategy in the development of materials for such applications is to create a substrate capable of responding to an external stimulus. Light as the external stimulus, combined with photocleavable caging molecules, has the advantage of high resolution - both spatial and temporal - and leaves no residue post reaction which could cause unexpected side effects on the development of these chips. Light-induced surface chemistry is crucial in the production of microchips for the microelectronics industry, so it is natural that such strategies be translated to the creation of chips for biological applications. Currently, a popular application of photocleavable surfaces is DNA screening and *in situ* synthesis. DNA probe arrays are useful tools in biomedical research and diagnostics because of their ability to simultaneously address large numbers of genes [46].

The photocleavable molecules in these designs aid in DNA synthesis; they are used to temporarily protect the terminal groups of nucleotide monomers, which are often assembled on glass or metal surfaces. They block the DNA polymerase enzyme from incorporating additional nucleotides to a growing nucleic acid strand after each "cycle" of nucleotide incorporation. Irradiation and the use of appropriate photomasks could allow the deprotection of specific terminal groups and control over the DNA sequence and size produced. Seo et al. reported a successful DNA sequencing approach based on DNA synthesis using photocleavable fluorescent nucleotides on a solid surface [46]. The researchers attached azido-labeled DNA onto alkyne-modified glass. The goal was to identify the DNA sequence of interest via the synthesis of its complimentary strand. A series of nucleotide analogues bound to different fluorescent dyes through an o-ntirobenzyl linker were exposed to the DNA to be sequenced. By observing which dye was incorporated, the identity of the unknown nucleotide could be established. They showed that near-UV irradiation leads to efficient release of the fluorophore and demonstrated the feasibility of performing the DNA polymerase reaction on the solid surface. They expected to be able to sequence at lease 25 bases per spot on their chip. Their DNA sequencing technology has the potential for use in whole genome sequencing and can be applied to pharmacogenetics. So far, limited success with photochemical approaches has been reported but several caging groups have been studied and demonstrated fast and efficient photo deprotection of the hydroxyl group for DNA microarray synthesis [47], such as coumarins [35], indoline and quinoline. The ability to produce high-density oligonucleotide arrays (GenChip®probe arrays) has been utilized in DNA sequencing technology [48]. Photocleavable protecting groups can also be used in the synthesis of other biomolecules on solid surfaces; Fodor et al. reported the generation of a successful array of 1024 peptides using an o-nitrobenzyl derivative [49].

The basic challenge for the fabrication of protein and cell microarrays is the ability to label proteins with high resolution on a substrate. To achieve this, a high density of individual, isolated reactive protein sites are required. Light as a trigger, can induce instantaneous photoreactions on surfaces to control immobilization of biomolecules without chemical reagents [50, 51]. Several studies have already been done to demonstrate the feasibility of applying different photocleavable molecules on different functionalized solid surfaces and have investigated their advantages. Basically, the caged molecules are attached to a solid substrate and irradiated through a mask or using laser lithography. Deprotection happens selectively on the molecules exposed to the light and free functional groups are generated ready to react with a second molecule, which is usually a biomolecule [51]. This technique can lead to the development of biosensors and offers certain advantages, including reduced operation time, parallel detection of multiple targets, and small sample requirements.

Sundberg et al fabricated a heterogeneous surface with two different antibodies on a solid substrate in 1996 [52]. An o-nitrobenzyl derivative was used to create a caged biotin analogue which was immobilized to a glass surface. UV light exposure through a photomask yielded regions of deprotected biotin. These regions can bind specifically to the molecule streptavidin to which biotinylated macromolecules could subsequently be linked. They achieved two different biotinylated antibodies immobilized on different regions of a planar substrate. Alonso et al. created a photosensitive silane based on NVOC chemistry [53]. They assembled tetraethylene glycol, a typical protein repellent, on a silica surface bound by a NVOC terminal group. By UV irradiation, they could achieve highly selective deprotection of amine groups and site-specific immobilization of tris-nitrilotriacetic acid (tris-NTA) for His-tagged protein attachment and biotin for streptavidin attachment (Figure 3). In their system, the oligoethylene glycol was introduced to the substrate to decrease non-specific binding between the substrate and undesirable proteins.

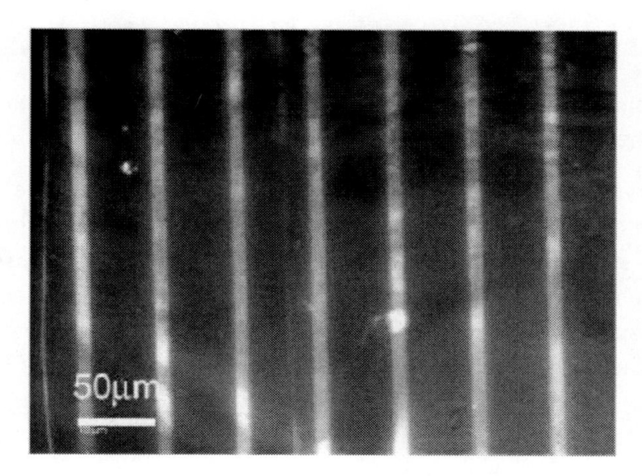

Figure 3. Fluorescence image of a patterned substrate after coupling biotin, then streptavidin, followed by fluorescently labelled biotin. Reprinted with permission from [53]. Copyright 2010 American Chemical Society.

Lee et al. developed a maskless photolithography process to control protein patterns with NVOC molecules [54]. They fabricated a two dimensional micro mirror array, where they used digital micro mirrors as a virtual photomask to create a biotin pattern. Most recently, Grunwald et al. applied photocleavable surfaces to capture and recognize viruses [55]. They created a photo-activatable tris-NTA via a nitrobenzyl linker. The photo-activatable tris-NTAs were self-inactivated by His-tagged proteins. Irradiation could activate the affinity by cleaving a tethered intramolecular ligand and arming a multivalent chelator head. In their study, different strategies were applied to create site-specific protein patterns, including mask patterning, laser lithography, and successive activation of different areas using *in situ* laser scanning lithography. Furthermore, they demonstrated the ability of their system to capture virus-specific very low-density lipoprotein receptor, which made it a highly flexible platform for detection and analysis of clinically relevant virus particles. Until now, many different solid surface-based systems using photocleavable caging groups have been developed and most of them used o-nitrobenzyl derivatives to control protein specific binding to the solid support [56-60]. Achieving control over the size of patterns and avoiding non-specific binding of proteins is still a big challenge for protein patterning techniques based on photochemistry.

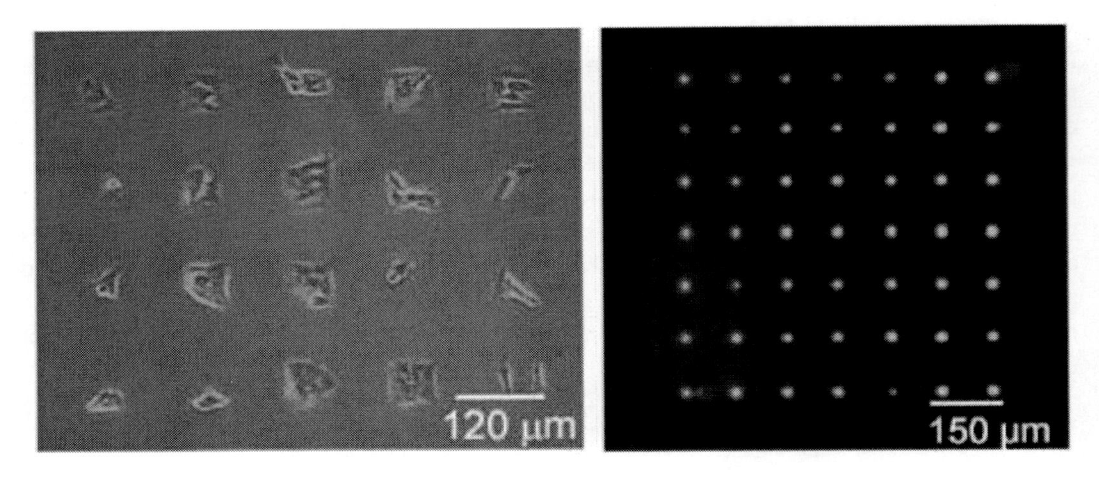

Figure 4. Swiss 3T3 fibroblast cells selectively attached to a SAM. Reprinted with permission from [63]. Copyright 2010 American Chemical Society.

Cell microarrays are important platforms for the study of cellular processes and cell behaviour. It is thus important to spatially control cell adhesion on substrates for cell culture. Using light as a stimulus and developing functional substrates that could respond to irradiation to switch surface properties from cell non-adhesive to adhesive is a powerful method to realize such technologies. Several researchers have already created cell patterns either on glass surfaces or silicon based on self-assembled monolayers (SAMs) [61, 62]. For example, Nakanishi et al. developed a method to control site-specific cell adhesion on a SAM spatiotemporally [45]. In their study, an o-nitrobenzyl derivative was used to create a photocleavable SAM surface based on an alkylsiloxane. Bovine serum albumin (BSA) was adsorbed onto the surface to prevent cell adhesion. Selective irradiation led to the binding of fibronectin to create cell adhesive regions. By controlling the size of the irradiated region, they could reduce the pattern to smaller than a single cell. Their results showed that the cells formed nodal structures corresponding to pattern size and shape. At the same time, by sequential irradiation of the substrate, they could control cell migration and proliferation from one patterned region to another. Their technique is a potential research tool for the manipulation of cells in biological studies. Dillmore et al. provided a method to pattern ligands and cells on SAMs made of alkane-thiol on gold using an o-nitrobenzyl derivate (Figure 4) [63]. Their method began with a NVOC-hydroquinone containing monolayer on gold. Irradiation revealed the hydroquinone, which following oxidation, provided a site for the immobilization of cell adhesive peptides. Finally, the designed surface could generate circular patterns of cell attachment corresponding to the photomask applied.

Park et al. also developed a protein and cell pattern on thiolated gold SAM surfaces using o-nitrobenzyl derivates [64]. Their strategy resulted in the immobilization of a variety of cell adhesive peptides containing the ketone group. At the same time, they showed the sequential immobilization of two fluorescent dyes in a pattern and also the immobilization of ligands in gradients. Our group also developed a photoactive substrate based on an alkane-thiol SAM on gold [65]. Poly(ethylene glycol) (PEG) was introduced to a gold SAM surface to initially create a cell repulsive surface via a photocleavable o-nitrobenzyl functional group. The cell repulsive surface was subsequently rendered cell adhesive by UV-irradiation to cleave the photoactive o-nitrobenzyl group and allow for the further immobilization of cell adhesive peptides onto the irradiated regions. Control of cell attachment was shown on the surfaces before and after UV-irradiation and the efficient control of cell attachment lasted up to 5 days in culture. Kikuchi et al. were able to photocage a variety of functional groups on a glass substrate and use their technology to pattern multiple cell types together on one surface at the single cell level [66]. Furthermore, there exist studies of photocleavable substrates to develop other technologies such as microfluidics devices [44] and devices for protein purification [67].

Photocleavable molecules applied on solid supports are a good platform to develop microarrays for DNA analysis and diagnostics, protein-protein interactions and the study of cell processes, drug screening and other biomedical applications. A solid flat support could minimize the topographical influence on the development of microarray chips and could also tolerate many harsher operating conditions or reaction conditions like organic solvent exposure and high temperatures, which are usually a problem for biomaterials.

3.2. Biomaterials for Controlled Release

The development of materials employing photolabile molecules destined for use in the human body is a much more recent area of research. One major application is for controlled release of drug molecules. Introducing photolabile molecules into the material design can potentially allow for release upon light exposure in a particular area of the body where it is needed and when it is needed. Previously, we had discussed the caging of actual drug molecules. These strategies, conversely, concentrate on caging a delivery vehicle which once developed, could hold any number of different drugs.

One such body of work has focused on the development of light-activated micelles for drug release. Micelles are aggregates formed in aqueous solutions such that a hydrophilic exterior surrounds and protects a mostly hydrophobic interior. Within this interior, drugs or bioactive molecules can be stored. In their pioneering work, Jiang et al. created a copolymer consisting of hydrophilic and hydrophobic blocks. The hydrophobic blocks, which ultimately form the micelle interior, contained a photolabile chromophore. Upon UV irradiation, the chromophore pendant group broke off from the polymer chain transforming its associated hydrophobic block to a hydrophilic one. This transformation ultimately led to the dissociation of the micelle and release of the hydrophobic cargo in its interior [68]. This group later modified their strategy to employ an o-nitrobenzyl derivative as the photolabile moiety in the co-block polymer seen in Fig 5. They were able to study the photo-controlled release of a model dye molecule and demonstrated release via two-photon irradiation, which unfortunately required longer irradiation times to achieve release [69]. In another drug delivery strategy, micelles were created containing a hydrophilic contrast agent in the interior detectable by MRI [70]. The hydrophobic component of the micelle contained an o-nitrobenzyl derivative. Upon UV-irradiation, the photolabile groups could leave resulting in the formation of exposed carboxylic acids whose polar nature increased the hydrophilicity of the polymer micelle causing it to undergo a rearrangement and releasing its cargo in the process. Xie et al. employed a similar strategy but designed the o-nitrobenzyl containing polymer micelles to be biodegradable allowing for effective elimination from the body after release [71]. Such a strategy shows great potential to control the delivery of hydrophilic drugs with light.

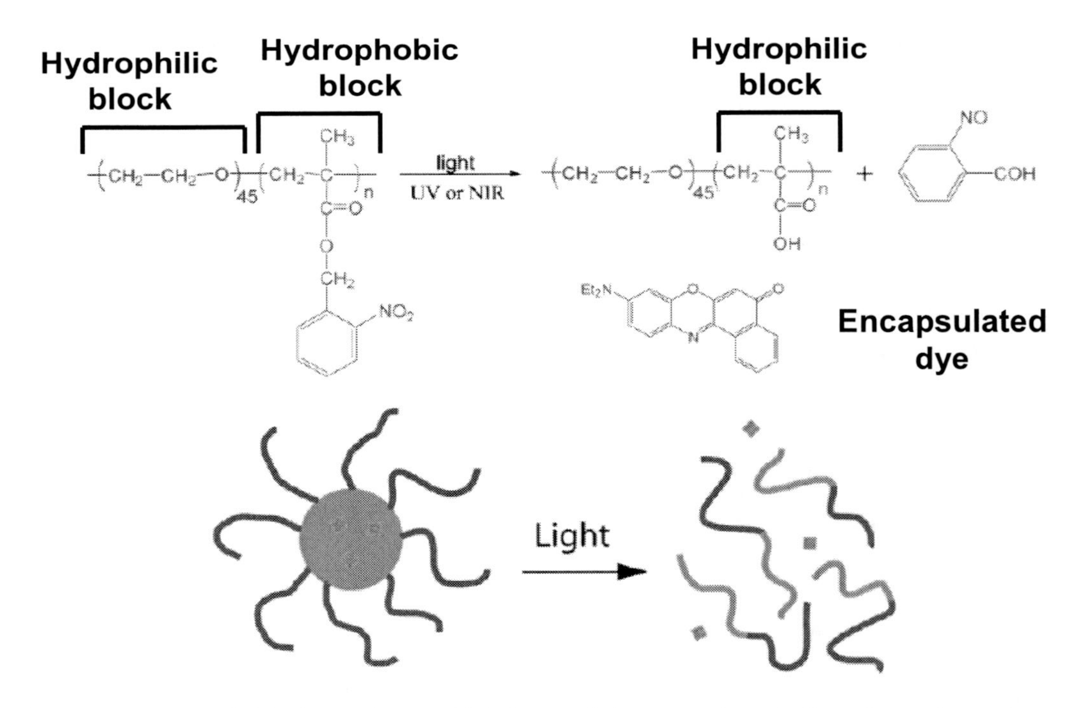

Figure 5. Structure of a block copolymer which forms a light-responsive micelle and the release of a representative encapsulated material after light exposure. Reprinted with permission from [69]. Copyright 2010 American Chemical Society.

Similar to the micelle systems, Li et al. employed dendrimers made of poly(amidoamine) bound to an o-nitrobenzyl group for photo-responsive drug delivery (Fig 6). Dendrimers are large synthetic molecules possessing nanoscale internal cavities and having surfaces with bound functional groups of controllable quantities. Drugs can be encapsulated in the internal cavities. In this study, the o-nitrobenzyl groups form the dendrimer shell keeping the drug molecules, salicylic acid and adriamycin, encased in the dendrimer. UV-induced cleavage of these photolabile groups allowed for drug escape [72]. Park et al. had previously designed dendrimers incorporating photo-activatable groups for light induced release of model dye molecules and developed strategies for encapsulating hydrophilic molecules in the watery interior of the dendrimer as well as hydrophobic molecules in the membrane of the particles [73].

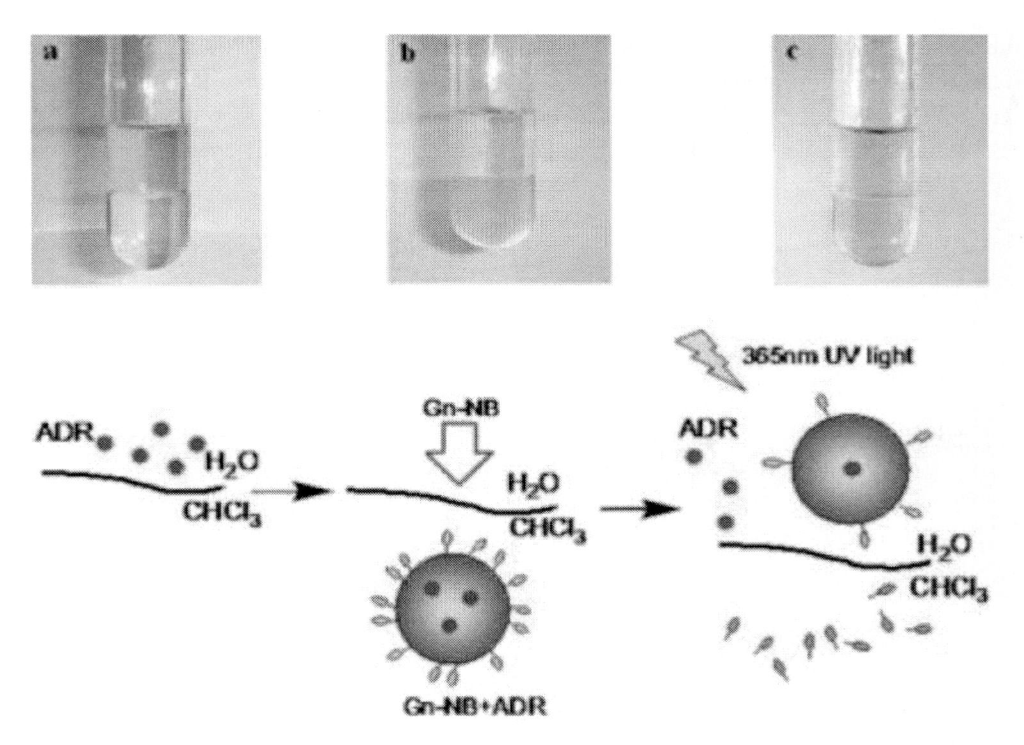

Figure 6. (A) Free Adriamycin (ADR) represented as dots is first dissolved in water and is encapsulated into the dendrimers (Gn-NB) in the chloroform phase upon addition. Note the presence of nitrobenzyl groups (NB) drawn on the surface shell of the dendrimer in (B). Irradiation causes the disintegration of the shell with freed NB groups shown in the chloroform causing release of the ADR molecules into aqueous solution in (C). Reprinted with permission from [72]. Copyright 2010 John Wiley & Sons, Inc.

In their work, Murayama and Kato employed hydrogels as a drug delivery system [74]. The hydrogels encapsulated proteins and degraded upon near-UV light exposure. Degradation was controlled by introducing a photolabile group in the hydrogel crosslinker causing the crosslinkers to break apart upon irradiation. The group demonstrated that enzyme activity could be preserved by encapsulation in the hydrogel whereby released enzymes presented strong activity levels after UV irradiation making their delivery system ideal for bioactive proteins, which are easily degraded in the body and thus often difficult to use as drugs.

Another interesting avenue of controlled drug release research focuses on directly binding the caging group to a drug material and embedding the drug in a polymeric scaffold to form a delivery device. In the previous studies with micelles, the bioactive molecules of interest were released at once in a relatively uncontrolled fashion upon micelle disintegration.

Binding the photocage directly to a drug molecule, by comparison, has the potential to better control the quantity of drug release. McCoy et al. bound several model drugs such as acetyl salicylic acid, ibuprofen, and ketoprofen to a photolabile group, 3,5-dimethyoxybenzoin. The bound compounds were quite hydrophobic and when incorporated into a methyl methacrylate hydrogel, they remained encased in the matrix. Upon irradiation, the photolabile groups were removed rendering the drug molecules more hydrophilic and able to diffuse through aqueous solution in the hydrogel and freeing the drugs for delivery [75].

In a unique strategy, Mizukami et al. created photoresponsive liposomes (Figure 7). However, instead of inserting the cage into the liposome structure, they caged an activator molecule that could create pores in the liposome upon UV irradiation [76]. This molecule was an antimicrobial peptide and they designed the liposome, which carries the biomolecules of interest in its interior, to be of similar composition to bacterial membranes. Using this methodology, they were able to visualize the release of a fluorescent dye from their system post irradiation.

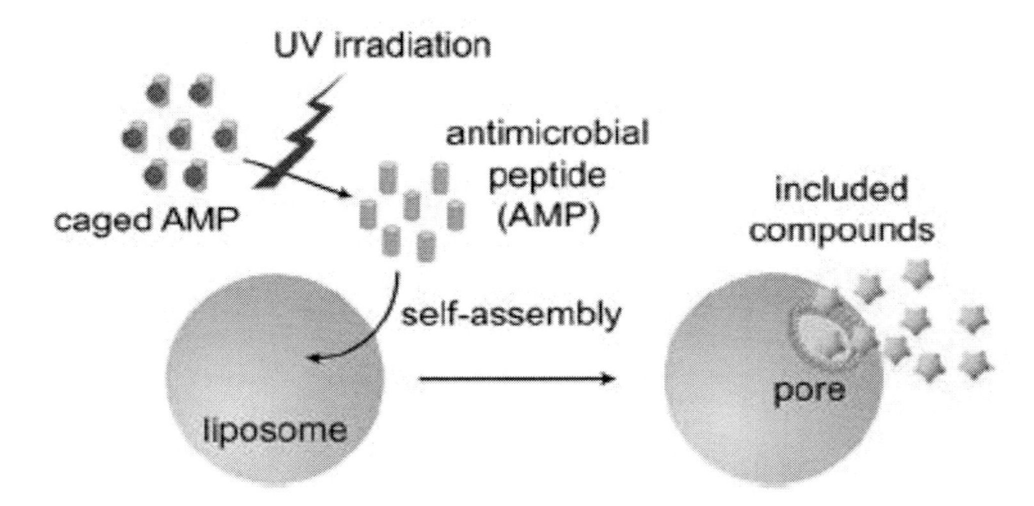

Figure 7. Diagram of a photorelease system based on a caged antimicrobial peptide which is activated upon irradiation to create pores in a liposome carrying molecules of interest. Reprinted with permission from [76]. Copyright 2010 American Chemical Society.

Targeted drug delivery devices are currently in great demand to avoid the side affects associated with active drug freely circulating the body, targeting healthy tissues and cells and causing unwanted side effects. This is an area where traditional caging chemistry can be incorporated into new designs with biomaterials to provide effective solutions. The designs discussed here have mostly been conceived over the past five years showing the beginnings of a new area of research which promises to grow stronger with time.

3.3. Controlling Biomaterial Physical Properties

In the previous section, we saw examples of how materials can be designed with traditional caging groups to control release of a bioactive molecule for drug delivery. For applications such as tissue engineering or cell studies, it would be ideal to have tools which can afford a high level of control over biomaterial physical properties, such as material degradation, formation, or strength.

In one study to control bulk gel degradation, an o-nitrobenzyl derivative was introduced into a peptide amphiphile [77]. Peptide amphiphiles contain a short peptide sequence, which is often bioactive and relatively hydrophilic, connected to a hydrophobic segment. In solution, the molecules act like surfactants presenting the hydrophilic peptide sequences on the surface and burying the hydrophobic faces. These amphiphiles assemble spontaneously to form nanofibres which can be interconnected to form gels [78]. Löwik et al. inserted a photolabile o-nitrobenzyl derivative between the hydrophilic peptide and hydrophobic block of an amphiphile. Near UV-light exposure served to cleave the hydrophobic block away destabilizing the amphiphile leading to degradation of the biomaterial. The group suggested that not only could bulk degradation be performed, but that photomasks could potentially be used to spatially control degradation [77].

Gels incorporating photolabile o-nitrobenzyl derivatives have also been designed to respond to more than one stimulus - both light and temperature. Gels were designed to undergo a transition from solution state to gel at higher temperatures and could be degraded upon exposure to near-UV light which cleaved a pendant 2-nitrobenzyl group leaving behind a free carboxylic acid. This caused a hydrophobic block of the polymer to become hydrophilic which served to increase the sol-gel transition temperature. Thus, the degraded gels could even be induced to gel again, but at a higher temperature [79]. This group had previously designed o-nitrobenzyl containing micelles also responsive to both light and temperature [80].

Gels have also now been designed to degrade in a controlled fashion in select regions. Kloxin et al. created a photodegradable PEG hydrogel by incorporating a novel light-responsive monomer into the material [81]. PEG formed the base of the monomer which was attached to a photolabile o-nitrobenzyl derivative and acrylate end groups to facilitate polymerization (Fig 8A). This modified PEG monomer was copolymerized with PEG acrylate monomers to create the final photoresponsive hydrogel. In regions exposed to light, the modified monomer breaks off from the polymer network causing gel degradation while leaving the material intact in non-exposed regions (Fig 8B). In this way, photomasks were used to control gel degradation for patterned erosion. The group demonstrated that live cells could be encapsulated in the gel and that controlled degradation could be used to manipulate cell movement by allowing for migration in eroded areas.

The group further studied this material system to characterize the degradation kinetics [82]. This ability to predict mass loss within a material can allow for control over the hydrogel's crosslinking density with irradiation time. The group was further able to vary the degree of degradation within the gel to create controlled gradients in both the x-y plane and z-direction. Material strength, which is related to the degree of crosslinking, is known to impact cell morphology and differentiation [83, 84]. Towards such applications, human mesenchymal stem cells were successfully encapsulated within their degradable gels which were irradiated to create a gradient in the z-direction. Cells were found to spread within the material in areas of decreased crosslinking density demonstrating the biocompatibility of their system in addition to the ability to control stem cell behavior with light.

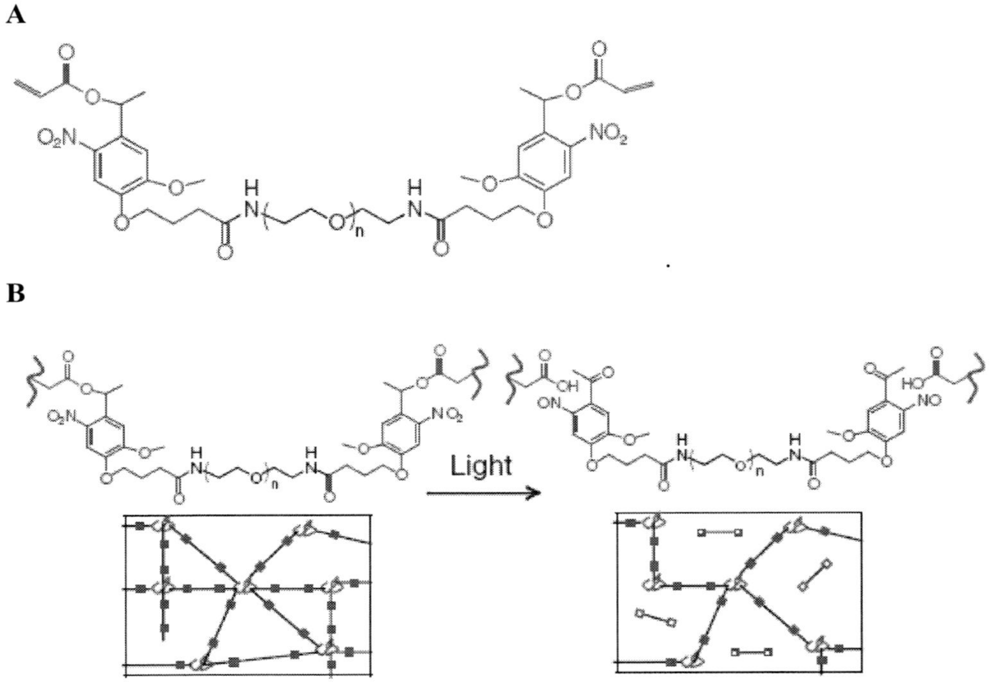

A

B

Light

Figure 8. (A) Monomer used in the formation of photodegradable hydrogels (B) Method of light catalyzed polymer degradation where PEG polymer chains (coils) are crosslinked with PEG (lines) containing photolabile nitrobenzyl groups (squares) From [81]. Reprinted with permission from AAAS.

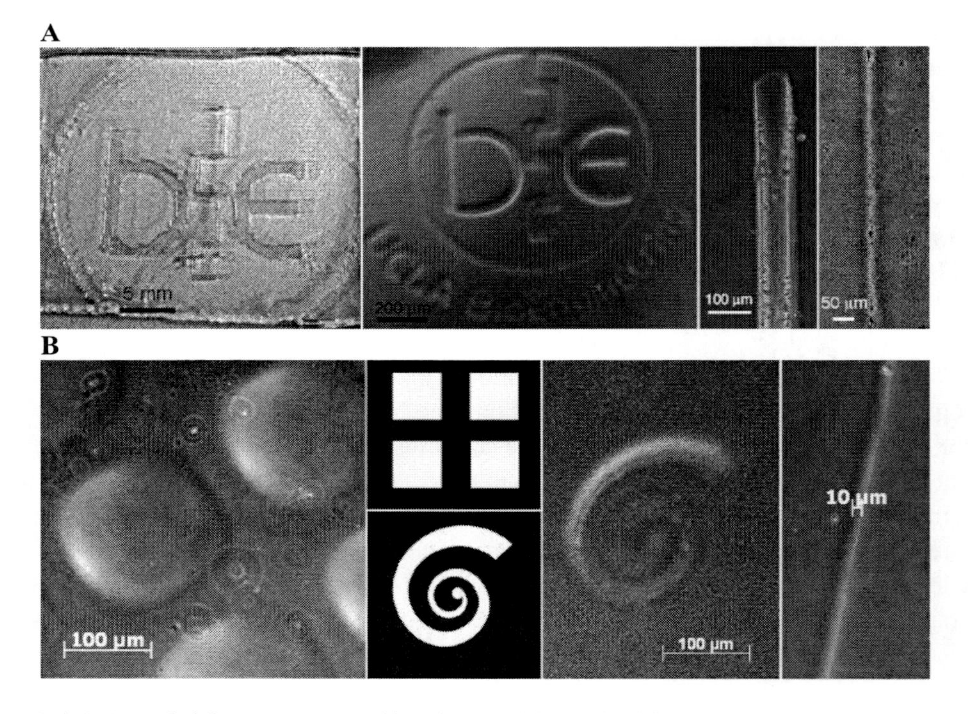

Figure 9. (A) Degraded features patterned in a hydrogel (B) Raised features patterned in a hydrogel. Reprinted with permission from [85]. Copyright 2010 American Chemical Society.

The controllable degradation of this material was further used to produce patterned features in the hydrogel [85]. Negative, or recessed, features were created by exposing the gel to light through a photomask resulting in degradation in the exposed areas (Fig 9 A) while positive, or raised, features were created by decreasing the crosslinking density in light-exposed regions without complete erosion of the material. This led to increased swelling of the hydrogel in these areas generating material patterns such as those seen in Figure 10 B. In addition, they were able to demonstrate degradation via two-photon photolysis. Since two-photon laser technology has the capability to focus light in three-dimensions, this leads to the possibility of forming a wide range of three dimensional patterns that do not necessarily need to be interconnected.

Materials with potential biological applications can not only be degraded with light, but also be induced to form with the help of photolabile switches. The ability to form hydrogels using light as a trigger can be very useful in biomedical applications. A liquid material could be injected into a body cavity and induced to gel at a pre-determined time point through irradiation allowing the gel to take the shape of the local environment. Haines et al. designed a peptide that can spontaneously fold in solution to self-assemble into a hydrogel. The peptide forms into a folded hairpin structure whereby one face of the hairpin is hydrophobic and the other hydrophilic. Since hydrophobic interactions were key to hydrogel self-assembly, the group hypothesized that introducing a negative charge on the hydrophobic side via a photolabile o-nitrobenzyl derivative could disrupt peptide folding. The group was able to successfully induce folding of the peptides upon near-UV light exposure [86]. Biocompatibility of the material was demonstrated by the adhesion and migration of 3T3 fibroblasts. Muraoka et al. similarly designed photoresponsive self-assembling peptides with the ability to gel upon irradiation and in one such peptide they were able to incorporate the well-known cell-adhesive peptide sequence Arg-Gly-Asp-Ser (RGDS) [87, 88]. Their material was also shown to support a 3T3 fibroblast culture in 3D where the cells developed focal adhesions on the gelled material indicative of cell attachment and spreading.

Traditional caging groups can thus be incorporated into soft biocompatible materials to control their degradation dynamically and spatially to produce high resolution patterns. Gelation can even be brought under temporal control using light as a switch. These techniques offer new tools for improved control over biomaterial properties for new tissue engineering solutions and applications in the study of cell biology.

3.4. Controlling Biomaterial Chemical Properties: Patterning Biomolecules and Cells

Another exciting application for traditional photolabile caging groups in biomaterial design is the generation of protein and cell patterns using light as a switch. One effective strategy to generate a peptide pattern for cell patterning is to cage an adhesive peptide and then bind it to a non-adhesive biomaterial. This creates a cell non-adhesive material prior to light exposure. Uncaging the peptide upon irradiation in a pattern leads to adhesive spots on the biomaterial to which cells can selectively bind. Our group utilized caged RGDS peptides bound to a non-adhesive hyaluronic acid hydrogel material to create such a cell pattern [89]. Figure 10 shows lined patterns of cells at various times post irradiation. We were later able to demonstrate the patterning of two cell types on the same surface to generate a patterned co-culture [90]. Petersen et al. and Ohmuro-Matsuyama utilized a similar caged peptide strategy to pattern 3T3s on solid-surfaces [91, 92].

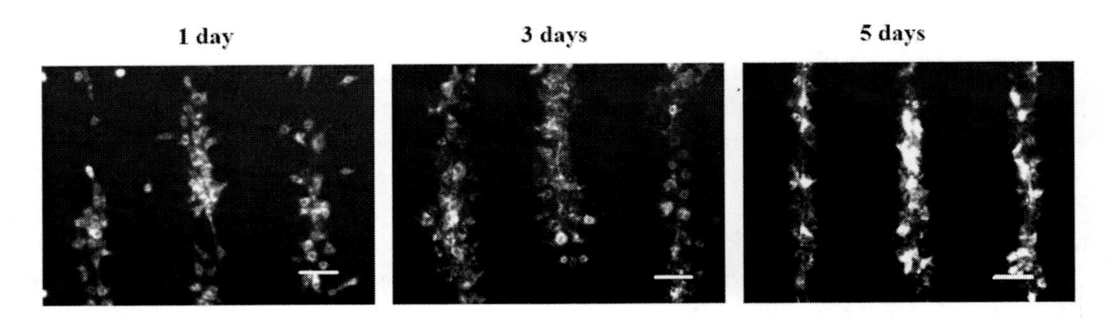

Figure 10. Fluorescence micrographs depicting line patterns of 3T3 fibroblasts dyed with CellTracker™ Red and grown on a HA hydrogel bound with patterned caged and uncaged R[G]DS peptide over five days post seeding. Scale bars = 100 μm. [90].

Gu and Yang developed a novel technique to create coordinated physical and chemical patterns in a biocompatible PEG hydrogel using a photolabile group and an enzyme [93]. Their technique relies on a unique crosslinker which contains a peptide sequence that can be cleaved by specific protease enzymes. However, the cleavage sequence is initially caged via an o-nitrobenzyl derivative so that it is inaccessible to the enzymes. Initial patterns are thus created by irradiating the crosslinked PEG gel to remove the photolabile group to create free peptide sequences in a desired pattern. Exposure to proteases then dissolves the gel in the irradiated regions to create a physical pattern. Protease degradation furthermore results in the formation of free reactive amine groups which can be used to bind bioactive molecules in the physically patterned regions to create chemical patterns as well. Using this methodology, the group was able to create cell patterns on the single cell level in a hydrogel. They patterned circular wells 20 μm in diameter separated by 20 μm and 500 μm in depth and chemically patterned antibodies in the wells to trap B-cells.

There is a great deal of interest in guiding cells through a 3D environment. Patterning cells in 3D would allow for improved re-creations of the natural cellular microenvironment. In an exciting study, Shoichet's group was able to guide neuronal cells in three dimensions employing photolabile o-nitrobenzyl derivatives. The base materials they developed included an optically transparent agarose hydrogel and hyaluronic acid hydrogel through which cells could migrate. They modified the hydrogels to contain sulfhydryl groups protected by the o-nitrobenzyl derivative. With a laser, they were able to create a channel of free sulfhydryl groups which could subsequently be used to bind bioactive, adhesive peptides. They chose RGDS as the peptide and discovered that neurons could indeed be guided along these adhesive channels through the bulk of the gel (Fig 11) [94, 95]. In subsequent studies, this group attempted 3D patterning via two photon microscopy [96, 97]. The two photon laser can potentially be used to form much more complex patterns to perhaps approach those complex biochemical cues found in natural tissue. Since nitrobenzyl-based photolabile groups have often been seen in the literature to be inefficiently removed with two-photon irradiation, the group decided to use a 6-bromo-7-hydroxycoumarin group in its place. They were able to demonstrate the formation of three dimensional patterns in the volume of the hydrogel (Fig 12) formed by two-photon irradiation by binding a dye to the freed sulfhydryl groups for pattern visualization. In a very recent study utilizing a similar patterning strategy, they were able to create gradients of a vascular endothelial growth factor in an agarose hydrogel bound with adhesive RGDS peptides to guide the migration of endothelial cells in 3D and induce the formation of tubule-like structures [98].

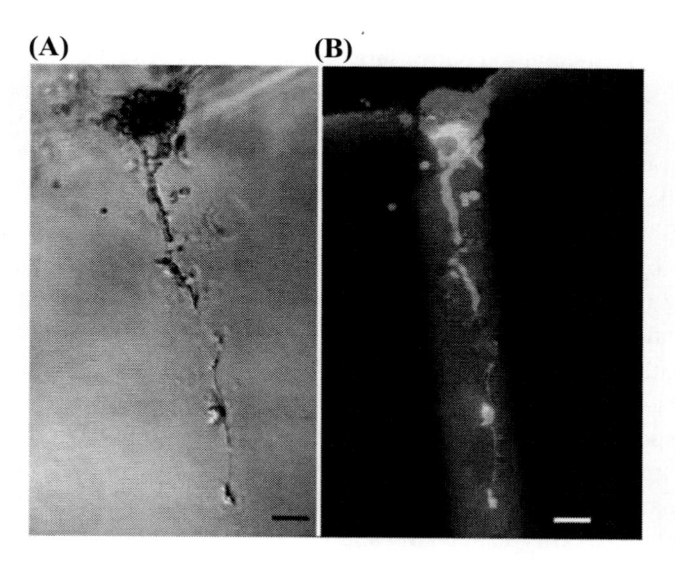

Figure 11. (A) Dorsal root ganglion cells growing within an agarose channel patterned with GRGDS peptides via photolabile nitrobenzyl groups (B) Fluorescent micrograph of patterned agarose channel dyed green with fluorescein and cells labeled red. Reprinted by permission from Macmillan Publishers Ltd.: Nature Materials [94], copyright 2010.

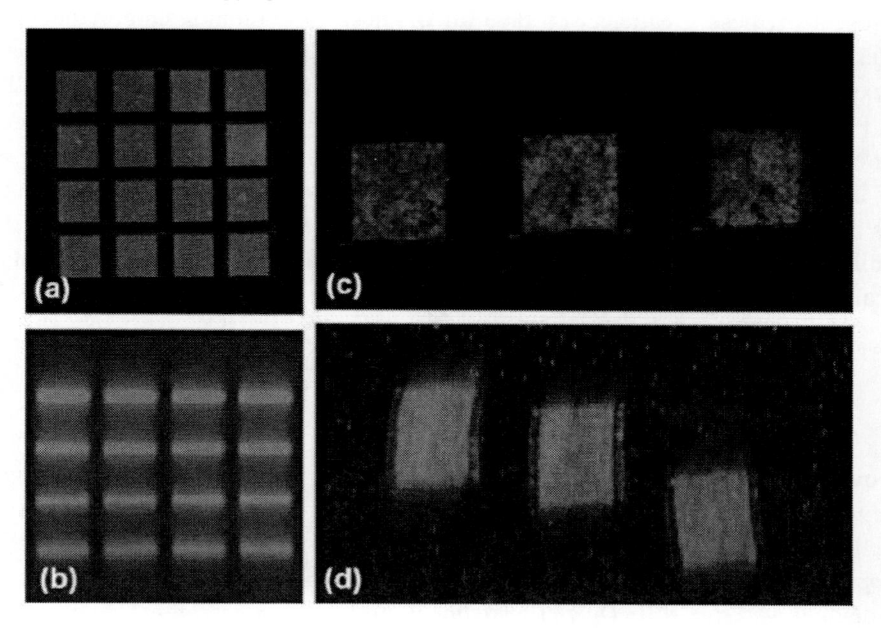

Figure 12. (A) Top view and (B) side view of two photon patterned squares in a hydrogel and (C) top view and (D) side view of rectangular volumes patterned in a hydrogel visualized with AF488-Mal dye. Reprinted with permission from [96]. Copyright 2010 American Chemical Society.

Conclusion

Since the late 1970s, the caging of biomolecules and the design of biomaterials incorporating traditional photolabile caging groups has grown into a great body of highly creative research with huge potential. With the technology to cage small activators, proteins, and nucleic acids, the potential to control virtually any gene product or biochemical pathway with a light switch is within our grasp. With this power, we can hope to expand our current knowledge in the biological sciences and come to improved understandings of a variety of disease states. However, we appear to be sitting in the beginning stages of this technology's development. While we are amassing growing libraries of caged biomolecules, the number of purely application-based research papers taking advantage of the unique spatial and temporal control afforded by these techniques is limited. The application of these traditional photolabile caging molecules for the light-activated control over the design of solid-surface biotechnologies and biomaterial properties is relatively newer. Many of the studies discussed were published within the last five years, or even current year at the time of this writing. Material engineers and scientists have harnessed these molecules to make truly creative dynamic, light-responsive materials which could be developed into biochips, drug delivery devices, tissue engineered constructs and more. We have the power to carve channels and features into biomaterials, modify their strength and degradation, dictate gel formation, and create cell and protein patterns all spatially and temporally controlled with light. On or within these materials, cells can be exposed to changing conditions under our control to monitor or direct their behavior. However, once again, these materials are mostly in purely developmental stages. Caging biomolecules and designing these complex materials often requires a high level of specialized technical skills and such training may not be available to the end-users of these creations. To take all of these technologies employing photocaged molecules to the next step, the chemists, material scientists and engineers who have become design experts in these areas must form close ties with biologists, biochemists, and medical experts. Physicists and electrical engineers should also be consulted in the design of optimal lasers or other light sources for these different applications, which may involve laser access to areas of the body for drug delivery, creating a patterned exposure, or irradiating small focal volumes in three dimensions for local uncaging. As long as we can work together, the use of photolabile caging molecules has the potential to bring about leaps in understanding in the biological sciences and a new generation of highly tunable and responsive biomaterials.

References

[1] Young DD, Deiters A. Photochemical control of biological processes *Organic and Biomolecular Chemistry*. 2007;5(7):999-1005.

[2] Pelliccioli AP, Wirz J. *Photoremovable protecting groups: Reaction mechanisms and applications Photochemical and Photobiological Sciences*. 2002;1(7):441-58.

[3] Goeldner M, Givens R. *Dynamic studies in biology : phototriggers, photoswitches and caged biomolecules*. Weinheim: Wiley-VCH 2005.

[4] Pelliccioli AP, Wirz J. Photoremovable protecting groups: Reaction mechanisms and applications. *Photochemical and Photobiological Sciences*. 2002;1(7):441-58.

[5] Yu H, Li J, Wu D, Qiu Z, Zhang Y. Chemistry and biological applications of photolabile organic molecules *Chemical Society Reviews*. 2010;39(2):464-73
[6] Denk W, Strickler JH, Webb WW. Two-Photon Laser Scanning Fluorescence Microscopy. *Science*. 1990;248(4951):73-6.
[7] Kaplan JH, Forbush B, Hoffman JF. Rapid photolytic release of adenosine 5'-triphosphate from a protected analog: utilization by the sodium:potassium pump of human red blood cell ghosts. *Biochemistry*. 1978 05/01;17(10):1929-35.
[8] Jun B, Kim S. Real-time Structural Transitions Are Coupled to Chemical Steps in ATP Hydrolysis by Eg5 Kinesin. *Journal of Biological Chemistry*. 2010 April 9, 2010;285(15):11073-7.
[9] Ellis-Davies GCR. Development and application of caged calcium. *Methods in Enzymology*: Academic Press 2003:226-38.
[10] Gomez TM, Spitzer NC. In vivo regulation of axon extension and pathfinding by growth-cone calcium transients. *Nature*. 1999;397(6717):350-5.
[11] Pavlos CM, Hua X, Toscano JP. Photosensitive Precursors to Nitric Oxide. *Current Topics in Medicinal Chemistry*. 2005 07;5(7):637-47.
[12] Dalva MB, Katz LC. Rearrangements of Synaptic Connections in Visual Cortex Revealed by Laser Photostimulation. *Science*. 1994;265(5169):255-8.
[13] Kramer RH, Fortin DL, Trauner D. New photochemical tools for controlling neuronal activity. *Current Opinion in Neurobiology*. 2009;19(5):544-52.
[14] Cambridge SB, Geissler D, Keller S, Cürten B. A caged doxycycline analogue for photoactivated gene expression. *Angewandte Chemie - International Edition* 2006;45(14):2229-31.
[15] Sauers DJ, Temburni MK, Biggins JB, Ceo LM, Galileo DS, Koh JT. Light-Activated Gene Expression Directs Segregation of Co-cultured Cells in Vitro. *ACS Chemical Biology*. 2010 01/05;5(3):313-20.
[16] Bayley H. Pore-Forming Proteins with Built-in Triggers and Switches. *Bioorganic Chemistry*. 1995;23(4):340-54.
[17] Chang C-y, Niblack B, Walker B, Bayley H. A photogenerated pore-forming protein. *Chemistry and Biology*. 1995 06/01;2(6):391-400.
[18] Marriott G. Caged Protein Conjugates and Light-Directed Generation of Protein Activity: Preparation, Photoactivation, and Spectroscopic Characterization of Caged G-Actin Conjugates. *Biochemistry*. 1994 08/01;33(31):9092-7.
[19] Nandy SK, Agnes RS, Lawrence DS. Photochemically-Activated Probes of Protein Protein Interactions. *Organic Letters*. 2007 05/17;9(12):2249-52.
[20] Bourgault S, Létourneau M, Fournier A. Development of photolabile caged analogs of endothelin-1. *Peptides*. 2007;28(5):1074-82.
[21] Wood JS, Koszelak M, Liu J, Lawrence DS. A Caged Protein Kinase Inhibitor. *Journal of the American Chemical Society*. 1998;120(28):7145-6.
[22] Tatsu Y, Nishigaki T, Darszon A, Yumoto N. A caged sperm-activating peptide that has a photocleavable protecting group on the backbone amide. *FEBS Letters*. 2002;525(1-3):20-4.
[23] Hiraoka T, Hamachi I. Caged RNase: photoactivation of the enzyme from perfect off-state by site-specific incorporation of 2-nitrobenzyl moiety. *Bioorganic & Medicinal Chemistry Letters*. 2003;13(1):13-5.

[24] Wu N, Deiters A, Cropp TA, King D, Schultz PG. A Genetically Encoded Photocaged Amino Acid. *Journal of the American Chemical Society.* 2004 10/19;126(44):14306-7.

[25] Gautier A, Nguyen DP, Lusic H, An W, Deiters A, Chin JW. Genetically Encoded Photocontrol of Protein Localization in Mammalian Cells. *Journal of the American Chemical Society.* 03/10/;132(12):4086-8.

[26] Chen PR, Groff D, Guo J, Ou W, Cellitti S, Geierstanger BH, et al. A facile system for encoding unnatural amino acids in mammalian cells. *Angewandte Chemie - International Edition.* 2009;48(22):4052-5

[27] Hahn ME, Muir TW. Photocontrol of Smad2, a Multiphosphorylated Cell-Signaling Protein, through Caging of Activating Phosphoserines. *Angewandte Chemie - International Edition.* 2004;43(43):5800-3.

[28] Pellois J-P, Muir TW. A ligation and photorelease strategy for the temporal and spatial control of protein function in living cells *Angewandte Chemie - International Edition.* 2005;44(35):5713-7.

[29] Chou C, Young DD, Deiters A. Photocaged T7 RNA polymerase for the light activation of transcription and gene function in pro- and eukaryotic cells. *ChemBioChem.* 2010;11(7):972-7.

[30] Sadovski O, Jaikaran ASI, Samanta S, Fabian MR, Dowling RJO, Sonenberg N, et al. A collection of caged compounds for probing roles of local translation in neurobiology. *Bioorganic & Medicinal Chemistry.* 2010;In Press, Corrected Proof.

[31] Goard M, Aakalu G, Fedoryak OD, Quinonez C, St. Julien J, Poteet SJ, et al. Light-Mediated Inhibition of Protein Synthesis. *Chemistry and Biology.* 2005 06/01;12(6):685-93.

[32] Miller DS, Chirayil S, Ball HL, Luebke KJ. Manipulating cell migration and proliferation with a light-activated polypeptide *ChemBioChem.* 2009;10(3):577-84.

[33] Priestman MA, Lawrence DS. Light-mediated remote control of signaling pathways. Biochimica et Biophysica Acta (BBA) - *Proteins & Proteomics.* 2010;1804(3):547-58.

[34] Sinha DK, Neveu P, Gagey N, Aujard I, Benbrahim-Bouzidi C, Le Saux T, et al. Photocontrol of protein activity in cultured cells and zebrafish with one- and two-photon illumination. *ChemBioChem.* 2009;11(5):653-63

[35] Furuta T, Watanabe T, Tanabe S, Sakyo J, Matsuba C. Phototriggers for Nucleobases with Improved Photochemical Properties. *Organic Letters.* 2007 10/11;9(23):4717-20.

[36] Ando H, Furuta T, Tsien RY, Okamoto H. Photo-mediated gene activation using caged RNA/DNA in zebrafish embryos. *Nature Genetics.* 2001 08;28(4):317.

[37] Yamaguchi S, Chen Y, Nakajima S, Furuta T, Nagamune T. Light-activated gene expression from site-specific caged DNA with a biotinylated photolabile protection group. *Chemical Communications.* 2010;46(13):2244-6.

[38] Casey JP, Blidner RA, Monroe WT. Caged siRNAs for Spatiotemporal Control of Gene Silencing. *Molecular Pharmaceutics.* 2009 04/16;6(3):669-85.

[39] Monroe WT, McQuain MM, Chang MS, Alexander JS, Haselton FR. Targeting Expression with Light Using Caged DNA. *Journal of Biological Chemistry.* 1999 July 23, 1999;274(30):20895-900.

[40] Thompson S, Self AC, Self CH. Light-activated antibodies in the fight against primary and metastatic cancer. *Drug Discovery Today.* 2010;15(11-12):468-73.

[41] Skwarczynski M, Noguchi M, Hirota S, Sohma Y, Kimura T, Hayashi Y, et al. Development of first photoresponsive prodrug of paclitaxel. *Bioorganic & Medicinal Chemistry Letters*. 2006;16(17):4492-6.

[42] Reinhard R, Schmidt BF. Nitrobenzyl-Based Photosensitive Phosphoramide Mustards: Synthesis and Photochemical Properties of Potential Prodrugs for Cancer Therapy. *Journal of Organic Chemistry*. 1998;63(8):2434-41.

[43] Li L-S, Babendure JL, Sinha SC, Olefsky JM, Lerner RA. Synthesis and evaluation of photolabile insulin prodrugs. *Bioorganic & Medicinal Chemistry Letters*. 2005;15(17):3917-20.

[44] Lim M, Rothschild KJ. Photocleavage-based affinity purification and printing of cell-free expressed proteins: Application to proteome microarrays. *Analytical Biochemistry*. 2008 Dec;383(1):103-15.

[45] Nakanishi J, Kikuchi Y, Takarada T, Nakayama H, Yamaguchi K, Maeda M. Spatiotemporal control of cell adhesion on a self-assembled monolayer having a photocleavable protecting group. *Analytica Chimica Acta*. 2006 Sep;578(1):100-4.

[46] Seo TS, Bai XP, Ruparel H, Li ZM, Turro NJ, Ju JY. Photocleavable fluorescent nucleotides for DNA sequencing on a chip constructed by site-specific coupling chemistry. *Proceedings of the National Academy of Sciences of the United States of America*. 2004 Apr;101(15):5488-93.

[47] Afroz F, Barone AD, Bury PA, Chen CA, Cuppoletti A, Kuimelis RG, et al. Photo-removable protecting groups for in situ DNA microarray synthesis. *Clinical Chemistry*. 2004 Oct;50(10):1936-9.

[48] Chee M, Yang R, Hubbell E, Berno A, Huang XC, Stern D, et al. Accessing genetic information with high-density DNA arrays. *Science*. 1996 Oct;274(5287):610-4.

[49] Fodor SPA, Read JL, Pirrung MC, Stryer L, Lu AT, Solas D. Light-Directed, Spatially Addressable Parallel Chemical Synthesis. *Science*. 1991;251(4995):767-73.

[50] Mrksich M. Using self-assembled monolayers to model the extracellular matrix. *Acta Biomaterialia*. 2009 Mar;5(3):832-41.

[51] Pelliccioli AP, Wirz J. Photoremovable protecting groups: reaction mechanisms and applications. *Photochemical & Photobiological Sciences*. 2002 Jul;1(7):441-58.

[52] Sundberg SA, Barrett RW, Pirrung M, Lu AL, Kiangsoontra B, Holmes CP. Spatially-addressable immobilization of macromolecules on solid supports. *Journal of the American Chemical Society*. 1995 Dec;117(49):12050-7.

[53] Alonso JM, Reichel A, Piehler J, del Campo A. Photopatterned surfaces for site-specific and functional immobilization of proteins. *Langmuir*. 2008 Jan;24(2):448-57.

[54] Lee KN, Shin DS, Chung WJ, Lee YS, Kim YK. Photochemical selective surface modification using micromirror array for biochip fabrication. In: Karam JM, Yasaitis J, eds. *Micromachining and Microfabrication Process Technology Vii* 2001:352-9.

[55] Grunwald C, Schulze K, Reichel A, Weiss VU, Blaas D, Piehler J, et al. In situ assembly of macromolecular complexes triggered by light. *Proceedings of the National Academy of Sciences of the United States of America*.107(14):6146-51.

[56] Mancini RJ, Li RC, Tolstyka ZP, Maynard HD. Synthesis of a photo-caged aminooxy alkane thiol. *Org Biomol Chem*. 2009;7(23):4954-9.

[57] Veiseh M, Zareie MH, Zhang MQ. Highly selective protein patterning on gold-silicon substrates for biosensor applications. *Langmuir*. 2002 Aug;18(17):6671-8.

[58] Banala S, Arnold A, Johnsson K. Caged substrates for protein labeling and immobilization. *Chembiochem*. 2008 Jan;9(1):38-41.

[59] Nakagawa M, Ichimura K. Photopatterning of self-assembled monolayers to generate aniline moieties. *Colloids and Surfaces a-Physicochemical and Engineering Aspects*. 2002 May;204(1-3):1-7.

[60] Ito Y, Nogawa M, Takeda M, Shibuya T. Photo-reactive polyvinylalcohol for photo-immobilized microarray. *Biomaterials*. 2005 Jan;26(2):211-6.

[61] Chen SY, Smith LM. Photopatterned Thiol Surfaces for Biomolecule Immobilization. *Langmuir*. 2009 Oct;25(20):12275-82.

[62] Nakayama H, Nakanishi J, Shimizu T, Yoshino Y, Iwai H, Kaneko S, et al. Silane coupling agent bearing a photoremovable succinimidyl carbonate for patterning amines on glass and silicon surfaces with controlled surface densities. *Colloids and Surfaces B-Biointerfaces*. Mar;76(1):88-97.

[63] Dillmore WS, Yousaf MN, Mrksich M. A photochemical method for patterning the immobilization of ligands and cells to self-assembled monolayers. *Langmuir*. 2004 Aug;20(17):7223-31.

[64] Park S, Yousaf MN. An interfacial oxime reaction to immobilize ligands and cells in patterns and gradients to photoactive surfaces. *Langmuir*. 2008 Jun;24(12):6201-7.

[65] Cheng N, Cao XD. Photoactive SAM surface for control of cell attachment. *Journal of Colloid and Interface Science*. Aug;348(1):71-9.

[66] Kikuchi Y, Nakanishi J, Shimizu T, Nakayama H, Inoue S, Yamaguchi K, et al. Arraying Heterotypic Single Cells on Photoactivatable Cell-Culturing Substrates. *Langmuir*. 2008 10/16/;24(22):13084-95.

[67] Besson E, Gue AM, Sudor J, Korri-Youssoufi H, Jaffrezic N, Tardy J. A novel and simplified procedure for patterning hydrophobic and hydrophilic SAMs for microfluidic devices by using UV photolithography. *Langmuir*. 2006 Sep;22(20):8346-52.

[68] Jiang J, Tong X, Zhao Y. A New Design for Light-Breakable Polymer Micelles. *Journal of the American Chemical Society*. 2005 05/17;127(23):8290-1.

[69] Jiang J, Tong X, Morris D, Zhao Y. Toward Photocontrolled Release Using Light-Dissociable Block Copolymer Micelles. *Macromolecules*. 2006 05/23;39(13):4633-40.

[70] Lepage M, Jiang J, Babin J, Qi B, Tremblay L, Zhao Y. MRI observation of the light-induced release of a contrast agent from photo-controllable polymer micelles. *Physics in Medicine and Biology*. 2007;52:N249–N55.

[71] Zhigang X, Xiuli H, Xuesi C, Guojun M, Jing S, Xiabin J. A Novel Biodegradable and Light-Breakable Diblock Copolymer Micelle for Drug Delivery. *Advanced Engineering Materials*. 2009;11(3):B7-B11.

[72] Li Y, Jia X, Gao M, He H, Kuang G, Wei Y. Photoresponsive nanocarriers based on PAMAM dendrimers with a o-nitrobenzyl shell. *Journal of Polymer Science Part A: Polymer Chemistry*. 2010;48(3):551-7.

[73] Park C, Lim J, Yun M, Kim C. Photoinduced Release of Guest Molecules by Supramolecular Transformation of Self-Assembled Aggregates Derived from Dendrons. *Angewandte Chemie - International Edition*. 2008;47(16):2959-63.

[74] Murayama S, Kato M. Photocontrol of Biological Activities of Protein by Means of a Hydrogel. *Analytical Chemistry*. 2010;82(6):2186-91.

[75] McCoy CP, Rooney C, Edwards CR, Jones DS, Gorman SP. Light-Triggered Molecule-Scale Drug Dosing Devices. *Journal of the American Chemical Society*. 2007 07/18/;129(31):9572-3.

[76] Mizukami S, Hosoda M, Satake T, Okada S, Hori Y, Furuta T, et al. Photocontrolled Compound Release System Using Caged Antimicrobial Peptide. *Journal of the American Chemical Society*. 2010 06/29;132(28):9524-5.

[77] Löwik DWPM, Meijer JT, Minten IJ, van Kalkeren H, Heckenmüller L, Schulten I, et al. Controlled disassembly of peptide amphiphile fibres. *Journal of Peptide Science*. 2008;14(2):127-33.

[78] Cui H, Webber MJ, Stupp SI. Self-assembly of peptide amphiphiles: From molecules to nanostructures to biomaterials. *Peptide Science*. 2010;94(1):1-18.

[79] Woodcock JW, Wright RAE, Jiang X, O'Lenick TG, Zhao B. Dually responsive aqueous gels from thermo- and light-sensitive hydrophilic ABA triblock copolymers. *Soft Matter*. 2010;6(14):3325-36.

[80] Jiang X, Lavender CA, Woodcock JW, Zhao B. Multiple Micellization and Dissociation Transitions of Thermo- and Light-Sensitive Poly(ethylene oxide)-b-poly(ethoxytri(ethylene glycol) acrylate-co-o-nitrobenzyl acrylate) in Water. *Macromolecules*. 2008 03/08;41(7):2632-43.

[81] Kloxin AM, Kasko AM, Salinas CN, Anseth KS. Photodegradable Hydrogels for Dynamic Tuning of Physical and Chemical Properties. *Science*. 2009 April 3, 2009;324(5923):59-63.

[82] Kloxin AM, Tibbitt MW, Kasko AM, Jonathan AF, J.A., Anseth KS. Tunable Hydrogels for External Manipulation of Cellular Microenvironments through Controlled Photodegradation. *Advanced Materials*. 2010;22(1):61-6.

[83] Lutolf MP, Gilbert PM, Blau HM. Designing materials to direct stem-cell fate. *Nature*. 2009;462(7272):433-41.

[84] Leipzig ND, Shoichet MS. The effect of substrate stiffness on adult neural stem cell behavior. *Biomaterials*. 2009;30(36):6867-78.

[85] Wong DY, Griffin DR, Reed J, Kasko AM. Photodegradable Hydrogels to Generate Positive and Negative Features over Multiple Length Scales. *Macromolecules*. 2010 02/24/;43(6):2824-31.

[86] Haines LA, Rajagopal K, Ozbas B, Salick DA, Pochan DJ, Schneider JP. Light-Activated Hydrogel Formation via the Triggered Folding and Self-Assembly of a Designed Peptide. *Journal of the American Chemical Society*. 2005 11/10;127(48):17025-9.

[87] Muraoka T, Cui H, Stupp SI. Quadruple Helix Formation of a Photoresponsive Peptide Amphiphile and Its Light-Triggered Dissociation into Single Fibers. *Journal of the American Chemical Society*. 2008 02/16;130(10):2946-7.

[88] Muraoka T, Koh C-Y, Cui H, Stupp SI. Light-Triggered Bioactivity in Three Dimensions. *Angewandte Chemie International Edition*. 2009;48(32):5946-9.

[89] Goubko C, Majumdar S, Basak A, Cao X. Hydrogel cell patterning incorporating photocaged RGDS peptides. *Biomedical Microdevices*. 2010;12(3):555-68.

[90] Goubko CA, Majumdar S, Basak A, Cao X. Novel cell patterning platform employing photocaged RGDS peptides on a hydrogel. *AIChE Annual Meeting, Conference Proceedings*. Nashville, TN 2009.

[91] Petersen S, Alonso JM, Specht A, Duodu P, Goeldner M, del Campo A. Phototriggering of Cell Adhesion by Caged Cyclic RGD Peptides. *Angewandte Chemie International Edition*. 2008;47(17):3192-5.

[92] Ohmuro-Matsuyama Y, Tatsu Y. Photocontrolled Cell Adhesion on a Surface Functionalized with a Caged Arginine-Glycine-Aspartate Peptide. *Angewandte Chemie International Edition*. 2008;47(39):7527-9.

[93] Gu Z, Tang Y. Enzyme-assisted photolithography for spatial functionalization of hydrogels. *Lab on a Chip*. 2010;10(15):1946-51.

[94] Luo Y, Shoichet MS. A photolabile hydrogel for guided three dimensional cell growth and migration. *Nature Materials*. 2004;3:249-53.

[95] Musoke-Zawedde P, Shoichet MS. Anisotropic three-dimensional peptide channels guide neurite outgrowth within a biodegradable hydrogel matrix *Biomedical Materials*. 2006;1(3):162-9.

[96] Wosnick JH, Shoichet MS. Three-dimensional Chemical Patterning of Transparent Hydrogels. *Chemistry of Materials*. 2008;20(1):55-60.

[97] Wylie RG, Shoichet MS. Two-photon micropatterning of amines within an agarose hydrogel. *Journal of Materials Chemistry*. 2008;18(23):2716-21.

[98] Aizawa Y, Wylie R, Shoichet M. Endothelial Cell Guidance in 3D Patterned Scaffolds. *Advanced Materials*. 2010.

In: Photochemistry
Editors: Karen J. Maes and Jaime M. Willems
ISBN: 978-1-61209-506-6

Chapter 5

PHOTOSYSTEM AT HIGH TEMPERATURE: MECHANISMS OF ADAPTATION AND DAMAGE

Yasuo Yamauchi[1] and Yukihiro Kimura[2]
[1]Graduate School of Agricultural Science, Kobe University
[2]Organization of Advanced Science and Technology, Kobe University

ABSTRACT

Recent global warming is presumed to reduce productivity of plants near future through deterioration of photosynthetic activity. In particular, photosystem II (PSII) of higher plants is the most sensitive site affected by heat stress. Elevation of temperature causes structural damages in the PSII pigment-protein complexes due to impairment of D1 proteins and release of extrinsic proteins accompanied with degradation of manganese-calcium clusters, resulting in the loss of photosynthetic activities. In contrast, some prokaryotic phototrophs are moderate thermophiles growing at high temperatures over 50°C. Although bacterial photosystems are primitive adapted for ancestral environments, they can provide some implications on stability and tolerance under heat stress condition. In this article, we focus on photosynthetic organisms with type II reaction centers (RC), and discuss a relationship between structures and thermal stability and/or tolerance of the pigment-protein complexes, to find possible countermeasures for overcoming the global warming condition.

PSII is evolutionally related to purple bacteria because of similarities in their heterodimeric RC and quinone-mediated electron transport system. The purple bacterial photosystem lacks extrinsic proteins which are ubiquitously residing in the periphery of PSII, but has a circular pigment-protein complex, light-harvesting 1 (LH1), which is closely associating with the RC to maintain structures and functions of the photosystem. Among purple bacteria, *Thermochromatium tepidum* is a unique thermophile. The enhanced thermal stability is attributed to Ca^{2+}-induced structural changes of the LH1 complex at the peripheral region. In cyanobacteria, extrinsic proteins PsbO, PsbU, PsbV at the lumenal side of PSII play significant roles for regulation and stabilization of the water oxidation machinery. Deletion of these proteins resulted in deterioration of oxygen–evolving ability. Also in higher plants, PSII core is stabilized by extrinsic proteins PsbO, PsbP, PsbQ, and liability of interaction between the extrinsic proteins and PSII core under heat stress condition is thought to be a primary cause of heat-derived damage of photosystem. In particular, PsbO is most important for stabilizing the manganese-calcium cluster, and thus, commonly exists in all oxygenic phototrophs. Accumulated knowledge suggests that release of PsbO is critical for loss of oxygen evolution, and biochemical modifications of PsbO by ROS and lipid peroxides enhance its release from PSII core, especially under heated condition in the presence of light.

These findings indicate that stabilizing of RC is crucial for protecting photosystem from heat damage, therefore regulation of light is a possible means to alleviate the heat damage because light is the primary determinant to regulate redox state. In addition, to induce heat tolerance of higher plants, application of artificial chemicals is expected to convenient and effective tools. Several small peroxidized products from lipid peroxidation are shown to induce gene expression involved in heat tolerance. We named these chemicals as "environmental elicitor", and malondialdehyde and ethylvinylketone were screened as heat tolerance-inducing environmental elicitors protecting PSII from heat damage. Taken together, improvement of structure of PSII to stabilize RC, control of light condition, and use of environmental elicitors might be potential countermeasures to overcome heat stress.

1. General Introduction

Vital energy for all organisms is ultimately depended on photosynthesis, i.e. conversion of photon energy of sunlight to chemical energy carried out by phototrophs. Thus global decrease of photosynthetic activity is a crisis to be avoided because it directly means deficiency of energy supply for ecosystem, however recent climate change foresees gloomy future. Prediction of surface temperature on earth simulated by the National Institute of Environmental Studies of Japan (2007) indicates that the global surface temperature likely increase 1.1 to 6.4°C during 21st century. Increase in average temperature could result in longer potential growing seasons at high latitudes, and often shorter seasons at low latitudes because of interactions with rainfall, evapotranspiration and soil moisture [1]. These climate changes would disturb agriculture on worldwide level and bring about unstable crop yield. Simplistic, but natural forecast can imagine frequent appearance of hot days above 40°C in midsummer over a wide area. On the other hand, accumulated research suggest that photosynthesis is potentially inhibited at 40-45°C. Taken together, inhibition of photosynthesis by heat-stressed condition will be inevitable in near future if no countermeasure is applied.

Against forthcoming warming environment, countermeasure must be taken to acquire heat tolerance of photosynthesis in higher plants. To the purpose, it is necessary to understand mechanisms of heat-derived damage in photosynthesis, adaptation and acclimation of various photosystems under high temperature conditions over 40°C. Therefore, in this article, we describe the relationship between photosystem and high temperature in terms of following issues, (i) structure and function of photosystems with type-II reaction center, (ii) thermal stability in bacterial photosystems, (iii) mechanism of damage of photosynthesis in higher plants, (iv) adaptation mechanism of photosynthesis against heat stress, (v) enhancement of heat tolerance by genetic engineering and application of chemicals, (vi) induction of heat tolerance by artificial treatments of chemicals.

2. Structure and Function of Photosystems with Type-II Reaction Center

Photosynthetic organisms with type-II RCs are evolutionally related each other since they have central domains of heterodimeric protein subunits, and a special pair of (bacterio)chlorophyll molecules which plays key roles for a primary charge separation in the photosynthetic reaction. These phototrophs are, however, largely different in the protein composition and several cofactors, which are closely related to the stability of photosystems under heat-stress condition. A number of biochemical and physicochemical studies have provided valuable information on the relation between structure and stability of type-II RCs, including LH1-RC complexes from purple bacteria, and PSII from cyanobacteria and higher plants.

2.1. Purple Bacteria

In purple photosynthetic bacteria, light energy is captured by two types of light-harvesting pigment-protein complexes (LH1 and LH2), and transferred to the RC where primary charge separation occurs across the membrane [2]. Two electrons from the RC are transferred to a ubiquinone to form a ubiquinol, which migrates to a quinone pool in the membrane. The oxidized RC complexes are reduced by cytochrome bc_1 complexes through soluble heme proteins to form the cyclic electron flow, and proton gradient generated across the membrane is utilized to drive ATP synthases (Figure 1).

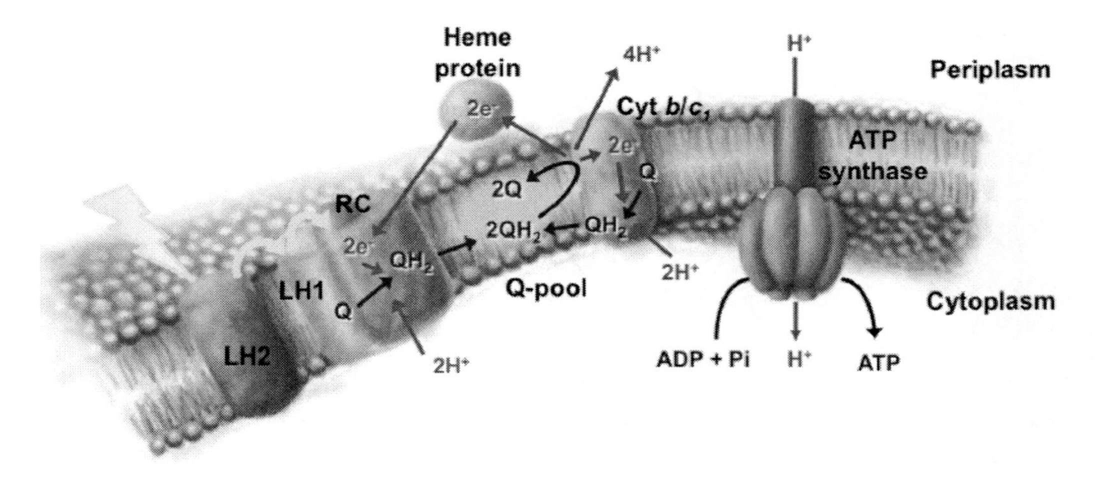

Figure 1. Photosynthetic apparatus and reaction scheme in the membrane of purple bacteria.

Ancestral purple bacteria are similar with PS II in the protein composition of the heterodimeric RC and the quinone-mediated electron transport system. High resolution three dimensional crystallographic structures of isolated RC complexes have been reported for both mosophilic [3, 4] and thermophilic [5] purple bacteria. The results demonstrated that RC,s are comprised of four intrinsic protein subunits (C, H, M and L), but the extrinsic proteins, ubiquitous for PS II, are not involved in the assembly of the bacterial RCs. In contrast, the X-ray structure of the RC associating with the LH1 has been limited to *Rhodopseudomonas* (*Rps.*) *palustris* at moderate resolution [6]. Generally, LH1 complexes are oligomers of minimal subunits composed of α- and β-polypeptides, *a*-type bacteriochlorophyll (BChl *a*) and carotenoid molecules. The RC is surrounded by LH1 with a one-to-one stoichiometric ratio to form core complexes (LH1-RC). Recent electron microscopic (EM) and atomic force microscopic (AFM) studies showed two-dimensional projection maps at 8.5 - 26 Å resolution, indicating a variety of LH1 complexes forming 12- to 16-meric assembly of $\alpha\beta$-subunits depending on the species [7-12]. However, the LH1-RC structure at atomic resolution is absolutely required for elucidating details of the function and structure of LH1 pigment-protein complexes. Since the RC tightly associates with the LH1, interaction modes between both complexes, and the size and/or shape of LH1 rings are significant to understand the structural functional consequences, including structural stabilities, quinone transport mechanisms, and excited state dynamics of the LH1-RC complex.

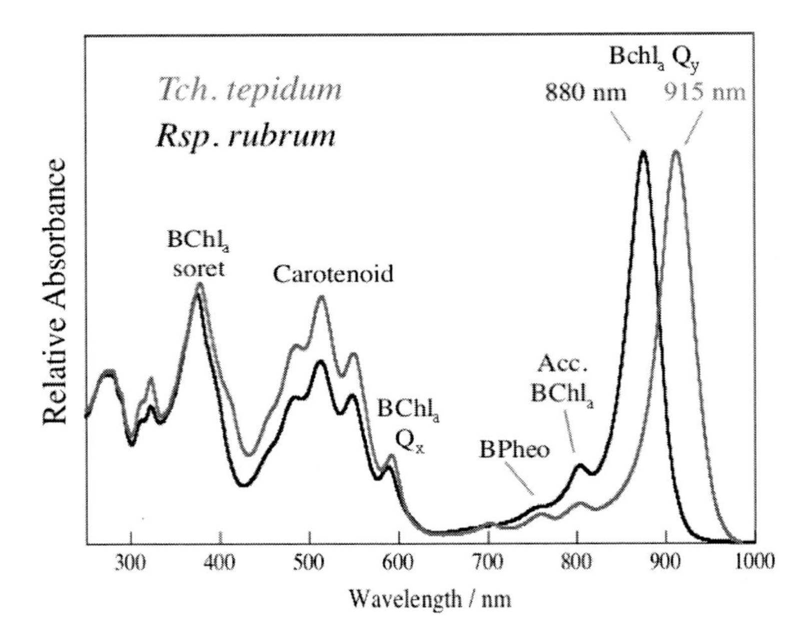

Figure 2. Absorption spectra of purified LH1-RC complexes from thermophilic *Tch. tepidum* (magenta) and mesophilic *Rsp. rubrum* (black).

The LH1 complex shows characteristic electronic absorption properties due to BChl *a* dimers and carotenoid molecules incorporated into the $\alpha\beta$-subunits. These pigments are useful as internal probes for monitoring structural and functional properties of the LH1-RC complexes. In a typical mesophilic purple bacterium *Rhodospilirum rubrum*, the LH1 complex is suggested to have a macrocycle with a 16-fold symmetry [7]. The BChl *a* – BChl *a* interactions within the LH1 ring resulted in the characteristic absorption property of the core complex as shown in Figure 2. The LH1-RC contains only *a*-type BChl molecule of which Q_y, Q_x and Soret bands appear at 880, 590 and 390 nm, respectively, as well as carotenoid bands around 450 – 550 nm and RC bands at 780 and 800 nm. Interestingly, thermophilic *Tch. tepidum* showed an unusual red-shift of the LH1 Q_y peak to 915 nm along with slight red-shifts of Soret and Q_x bands although peak positions for carotenoid and RC bands remained to be unchanged [13]. The large red-shift of the Q_y band seems to be disadvantage in terms of the uphill energy transfer from the LH1 to the RC [14]. Taking into account that *Tch. tepidum* is a sole thermophile among purple bacteria and has the abnormal LH1 Q_y band, the relationship between thermostability and low-energy Q_y transition is a matter of interest. It is of note that mesophilic *Roseospirillum parvum* 930I [15] and strain 970 [16] have been reported to show largely red-shifted LH1 Q_y transitions at 909 nm and 963 nm, respectively. The reason for unusual red-shifts were supposed as interactions between BChl *a* and specific residues of the LH1 polypeptides and/or enhanced exciton couplings within the ring-like assembly of BChl *a* molecules [15, 16]. Therefore, the Q_y red-shift is not necessarily responsible for the enhanced thermal stability but another mechanism is suggested to be involved in acquiring thermal stability of *Tch. tepidum*.

2.2. Cyanobacteria

Cyanobacteria are oxygenic photosynthetic prokaryotes which acquired the oxygen evolving ability through evolution. The most notable difference with purple bacteria is the water oxidation machinery, Mn_4Ca cluster, which can utilize water molecules as electron donors (Figure 3). The photosynthetic water oxidation is initiated by the light-induced charge separation of P680 special pairs located at the central part of heterodimeric RCs, followed by electron transfer reactions to cytochrome b_6f and PS I to reduce $NADP^+$ for CO_2 assimilation [17]. In the donor side of the PSII, the light-driven water oxidation occurs at the Mn_4Ca cluster through the five intermediate states labeled S_n (n = 0 – 4). The oxidized P680 is reduced by the cluster, which accumulates oxidizing equivalents necessary for the water oxidation during the S-state cycling. Both Ca^{2+} and Cl^- are indispensable cofactors for O_2 evolution. Although several pathways for water, proton, and O_2 channels have been proposed based on the crystallographic structure [18, 19], details of the water oxidation mechanism are still a matter of debate.

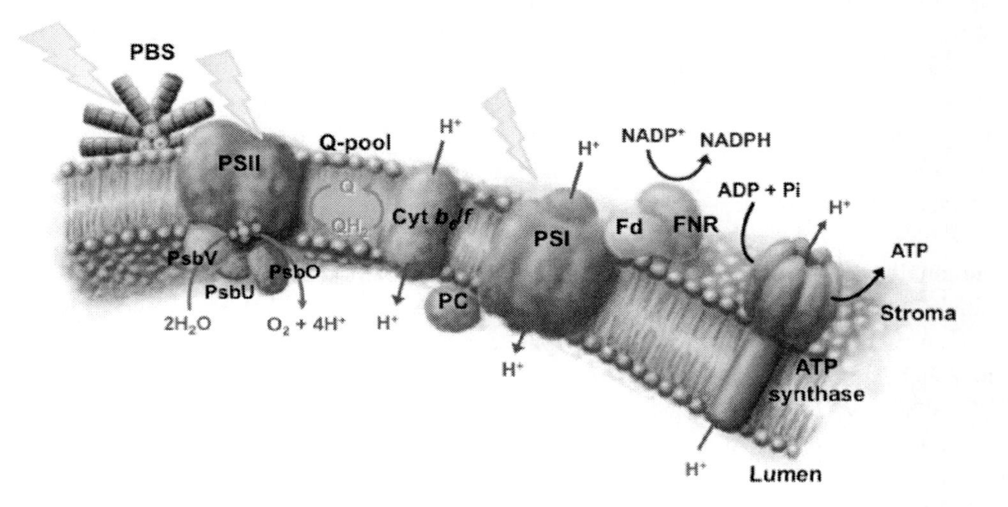

Figure 3. Photosynthetic apparatus and reaction scheme in thylakoid membrane of cyanobacteria.

In this decade, X-ray crystallographic structures of PSII from thermophilic cyanobacteria have been improved step by step with starting from 3.8 Å to 2.9 Å resolution [18, 20-22]. Recently, the resolution has been achieved to atomic level, revealing detailed structure of the oxygen-evolving complex (OEC) including the Mn_4Ca cluster (Shen et al., the 15th international congress of photosynthesis). The PSII core is comprised of several intrinsic proteins including PsbA (D1), PsbB (CP47), PsbC (CP43) and PsbD (D2), and the Mn_4Ca cluster, which are highly conserved from cyanobacteria to plants, keeping the essential function of oxygenic phototrophs. The cluster is a pentanuclear metal complex comprised of four manganese and one calcium ions which are bridged by several oxygen atoms and coordinated by water molecules as well as Asp, Glu, Ala, and His residues from PsbA and/or PsbC proteins.

Another interest in PSII is the presence of gene-encoded extrinsic proteins which are lacking in purple bacteria (Figures 1 and 3). Extrinsic proteins are thought to play essential roles for stabilizing and protecting PSII cores [23]. Bacterial PSII mainly possesses PsbO, PsbV, and PsbU whereas plant PSII contains PsbO, PsbP, and PsbQ. PsbO protein commonly exists in every oxygenic phototroph with different stoichiometry; one and two PsbO per PSII for cyanobacteria and higher plants, respectively [24]. PsbO and PsbV independently bind to the PSII lacking extrinsic proteins [25]. The later was more stabilized in the presence of PsbO and PsbU, and the full binding of PsbU to the PSII core requires both PsbO and PsbV [25, 26].

High-resolution X-ray structures of PsbO protein associated with the PSII core are available for *Thermosynechococcus elongatus* [18, 21] and *vulcanus* [27], in which PsbO is comprised of a β-barrel core and an extended α-helix domain. The structure revealed that none of amino acid residues from the PsbO were served as direct ligands for the Mn_4Ca cluster. However, a number of studies strongly indicated that this protein plays significant roles for protecting and stabilizing the catalytic center and modulates the Ca^{2+} and Cl^- requirements for the water oxidation [24, 28]. A site-directed mutagenesis studies indicated that mutations at D159 and R163 in PsbO in the absence of PsbV failed photoautotrophic growth [29]. In addition, several studies using isolated PsbO revealed different Ca^{2+}-binding properties between higher plants and cyanobacteria. In plants PsbO, spectroscopic studies suggested that Ca^{2+}-induced structural changes facilitate the association of PsbO with PSII core [30, 31] although an EPR analysis indicated that the functional Ca^{2+} ion was not involved in the binding to the PsbO [32]. In cyanobacteria, spectroscopic analyses concluded no significant conformational change induced by Ca^{2+} [33], whereas the low-affinity Ca^{2+}-binding site in PsbO locating at the luminal exit of a proton channel has been implicated to be responsible for the water oxidation [34, 35].

PsbV and PsbU proteins are specific for bacterial PSII [23] and proposed to be functionally replaced with PsbP and PsbQ in plants PSII through evolution from ancestral cyanobacteria to higher plants [25, 36]. However, there is very little structural homology between them based on the crystallographic structures [18, 21, 27, 37, 38]. Recently, PsbP and PsbQ homologues associated with the PSII core were discovered also in *Synechocystis* PCC6803 [39]. Phylogenetic studies suggested that PsbP and PsbQ homologues in cyanobacteria were evolved into plants PsbP and PsbQ, respectively, through intensive genetic modification during endosymbiosis and subsequent gene transfer to the host nucleus [40-42]. These results indicated that the origins of PsbV/PsbU and PsbP/PsbQ are different each other regardless of their functional similarities in the OEC.

PsbV (Cyt c_{550}) is a soluble *c*-type monoheme cytochrome. This protein has variable redox potential, -240 mV or -80 mV in a free or PSII-bound form although participation in the light-induced electron transport chain was suggested to be unlikely [43]. The redox poteintial of PsbV was also altered by specific mutations, but was not important for PSII activities under normal growth conditions [44]. However, PsbV deletion mutants were unable to grow photoautotrophically in the absence of Ca^{2+} or Cl^- and inactivated the oxygen-evolving ability upon the dark incubation [45]. Interestingly, PsbV was possible to support the water oxidation even in the absence of PsbO, in contrast to the function of PsbP in higher plants [26]. Therefore, the role of PsbV protein was suggested to maintain a proper ion environment within the OEC for optimal oxygen-evolving activity [25, 26, 45-47]. Although none of residue from PsbV was served as a direct ligand for the Mn_4Ca cluster, this protein was suggested to participate in stabilizing PsbA protein through electrostatic interactions [48].

The general function of PsbU is assigned to enhancing the structural stability and shielding of the Mn_4Ca cluster. PsbU deletion mutants were capable of photoautotrophic growth in the absence of Ca^{2+} or Cl^- but at reduced rates [45]. The oxygen-evolving ability was lowered by the removal of PsbU and restored in part by adding Cl^- but not Ca^{2+}, indicating that PsbU regulates Cl^- requirement [49, 50]. Based on the results, this protein was suggested to optimize the water oxidation process by forming the stable architecture in the OEC [45, 49, 50]. Another functions including suppression of light-induced D1 degradation [50], and protection of the PSII core from reactive oxygen species (ROS) [51] were also proposed.

2.3. Higher Plants

The photosynthetic apparatus and reaction mechanisms in higher plants are largely similar with those of cyanobacteria as illustrated in Figure 4. However, high-resolution structural analysis has been delayed in higher plants due to instability of each membrane protein complex compared with those of thermophilic cyanobacteria. As for the PSII complex, the structural information is limited to a low-resolution data collected by electron microscopy [52, 53]. However, accumulated knowledge strongly indicated that the PS II core assembly and their function are almost identical to those of prokaryotic counterparts except for the critical difference in the composition of extrinsic proteins, which provides valuable insight to understand evolution of photosynthetic organisms [54].

Higher plants possess gene-encoded extrinsic proteins, PsbP, PsbQ, and PsbR as well as PsbO which commonly exists in oxygenic phototrophs. These proteins play a key role for maintaining oxygen-evolving activity at physiological rates [23, 24]. They have a strict binding order of PsbO > PsbP > PsbQ. PsbO independently associates with the PSII core [55, 56], whereas binding of PsbP occurred through electrostatic interactions with PsbO [57, 58]. PsbQ requires both PsbO and PsbP for its binding [56, 57]. PsbP protein is related to the stability of the Mn_4Ca cluster [42] as well as the binding affinity of Ca^{2+} and Cl^- ions [28], which are essential for the water oxidation chemistry [17]. In addition, *psbP*-deletion mutants revealed that PsbP is indispensable for the normal PSII function in higher plants [59]. Another function of this protein was indicated that PsbP has a Mn^{2+}-binding ability as a reservoir to keep or deliver manganese ions [60]. PsbQ protein is related to be binding of Cl ions [38]. Studies on transgenic tobacco [59] and Arabidopsis [61] revealed that PsbQ protein is not necessary under normal growth condition [59, 61] but is required for photoautotrophic growth under low-light condition [61]. The role of PsbR has not been fully understood. This protein was suggested to play an important link for stable assembly of PsbP protein, whereas not closely related to the assembly of PsbQ [62]. However, a recent bioinfomatic analysis suggested a relationship between PsbR and the Mn_4Ca cluster [63].

Most important physiological role of PsbO is to stabilize the binding of the Mn_4Ca cluster that is essential for oxygen-evolving activity [64]. Binding interaction between PsbO and PSII core is thought to be electrostatic and hydrophobic interaction because PsbO can be dissociated from the PSII core by a number of chemical treatments including washing with alkaline Tris, 1 M $CaCl_2$, and chaotropic agents, or by physical means such as heat treatment [65, 66]. Especially, Lys residue-modifying chemicals such as *N*-succinimidyl propionate and 2,4,6-trinitrobenzene sulfonic acid caused release of PsbO from PSII and loss of oxygen-evolution activity [67], suggesting that the positive charge of Lys is important for the electrostatic interaction between PsbO and PSII.

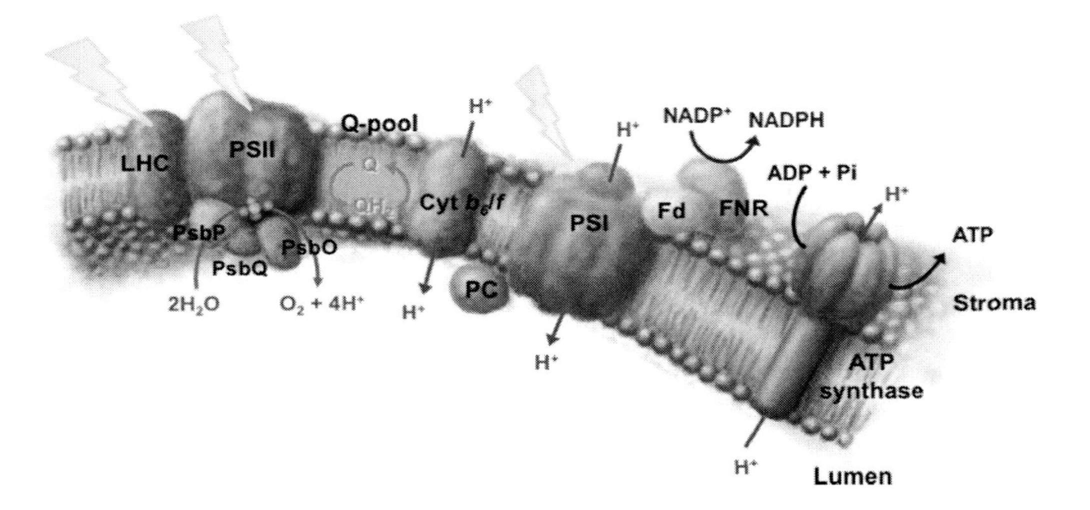

Figure 4. Photosynthetic apparatus and reaction scheme in thylakoid membrane from higher plants.

THERMAL STABILITY OF TYPE-II REACTION CENTERS

Thermal stability of proteins has been extensively investigated by comparing thermophiles with the mesophilic analogues. Several factors including electrostatic and hydrophobic interactions, solvent-accessible surface area, oligomerization, shortning of loops, and buried molecules have been proposed to be responsible for the thermostability of the proteins [68-74]. In photosynthetic organisms, one of the sensitive sites affected by heat stress is type-II RCs due to the impairment and/or decomposition of large pigment-protein complexes. However, several cofactors responsible for enhancing thermal stability of type-II RCs have been proposed, which can provide some implications on stability and tolerance under heat-stress condition.

3.1. Purple Bacteria

Thermophilic purple sulfur bacterium, *Tch. tepidum,* can grow at optimum temperatures at 48 – 50 °C, up to 56 °C, the highest temperature among purple bacteria ever known [75]. Another purple sulfur bacterium, *Ach. Vinosum,* is phylogenetically related with *Tch. tepidum*, however, this mesophilic analogue favors a much lower growth temperature by ~ 20 °C than *Tch. tepidum*. Based on reconstitution experiments in liposomes, it has been supposed that thermal stability of the *tepidum* RC was comparable to those of mesophilic RCs, but was enhanced by the interaction with the LH1 complex [76]. In addition, the high-resolution crystallographic structure of the *tepidum* RC demonstrated that characteristic three arginine residues (two from L-subunit and one from M-subunit) were located at the membrane interface [5]. Based on the study of site-directed mutations into structurally homologue positions of *Rba. sphaeroides* RC, electrostatic interactions between acidic residues at the C-terminal region of LH1 $\alpha\beta$-polypeptides and the specific basic residues in the RC are suggested to be involved in the enhanced thermal stability of *Tch. tepidum* [77]. The results described above strongly indicated that enhanced thermal stability of *Tch. tepidum* is caused by the interaction between RC and LH1 complexes, and the interaction modes seem to be different among the species resulting from the structural properties of the LH1 complex [2]. However, the critical factor responsible for the thermostability had been unclear.

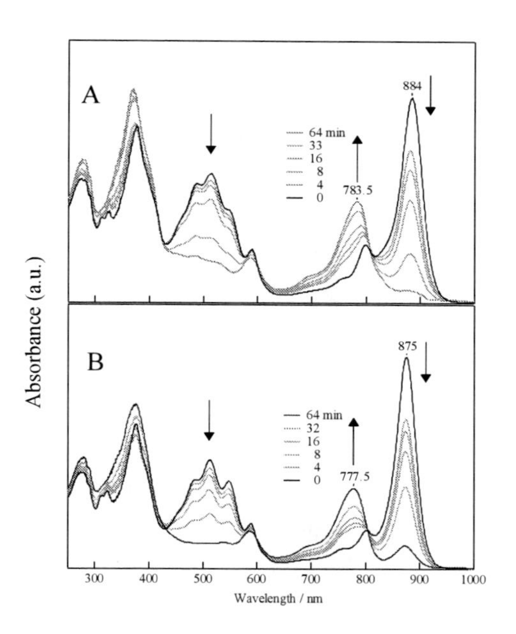

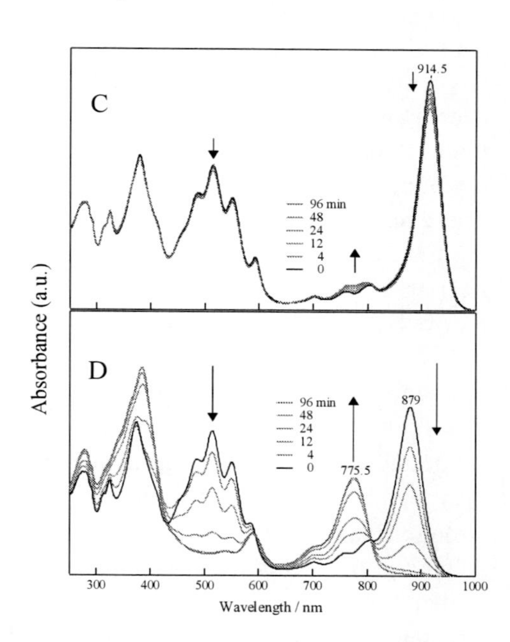

Figure 5. Spectral changes of absorption spectra for the LH1-RC complexes upon incubation at 50°C. (A) *Avn. vinosum*, (B) *Rsp. rubrum*, (C) native and (D) Ca^{2+}-depleted *Tch. tepidum*.

Recently, it was revealed that an inorganic cofactor, calcium ion, is closely related to the enhanced thermal stability of the *tepidum* LH1-RC complex [78] as well as its unusual spectroscopic properties [13]. Figure 5 shows thermal stability of LH1-RC complexes from purple bacteria monitored by electronic absorption spectra of the pigments incorporated into the complex. In mesophilic *vinosum* (A) and *rubrum* (B), the LH1 Q_y band intensity around

880 nm largely decreased upon 50 °C incubation, and new bands around 780 nm appeared with showing several isosbestic points. Similar deterioration was observed in 450 – 550 nm bands for carotenoid molecules which are thought to be participating in the thermal stability of the LH1-RC complex. Since a BChl *a* monomer bound to α- or β-polypeptide exhibits the absorption band at 780 nm, the spectral change indicates that the LH1 complex directly decomposed to the BChl-bound α- or β-polypeptide. In contrast, LH1 Q_y and carotenoid bands of *Tch. tepidum* were retained almost completely under the incubation at 30 – 40 °C (data not shown) and ~ 90 % even at 50 °C (Figure 5C), showing much higher thermal tolerance. However, after the removal of Ca^{2+} from the *tepidum* LH1-RC, the thermostability was largely deteriorated (Figure 5D) with unusual Q_y blue-shifts as comparable to those of the mesophilic counterparts. The results clearly indicated that Ca ions play key roles in stabilizing the pigment-protein assembly of the *tepidum* LH1-RC complex.

Furthermore, the thermal stability was examined by monitoring the secondary or tertiary structure of the protein using UV CD spectroscopy or differential scanning calorimetry (DSC) [13]. The CD bands reflecting secondary structure of *tepidum* LH1-RC were little influenced upon the Ca^{2+}-depletion, indicating that the secondary structure was maintained to be normal. However, the CD bands largely decreased upon 50 °C incubation after the removal of Ca^{2+}. These results suggested that small but distinctive changes were induced by the Ca^{2+}-binding at the peripheral region to strengthen the structural coupling between LH1 and RC complexes and enhancing the thermal stability. Further inspection using DSC revealed that peak temperatures for the denaturation were strongly dependent on Ca^{2+} as shown in Figure 6. In the native state (B915), the LH1 core complex was firmly retained but was rapidly decomposed at 75 °C as revealed by the intensive and sharp DSC band. In contrast, after the removal of Ca^{2+} partially by a size-exclusion column chromatography or completely by chelating agents, denaturing temperatures of the DSC bands lowered with much broadening. These results clearly indicate that the stability of tertiary structures was strongly dependent on the Ca^{2+} concentration in *Tch. tepidum*. It was reported that electrostatic interactions of the LH1 with its RC are responsible for the enhanced thermal stability of this organism [77]. However, the stability of the core complex was lost upon the Ca^{2+}-depletion even in the presence of LH1 complexes. Therefore, not only pure electrostatic interaction but the inorganic cofactor, Ca^{2+} is required for the marked enhancement of the thermal stability. Based on the primary structure of *tepidum* LH1 complex, putative Ca^{2+} binding sites were suggested to be at the *C*-terminal region of α-polypeptide or the *N*-terminal region of β-polypeptide, where several neighboring acidic residues are located (Figure 7). Recent study strongly indicated the location of the binding site to be at the *C*-terminal region of the $\alpha\beta$-subunit [79]. Therefore, a possible mechanism for acquiring heat tolerance of this thermophile is that Ca^{2+}-induced structural modifications at the LH1 membrane interface increased the structural stability of the LH1 complex itself and/or the interaction between LH1 and RC complexes, resulting in the protection of the RC showing the enhanced thermal stability of the core complex. *Tch. tepidum* was isolated from the living environment containing rich mineral calcium carbonate. Thus, it is presumed that this bacterium enhanced heat tolerance by utilizing this inorganic cofactor to survive in the extreme environment.

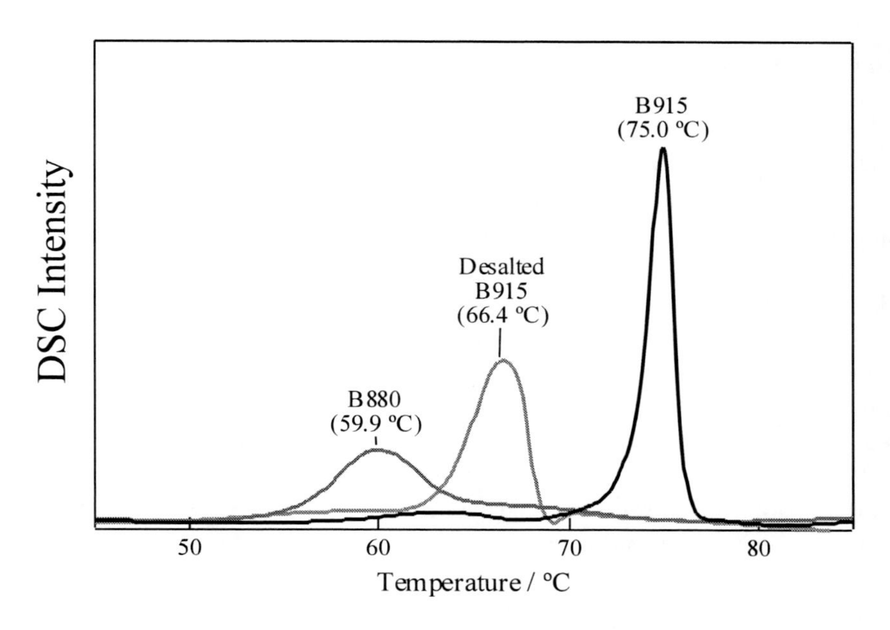

Figure 6. DSC profiles of *tepidum* LH1-RC complexes in Ca^{2+}-bound form (black), partially (green) and completely (magenta) Ca^{2+}-depleted form.

LH1 α-polypeptide

TTP	MFTMNANLYKIWLILDPRRVLVSIVAFQIVLGLLIHMIVLST-DLNWLDDNIPVSYQALGKK
AVN	MSPDLWKIWLLVDPRRILIAVFAFLTVLGLAIHMILLSTAEFNWLEDGVPAA

LH1 β-polypeptide

TTP	AEQKSLTGLTDDEAKEFHAIFMQSMYAWFGLVVIAHLLAWLYRPWL
AVN	NSSMTGLTEQEAQEFHGIFVQSMTAFFGIVVIAHILAWLWRPWL

Figure 7. Amino acid sequences of LH1 α- and β-polypeptides from *Tch. tepidum* and *Ach. vinosum*. (Red: acidic residues forming putative metal binding sites. Green: BChl a binding site.)

3.2. Cyanobacteria

PSII is the most sensitive photosynthetic apparatus affected by heat stress, due to impairment of D1 protein, release of extrinsic proteins, and disruption of the Mn_4Ca cluster. Thus, thermal stability of the OEC is closely related to acclimation of cellular thermal tolerance in oxyphototrophs. In cyanobacteria, extrinsic proteins, PsbO, PsbV, and PsbU, were reported to be a crucial factor responsible for enhancing thermostability [23, 24, 32]. Systematic analyses for targeted *Synechocystis* mutants revealed that cells lacking either PsbO or PsbV decreased their thermal stability of the OEC, and therefore, failed to enhance the cellular thermotolerace when grown at moderately high temperatures [80]. The isolated PsbO from *Thermosynechococcus elongatus* showed little structural alteration induced by the Ca^{2+}-binding [33]. In contrast, plants PsbO was thermally destabilized upon the binding of Ca^{2+} due to slight changes in secondary structure from β-sheet to loop or nonorderd structure [30]. Interestingly, the thermal stability was enhanced when plants PsbO proteins were reconstituted with thermally stable homologues from the thermophilic *Phormidium laminosum* [81], confirming structural and functional significance of PsbO protein. PsbV was also involved in the thermal stability of cyanobacteria since this protein was reported to be responsible for decrease/increase in the heat stability of the oxygen evolution in *Synechococcus* sp. PCC7002 [46]. Additionally, targeted mutagenesis studies in *Synechococcus* species revealed that PsbU mutants lost the ability to acclimate against high temperatures [82, 83] and additional PsbU enhanced their thermostability [82]. The heat-shock response under high temperatures was not affected by this mutation [83], indicating that PsbU contributes to enhancement of the thermal stability of the oxygen-evolving machinery to protect PSII against heat-induced inactivation.

Another factor related to thermal stability and/or tolerance was reported to be saturated fatty acids [40, 42, 45, 49, 51, 84-89]. It was suggested that saturation of membrane lipids enhanced thermal tolerance of plants PS II [84, 85, 90]. However, in *Synechocystis* sp. PCC6803, unsaturation of membrane lipids enhanced tolerance of the OEC toward to chilling stress, but not to heat stress and photosynthetic activities [86, 87, 91]. No appreciable difference in protein composition was observed between PSII complexes (or thylakoids) grown at 25°C and 35°C [89]. Instead, decreased desaturation level of the lipids under high temperature results in deceleration of the pH change, and reduction of PSII lacking Q_B binding ability, which was proposed to be one of the main cause for the acclimation of PSII [89]. Additionally, several factors including some specific lipids [92-95], heat shock proteins [96-98], sigma factors [99, 100] were also proposed to be responsible for the thermal stability of PSII and/or cellular thermotolerance in cyanobacteria.

3.3. Higher Plants

In higher plants, PSII is much susceptible to high temperature than PSI [101]. Thermosensitivity of oxygen evolution in higher plants has been studied by simplified experiments using PSII particles or isolated thylakoid membranes. Earlier works suggested that heat-labile properties of the OEC as revealed by in vivo and in vitro studies [86, 101, 102]. These studies demonstrated that the release of the PsbO occurs first, followed by liberation of two out of four Mn ions from the Mn_4Ca cluster of OEC [66, 103, 104], and finally the loss of oxygen evolution takes place at high temperature (Figure 8) [66, 105]. Unstable condition of Mn_4Ca cluster by disturb of interaction between PSII core and PsbO. The release of PsbO is thought to be due to thermal denaturation of PsbO; however, as described below, the possibility that other factors are also involved in the release of PsbO remains because PsbO itself is a thermostable protein [106].

4. Heat-Derived Damage of Photosystems in Higher Plant

Temperature is a critical factor for determining photosynthetic activity by means of both catalytic and biochemical properties. Net photosynthesis is highly dependent on surrounding temperature. Below 35-40°C, CO_2 assimilation did not affected by temperature or it increases as temperature rises, however over 40°C CO_2 assimilation drastically decreases [107, 108]. In the heated condition, PSII is a primary site of heat-drived damage, and thermal stress also leads to an increase of the relative fraction of QB-restoring RC [109, 110], loss of thylakoid membrane integrity, especially destacking of the thylakoid membranes [111, 112], and dysfunction in the system of CO_2 assimilation [113].

Accumulated knowledge suggests that light is a critical factor to determine heat-derived damage of photosynthesis, that is, irreversible damage of photosystems in heat-stressed plants can divide into light-mediated and light-independent mechanisms.

4.1. Light-Mediated Damage of Photosystem in Heat-Stressed Higher Plants

In nature, heat damage of photosystem usually occurs under both high temperature and high light intensity conditions. However, the activation of cyclic electron flow at the expense of linear flow is proposed to protect PSII from damage at high temperature by inducing photoprotective quenching, stabilizing the thylakoid membrane through enhanced zeaxanthin formation, and reducing the size of the PSII light-harvesting antennae [114-116]. The distribution of absorbed light energy between the two photosystems is dynamically balanced and regulated by a process termed ''state transitions.'' Light absorbed by PSII gets funneled to PSI in State 2 but not in State 1. Energy transfer from LHCII to the core antenna complex of PSII is affected at elevated temperatures. Even at 35°C, the migration of LHCII from the intergranal space toward PSI gets arrested affecting the balance of light absorption between the two photosystems [117]. At temperature above 40–42°C the loss of photosynthetic activity is partly caused by the inactivation of the acceptor side of PSII and reduction of the rate of electron transport in the chloroplasts.

Accumulated research suggest that heat-derived damage of PSII in the presence of light is caused by reactive oxygen species generated under high light conditions (Figure 8). Yamashita et al. indicated that the reactive oxygen species are generated by heat-induced inactivation of a water-oxidizing manganese complex and through lipid peroxidation [118]. The damages caused by the moderate heat stress to PSII are quite similar to those induced by excessive illumination where reactive oxygen species are involved; reactive oxygen species (ROS) damage D1 protein and inhibit the repair of photodamaged PSII by suppressing the synthesis of D1 proteins *de novo*[119]. On the other hand, saturation of polyunsaturated fatty acids (PUFAs) contributes to acquisition of heat tolerance of photosynthesis, as reported in isolated chloroplasts [90], cell culture [120], and intact plants [121]. It is thought that saturation contributes to heat tolerance by altering physicochemical properties, i.e., increased saturation of PUFAs raises the temperature at which lipids phase-separate into non-bilayer structures, which disrupts membrane organization [120, 121] and thereby provides the proper assembly and dynamics of PSII at higher temperature [122]. In addition to this physicochemical explanation for damage in heat-stressed plants, we provide a biochemical explanation in that the biological effect of reactive carbnyls such as malondialdehyde (MDA) and aclorein, i.e., its protein-binding ability, is greatly enhanced in heat-stressed conditions. Yamauchi and Sugimoto indicated that protein modification of malondialdehyde by peroxidized fatty acids causes loss of binding activity of PsbO to PSII core [123]. Protein modification by MDA is dependent on temperature, and highly enhanced over 35°C [123].

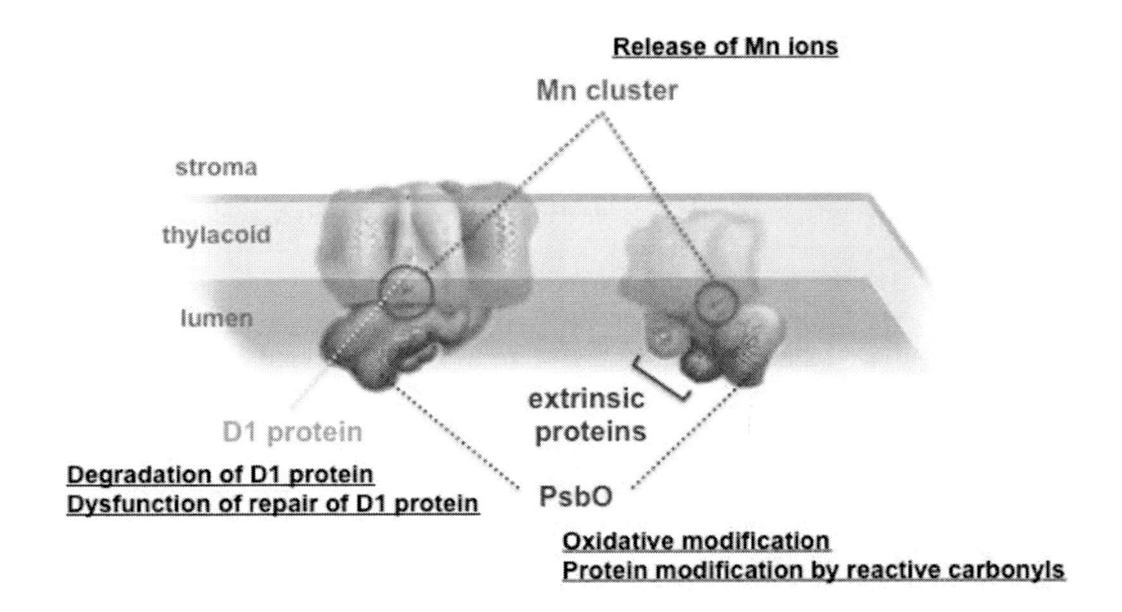

Figure 8. Diagram of structure of higher plant PSII and predicted components damaged by heat stress in the presence of light. Hypothesized mechanisms of the damage are underlined.

4.2. Light-Independent Mechanisms of Damage of Photosystems in Heat-Stressed Higher Plants

Photosystems are damaged by heat treatment in the dark. In Figure 9, maximal quantum yield of PSII (Fv/Fm) of Arabidopsis plants rapidly declined by 40°C treatment in the dark, whereas 40°C treatment in the presence of light did not reduce Fv/Fm. Resemble phenomena have been reported by several literatures [109, 124, 125]. This phenomena can be explained by activated cyclic electron flow around PSI, that is, stability of electron flow through PSI compared to PSII above 40°C.

The mechanism causing the decline in the electron transport rate above the thermal optimum remains uncertain [126]. A leading proposal is that cyclic electron transport is activated at elevated temperature at the expense of linear electron transport, thereby causing a shortage of NADPH [115]. In barley, elevated temperature is proposed to activate cyclic electron flow by diverting electrons from the NADPH pool to the plastoquinone pool [127-129]. In pima cotton, enhanced cyclic photophosphorylation may explain reductions in the stromal oxidation state observed above the thermal optimum [130]. The rise in cyclic electron flow above the thermal optimum increases the thylakoid pH gradient, resulting in activation of photoprotective quenching and a dissociation of the outer light harvesting antennae from PSII. Electron flow through PSII decreases above the thermal optimum in a pattern that mimics a decline in whole chain electron transport [131]; by contrast, electron flow rate through PSI is stable between the thermal optimum for photosynthesis and 40 °C, indicating that it has high capacity to support enhanced cyclic electron flow at elevated temperature [101, 124, 131]. Non-photochemical quenching of PSII is widely observed at temperatures where electron transport capacity slows with rising temperature, indicating that the reduction in electron flow through PSII is a regulatory response to limitations further down the electron transport chain [130-132]. Consistently, Yamasaki *et al.* [131] demonstrated that the capacity for electron transfer from plastoquinone to P700 declines above the thermal optimum, implicating electron flow between the photosystems as a possible cause for the decline in the electron transport rate.

Introduction of stromal reducing power into thylakoid membranes is mediated by two distinct pathways; one is a pathway through NAD(P)H dehydrogenase (NDH) catalyzing reduction of oxidized plastquinone (PQ) by NAD(P)H and another is a pathway through ferredoxin-quinone oxidoreductase (FQR) catalyzing reduction of oxidized PQ by reduced ferredoxin. Which pathway is involved in the light-independent damage of photosystem under heat conditions, is confused. Reduction of PQ was suppressed in heat-stressed NDH-deficient tobacco plants in the dark compare to wild type, suggesting that NDH mediates introduce of reducing power in stroma into thylakoid membrane[125]. However, Yamane et al showed that reduction of PQ under high temperature in the dark was not different between NDH-deficient tobacco plants and wild type. Delayed luminescence analysis using NDH and FQR mutants indicated that the predominant pathway responsible for the afterglow in tobacco was mainly due to FQR while the NDH complex is involved in the predominant pathway in Arabidopsis. These results imply that the pathway varies depend on the plant species and developing conditions.

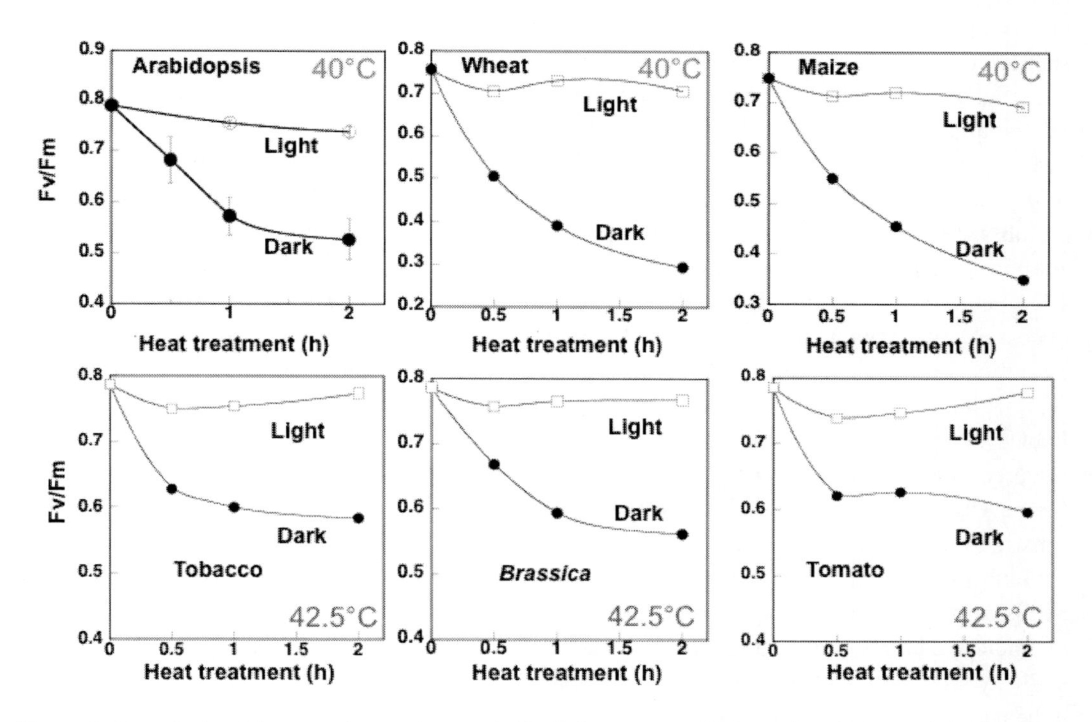

Figure 9. Heat-derived decrease in maximum yield of photosystem II photochemical reactions (Fv/Fm) in various plants. Each plant was incubated in the presence (100 μmol m^{-2} s^{-1}) or absence of light at indicated temperature, and measured Fv/Fm by junior-PAM (Walz, Germany).

Though detailed molecular mechanism of damage of photosystem components derived from over-reducing by stromal reductants is unknown, several possibilities involved in the oxidative damage can be predicted. Backflow of electron from stromal reductants to plastquinone [133] can potentially produce ROS around PSII [134] and over-reduced PQ can produce superoxide radicals in thylakoid membranes [135], resulting in oxidative damage to photosystem components.

5. Adaptation Mechanism of Photosynthesis Against Heat Stress

Heat-shock proteins (HSPs) are involved in protecting cells against high temperature and other stresses [136]. Also in chloroplasts, HSPs can potentially act as protectants for heat stress. Chloroplast-localized HSPs contributed heat tolerance to the photosynthetic electron transport chain in isolated chloroplasts, because small HSPs could associate with thylakoids and protect O_2 evolution and OEC proteins of PSII against heat stress. Evidence for the significance of chloroplast-localized HSPs for thermotolerance was obtained in tomato species [137, 138]. Neta-Sharir et al. [139] demonstrated that chloroplasts small heat-shock protein, HSP21, induced by heat treatment in tomato leaves protected PSII from temperature-dependent oxidative stress, and chloroplast HSPs are thought to prevent heat damage rather than repair stress-related damage [138, 140]. HSPs can function in protection as molecular chaperons to prevent but no reverse protein denaturation and aggregation, as membrane stabilizers and possibly, as site-specific antioxidants [136]. HSP101 has been shown to be essential for thermotolerance by genetic analysis [141]. By genetic manipulation, the role and sub-cellular localization of the small (16 kDa) HSP (HspA) was investigated comparing the cyanobacterium *Synechococcus* strain ECT16-1, with constitutively expressed HspA, with the reference strain [142]. The protein possesses the unique property to associate with thylakoid membranes during heat stress and support stability of thylakoid membranes. It is also supposed that other compounds different from HSPs can contribute to heat tolerance [83, 143]. Numerous studies have shown that the extrinsically associated proteins are not necessary for oxygen evolution activity in vitro, but they are required to enhance oxygen evolution and play important roles in vivo [32]. In the cyanobacterium *Synechococcus* sp. PCC 7002, the oxygen-evolving machinery is stabilized against heat-induced inactivation by PsbU, which is an extrinsic protein of the PSII complex [83]. Inactivation of psbU gene lowered the thermal stability of OEC in the mutant in comparison with the WT cells. However, the levels of HSPs, namely, the homologs of HSP70, HSP 60, and HSP 17 remained unaffected by mutation in psbU gene suggesting that HSPs are not involved in cell acclimation to high temperature [83]. Similarly, Tanaka et al. [143] reported that thermal acclimation of OEC of *Chlamydomonas reinhardtii* did not change the levels of well-characterized HSPs with molecular masses 70, 60, or 22 kDa. It is possible that OEC thermal stability and acclimation are mostly regulated by PsbO gene that produces the 33-kDa extrinsic protein [21, 81, 104].

Protection against photo-oxidation under heat stressed condition is assumed to occur through cyclic phosphorylation by heat inducible NAD(P)H dehydrogenase [144]. Our results shown in Figure 9 might support the involvement of cyclic electron flow around PSI in protection against heat stress because complete cyclic electron flow is accomplished in the presence of the light, but not in the dark.

6. Improvement of Heat Tolerance by Genetic Engineering

Attempts to acquire heat tolerance by overexpression of HSPs have been made, and heat shock transcription factors (HSFs), those are the central regulators of the heat shock (HS) stress response in all eukaryotic organisms [145]. Overexpression of tomato (*Lycopersicon esculentum*) chloroplast small heat shock protein (sHSP), HSP21 enhanced heat tolerance of transgenic tomato plants [139]. The effect of the transgene was not on PSII thermotolerance, but the protein protects PSII from temperature-dependent oxidative stress. In transgenic Arabidopsis overexpressing HSFA2, which is a dominant HSF in Arabidopsis, showed enhanced thermotolerance of PSII [146] and basal and acquired thermotolerance was significantly enhanced in high-level HsfA2-overexpressed transgenic lines in comparison with wild-type plants [147]. HSPs can contribute improvement of heat tolerance although effects were a little with undesirable side effects such as dwarf phenotype and delay of development [147].

Overexpression of multiprotein bridging factor 1c (MBF1c) which is a stress-response transcriptional coactivator enhanced the tolerance of transgenic plants to heat stress [148]. The expression of MBF1c in transgenic plants augmented the accumulation of a number of defense transcripts in response to heat stress, resulting in enhanced tolerance to heat stress. Transcription factor DREB2A interacts with a cis-acting dehydration-responsive element (DRE) sequence and activates expression of downstream genes involved in drought- and salt-stress response in *Arabidopsis thaliana*. Transgenic*Arabidopsis*-overexpressing *DREB2A* induces not only drought- and salt-responsive genes but also heat-shock (HS)-related genes [149]. Thermotolerance was significantly increased in plants overexpressing *DREB2A* and decreased in *DREB2A* knockout plants, indicating that DREB2A functions in both water and HS-stress responses. Thermotolerance of photosystems in these two transgesgenic lines are unknown, however these research exhibit that overexpression of proteins involved in signal transduction can contribute acquisition of thermotolerance.

Saturation of fatty acid in biomembranes contributes to improve tolerance of photosynthesis against heat stress [121]. Oxygen evolution of linolenic acid deficient mutants after heat treatment maintained higher level than those of wild types. Detailed mechanism is unknown, however it is expected that reduced lipid peroxidation might contribute the heat tolerance of OEC.

Overexpression of betaine aldehyde dehydrogenase contributed to an improvement on thermostability of the oxygen evolving complex and the reaction center of PSII [150]. The increased thermotolerance induced by accumulation of glycinebetaine, a compatible slute, in vivo was associated with the enhancement of the repair of PSII from heat-enhanced photo inhibition, which might be due to less accumulation of reactive oxygen species in transgenic plants.

7. Induction of Heat Tolerance by Artificial Treatments of Chemicals

In presence, many countries permit to cultivate and/or use of genetically modified organisms (GMO). Actually, production of several crops such as soybean, cotton and maize occupies almost part of net production in agricultural countries such as America, Brazil, Argentina, India and Canada (James, ISAAA Briefs, 2009). In future, development of heat-tolerant GMO will be accomplished, however it is uncertain whether citizens in whole world can accept GMO. Thus it is important to develop alternative solution to overcome heat-damage of plants. One solution is to use artificial inducer of heat tolerant of plants. To overcome loss of photosynthetic activity under warming condition, establishment of heat tolerance by artificial treatment is an important mission.

7.1. Ca^{2+}

Ca is not only a macronutrient, but also a major intracelluar messenger involved in the mediation of many physiological processes in plants, and calmodulin plays a pivotal role in the calcium messenger system. Gong et al. [151] indicated that Ca^{2+} and calmodulin may be involved in the acquisition of the heat-shock induced thermotolerance, and exogenous application of Ca^{2+} enhanced thermotolerance in maize.

7.2. H_2O_2

Signal transduction of HSP induction remain unclear, however ROS could also play a key role in mediating important signal transduction events [152]. H_2O_2 can induce HSFA2 gene expression, thus H_2O_2 is potentially a signal molecule of heat stress [153]. Low levels (<10 μM) of H_2O_2 to rice plants permitted the survival of more green leaf tissue, and of higher quantum yield for PSII, than in non-treated controls, under salt and heat stresses [154]. The H_2O_2 treatment induced not only active oxygen scavenging enzymes activities, but also expression of transcripts for stress-related genes encoding sucrose-phosphate synthase, Δ'-pyrroline-5-carboxylate synthase, and small heat shock protein 26.

7.3. Phytohormone

Plant hormones such as salicylic acid (SA), abscisic acid (ABA), and ethylene have been shown to play an important role in mediating ROS and temperature stress signals [155]. ABA, in particular, was shown to activate Ca^{2+} channels during drought stresses via the function of ROS and Rboh proteins [156]. However, the cause-and-effect relationship(s) between ROS, SA, ABA, and ethylene during temperature stress is not clear. Further studies are needed to elucidate the role of ROS in mediating the action of different plant hormones during temperature stress.

Exogenous application of ABA can improved the recovery growth of maize seedlings after heat treatment [157], and enhanced the survival rate of bromegrass cell suspension cultures at 42.5°C [158]. This ABA-induced thermotolerance might be mediated by Ca, and ABA application might maintain antioxidant enzyme systems at higher levels [159].

SA also can induce heat tolerance. Compared with the control cucumbers (foliar spray of distilled water), a foliar spray of 1 mM SA decreased electrolyte leakage and the concentration of H_2O_2 and lipid peroxidation by heat stress treatment with 36 h after heat stress and 24 h after recovery [160]. This treatment also enhanced maximum yield of PSII photochemical reactions (Fv/Fm) and the quantum yield of the PSII electron transport (ϕPSII) after both heat stress and recovery.

Increase of temperature influences the early phase of anther development, causing premature progression through meiosis of pollen mother cells and proliferation arrest and premature degradation of anther wall cells [161]. Complete male sterility resulted from elevated temperatures for 4 days or longer days are observed widely among heat-stressed plant species such as wheat, tomato, cowpea and Arabidopsis. Endogenous auxin levels specifically decreased in the developing anthers of barley and Arabidopsis, and expression of the *YUCCA* auxin biosynthesis genes was repressed by increasing temperatures. Thus, application of auxin completely reversed male sterility in both plant species [161].

7.4. Environmental Elicitor

Originally, elicitor is defined as a substance that triggers the hypersensitive response against biotic stresses in a plant. Most elicitors are polysaccharides, small proteins, or lipids associated with the fungal or bacterial cell wall. However, pectic fragments resulting from microbial damage to the plant's own cell walls may also act as elicitors. The elicitors interact with the plasma membrane of undamaged cells and trigger activation of genes involved in the defense response.

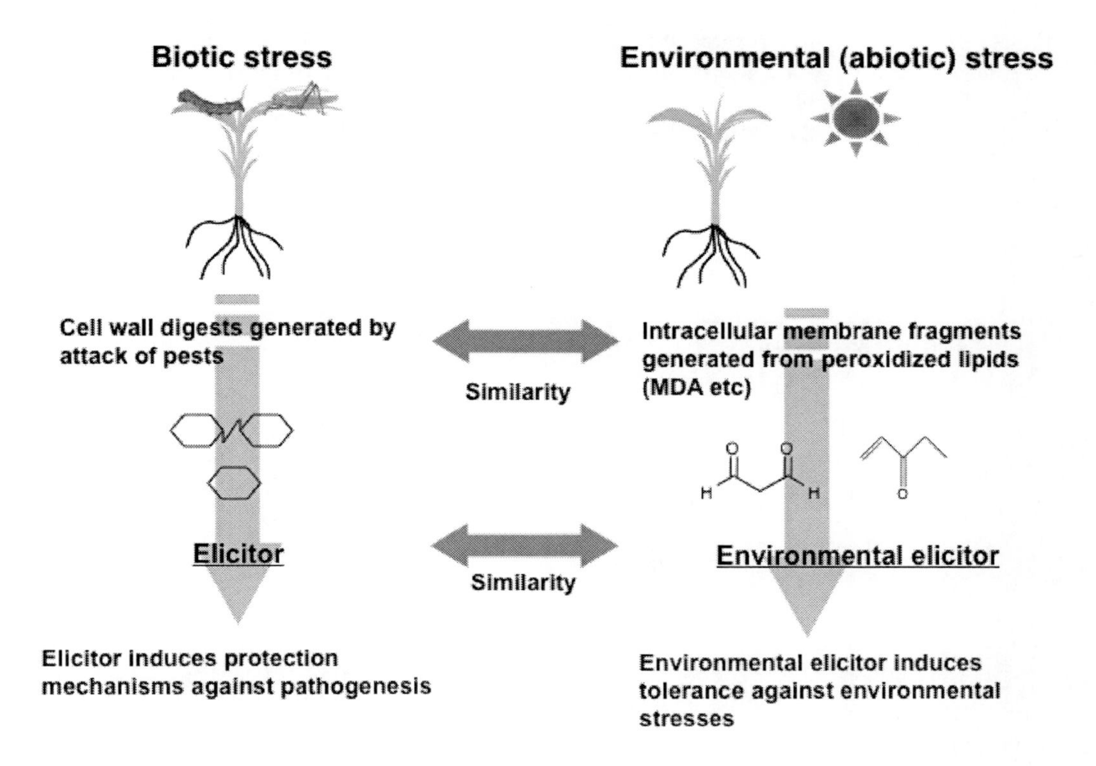

Figure 10. Concept of environmental elicitor.

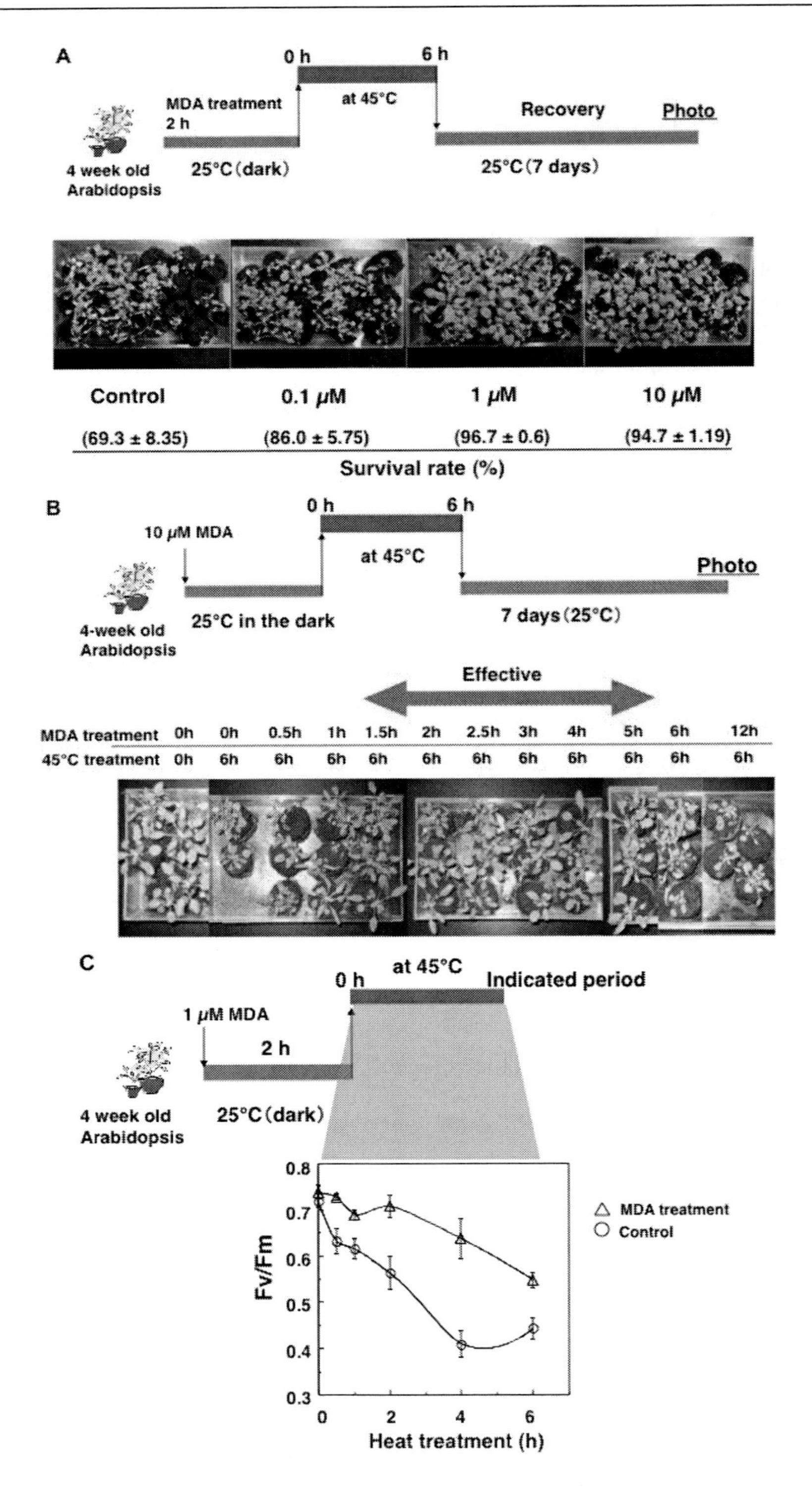

Figure 11. Enhanced thermotolerance of Arabidopsis plants treated with MDA as an environmental elicitor. (A) Determination of effective period of MDA for acquirement of thermotolerance in Arabidopsis plants. (B) Determination of effective concentration of MDA. (C) MDA treatment enhance thermotolerance of PSII during heat treatment. Diagrams of each treatment are also shown in the figure.

In the case of environmental stresses, it is widely believed that ROS-mediated breakdown of intracellular components is involved in many aspects of biological processes. Intracellular concentration of ROS is dependent on balance between generation and scavenge of ROS. In healthy photosynthetic tissues, generation of ROS is maintained at low level, and ROS scavenging system is well functional. As a result, concentration of ROS keeps low level and photosynthetic apparatus remains healthy. However in stressed conditions, generation of ROS is enhanced because consumption of NADPH by carbon-fixation reaction reduced under heat stress and significant amount of the reducing power flows oxygen molecules. Also ROS-scavenging system might become not functional at such conditions because of heat-inactivation of enzymes.

In such condition, ROS trigger peroxidation of fatty acids. Recently, it has been shown that degraded products including MDA and methylviniylketone from peroxidized fatty acids can regulate gene expression involved in environmental stresses [162]. This result leads an idea that small reactive carbonyls produced by lipid peroxidation act as signal molecules to facilitate environmental stress tolerance, mimicking to conventional elicitors in the case of pathogenesis (Figure 10). We named these reactive carbonyls as "environmental elicitor", and effect of environmental elicitors to heat stressed plants was investigated.

First, we determined effective MDA concentration for enhancement of heat tolerance of Arabidopsis plants. MDA solution (0.1, 1.0, or 10 μM) was foliar-sprayed to 4 week-old Arabidopsis plants, and then the plants were incubated for 2 hours at 25°C. The MDA-treated Arabidopsis were heat-stressed at 45°C for 6 hours and recovered at 25°C for a week. As shown in Figure 11A, Arabidopsis plants treated with 1.0 and 10 μM MDA significantly survived by the heat treatment. When 10 μM MDA was used as environmental elicitor, effect of MDA appeared after 1.5 hours then kept until 5 hours of treatment (Figure 11B). MDA treatment contributed to stabilize PSII under heat stress (Figure 11C). These results indicate that MDA can use as environmental elicitor to enhance heat tolerance of plants.

CONCLUSION

In this article, we focused on photosystems with type II RCs which are the highly sensitive sites affected by heat stress, and discussed their stability, tolerance, and acclimation under high-temperature conditions. The integrated pigment-protein complexes are embedded into the membranes, therefore, structures and interactions of the proteins at the membrane interface are significant for protecting the RC from invasion by endogenous or exogenous chemicals in functional photosystems. Based on the results presented here in combination with numbers of biochemical and physicochemical studies strongly indicated that several organic and inorganic cofactors are available for regulation of the proteins interacting with the intrinsic proteins to maintain the functional structure of type-II RCs and enhance the thermal stability and tolerance. Application of the knowledge obtained from bacterial photosystems to plants by biochemical technology and genetic engineering is one of potential countermeasures against the global warming.

Heat damage and resultant decrease of crop yields forecasted in near future are important matter of great urgent. However, as described in this article, heat damage and tolerant mechanisms are very complicated including thermodynamic, biochemical aspects. Many studies give potentially useful implication, nevertheless, each countermeasure is insufficient to solve the problem, therefore combination of multiple countermeasures are necessary to contribute the solution of the global problem. Thus, cooperative studies of diverse researchers among different fields and sophisticated integration of their knowledge are important to overcome the problems of global warming.

ACKNOWLEDGMENT

We thank Ms. Chihiro Nakamoto (Kobe University) for technical assistance and Dr. Kotomi Ueno (Kobe University) for illustrating scientific figures used in this article. We also thank Dr. Zheng-Yu Wang (Ibaraki University) for the generous gift of *Rsp. rubrum* and *Ach. vinosum,* and valuable comments.

REFERENCES

[1] Long, S. P. & Ort, D. R. (2010) More Than Taking The Heat: Crops And Global Change, *Current Opinion In Plant Biology 13*, 241-248.

[2] Cogdell, R. J., Gall, A. & Kohler, J. (2006) The Architecture And Function Of The Light-Harvesting Apparatus Of Purple Bacteria: From Single Molecules To In Vivo Membranes, *Quarterly Reviews Of Biophysics 39*, 227-324.

[3] Deisenhofer, J., Epp, O., Miki, K., Huber, R. & Michel, H. (1984) X-Ray Structure-Analysis Of A Membrane-Protein Complex - Electron-Density Map At 3a Resolution And A Model Of The Chromophores Of The Photosynthetic Reaction Center From Rhodopseudomonas-Viridis, *Journal Of Molecular Biology 180*, 385-398.

[4] Allen, J. P., Feher, G., Yeates, T. O., Rees, D. C., Deisenhofer, J., Michel, H. & Huber, R. (1986) Structural Homology Of Reaction Centers From Rhodopseudomonas-Sphaeroides And Rhodopseudomonas-Viridis As Determined By X-Ray-Diffraction, *Proceedings Of The National Academy Of Sciences Of The United States Of America 83*, 8589-8593.

[5] Nogi, T., Fathir, I., Kobayashi, M., Nozawa, T. & Miki, K. (2000) Crystal Structures Of Photosynthetic Reaction Center And High-Potential Iron-Sulfur Protein From Thermochromatium Tepidum: Thermostability And Electron Transfer, *Proceedings Of The National Academy Of Sciences Of The United States Of America 97*, 13561-13566.

[6] 6. Roszak, A. W., Howard, T. D., Southall, J., Gardiner, A. T., Law, C. J., Isaacs, N. W. & Cogdell, R. J. (2003) Crystal Structure Of The RC-LH1 Core Complex From Rhodopseudomonas Palustris, *Science 302*, 1969-1972.

[7] 7. Karrasch, S., Bullough, P. A. & Ghosh, R. (1995) The 8.5-Angstrom Projection Map Of The Light-Harvesting Complex-I From Rhodospirillum-Rubrum Reveals A Ring Composed Of 16 Subunits, *Embo Journal 14*, 631-638.

[8] Jamieson, S. J., Wang, P. Y., Qian, P., Kirkland, J. Y., Conroy, M. J., Hunter, C. N. & Bullough, P. A. (2002) Projection Structure Of The Photosynthetic Reaction Centre-Antenna Complex Of Rhodospirillum Rubrum At 8.5 Angstrom Resolution, *Embo Journal 21*, 3927-3935.

[9] Fotiadis, D., Qian, P., Philippsen, A., Bullough, P. A., Engel, A. & Hunter, C. N. (2004) Structural Analysis Of The Reaction Center Light-Harvesting Complex I Photosynthetic Core Complex Of Rhodospirillum Rubrum Using Atomic Force Microscopy, *Journal Of Biological Chemistry 279*, 2063-2068.

[10] Scheuring, S., Francia, F., Busselez, J., Melandri, B. A., Rigaud, J. L. & Levy, D. (2004) Structural Role Of Pufx In The Dimerization Of The Photosynthetic Core Complex Of Rhodobacter Sphaeroides, *Journal Of Biological Chemistry 279*, 3620-3626.

[11] Qian, P., Hunter, C. N. & Bullough, P. A. (2005) The 8.5 Angstrom Projection Structure Of The Core RC-LH1-Pufx Dimer Of Rhodobacter Sphaeroides, *Journal Of Molecular Biology 349*, 948-960.

[12] Scheuring, S., Busselez, J. & Levy, D. (2005) Structure Of The Dimeric Pufx-Containing Core Complex Of Rhodobacter Blasticus By In Situ Atomic Force Microscopy, *Journal Of Biological Chemistry 280*, 1426-1431.

[13] Kimura, Y., Hirano, Y., Yu, L. J., Suzuki, H., Kobayashi, M. & Wang, Z. Y. (2008) Calcium Ions Are Involved In The Unusual Red Shift Of The Light-Harvesting 1 Qy Transition Of The Core Complex In Thermophilic Purple Sulfur Bacterium Thermochromatium Tepidum, *Journal Of Biological Chemistry 283*, 13867-13873.

[14] Ma, F., Kimura, Y., Zhao, X. H., Wu, Y. S., Wang, P., Fu, L. M., Wang, Z. Y. & Zhang, J. P. (2008) Excitation Dynamics Of Two Spectral Forms Of The Core Complexes From Photosynthetic Bacterium Thermochromatium Tepidum, *Biophysical Journal 95*, 3349-3357.

[15] Tuschak, C., Beatty, J. T. & Overmann, J. (2004) Photosynthesis Genes And LH1 Proteins Of Roseospirillum Parvum 930I, A Purple Non-Sulfur Bacterium With Unusual Spectral Properties, *Photosynthesis Research 81*, 181-199.

[16] Permentier, H. P., Neerken, S., Overmann, J. & Amesz, J. (2001) A Bacteriochlorophyll A Antenna Complex From Purple Bacteria Absorbing At 963 Nm, *Biochemistry 40*, 5573-5578.

[17] Renger, G. & Renger, T. (2008) Photosystem II: The Machinery Of Photosynthetic Water Splitting, *Photosynth Res 98*, 53-80.

[18] Guskov, A., Kern, J., Gabdulkhakov, A., Broser, M., Zouni, A. & Saenger, W. (2009) Cyanobacterial Photosystem II At 2.9-Angstrom Resolution And The Role Of Quinones, Lipids, Channels And Chloride, *Nature Structural & Molecular Biology 16*, 334-342.

[19] Gabdulkhakov, A., Guskov, A., Broser, M., Kern, J., Muh, F., Saenger, W. & Zouni, A. (2009) Probing The Accessibility Of The Mn4Ca Cluster In Photosystem II: Channels Calculation, Noble Gas Derivatization, And Cocrystallization With DMSO, *Structure 17*, 1223-1234.

[20] Zouni, A., Witt, H. T., Kern, J., Fromme, P., Krauss, N., Saenger, W. & Orth, P. (2001) Crystal Structure Of Photosystem II From Synechococcus Elongatus At 3.8 Angstrom Resolution, *Nature 409*, 739-743.

[21] Ferreira, K. N., Iverson, T. M., Maghlaoui, K., Barber, J. & Iwata, S. (2004) Architecture Of The Photosynthetic Oxygen-Evolving Center, *Science 303*, 1831-1838.

[22] Loll, B., Kern, J., Saenger, W., Zouni, A. & Biesiadka, J. (2005) Towards Complete Cofactor Arrangement In The 3.0 Angstrom Resolution Structure Of Photosystem II, *Nature 438*, 1040-1044.

[23] Roose, J. L., Wegener, K. M. & Pakrasi, H. B. (2007) The Extrinsic Proteins Of Photosystem II, *Photosynthesis Research 92*, 369-387.

[24] Williamson, A. K. (2008) Structural And Functional Aspects Of The MSP (Psbo) And Study Of Its Differences In Thermophilic Versus Mesophilic Organisms, *Photosynth Res 98*, 365-389.

[25] Shen, J. R. & Inoue, Y. (1993) Binding And Functional Properties Of Two New Extrinsic Components, Cytochrome C-550 And A 12-Kda Protein, In Cyanobacterial Photosystem II, *Biochemistry 32*, 1825-1832.

[26] Shen, J. R., Burnap, R. L. & Inoue, Y. (1995) An Independent Role Of Cytochrome C-550 In Cyanobacterial Photosystem-Ii As Revealed By Double-Deletion Mutagenesis Of The Psbo And Psbv Genes In Synechocystis Sp Pcc-6803, *Biochemistry 34*, 12661-12668.

[27] Kawakami, K., Umena, Y., Kamiya, N. & Shen, J. R. (2009) Location Of Chloride And Its Possible Functions In Oxygen-Evolving Photosystem II Revealed By X-Ray Crystallography, *Proceedings Of The National Academy Of Sciences Of The United States Of America 106*, 8567-8572.

[28] Seidler, A. (1996) The Extrinsic Polypeptides Of Photosystem II, *Biochimica Et Biophysica Acta-Bioenergetics 1277*, 35-60.

[29] Al-Khaldi, S. F., Coker, J., Shen, J. R. & Burnap, R. L. (2000) Characterization Of Site-Directed Mutants In Manganese-Stabilizing Protein (MSP) Of Synechocystis Sp PCC6803 Unable To Grow Photoautotrophically In The Absence Of Cytochrome C-550, *Plant Molecular Biology 43*, 33-41.

[30] Heredia, P. & De Las Rivas, J. (2003) Calcium-Dependent Conformational Change And Thermal Stability Of The Isolated Psbo Protein Detected By FTIR Spectroscopy, *Biochemistry 42*, 11831-11838.

[31] Kruk, J., Burda, K., Jemiola-Rzeminska, M. & Strzalka, K. (2003) The 33 Kda Protein Of Photosystem II Is A Low-Affinity Calcium- And Lanthanide-Binding Protein, *Biochemistry 42*, 14862-14867.

[32] Seidler, A. & Rutherford, A. W. (1996) The Role Of The Extrinsic 33 Kda Protein In Ca2+ Binding In Photosystem II+, *Biochemistry 35*, 12104-12110.

[33] Loll, B., Gerold, G., Slowik, D., Voelter, W., Jung, C., Saenger, W. & Irrgang, K. D. (2005) Thermostability And Ca2+ Binding Properties Of Wild Type And Heterologously Expressed Psbo Protein From Cyanobacterial Photosystem II, *Biochemistry 44*, 4691-4698.

[34] 34. Rutherford, A. W. & Faller, P. (2001) The Heart Of Photosynthesis In Glorious 3D, *Trends Biochem Sci 26*, 341-344.

[35] Murray, J. W. & Barber, J. (2006) Identification Of A Calcium-Binding Site In The Psbo Protein Of Photosystem II, *Biochemistry 45*, 4128-4130.

[36] Enami, I., Suzuki, T., Tada, O., Nakada, Y., Nakamura, K., Tohri, A., Ohta, H., Inoue, I. & Shen, J. R. (2005) Distribution Of The Extrinsic Proteins As A Potential Marker For The Evolution Of Photosynthetic Oxygen-Evolving Photosystem II, *Febs Journal 272*, 5020-5030.

[37] Ifuku, K., Nakatsu, T., Kato, H. & Sato, F. (2004) Crystal Structure Of The Psbp Protein Of Photosystem II From Nicotiana Tabacum, *Embo Reports 5*, 362-367.

[38] Balsera, M., Arellano, J. B., Revuelta, J. L., De Las Rivas, J. & Hermoso, J. A. (2005) The 1.49 Angstrom Resolution Crystal Structure Of Psbq From Photosystem II Of

Spinacia Oleracea Reveals A PPII Structure In The N-Terminal Region, *Journal Of Molecular Biology 350*, 1051-1060.

[39] Thornton, L. E., Ohkawa, H., Roose, J. L., Kashino, Y., Keren, N. & Pakrasi, H. B. (2004) Homologs Of Plant Psbp And Psbq Proteins Are Necessary For Regulation Of Photosystem II Activity In The Cyanobacterium Synechopystis 6803, *Plant Cell 16*, 2164-2175.

[40] De Las Rivas, J. & Roman, A. (2005) Structure And Evolution Of The Extrinsic Proteins That Stabilize The Oxygen-Evolving Engine, *Photochem Photobiol Sci 4*, 1003-1010.

[41] Ishihara, S., Takabayashi, A., Ido, K., Endo, T., Ifuku, K. & Sato, F. (2007) Distinct Functions For The Two Psbp-Like Proteins PPL1 And PPL2 In The Chloroplast Thylakoid Lumen Of Arabidopsis, *Plant Physiol 145*, 668-679.

[42] Ifuku, K., Ishihara, S., Shimamoto, R., Ido, K. & Sato, F. (2008) Structure, Function, And Evolution Of The Psbp Protein Family In Higher Plants, *Photosynth Res 98*, 427-437.

[43] Roncel, M., Boussac, A., Zurita, J. L., Bottin, H., Sugiura, M., Kirilovsky, D. & Ortega, J. M. (2003) Redox Properties Of The Photosystem II Cytochromes B559 And C550 In The Cyanobacterium Thermosynechococcus Elongatus, *J Biol Inorg Chem 8*, 206-216.

[44] Kirilovsky, D., Roncel, M., Boussac, A., Wilson, A., Zurita, J. L., Ducruet, J. M., Bottin, H., Sugiura, M., Ortega, J. M. & Rutherford, A. W. (2004) Cytochrome C(550) In The Cyanobacterium Thermosynechococcus Elongatus - Study Of Redox Mutants, *Journal Of Biological Chemistry 279*, 52869-52880.

[45] Shen, J. R., Qian, M., Inoue, Y. & Burnap, R. L. (1998) Functional Characterization Of Synechocystis Sp. PCC 6803 Delta Psbu And Delta Psbv Mutants Reveals Important Roles Of Cytochrome C-550 In Cyanobacterial Oxygen Evolution, *Biochemistry 37*, 1551-1558.

[46] Nishiyama, Y., Hayashi, H., Watanabe, T. & Murata, N. (1994) Photosynthetic Oxygen Evolution Is Stabilized By Cytochrome C(550) Against Heat Inactivation In Synechococcus Sp Pcc-7002, *Plant Physiology 105*, 1313-1319.

[47] Shen, J. R., Vermaas, W. & Inoue, Y. (1995) The Role Of Cytochrome C-550 As Studied Through Reverse Genetics And Mutant Characterization In Synechocystis Sp Pcc-6803, *Journal Of Biological Chemistry 270*, 6901-6907.

[48] Sugiura, M., Iwai, E., Hayashi, H. & Boussac, A. (2010) Differences In The Interactions Between The Subunits Of Photosystem II Dependent On D1 Protein Variants In The Thermophilic Cyanobacterium Thermosynechococcus Elongatus, *J Biol Chem 285*, 30008-30018.

[49] Shen, J. R., Ikeuchi, M. & Inoue, Y. (1997) Analysis Of The Psbu Gene Encoding The 12-Kda Extrinsic Protein Of Photosystem II And Studies On Its Role By Deletion Mutagenesis In Synechocystis Sp. PCC 6803, *J Biol Chem 272*, 17821-17826.

[50] Inoue-Kashino, N., Kashino, Y., Satoh, K., Terashima, I. & Pakrasi, H. B. (2005) Psbu Provides A Stable Architecture For The Oxygen-Evolving System In Cyanobacterial Photosystem II, *Biochemistry 44*, 12214-12228.

[51] Balint, I., Bhattacharya, J., Perelman, A., Schatz, D., Moskovitz, Y., Keren, N. & Schwarz, R. (2006) Inactivation Of The Extrinsic Subunit Of Photosystem II, Psbu, In Synechococcus PCC 7942 Results In Elevated Resistance To Oxidative Stress, *FEBS Lett 580*, 2117-2122.

[52] Nield, J., Balsera, M., De Las Rivas, J. & Barber, J. (2002) Three-Dimensional Electron Cryo-Microscopy Study Of The Extrinsic Domains Of The Oxygen-Evolving Complex Of Spinach - Assignment Of The Psbo Protein, *Journal Of Biological Chemistry 277*, 15006-15012.

[53] Nield, J. & Barber, J. (2006) Refinement Of The Structural Model For The Photosystem II Supercomplex Of Higher Plants, *Biochimica Et Biophysica Acta-Bioenergetics 1757*, 353-361.

[54] De Las Rivas, J., Balsera, M. & Barber, J. (2004) Evolution Of Oxygenic Photosynthesis: Genome-Wide Analysis Of The OEC Extrinsic Proteins, *Trends In Plant Science 9*, 18-25.

[55] Miyao, M. & Murata, N. (1983) Partial Reconstitution Of The Photosynthetic Oxygen Evolution System By Rebinding Of The 33-Kda Polypeptide, *Febs Letters 164*, 375-378.

[56] Miyao, M. & Murata, N. (1989) The Mode Of Binding Of 3 Extrinsic Proteins Of 33-Kda, 23-Kda And 18-Kda In The Photosystem-Ii Complex Of Spinach, *Biochimica Et Biophysica Acta 977*, 315-321.

[57] Miyao, M. & Murata, N. (1983) Partial Disintegration And Reconstitution Of The Photosynthetic Oxygen Evolution System - Binding Of 24 Kilodalton And 18 Kilodalton Polypeptides, *Biochemical Et Biophysica Acta 725*, 87-93.

[58] Tohri, A., Dohmae, N., Suzuki, T., Ohta, H., Inoue, Y. & Enami, I. (2004) Identification Of Domains On The Extrinsic 23 Kda Protein Possibly Involved In Electrostatic Interaction With The Extrinsic 33 Kda Protein In Spinach Photosystem II, *European Journal Of Biochemistry 271*, 962-971.

[59] Ifuku, K., Yamamoto, Y., Ono, T., Ishihara, S. & Sato, F. (2005) Psbp Protein, But Not Psbq Protein, Is Essential For The Regulation And Stabilization Of Photosystem II In Higher Plants, *Plant Physiology 139*, 1175-1184.

[60] Bondarava, N., Beyer, P. & Krieger-Liszkay, A. (2005) Function Of The 23 Kda Extrinsic Protein Of Photosystem II As A Manganese Binding Protein And Its Role In Photoactivation, *Biochimica Et Biophysica Acta-Bioenergetics 1708*, 63-70.

[61] Yi, X. P., Hargett, S. R., Frankel, L. K. & Bricker, T. M. (2006) The Psbq Protein Is Required In Arabidopsis For Photosystem II Assembly/Stability And Photoautotrophy Under Low Light Conditions, *Journal Of Biological Chemistry 281*, 26260-26267.

[62] Suorsa, M., Sirpio, S., Allahverdiyeva, Y., Paakkarinen, V., Mamedov, F., Styring, S. & Aro, E. M. (2006) Psbr, A Missing Link In The Assembly Of The Oxygen-Evolving Complex Of Plant Photosystem II, *Journal Of Biological Chemistry 281*, 145-150.

[63] De Las Rivas, J., Heredia, P. & Roman, A. (2007) Oxygen-Evolving Extrinsic Proteins (Psbo,P,Q,R): Bioinformatic And Functional Analysis, *Biochimica Et Biophysica Acta-Bioenergetics 1767*, 575-582.

[64] Debus, R. J. (2001) Amino Acid Residues That Modulate The Properties Of Tyrosine Y-Z And The Manganese Cluster In The Water Oxidizing Complex Of Photosystem II, *Biochimica Et Biophysica Acta-Bioenergetics 1503*, 164-186.

[65] Ghanotakis, D. F. & Yocum, C. F. (1990) Photosystem-Ii And The Oxygen-Evolving Complex, *Annual Review Of Plant Physiology And Plant Molecular Biology*

[66] Enami, I., Kitamura, M., Tomo, T., Isokawa, Y., Ohta, H. & Katoh, S. (1994) Is The Primary Cause Of Thermal Inactivation Of Oxygen Evolution In Spinach Ps-Ii

Membranes Release Of The Extrinsic 33 Kda Protein Or Of Mn, *Biochimica Et Biophysica Acta-Bioenergetics 1186*, 52-58.

[67] Miura, T., Shen, J. R., Takahashi, S., Kamo, M., Nakamura, E., Ohta, H., Kamei, A., Inoue, Y., Domae, N., Takio, R., Nakazato, K., Inoue, Y. & Enami, I. (1997) Identification Of Domains On The Extrinsic 33-Kda Protein Possibly Involved In Electrostatic Interaction With Photosystem II Complex By Means Of Chemical Modification, *Journal Of Biological Chemistry 272*, 3788-3798.

[68] Spassov, V. Z., Karshikoff, A. D. & Ladenstein, R. (1995) The Optimization Of Protein-Solvent Interactions - Thermostability And The Role Of Hydrophobic And Electrostatic Interactions, *Protein Science 4*, 1516-1527.

[69] Tanner, J. J., Hecht, R. M. & Krause, K. L. (1996) Determinants Of Enzyme Thermostability Observed In The Molecular Structure Of Thermus Aquaticus D-Glyceraldehyde-3-Phosphate Dehydrogenase At 2.5 Angstrom Resolution, *Biochemistry 35*, 2597-2609.

[70] Tahirov, T. H., Oki, H., Tsukihara, T., Ogasahara, K., Yutani, K., Ogata, K., Izu, Y., Tsunasawa, S. & Kato, I. (1998) Crystal Structure Of Methionine Aminopeptidase From Hyperthermophile, Pyrococcus Furiosus, *Journal Of Molecular Biology 284*, 101-124.

[71] Villeret, V., Clantin, B., Tricot, C., Legrain, C., Roovers, M., Stalon, V., Glansdorff, N. & Van Beeumen, J. (1998) The Crystal Structure Of Pyrococcus Furiosus Ornithine Carbamoyltransferase Reveals A Key Role For Oligomerization In Enzyme Stability At Extremely High Temperatures, *Proceedings Of The National Academy Of Sciences Of The United States Of America 95*, 2801-2806.

[72] Kumar, S., Tsai, C. J. & Nussinov, R. (2000) Factors Enhancing Protein Thermostability, *Protein Engineering 13*, 179-191.

[73] Li, T., Sun, F., Ji, X., Feng, Y. & Rao, Z. H. (2003) Structure Based Hyperthermostability Of Archaeal Histone Hpha From Pyrococcus Horikoshii, *Journal Of Molecular Biology 325*, 1031-1037.

[74] Kaushik, J. K., Iimura, S., Ogasahara, K., Yamagata, Y., Segawa, S. I. & Yutani, K. (2006) Completely Buried, Non-Ion-Paired Glutamic Acid Contributes Favorably To The Conformational Stability Of Pyrrolidone Carboxyl Peptidases From Hyperthermophiles, *Biochemistry 45*, 7100-7112.

[75] Madigan, M. T. (2003) Anoxygenic Phototrophic Bacteria From Extreme Environments, *Photosynthesis Research 76*, 157-171.

[76] Kobayashi, M., Fujioka, Y., Mori, T., Terashima, M., Suzuki, H., Shimada, Y., Saito, T., Wang, Z. Y. & Nozawa, T. (2005) Reconstitution Of Photosynthetic Reaction Centers And Core Antenna-Reaction Center Complexes In Liposomes And Their Thermal Stability, *Bioscience Biotechnology And Biochemistry 69*, 1130-1136.

[77] Watson, A. J., Hughes, A. V., Fyfe, P. K., Wakeham, M. C., Holden-Dye, K., Heathcote, P. & Jones, M. R. (2005) On The Role Of Basic Residues In Adapting The Reaction Centre - LH1 Complex For Growth At Elevated Temperatures In Purple Bacteria, *Photosynthesis Research 86*, 81-100.

[78] Kimura, Y., Yu, L. J., Hirano, Y., Suzuki, H. & Wang, Z. Y. (2009) Calcium Ions Are Required For The Enhanced Thermal Stability Of The Light-Harvesting-Reaction Center Core Complex From Thermophilic Purple Sulfur Bacterium Thermochromatium Tepidum, *Journal Of Biological Chemistry 284*, 93-99.

[79] Yu, L. J., Kato, S. & Wang, Z. Y. (In Press) Examination Of The Putative Ca2+-Binding Site In The Light-Harvesting Complex 1 Of Thermophilic Purple Sulfur Bacterium Thermochromatium Tepidum, *Photosynthesis Research*.

[80] Kimura, A., Eaton-Rye, J. J., Morita, E. H., Nishiyama, Y. & Hayashi, H. (2002) Protection Of The Oxygen-Evolving Machinery By The Extrinsic Proteins Of Photosystem II Is Essential For Development Of Cellular Thermotolerance In Synechocystis Sp PCC 6803, *Plant And Cell Physiology 43*, 932-938.

[81] Pueyo, J. J., Alfonso, M., Andres, C. & Picorel, R. (2002) Increased Tolerance To Thermal Inactivation Of Oxygen Evolution In Spinach Photosystem II Membranes By Substitution Of The Extrinsic 33-Kda Protein By Its Homologue From A Thermophilic Cyanobacterium, *Biochimica Et Biophysica Acta-Bioenergetics 1554*, 29-35.

[82] Nishiyama, Y., Los, D. A., Hayashi, H. & Murata, N. (1997) Thermal Protection Of The Oxygen-Evolving Machinery By Psbu, An Extrinsic Protein Of Photosystem II, In Synechococcus Species PCC 7002, *Plant Physiology 115*, 1473-1480.

[83] Nishiyama, Y., Los, D. A. & Murata, N. (1999) Psbu, A Protein Associated With Photosystem II, Is Required For The Acquisition Of Cellular Thermotolerance In Synechococcus Species PCC 7002, *Plant Physiology 120*, 301-308.

[84] Raison, J. K., Roberts, J. K. M. & Berry, J. A. (1982) Correlations Between The Thermal Stability Of Chloroplast (Thylakoid) Membranes And The Composition And Fluidity Of Their Polar Lipids Upon Acclimation Of The Higher Plant, Nerium Oleander, To Growth Temperature, *Biochimica Et Biophysica Acta (BBA) - Biomembranes 688*, 218-228.

[85] Hugly, S., Kunst, L., Browse, J. & Somerville, C. (1989) Enhanced Thermal Tolerance Of Photosynthesis And Altered Chloroplast Ultrastructure In A Mutant Of Arabidopsis Deficient In Lipid Desaturation, *Plant Physiol 90*, 1134-1142.

[86] Mamedov, M., Hayashi, H. & Murata, N. (1993) Effects Of Glycinebetaine And Unsaturation Of Membrane-Lipids On Heat-Stability Of Photosynthetic Electron-Transport And Phosphorylation Reactions In Synechocystis Pcc6803, *Biochimica Et Biophysica Acta 1142*, 1-5.

[87] Wada, H., Gombos, Z. & Murata, N. (1994) Contribution Of Membrane Lipids To The Ability Of The Photosynthetic Machinery To Tolerate Temperature Stress, *Proc Natl Acad Sci U S A 91*, 4273-4277.

[88] Nelson, N. & Yocum, C. F. (2006) Structure And Function Of Photosystems I And II, *Annu Rev Plant Biol 57*, 521-565.

[89] Aminaka, R., Taira, Y., Kashino, Y., Koike, H. & Satoh, K. (2006) Acclimation To The Growth Temperature And Thermosensitivity Of Photosystem II In A Mesophilic Cyanobacterium, Synechocystis Sp PCC6803, *Plant And Cell Physiology 47*, 1612-1621.

[90] Thomas, P. G., Dominy, P. J., Vigh, L., Mansourian, A. R., Quinn, P. J. & Williams, W. P. (1986) Increased Thermal-Stability Of Pigment-Protein Complexes Of Pea Thylakoids Following Catalytic-Hydrogenation Of Membrane-Lipids, *Biochimica Et Biophysica Acta 849*, 131-140.

[91] Gombos, Z., Wada, H. & Murata, N. (1991) Direct Evaluation Of Effects Of Fatty-Acid Unsaturation On The Thermal-Properties Of Photosynthetic Activities, As Studied By Mutation And Transformation Of Synechocystis Pcc6803, *Plant And Cell Physiology 32*, 205-211.

[92] Sato, N., Aoki, M., Maru, Y., Sonoike, K., Minoda, A. & Tsuzuki, M. (2003) Involvement Of Sulfoquinovosyl Diacylglycerol In The Structural Integrity And Heat-Tolerance Of Photosystem II, *Planta 217*, 245-251.

[93] Balogi, Z., Torok, Z., Balogh, G., Josvay, K., Shigapova, N., Vierling, E., Vigh, L. & Horvath, L. (2005) "Heat Shock Lipid" In Cyanobacteria During Heat/Light-Acclimation, *Archives Of Biochemistry And Biophysics 436*, 346-354.

[94] Sakurai, I., Mizusawa, N., Wada, H. & Sato, N. (2007) Digalactosyldiacylglycerol Is Required For Stabilization Of The Oxygen-Evolving Complex In Photosystem II1[C][OA], *Plant Physiology 145*, 1361-1370.

[95] Mizusawa, N., Sakata, S., Sakurai, I., Sato, N. & Wada, H. (2009) Involvement Of Digalactosyldiacylglycerol In Cellular Thermotolerance In Synechocystis Sp. PCC 6803, *Arch Microbiol 191*, 595-601.

[96] Eriksson, M. J. & Clarke, A. K. (1996) The Heat Shock Protein Clpb Mediates The Development Of Thermotolerance In The Cyanobacterium Synechococcus Sp. Strain PCC 7942, *J Bacteriol 178*, 4839-4846.

[97] Clarke, A. K. & Eriksson, M. J. (2000) The Truncated Form Of The Bacterial Heat Shock Protein Clpb/HSP100 Contributes To Development Of Thermotolerance In The Cyanobacterium Synechococcus Sp. Strain PCC 7942, *J Bacteriol 182*, 7092-7096.

[98] Nakamoto, H., Suzuki, N. & Roy, S. K. (2000) Constitutive Expression Of A Small Heat-Shock Protein Confers Cellular Thermotolerance And Thermal Protection To The Photosynthetic Apparatus In Cyanobacteria, *Febs Letters 483*, 169-174.

[99] Tuominen, I., Pollari, M., Tyystjarvi, E. & Tyystjarvi, T. (2006) The Sigb Sigma Factor Mediates High-Temperature Responses In The Cyanobacterium Synechocystis Sp. PCC6803, *Febs Letters 580*, 319-323.

[100] Singh, A. K., Summerfield, T. C., Li, H. & Sherman, L. A. (2006) The Heat Shock Response In The Cyanobacterium Synechocystis Sp Strain PCC 6803 And Regulation Of Gene Expression By Hrca And Sigb, *Archives Of Microbiology 186*, 273-286.

[101] Berry, J. & Bjorkman, O. (1980) Photosynthetic Response And Adaptation To Temperature In Higher-Plants, *Annual Review Of Plant Physiology And Plant Molecular Biology 31*, 491-543.

[102] Havaux, M. & Tardy, F. (1996) Temperature-Dependent Adjustment Of The Thermal Stability Of Photosystem II In Vivo: Possible Involvement Of Xanthophyll-Cycle Pigments, *Planta 198*, 324-333.

[103] Nash, D., Miyao, M. & Murata, N. (1985) Heat Inactivation Of Oxygen Evolution In Photosystem-Ii Particles And Its Acceleration By Chloride Depletion And Exogenous Manganese, *Biochimica Et Biophysica Acta 807*, 127-133.

[104] Enami, I., Kamo, M., Ohta, H., Takahashi, S., Miura, T., Kusayanagi, M., Tanabe, S., Kamei, A., Motoki, A., Hirano, M., Tomo, T. & Satoh, K. (1998) Intramolecular Cross-Linking Of The Extrinsic 33-Kda Protein Leads To Loss Of Oxygen Evolution But Not Its Ability Of Binding To Photosystem II And Stabilization Of The Manganese Cluster, *Journal Of Biological Chemistry 273*, 4629-4634.

[105] Yamane, Y., Kashino, Y., Koike, H. & Satoh, K. (1998) Effects Of High Temperatures On The Photosynthetic Systems In Spinach: Oxygen-Evolving Activities, Fluorescence Characteristics And The Denaturation Process, *Photosynthesis Research 57*, 51-59.

[106] Lydakis-Simantiris, N., Hutchison, R. S., Betts, S. D., Barry, B. A. & Yocum, C. F. (1999) Manganese Stabilizing Protein Of Photosystem II Is A Thermostable, Natively Unfolded Polypeptide, *Biochemistry 38*, 404-414.

[107] Sage, R. F. & Kubien, D. S. (2007) The Temperature Response Of C-3 And C-4 Photosynthesis, *Plant Cell And Environment 30*, 1086-1106.

[108] Kana, R., Kotabova, E. & Prasil, O. (2008) Acceleration Of Plastoquinone Pool Reduction By Alternative Pathways Precedes A Decrease In Photosynthetic CO2 Assimilation In Preheated Barley Leaves, *Physiologia Plantarum 133*, 794-806.

[109] Havaux, M., Greppin, H. & Strasser, R. J. (1991) Functioning Of Photosystem-I And Photosystem-Ii In Pea Leaves Exposed To Heat-Stress In The Presence Or Absence Of Light - Analysis Using Invivo Fluorescence, Absorbency, Oxygen And Photoacoustic Measurements, *Planta 186*, 88-98.

[110] Kreslavskii, V. D. & Khristin, M. S. (2003) [After Effect Of Heat Shock On Induction Of Fluorescence And Low Temperature Fluorescence Spectra Of Wheat Leaves], *Biofizika 48*, 865-872.

[111] Gounaris, K., Brain, A. R. R., Quinn, P. J. & Williams, W. P. (1984) Structural Reorganization Of Chloroplast Thylakoid Membranes In Response To Heat-Stress, *Biochimica Et Biophysica Acta 766*, 198-208.

[112] Semenova, G. A. (2004) Structural Reorganization Of Thylakoid Systems In Response To Heat Treatment, *Photosynthetica 42*, 521-527.

[113] Sharkey, T. D. (2005) Effects Of Moderate Heat Stress On Photosynthesis: Importance Of Thylakoid Reactions, Rubisco Deactivation, Reactive Oxygen Species, And Thermotolerance Provided By Isoprene, *Plant Cell And Environment 28*, 269-277.

[114] Tardy, F. & Havaux, M. (1997) Thylakoid Membrane Fluidity And Thermostability During The Operation Of The Xanthophyll Cycle In Higher-Plant Chloroplasts, *Biochimica Et Biophysica Acta-Biomembranes 1330*, 179-193.

[115] Bukhov, N. G., Wiese, C., Neimanis, S. & Heber, U. (1999) Heat Sensitivity Of Chloroplasts And Leaves: Leakage Of Protons From Thylakoids And Reversible Activation Of Cyclic Electron Transport, *Photosynthesis Research 59*, 81-93.

[116] Schrader, S. M., Kane, H. J., Sharkey, T. D. & Von Caemmerer, S. (2006) High Temperature Enhances Inhibitor Production But Reduces Fallover In Tobacco Rubisco, *Functional Plant Biology 33*, 921-929.

[117] Pastenes, C. & Horton, P. (1996) Effect Of High Temperature On Photosynthesis In Beans .1. Oxygen Evolution And Chlorophyll Fluorescence, *Plant Physiology 112*, 1245-1251.

[118] Yamashita, A., Nijo, N., Pospisil, P., Morita, N., Takenaka, D., Aminaka, R. & Yamamoto, Y. (2008) Quality Control Of Photosystem II: Reactive Oxygen Species Are Responsible For The Damage To Photosystem II Under Moderate Heat Stress, *J Biol Chem 283*, 28380-28391.

[119] Murata, N., Takahashi, S., Nishiyama, Y. & Allakhverdiev, S. I. (2007) Photoinhibition Of Photosystem II Under Environmental Stress, *Biochim Biophys Acta 1767*, 414-421.

[120] Alfonso, M., Yruela, I., Almarcegui, S., Torrado, E., Perez, M. A. & Picorel, R. (2001) Unusual Tolerance To High Temperatures In A New Herbicide-Resistant D1 Mutant From Glycine Max (L.) Merr. Cell Cultures Deficient In Fatty Acid Desaturation, *Planta 212*, 573-582.

[121] Murakami, Y., Tsuyama, M., Kobayashi, Y., Kodama, H. & Iba, K. (2000) Trienoic Fatty Acids And Plant Tolerance Of High Temperature, *Science 287*, 476-479.

[122] Alfonso, M., Collados, R., Yruela, I. & Picorel, R. (2004) Photoinhibition And Recovery In A Herbicide-Resistant Mutant From Glycine Max (L.) Merr. Cell Cultures Deficient In Fatty Acid Unsaturation, *Planta 219*, 428-439.

[123] Yamauchi, Y. & Sugimoto, Y. (2010) Effect Of Protein Modification By Malondialdehyde On The Interaction Between The Oxygen-Evolving Complex 33 Kda Protein And Photosystem II Core Proteins, *Planta 231*, 1077-1088.

[124] Havaux, M. (1993) Characterization Of Thermal-Damage To The Photosynthetic Electron-Transport System In Potato Leaves, *Plant Science 94*, 19-33.

[125] Sazanov, L. A., Burrows, P. A. & Nixon, P. J. (1998) The Chloroplast Ndh Complex Mediates The Dark Reduction Of The Plastoquinone Pool In Response To Heat Stress In Tobacco Leaves, *FEBS Lett 429*, 115-118.

[126] June, T., Evans, J. R. & Farquhar, G. D. (2004) A Simple New Equation For The Reversible Temperature Dependence Of Photosynthetic Electron Transport: A Study On Soybean Leaf, *Functional Plant Biology 31*, 275-283.

[127] Egorova, E. A. & Bukhov, N. G. (2002) Effect Of Elevated Temperatures On The Activity Of Alternative Pathways Of Photosynthetic Electron Transport In Intact Barley And Maize Leaves, *Russian Journal Of Plant Physiology 49*, 575-584.

[128] Egorova, E. A., Bukhov, N. G., Heber, U., Samson, G. & Carpentier, R. (2003) Effect Of The Pool Size Of Stromal Reductants On The Alternative Pathway Of Electron Transfer To Photosystem I In Chloroplasts Of Intact Leaves, *Russian Journal Of Plant Physiology 50*, 431-440.

[129] Bukhov, N. G., Dzhibladze, T. G. & Egorova, E. A. (2005) Elevated Temperatures Inhibit Ferredoxin-Dependent Cyclic Electron Flow Around Photosystem I, *Russian Journal Of Plant Physiology 52*, 578-583.

[130] Schrader, S. M., Wise, R. R., Wacholtz, W. F., Ort, D. R. & Sharkey, T. D. (2004) Thylakoid Membrane Responses To Moderately High Leaf Temperature In Pima Cotton, *Plant Cell And Environment 27*, 725-735.

[131] Yamasaki, T., Yamakawa, T., Yamane, Y., Koike, H., Satoh, K. & Katoh, S. (2002) Temperature Acclimation Of Photosynthesis And Related Changes In Photosystem II Electron Transport In Winter Wheat, *Plant Physiology 128*, 1087-1097.

[132] Salvucci, M. E. & Crafts-Brandner, S. J. (2004) Relationship Between The Heat Tolerance Of Photosynthesis And The Thermal Stability Of Rubisco Activase In Plants From Contrasting Thermal Environments, *Plant Physiology 134*, 1460-1470.

[133] Havaux, M., Rumeau, D. & Ducruet, J. M. (2005) Probing The FQR And NDH Activities Involved In Cyclic Electron Transport Around Photosystem I By The 'Afterglow' Luminescence, *Biochimica Et Biophysica Acta-Bioenergetics 1709*, 203-213.

[134] Hideg, E., Spetea, C. & Vass, I. (1994) Singlet Oxygen And Free-Radical Production During Acceptor-Induced And Donor-Side-Induced Photoinhibition - Studies With Spin-Trapping Epr Spectroscopy, *Biochimica Et Biophysica Acta-Bioenergetics 1186*, 143-152.

[135] Khorobrykh, S. A. & Ivanov, B. N. (2002) Oxygen Reduction In A Plastoquinone Pool Of Isolated Pea Thylakoids, *Photosynthesis Research 71*, 209-219.

[136] Barua, D., Downs, C. A. & Heckathorn, S. A. (2003) Variation In Chloroplast Small Heat-Shock Protein Function Is A Major Determinant Of Variation In Thermotolerance Of Photosynthetic Electron Transport Among Ecotypes Of Chenopodium Album, *Functional Plant Biology 30*, 1071-1079.

[137] Heckathorn, S. A., Downs, C. A., Sharkey, T. D. & Coleman, J. S. (1998) The Small, Methionine-Rich Chloroplast Heat-Shock Protein Protects Photosystem II Electron Transport During Heat Stress, *Plant Physiology 116*, 439-444.

[138] Heckathorn, S. A., Ryan, S. L., Baylis, J. A., Wang, D. F., Hamilton, E. W., Cundiff, L. & Luthe, D. S. (2002) In Vivo Evidence From An Agrostis Stolonifera Selection Genotype That Chloroplast Small Heat-Shock Proteins Can Protect Photosystem II During Heat Stress, *Functional Plant Biology 29*, 933-944.

[139] Neta-Sharir, I., Isaacson, T., Lurie, S. & Weiss, D. (2005) Dual Role For Tomato Heat Shock Protein 21: Protecting Photosystem II From Oxidative Stress And Promoting Color Changes During Fruit Maturation, *Plant Cell 17*, 1829-1838.

[140] Downs, C. A., Coleman, J. S. & Heckathorn, S. A. (1999) The Chloroplast 22-Ku Heat-Shock Protein: A Lumenal Protein That Associates With The Oxygen Evolving Complex And Protects Photosystem II During Heat Stress, *Journal Of Plant Physiology 155*, 477-487.

[141] Hong, S. W. & Vierling, E. (2001) Hsp101 Is Necessary For Heat Tolerance But Dispensable For Development And Germination In The Absence Of Stress, *Plant Journal 27*, 25-35.

[142] Nitta, K., Suzuki, N., Honma, D., Kaneko, Y. & Nakamoto, H. (2005) Ultrastructural Stability Under High Temperature Or Intensive Light Stress Conferred By A Small Heat Shock Protein In Cyanobacteria, *Febs Letters 579*, 1235-1242.

[143] Tanaka, Y., Nishiyama, Y. & Murata, N. (2000) Acclimation Of The Photosynthetic Machinery To High Temperature In Chlamydomonas Reinhardtii Requires Synthesis De Novo Of Proteins Encoded By The Nuclear And Chloroplast Genomes, *Plant Physiology 124*, 441-449.

[144] Wang, P., Duan, W., Takabayashi, A., Endo, T., Shikanai, T., Ye, J. Y. & Mi, H. L. (2006) Chloroplastic NAD(P)H Dehydrogenase In Tobacco Leaves Functions In Alleviation Of Oxidative Damage Caused By Temperature Stress, *Plant Physiology 141*, 465-474.

[145] Baniwal, S. K., Bharti, K., Chan, K. Y., Fauth, M., Ganguli, A., Kotak, S., Mishra, S. K., Nover, L., Port, M., Scharf, K. D., Tripp, J., Weber, C., Zielinski, D. & Von Koskull-Doring, P. (2004) Heat Stress Response In Plants: A Complex Game With Chaperones And More Than Twenty Heat Stress Transcription Factors, *J Biosci 29*, 471-487.

[146] Nishizawa, A., Yabuta, Y., Yoshida, E., Maruta, T., Yoshimura, K. & Shigeoka, S. (2006) Arabidopsis Heat Shock Transcription Factor A2 As A Key Regulator In Response To Several Types Of Environmental Stress, *Plant Journal 48*, 535-547.

[147] Ogawa, D., Yamaguchi, K. & Nishiuchi, T. (2007) High-Level Overexpression Of The Arabidopsis Hsfa2 Gene Confers Not Only Increased Themotolerance But Also Salt/Osmotic Stress Tolerance And Enhanced Callus Growth, *Journal Of Experimental Botany 58*, 3373-3383.

[148] Suzuki, N., Rizhsky, L., Liang, H. J., Shuman, J., Shulaev, V. & Mittler, R. (2005) Enhanced Tolerance To Environmental Stress In Transgenic Plants Expressing The

Transcriptional Coactivator Multiprotein Bridging Factor 1c, *Plant Physiology 139*, 1313-1322.
[149] Sakuma, Y., Maruyama, K., Qin, F., Osakabe, Y., Shinozaki, K. & Yamaguchi-Shinozaki, K. (2006) Dual Function Of An Arabidopsis Transcription Factor DREB2A In Water-Stress-Responsive And Heat-Stress-Responsive Gene Expression, *Proc Natl Acad Sci U S A 103*, 18822-18827.
[150] Yang, X. H., Wen, X. G., Gong, H. M., Lu, Q. T., Yang, Z. P., Tang, Y. L., Liang, Z. & Lu, C. M. (2007) Genetic Engineering Of The Biosynthesis Of Glycinebetaine Enhances Thermotolerance Of Photosystem II In Tobacco Plants, *Planta 225*, 719-733.
[151] Gong, M., Li, Y. J., Dai, X., Tian, M. & Li, Z. G. (1997) Involvement Of Calcium And Calmodulin In The Acquisition Of Heat-Shock Induced Thermotolerance In Maize Seedlings, *Journal Of Plant Physiology 150*, 615-621.
[152] Suzuki, N. & Mittler, R. (2006) Reactive Oxygen Species And Temperature Stresses: A Delicate Balance Between Signaling And Destruction, *Physiologia Plantarum 126*, 45-51.
[153] Miller, G. & Mittler, R. (2006) Could Heat Shock Transcription Factors Function As Hydrogen Peroxide Sensors In Plants?, *Annals Of Botany 98*, 279-288.
[154] Uchida, A., Jagendorf, A. T., Hibino, T., Takabe, T. & Takabe, T. (2002) Effects Of Hydrogen Peroxide And Nitric Oxide On Both Salt And Heat Stress Tolerance In Rice, *Plant Science 163*, 515-523.
[155] Kotak, S., Larkindale, J., Lee, U., Von Koskull-Doring, P., Vierling, E. & Scharf, K. D. (2007) Complexity Of The Heat Stress Response In Plants, *Current Opinion In Plant Biology 10*, 310-316.
[156] Torres, M. A. & Dangl, J. L. (2005) Functions Of The Respiratory Burst Oxidase In Biotic Interactions, Abiotic Stress And Development, *Current Opinion In Plant Biology 8*, 397-403.
[157] Bonhamsmith, P. C., Kapoor, M. & Bewley, J. D. (1988) Exogenous Application Of Abscisic-Acid Or Triadimefon Affects The Recovery Of Zea-Mays Seedlings From Heat-Shock, *Physiologia Plantarum 73*, 27-30.
[158] Robertson, A. J., Ishikawa, M., Gusta, L. V. & Mackenzie, S. L. (1994) Abscisic Acid-Induced Heat Tolerance In Bromus-Inermis Leyss Cell-Suspension Cultures - Heat-Stable, Abscisic Acid-Responsive Polypeptides In Combination With Sucrose Confer Enhanced Thermostability, *Plant Physiology 105*, 181-190.
[159] Gong, M., Li, Y. J. & Chen, S. Z. (1998) Abscisic Acid-Induced Thermotolerance In Maize Seedlings Is Mediated By Calcium And Associated With Antioxidant Systems, *Journal Of Plant Physiology 153*, 488-496.
[160] Shi, Q., Bao, Z., Zhu, Z., Ying, Q. & Qian, Q. (2006) Effects Of Different Treatments Of Salicylic Acid On Heat Tolerance, Chlorophyll Fluorescence, And Antioxidant Enzyme Activity In Seedlings Of Cucumis Sativa L., *Plant Growth Regulation 48*, 127-135.
[161] Sakata, T., Oshino, T., Miura, S., Tomabechi, M., Tsunaga, Y., Higashitani, N., Miyazawa, Y., Takahashi, H., Watanabe, M. & Higashitani, A. (2010) Auxins Reverse Plant Male Sterility Caused By High Temperatures, *Proceedings Of The National Academy Of Sciences Of The United States Of America 107*, 8569-8574.

[162] Weber, H., Chetelat, A., Reymond, P. & Farmer, E. E. (2004) Selective And Powerful Stress Gene Expression In Arabidopsis In Response To Malondialdehyde, *Plant Journal 37*, 877-888.

In: Photochemistry
Editors: Karen J. Maes and Jaime M. Willems

ISBN: 978-1-61209-506-6

Chapter 6

The Little-Known Wavelength Effect in Provitamin D Photochemistry: The Ambiguous Role of the Weak Irreversible Channel

Irina Terenetskaya
Institute of Physics, National Academy of Sciences of Ukraine,
Kiev, Ukraine

Abstract

In this chapter brief description of the well-known wavelength effects in the provitamin D photochemistry which is the first stage of vitamin D synthesis initiated by the UV irradiation is presented. Unusual spectral kinetics revealed under irradiation of provitamin D with XeCl excimer laser at $\lambda = 308$ nm which conflicted with standard neglect of the weak irreversible channel is shown, and the origin of the observed anomaly is described in detail using simplified model and computer simulations. Particular attention is given to the original spectrophotometric analysis of the multi-component mixture of the vitamin D photoisomers which takes into account the irreversible photodegradation. Significant consequences of the effect revealed for the industrial synthesis of vitamin D and for biological UV dosimetry are discussed.

Introduction

The wide range of monomolecular photoisomerizations in the synthesis of vitamin D has attracted the attention of investigators over many years. The synthesis of vitally important vitamin D_3 in skin induced by solar ultraviolet (UV) irradiation is a two-stage process which begins with the production of previtamin D_3 from the steroidal precursor 7-dehydrocholesterol (7-DHC). Then vitamin D_3 itself is formed upon reversible thermal rearrangement of the previtamin D molecule. However, previtamin D is not stable to UV radiation and undergoes a number of side photoconversions.

It is generally accepted that reversible photoreactions play a determining role in the kinetics of a photoreaction, i.e. that UV irradiation of the initial provitamin D at a temperature excluding the formation of vitamin D leads to the formation of a mixture of the four principal photoisomers, between which a dynamic equilibrium – so-called photostationary state (PS) – is established after a specific time, and its composition strongly depends on the irradiation spectrum applied [1]. This well-known wavelength effect is caused mainly by the different absorbencies of the photoisomers involved in the reaction network [2,3].

Using tunable laser irradiation within 295-305 nm another, more complicated, wavelength effect in previtamin D photochemistry has been revealed, namely, a sudden increase in the efficiency of ring closure into lumisterol within narrow spectral range between 302 and 305 nm [4] that is still not fully understood.

In spite of the fact that irreversible photoconversions into so-called "over-irradiation products" toxisterols have been known for a long time [2,5], it is usually assumed that due to the low quantum yield toxisterols accumulate on prolonged exposure after complete conversion of the initial provitamin D. Therefore, the irreversible channel does not as a rule appear on the reaction scheme, and the presence of toxisterols is disregarded when analyzing the photoisomer mixture, i.e. the total concentration of the four main photoisomers is taken as 100%.

In this chapter based on our studies on laser initiation of provitamin D photoisomerization we will show the limitations of the photostationary state approximation and explain the little-known wavelength dependence of the irreversible channel efficiency using a simplified model and computer simulations of the photoreaction kinetics. The data on the excimer UV laser irradiation will illustrate the importance of the wavelength effect revealed. With due consideration for the irreversible photoconversions the ways of the photochemical stage optimization in the industrial synthesis of vitamin D will be discussed.

Particular attention is given to original spectrophotometric analysis of the multicomponent mixture of the vitamin D photoisomers which does take into account irreversible degradation of the photoisomer mixture and has most relevant applications in an *in situ* monitoring of biologically active antirachitic solar UV radiation. Characterization of the 'Vitamin D' biodosimeter using special laboratory facility is described in detail with focus on the difference between the *in vivo* and the *in vitro* action spectra of the vitamin D synthesis.

1. Commonly Accepted Interpretation of Provitamin D Photoisomerization

The complex network of vitamin D[1] synthesis *in vitro* has been intensively studied by E. Havinga and co-authors [1,3]. UV irradiation of provitamin D (Pro) within its absorption band (240-315 nm) yields previtamin D (Pre) by hexadiene ring opening (Figure 1). Once

[1] The terminology vitamin D is employed here in a general sense. Vitamin D2, or ergocalciferol ($C_{28}H_{44}O$), is synthesized upon UV irradiation mainly in plants from ergosterol (provitamin D2), similar to vitamin D3, or cholecalciferol ($C_{27}H_{44}O$) which is photochemically produced in animal and human skin from 7-dehydrocholesterol (7-DHC, provitamin D3). It is significant that basic monomolecular isomerizations of the two steroid species occur in perfect analogy.

formed, previtamin D converts into vitamin D (D) by the thermoinduced intramolecular hydrogen shift[2].

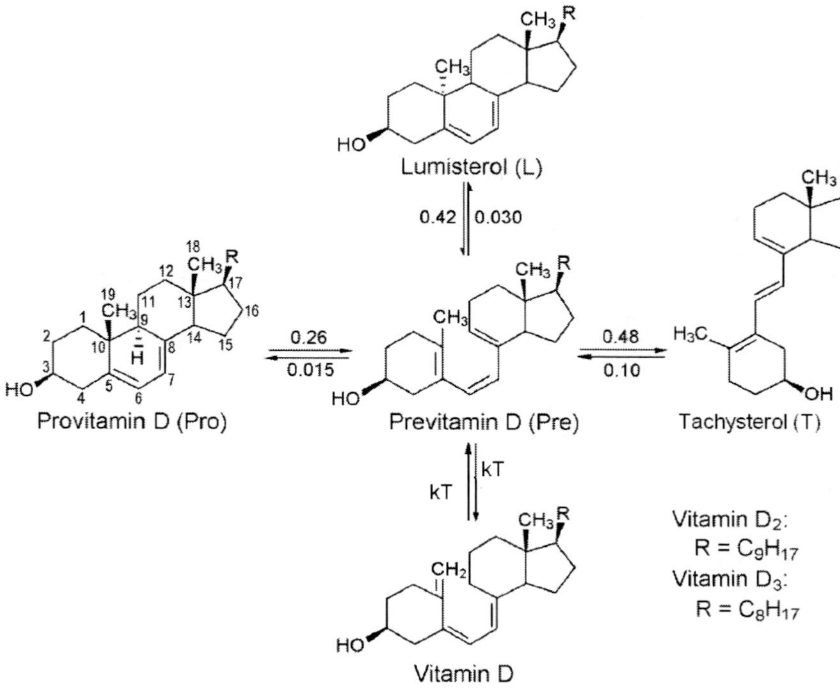

Lumisterol (L)

0.42 0.030

Provitamin D (Pro) 0.26 0.015 Previtamin D (Pre) 0.48 0.10 Tachysterol (T)

kT kT

Vitamin D

Vitamin D_2: R = C_9H_{17}
Vitamin D_3: R = C_8H_{17}

Figure 1. Commonly used scheme of vitamin D synthesis.

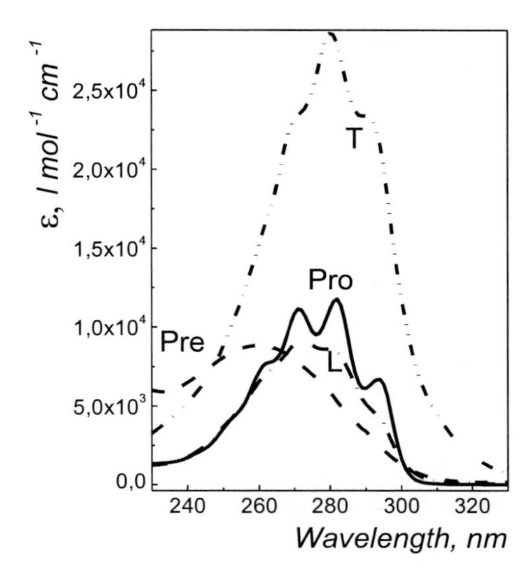

Figure 2. UV absorption spectra of provitamin D and its main photoisomers.

[2] At 20C, *in vitro*, it takes 20 hours to form 10% of vitamin D from previtamin D, and Pre<=>D equilibrium is established in 30 days. At 40C these times reduce to 2,3 hours and 3,5 days correspondingly [6].

However, the photoreaction is considerably complicated by the side photoconversions of previtamin D which absorption band lies in the same spectral range as is provitamin D (Figure 2). As a result, upon UV irradiation previtamin D undergoes a series of the photoconversions: reversible *cis-trans* isomerization into tachysterol (T) exhibits the highest quantum yield, and reversible ring-closures into initial provitamin D or its diastereomer lumisterol (L) are less probable. Hereupon UV irradiation of provitamin D gives rise to the mixture of four photoisomers (Pro, Pre, T and L) at a temperature preventing formation of vitamin D.

In due course, a dynamic equilibrium between the photoisomers, so-called photostationary state (PS), is established. Previtamin D and tachysterol usually dominate in the photostationary state, and the Pre/T ratio is strongly dependent on the irradiation wavelength [2,3]. It has been found that irradiation at 254 nm ensures high yield of tachysterol, and its amount decreases as the initiation radiation wavelength becomes longer. This wavelength effect known long ago is caused mainly by the different absorbencies of the photoisomers involved in the reaction network.

At the same time, it has long been noted that besides the fundamental photoreactions, irreversible photoconversions also take place with low rates in the system, that after prolonged lamp irradiation at $\lambda = 254$ nm result in the irreversible photodegradation of the isomeric mixture, and at $\lambda = 302{,}5/313$ nm in the accumulation of the 'overirradiation products', the toxisterols (Tox). In [5] 13 toxisterols were isolated from the photoisomeric mixture formed during irradiation in various solvents, their UV, IR, and NMR spectra were studied in detail, and their structure was established.

Nevertheless in the majority of the investigations the authors restricted themselves to the simple scheme of the reversible photoreactions shown in Figure 1 as far as the quantum yield of the formation of toxisterols is low ($\varphi \approx 0.039$) [2] and toxisterols accumulate in the photoisomer mixture during prolonged exposure after complete conversion of the initial Pro. Therefore, as a rule, the irreversible channel is not shown in the reaction scheme, and the presence of Tox is disregarded during concentration analysis of the photoisomeric mixture. In other words, it is generally assumed that accumulation of toxisterols in the photostationary state can be neglected and the total concentration of the four main photoisomers can be kept constant (equal to 100%) [4,7-9].

It was within the scope of the PS approximation [4] that a dramatic change was found in the calculated quantum yield of the photocyclization Pre → L within the limits of the narrow spectral range of 300-305 nm. Later on the 'dramatic' increase in the quantum yields of photocyclization and decrease in the quantum yield of *cis–trans* isomerisation of previtamin D received several interpretations [10-13], but the question of sharp changes around 300 nm in the quantum yields of previtamin D photoconversions remains still open [14].

2. Specific Features of Provitamin D Photoisomerization Under XeCl Laser Excitation at Λ = 308 Nm

Inadequacy of the photostationary state approximation has been revealed when we used laser initiation of provitamin D photoisomerization and spectral monitoring of the photoreaction kinetics [15-17].

The kinetics of 7-DHC photoisomerization was studied upon initiation by excimer XeCl laser radiation at $\lambda = 308$ nm (average power $P_{av} \approx 300$ mW, pulse duration $\tau_{pulse} \approx 15$ ns, pulse repetition frequency $f = 20$ Hz). A high purity 7-DHC specimen (melting temperature $t_m = 148\text{-}150^0$C, $\varepsilon_{282} = 12100$) had been prepared at the Institute of Vitamins (Moscow). Irradiation of 7-DHC in ethanol solution at a concentration $C \approx 10^{-4}$mol/l ensuring an absorbance $A_{282} = 1$ at the absorption maximum at $\lambda = 282$ nm was carried out in a standard quartz cuvette of 1 cm thickness. Before irradiation argon was bubbled through the solution to remove oxygen, and then the cuvette was sealed hermetically. The emission of the XeCl laser was focused on a cuvette surface onto a 2 x 1 cm^2 spot using a cylindrical lens, and the photoreaction kinetics was studied by the recording ultraviolet absorption spectra of the solution with Specord UV-VIS spectrophotometer. At the concentration used, the absorbance of the solution at the wavelength of the XeCl laser generation was less than 10^{-2}, i.e. the illumination along the cuvette volume could be considered constant (only 1% of the incident photons were absorbed), and the photoreaction proceeded according to the first order.

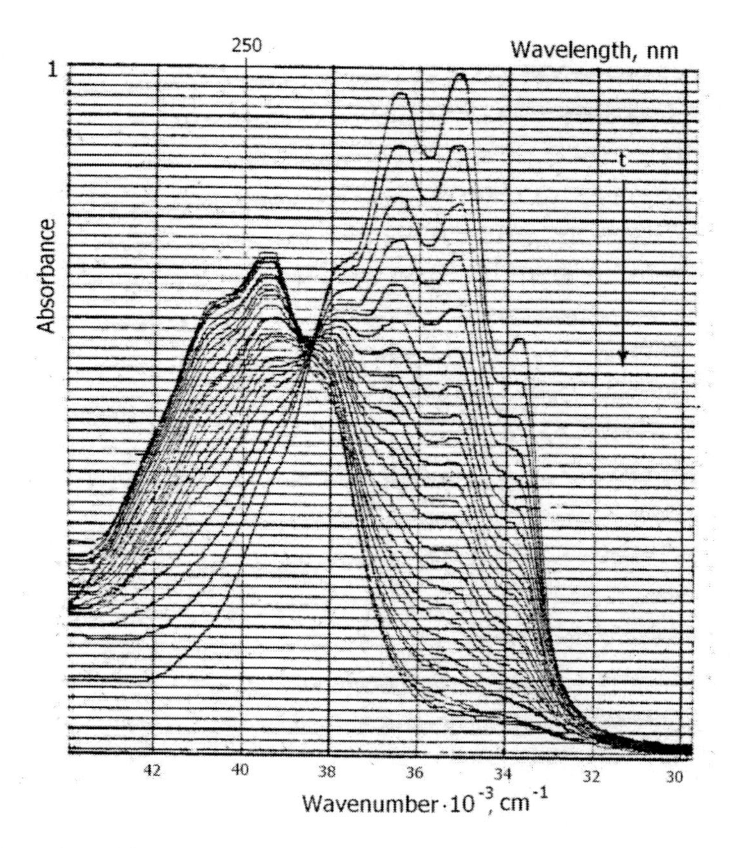

Figure 3. Transformation of the initial spectrum of 7-DHC solution in ethanol upon irradiation with excimer XeCl laser at 308 nm.

Transformation of the spectrum of 7-DHC with exposure showed unexpected behavior (Figure 3). In conflict with the generally accepted idea on the establishment of a photostationary state the absorbance near to maximum at 282 nm dropped exponentially from the very beginning of laser irradiation indicating irreversible photodegradation. In addition, comparison of the final spectrum in Figure 3 with the spectra of the main photoisomers showed that the resultant photoproduct was none of those given in Figure.2.

The result obtained did not correlate with the data published in [7] in which the establishment of a photostationary state during a XeCl laser irradiation was observed.

The resultant photoproducts were identified by thin-film chromatography (a benzene-acetone mixture was used as a mobile phase) and by NMR spectroscopy ('Bruker-200', a solution in deuterochloroform, using tetramethylsilane as an internal standard) [16]. From the comparison of the data obtained with the results published in [5], it was safe to say that the main component of the photoisomer mixture formed after laser irradiation was toxisterol B_3, produced by the addition of one ethanol molecule.

Next the photoreaction under XeCl laser irradiation was studied in various solvents, such as ether, isopropanol, hexane, and octane. In all cases the irreversible transformation of 7-DHC was observed into the products absorbing in the short-wave region of the spectrum (with a maximum at ~250 nm), but not into the main photoisomers [17].

Hence our experiments have demonstrated that the usual range of reversible photoreactions, shown in Figure 1, was blocked by the irreversible photoconvertions with low quantum yield, and the concept of their secondary role is no longer correct when the photoreaction is initiated by laser radiation at 308 nm. We were able to detect this feature due to use of UV absorption spectroscopy in contrast to the widely used chromatographic analysis.

It was required to explain why at a laser irradiation at λ = 308 nm the photoreaction occurs by a nonconventional way, i.e. the establishment of a PS state with formation of the basic products is not observed and there is an irreversible degradation of the initial provitamin D into toxisterols?

2.1. Influence of Two-Photon Processes

It was natural to relate such anomalous behavior of the photoreaction to specificity of pulsed laser excitation because it is known that when the intensity of the exciting light is changed, not only the reaction yields may change, but also the type of the predominant photoreaction. In our case of pulsed XeCl laser excitation with a relatively low average power of 300 mW the peak power reached 1 MW.

Thus, at first we assumed that the formation of previtamin D occurs at a one-photon excitation of Pro, while the formation of toxisterols takes place as a result of two-quantum reactions (with the absorption of the second photon by the excited molecule of Pro), which become prevalent at high power density of the exciting radiation.

To clarify the influence of the intensity on the photoreaction kinetics, we studied its behavior under changes of power density of the laser irradiation [15]. This parameter was increased tenfold (to 10 MW/cm^2) by additional focusing, and decreased by a factor of 100 by reduction with filters. It was found that the qualitative picture of the spectrum transformation did not change; only the rate of the absorbance decrease at 282 nm was linearly dependent on the power density that at first glance excluded the participation of two quantum processes.

Thorough analysis of the kinetic equations describing the formation of toxisterols as the product of two-quantum reactions has shown that only the improbably large value of 7-DHC absorption from the excited state can assure such linear dependence on the power density of laser irradiation [15].

Finally, to determine the possible role of the two-quantum processes we measured transmission of a 7-DHC solution at 308 nm varying the power density of the incident radiation in a wide range 10^2-10^8 W/cm^2. An ethanol solution of 7-DHC with initial transmission ~40% was placed in a flow-through cell that ensured replacing of the irradiated solution between the laser pulses. A beam from the excimer laser was focused by a long-focus lens to form a spot of 0,4 mm x 2,5 mm, and the power density of the incident radiation was varied with calibrated filters. A two-channel system for recording the transmission of a sample, as well as the intensities of transmitted and reference XeCl laser beams was used in conjunction with a precision pulse photometer (the irreproducibility in the measurements did not exceed 1% with averaging over 100 pulses).

Within these limits of the power density variations the 7-DHC solution transmission remained constant. This demonstrated convincingly the absence of any induced absorption and particularly of any saturation. Consequently, two photon processes are not involved in the discovered abnormality of the provitamin D photoisomerization on excitation with XeCl laser at 308 nm[3].

2.2. Influence of the Radiation Wavelength

In our subsequent work for excitation of 7-DHC within the wavelength range 280-305 nm that covers the maximum and the long-wavelength wing of its absorption band, we used dye-laser utilizing rhodamine dyes solutions pumped by a XeCl laser. Laser radiation of width ~10^{-2}nm had an average power of ~5 mW for $\tau_p = 10$ ns and $f = 20$ Hz. The radiation power density in the 7-DHC solution was 200 kW/cm^2, i.e. it corresponded to the average power density during irradiation with the excimer XeCl laser.

Our spectral observations of the photoreaction kinetics initiated by the second harmonic of a tunable dye-laser revealed substantial differences in the kinetics in relation to the irradiation wavelength [15].

Irradiation at 282 nm showed that the reaction occurred in compliance with the traditional scheme: a photostationary state was established rapidly, as demonstrated by stabilization of the absorption spectrum in 18 min after the beginning of irradiation and by its constancy during the subsequent irradiation period.

When excitation was provided at 295 nm, fast transformation of the spectrum during the initial stage of the irradiation process later became slower, but the changes during the subsequent irradiation did not stop. This was manifestation of the irreversible photoreactions, so that the establishment of quasi-stationary state of a photoisomer mixture could be observed.

These features of the photoreaction kinetics turned to be much more noticeable as the wavelength of the laser radiation was shifted to the red edge of the 7-DHC absorption band, and irradiation at $\lambda = 305$ nm caused irreversible photoreactions which resulted in such

[3] It is pertinent to note that the intensity effect on the photoreaction kinetics has been revealed later upon picosecond laser irradiation [19-21].

changes in the spectral pattern that could not be explained under any circumstances by a photostationary state.

We thus established experimentally pronounced wavelength dependence of the photoreaction kinetics when a solution of 7-DHC was irradiated by laser pulses of nanosecond duration with a power density of the order of 10^4-10^6 W/cm^2; besides, the rate of the photodegradation increased considerably with a wavelength. On excitation at the center of the absorption band (280-295 nm) the reaction still occurred in a traditional manner, but in the case of excitation at the long-wavelength edge ($\lambda > 305$ nm) the reaction became "anomalous": the initial product was converted into toxisterols and the accumulation of the main photoproducts -previtamin and tachysterol - was nearly suppressed.

The reasons for changes in the photoreaction pattern as a result of long-wavelength irradiation were identified by numerical calculations of the reaction kinetics at different wavelengths in accordance with two schemes: that including only reversible photoconvesrtions (Figure 1) and that supplemented by the Pre → Tox irreversible photoreaction channel (Figure 4) [17].

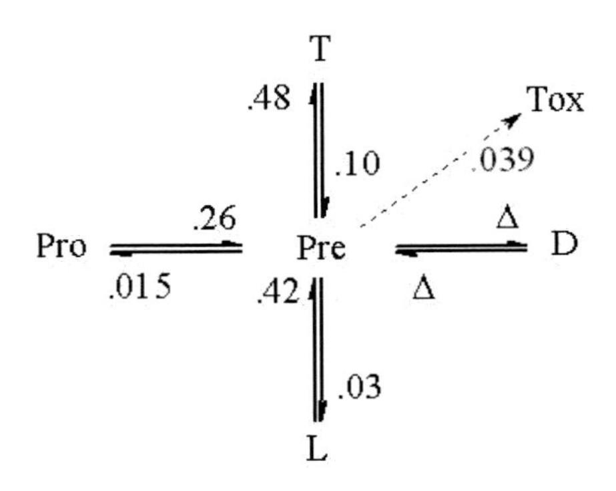

Figure 4. Schematic representation of provitamin D photoisomerization with regard to the irreversible channel.

The scheme with the reversible photoreactions only is described by the following familiar system of differential equations:

$$\begin{cases} \dfrac{dc_1}{dt} = \dfrac{I}{\sum_{i=1}^{4} \varepsilon_i c_i} (\varphi_{21}\varepsilon_2 c_2 - \varphi_{12}\varepsilon_1 c_1); & (2.1) \\ \dfrac{dc_2}{dt} = \dfrac{I}{\sum_{i=1}^{4} \varepsilon_i c_i} [(\varphi_{12}\varepsilon_1 c_1 + \varphi_{32}\varepsilon_3 c_3 + \varphi_{42}\varepsilon_4 c_4) - (\varphi_{21} + \varphi_{23} + \varphi_{24})\varepsilon_2 c_2]; & (2.2) \\ \dfrac{dc_3}{dt} = \dfrac{I}{\sum_{i=1}^{4} \varepsilon_i c_i} (\varphi_{23}\varepsilon_2 c_2 - \varphi_{32}\varepsilon_3 c_3); & (2.3) \\ \dfrac{dc_4}{dt} = \dfrac{I}{\sum_{i=1}^{4} \varepsilon_i c_i} (\varphi_{24}\varepsilon_2 c_2 - \varphi_{42}\varepsilon_4 c_4). & (2.4) \end{cases}$$

Here, the indices 1,2,3, and 4 denote Pro, Pre, T, and L, respectively; c_i - are the concentrations; ε_i are the molar extinction coefficients; φ_{ij} – are the quantum yields of $i \rightarrow j$ photoconversions. It is assumed that the principle of additivity of the optical densities applies, so that the intensity of the radiation absorbed by the i -th component of the mixture is

$$I_i = I_0 \frac{\varepsilon_i c_i}{\sum_{i=1}^{n} \varepsilon_i c_i}(1-10^{-\sum_{i=1}^{n}\varepsilon_i c_i d}) = I\frac{\varepsilon_i c_i}{\sum_{i=1}^{n} \varepsilon_i c_i}, \tag{2.5}$$

where $I = I_0(1-10^{\sum_{i=1}^{n}\varepsilon_i c_i d})$ is the intensity of the radiation absorbed by a system with n components; d is the thickness of the irradiated layer; I_0 is the intensity of the incident radiation.

The scheme including the Pre → Tox (2→ 5) irreversible photoconversion channel can be described if the second equation in the above system is supplemented by an appropriate term:

$$\frac{dc_2}{dt} = \frac{I}{\sum_{i=1}^{4} \varepsilon_i c_i}[(\varphi_{12}\varepsilon_1 c_1 + \varphi_{32}\varepsilon_3 c_3 + \varphi_{42}\varepsilon_4 c_4) - (\varphi_{21} + \varphi_{23} + \varphi_{24})\varepsilon_2 c_2 - \varphi_{25}\varepsilon_2 c_2]. \tag{2.6}$$

Then the rate of change of the concentration of Tox becomes

$$\frac{dc_5}{dt} = \frac{I}{\sum_{i=1}^{4} \varepsilon_i c_i}\varphi_{25}\varepsilon_2 c_2 \tag{2.7}$$

(the absorbance of radiation by Tox and, consequently, their phototransformations are ignored). In these calculations we used numerical values of ε_i from Ref.22, φ_{ij} from Ref.3 and φ_{25} from Ref.2.

The results of a numerical analysis of the kinetics of provitamin D photoisomerization in accordance with the above two schemes at λ_{irr} = 295 nm and 308 nm are shown in Figure 5. A comparison of Figures 5a) and 5c), corresponding to the reversible photoreaction scheme shows different compositions of the photoisomer mixture in the PS state for different irradiation wavelengths: in the case of λ_{irr} = 282 nm the dominant products are Pre and T, whereas at λ_{irr} = 308 nm the concentrations of all four photoisomers are comparable.

A comparison of Figures 5a) and 5c) with Figures 5b) and 5d), where the kinetics are shown for the same irradiation conditions but allowing for the Pre → Tox irreversible channel, shows that inclusion of the irreversible channel has little effect on the kinetics upon irradiation at 295 nm: there is only a slight reduction in the concentrations of Pre and T in the PS state, and neglect of the small amount of Tox accumulated up to the moment of establishments of the PS state does not result in significant error in the concentration analysis.

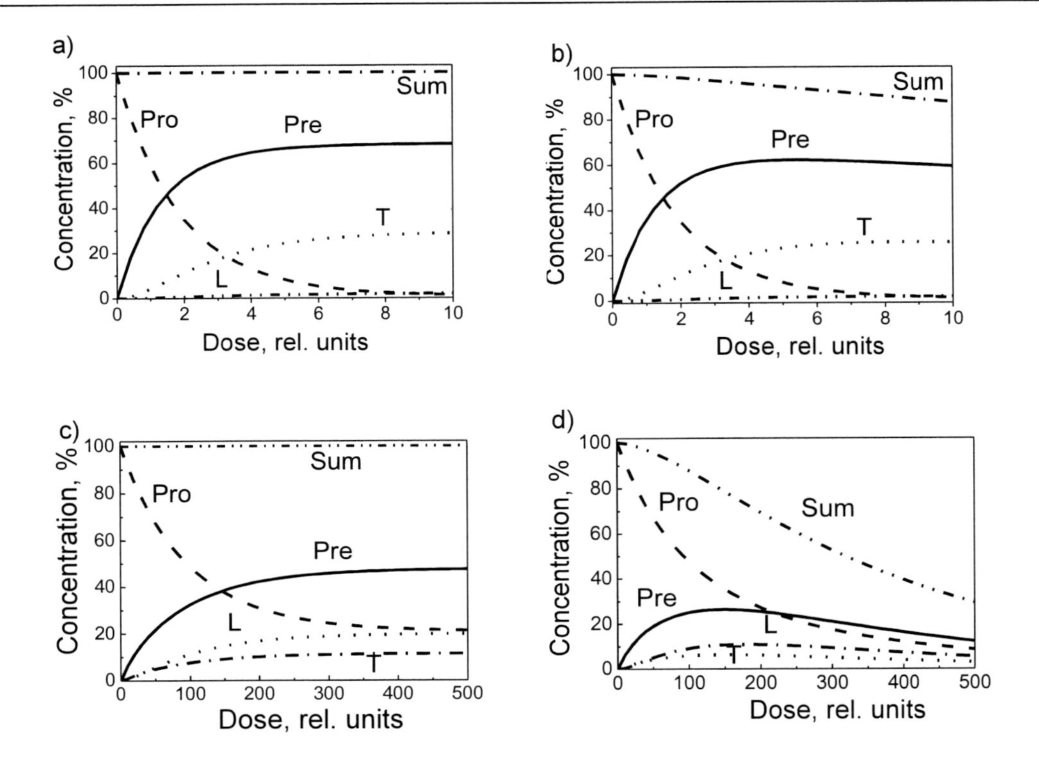

Figure 5. Simulated kinetics of provitamin D photoisomerization upon UV irradiation at 295 nm (a,b) and 308 nm (c,d) without the irreversible channel (a,c) and with due regard to the irreversible photoconversions Pre → Tox (b,d). 'Sum' is summary concentration of the four main photoisomers.

But at λ_{irr} = 308 nm an allowance for the Pre → Tox conversions changes dramatically the reaction kinetics: Tox accumulate rapidly from the earliest stage of irradiation significantly reducing (by a factor of almost 2) the maximum concentrations of Pre and T, so that the content of Tox by that moment becomes comparable with the concentration of Pre and exceeds the concentrations of the remaining photoisomers.

Similar computer simulations were then made for different excitation wavelengths, and they showed that account for the channel of irreversible Pre → Tox phototransformations had little effect on the kinetics for λ_{irr} in the range of 270-295 nm.

Thus, the anomalous kinetics of the photoreaction during irradiation in the region of λ_{ir} > 300 nm required explanation. It was necessary to determine why the establishment of a PS state was not observed while the initial provitamin D converted irreversibly into toxisterols.

2.3. Simplified Model of Reversible Photoconversions with a Weak Irreversible Channel

For a better understanding of the reasons for the increase effectiveness of the irreversible Pre → Tox channel in the region of 295-305 nm we have examined the following model [18].

Let us suppose that there are two photoisomers **A** and **B** with overlapping absorption spectra (Figure 6), between which reversible photoconversions A ↔ B with identical quantum yields $\varphi_{AB} = \varphi_{BA} = 0.5$ are possible. The initial concentrations are $C_A = 1$ and $C_B = 0$. Since the photoreaction is reversible, a photostationary state is clearly realized at some time after the beginning of irradiation in which the concentrations of **A** and **B** adopt specific values depending on the irradiation wavelength.

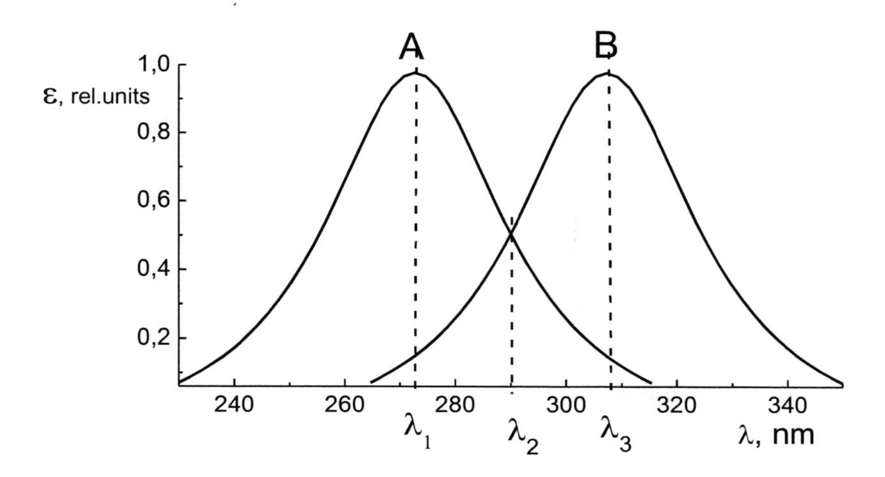

Figure 6. Model absorption spectra of the photoisomers **A** and **B**.

We have examined three different cases of irradiation at the wavelengths with the following ratios of the absorption of **A** and **B**:

at λ_1: $\varepsilon_A/\varepsilon_B = 10$,

at λ_2: $\varepsilon_A/\varepsilon_B = 1$,

at λ_3: $\varepsilon_A/\varepsilon_B = 0.1$.

The concentrations of the photoisomers in the photostationary state are determined by the following relation:

$$\frac{C_A}{C_B} = \frac{\varepsilon_B \, \varphi_{BA}}{\varepsilon_A \, \varphi_{AB}} \tag{2.8}$$

Thus, the ratios of the absorption parameters at the three selected wavelengths (at equality of quantum yields of reciprocal photoconversions) predetermine the following concentrations of **A** and **B** in the photostationary state during monochromatic irradiation:

$$\begin{aligned} &\text{at } \lambda_1: C_A/C_B = 0.1, \\ &\text{at } \lambda_2: C_A/C_B = 1, \\ &\text{at } \lambda_3: C_A/C_B = 10. \end{aligned} \tag{2.9}$$

Then we will have assumed that an irreversible photoreaction **B** $\rightarrow$ **X** is possible from **B** with the quantum yield 10 times smaller than the quantum yields of the reversible photoconversions (i.e., $\varphi_{BX} = 0.05$),.

For the photoreaction **A** $\leftrightarrow$ **B** $\rightarrow$ **X** the variation of the concentrations of **A** and **B** with the irradiation time is described by the following system of equations:

$$\frac{dC_A}{dt} = Il(\varepsilon_B \varphi_{BA} C_B - \varepsilon_A \varphi_{AB} C_A), \tag{2.10}$$

$$\frac{dC_B}{dt} = Il[\varepsilon_A \varphi_{AB} C_A - (\varepsilon_B \varphi_{BA} - \varepsilon_B \varphi_{BX})C_B], \tag{2.11}$$

where ***I*** is the irradiation intensity; *l* is the thickness of the irradiated layer.

In this case the ratio of the rates V_{BX} for the decomposition of **B** (the effectiveness of the irreversible channel **B** $\rightarrow$ **X**) and V_{AB} for the formation of B (by the phototransformation **A** $\rightarrow$ **B**) is determined by the equation

$$\frac{V_{BX}}{V_{AB}} = \frac{\varepsilon_B \varphi_{BX}}{\varepsilon_A \varphi_{AB}}. \tag{2.12}$$

The simplest calculation shows that the effectiveness of the irreversible channel for the three wavelengths will differ substantially:

$$\text{At } \lambda_1\text{: } \frac{V_{BX}}{V_{AB}} = \frac{1}{10} \cdot \frac{0.05}{0.5} = 0.01, \tag{2.13}$$

$$\text{at } \lambda_2\text{: } \frac{V_{BX}}{V_{AB}} = 0.1, \tag{2.14}$$

$$\text{at } \lambda_3\text{: } \frac{V_{BX}}{V_{AB}} = 1. \tag{2.15}$$

As evidenced by comparison of the expressions (2.13) - (2.15), the secondary role of the irreversible channel during irradiation at λ_1 can increase substantially with initiation of the photoreaction at other wavelengths even in spite of the low value of its quantum yield. Thus, during irradiation at λ_3 the rate of the irreversible decay **B** $\rightarrow$ **X** is already equal to the rate of formation of **B** from **A**, as a result of which the almost direct photoconversion **A** $\rightarrow$ **X** occurs, and the accumulation of **B** here is greatly reduced.

The situation described above (with identification of the hypothetical photoisomers **A**, **B**, and **X** with Pro, Pre, and Tox) is realized at the long-wave edge of the absorption spectrum of Pro from the long-wave side of the intersection of the spectra of Pro and Pre (Figure 2).

As it is seen in Figure 7, the calculated rate of the irreversible photoreactions Pre → Tox in relation to the rate of formation of previtamin D increases greatly when irradiation is provided at the long-wave edge of the absorption spectrum.

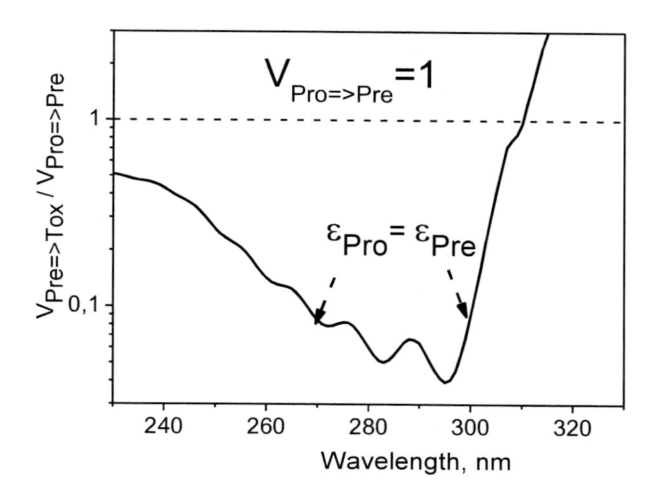

Figure 7. Calculated wavelength dependence of the rate of the previtamin D decay due to the irreversible photoconverstions Pre → Tox in relation to the rate of its formation from the provitamin D.

It is seen that only in the range of the wavelengths of 266-300 nm between the isosbestic points, at which the absorption coefficients of Pro and Pre are equal to each other, the rate of the irreversible phototransformations of previtamin D into the toxisterols is ten times less than the rate of its formation from provitamin D. Therefore, only at irradiation in this spectral region, in which the absorption coefficient of Pro is higher than for Pre, does the PS approximation adequately describe the kinetics of the photoreaction, and the traditional disregard of the irreversible channel is justified.

However, on account of the significant increase in the relative rate of the Pre → Tox at the long-wave edge at wavelength 310 nm the rate of formation of previtamin D (Pro → Pre) becomes equal to its irreversible photodegradation Pre → Tox, and as a result, the PS approximation loses any meaning.

Thus, it is the change in the ratio of the absorption coefficients of Pro and Pre in the region of $\lambda > 300$ nm ($\varepsilon_{Pre} >> \varepsilon_{Pro}$) that is the reason for the considerable increase in the effectiveness of the irreversible channel, in spite of its comparatively low quantum yield (an order of magnitude lower than for the Pro → Pre and Pre → T reactions). As a result, the probability of absorption of light by the Pre molecules and their phototransformations (including Pre → Tox) is substantially increased at the initial stage of the photoreaction. Moreover, the relation $\varepsilon_T >> \varepsilon_{Pre} >> \varepsilon_{Pro}$ compensates for the difference in the quantum effectiveness of the photoreactions Pre → T and T → Pre and leads to the rapid establishment of the equilibrium in the photoreaction Pre ↔ T and to the previtamin D "escape" due to the irreversible transitions in Tox.

It is clear from the above that the usual concept of the unimportance of the irreversible photoreactions breaks down during long-wave irradiation. Under these conditions the toxisterols stop conforming to the name "overirradiation products" since they are already accumulating at the initial stage of irradiation and their rapid accumulation restricts the growth in the concentrations of the main photoisomers. As a result, the maximum attainable concentrations of Pre and T differ radically from those calculated in terms of the PS approximation.

As evidenced by our calculations, if the total concentration of the four photoisomers Pro, Pre, T, and L has traditionally been taken as equal to 100% and the Tox content has been disregarded in concentration analysis, this introduces error which increases sharply as the irradiation wavelength is shifted toward the red region from 300 nm, and an adequate analysis should take into account the mixture photodegradation (Figure 5d). Accumulation of toxisterols that in fact represent 13 photoisomers [5], poses major problem for the commonly used method of high performance liquid chromatography (HPLC). In our opinion it is just the inadequacy of the concentration analysis that led to the "dramatic changes" of the calculated quantum yields in the region of 295-305 nm [4].

In the next sections we will show that these conclusions are supported fully by the experimental research [23] carried out using specially designed spectrophotometric analysis, which does take into account the photoisomeric mixture degradation of [24].

3. Spectrophotometric Analysis of the Multicomponent Photoisomeric Mixture

Spectrophotometric analysis of a multicomponent mixture is based on the well known equation (3.1), which is valid when the requirements of Bouguer's law are fulfilled for all m components and the additivity principle holds for their mixture:

$$A(\lambda) = \Sigma A_{\varphi}(\lambda) = \lambda \Sigma \varepsilon_{\varphi}(\lambda) C_j (j = 1, \ldots . m) \qquad (3.1)$$

Here $A(\lambda)$and $A_j(\lambda)$ are the total and the partial absorbance at the wavelength λ; $\varepsilon_j(\lambda)$ is molar extinction coefficient of individual component and C_j is corresponding concentration; l is the thickness of the absorbing layer.

As a rule, spectrophotometric analysis of a multicomponent mixture uses curve-fitting techniques to their spectra, utilizing the available spectral data for the components. However, in the case of vitamin D photoisomers the problem is complicated because the individual absorption spectra overlap in the same spectral region and show partial structural similarity (Figure 2). Besides, the total concentration of the four main photoisomers is not kept constant during UV exposure due to the irreversible photodegradation into toxisterols. As it is evident from Figure 3, the absorption spectra of the toxisterols are shifted to the shorter wavelengths relative to the spectra of the main photoisomers.

First proposed in [22], the spectrophotometric analysis of the mixtures of vitamin D isomers was based on the least-squares method in matrix form and operated at twelve evenly spaced analytical wavelengths in the range from 252 nm to 296 nm. The mixtures of known compositions were prepared and tested, and the calculated compositions of the synthetic mixtures were compared with known compositions. A standard deviation of ±4% in the percentage of each component was obtained. However, appearance of negative concentrations for the components presented in small amounts (or missing at all) has been observed (that is meaningless). Besides, the method was designed on the assumption that the total concentration of the components is kept constant during irradiation.

To overcome the above problems the new version of spectrophotometric analysis had been designed [24]. To avoid negative values the concentrations were expressed in the form of the square values $C_i = \chi_i^2$ $(i = 1,2,3,4)$.

Values χ_i were defined as giving a minimum to the functional

$$\chi^2 \equiv \chi_A^2 + \chi_C^2, \tag{3.2}$$

where the first summand χ_A^2 represents the usual [25] statistical criterion describing difference of calculated values from the experimentally measured ones and is defined by the root-mean-square deviation between the experimental absorbancy curve and an absorbancy curve calculated by combining the extinction curves of the individual components with selected weighting factors (the concentrations) at N wavelengths λ:

$$\chi_A^2 = \frac{1}{N+1}\sum_{j=1}^{N}\left(\frac{A_c^{\lambda_j} - A_{ex}^{\lambda_j}}{\Delta_{A_{ex}}^{\lambda_j}}\right), \tag{3.3}$$

Here $A_{ex}^{\lambda_j}$ - experimentally measured absorbance at the wavelength λ_j , $\Delta_{A_{ex}}^{\lambda_j}$ - experimental error of its definition, and $A_c^{\lambda_j}$ - calculated absorbance at the wavelength λ_j:

$$A_c^{\lambda_j} = \sum_{J=1}^{4} \chi_i^2 \varepsilon_i^{\lambda_j}. \tag{3.4}$$

The fit between the calculated and measured spectrum is considered satisfactory if for the given set of values $\{\chi_i\}$ all calculated and experimental absorbance values coincide within an error of measurement of the last, i.e.

$$\chi_A^2(\min) \leq \frac{N}{N+1}. \tag{3.5}$$

Proper account of the irreversible photodegradation of the 4-component photoisomer mixture is made by introducing an additional term χ_C^2 into the functional χ^2 to be minimized:

$$\chi_C^2 = \frac{1}{N+1} \cdot \frac{1 - \exp\left[-\left(1 - \sum_{J=1}^{4} \chi_i^2\right)^2\right]}{\Delta_C^2}. \tag{3.6}$$

This so-called "penalty" function controls the difference between the 100% and the total real concentration of the four main photoisomers (Pro, Pre, T and L) within uncertainty $\Delta_c << 1$. This function rises steeply in response to toxisterols accumulation that remarkably distorts the short wavelength edge of the absorption spectrum and achievement of the minimum value of the functional (3.2) becomes impossible along the set interval of the wavelengths. To achieve the χ^2 minimum the wavelengths interval should be narrowed from the short-wave side. Thus, a distinguishing feature of the developed program is the fact that the program itself "signals" when degradation of the photoisomeric mixture becomes greater than the 10% level.

Another important point is that the search for the local extremum of χ^2 is done using powerful minimization method - the Fletcher's variable metric algorithm [26].

The starting parameters for the minimization procedure are:

- Experimentally measured absorption spectrum of the photoisomer mixture within the range 230-330 nm with the 1 nm step;
- four reference spectra of individual photoisomers (Pro, Pre, T and L) within the same spectral range taken from [22];
- the "errors" table calculated within 230-330 nm as deviations between the known spectrum of provitamin D [22] and the measured initial spectrum of provitamin D solution[4];
- four initial values for the concentrations;
- the interval of analytical wavelengths;
- admissible value of a deviation of total concentration of 4 photoisomers from unit.

Several tests were performed to check adequacy, accuracy and reproducibility of the spectrophotometric analysis. At first, the method tolerance had been checked in relation to the initial conditions, i.e. initial values of four concentrations and the number N of analytical wavelengths.

Particular attention was given to comparison with HPLC analysis, and the results of the spectrophotometric and HPLC analyses are presented in Figure 8. In view of different initial concentrations and, as a result, of different time scale, the concentrations of Pre, T and summary concentration of four main photoisomers are shown in relation to the converted provitamin D. As it follows from Figure 8, close correspondence (within ± 2%) between the two sets of the concentrations is strong evidence in favor of the spectrophotometric analysis adequacy.

[4] For purpose of the "errors" calculation, it is convenient to join together the reference and experimental spectrum (before irradiation) at λ_{max} = 282 nm; to do this the ratio 306/ A_{282} is to be calculated ($E^{1\%}_{1cm}$ =306). All subsequent spectra after exposures should be multiplied by this ratio before computer processing.

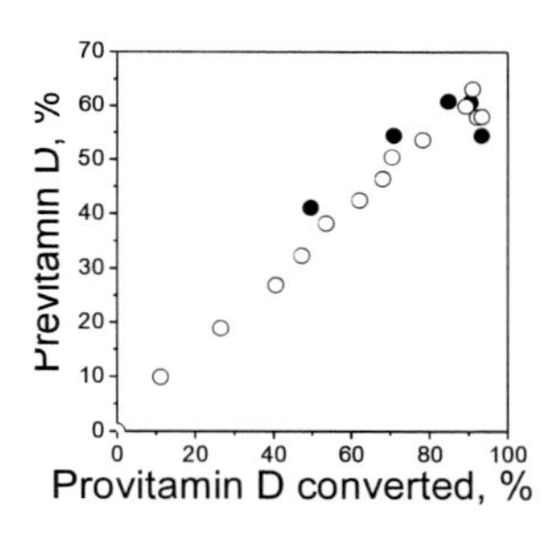

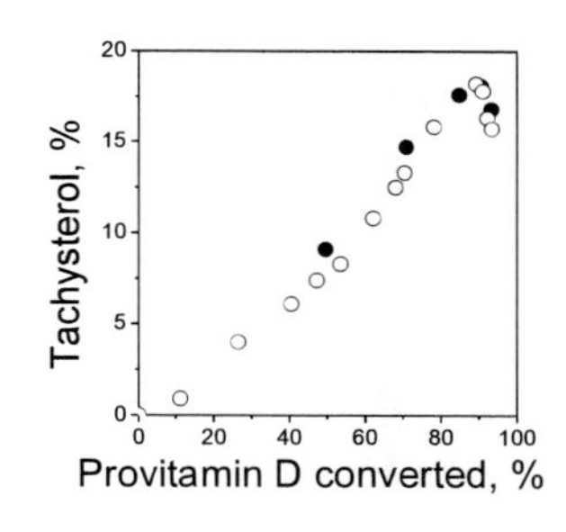

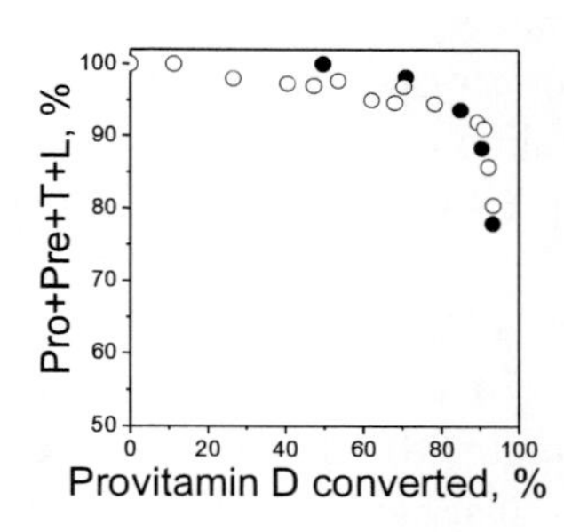

Figure 8. (a,b,c) Comparison of the results of spectrophotometric (open symbols) and HPLC (solid symbols) analyses of the photoisomer mixture formed upon irradiation with El-30 lamp.

4. Irradiation with KrF and XeCl Excimer Lasers

For initiation of the photoreaction pulsed (τ_p = 16 ns) excimer UV lasers has been used: KrF (λ_{irr} = 248 nm) and XeCl (λ_{irr} = 308 nm) (Lambda Physics). The experiments were carried out at the National Institute of Material Technology and Chemical Research (Tsukuba, Japan) [23]. The energy density of the laser pulses at the fully illuminated front side of rectangular cuvette (4 cm x 1 cm) with a solution of 7-DHC was measured before the beginning and upon termination of each series of a laser irradiation and fluctuated in different series within the limits of 40-60 mJ/cm^2 for the XeCl laser at frequency f= 10 Hz and 70-75 mJ/cm^2 for the KrF laser at frequency f= 1 Hz. Under such conditions the power density of laser radiation in an irradiated solution was 2,5÷3.8·10^6 W/cm^2 and 4.5÷4.8·10^6 W/cm^2, accordingly, that, in view of the short lifetime of excited singlet electronic state 7-DHC (<1 ps) [27], excluded the nonlinear effects resulted from the two-photon absorption.

The laser irradiation of solutions 7-DHC in different solvents (ethanol and deuteroethanol, methanol and deuteromethanol, isopropanol and cyclohexane) was carried out in the hermetically sealed quartz cuvettes of 1 cm thickness. Before an irradiation the solutions were bubbled with argon to exclude oxygen. The solvents were used of spectral purity, and 7-DHC of high purity (98%) and ergosterol (95%) were purchased from Sigma. Concentration of 7-DHC was 22 μg/ml (at irradiation with XeCl laser) and 14 μg/ml (at irradiation with KrF laser) that provided absorbance at the irradiation wavelengths 0.03 and 0.13, accordingly. It provided almost uniform illumination of a solution on the depth and excluded necessity of its stirring in the course of a laser irradiation.

Before the beginning of an irradiation and after the fixed exposures the absorption spectra of the solutions were registered with spectrophotometer Perkin Elmer Lambda 900 within a range of 230-330 nm and then were processed with computer for concentration analysis using the method described in the previous section.

4.1. The Provitamin D Photoisomerization upon Excitation with Krf Laser at the Wavelength Λ = 248 Nm

As it was mentioned, in the conditions of a short-wave irradiation of provitamin D accumulation of previtamin D at an initial stage doesn't exceed 20 % and at further irradiation it undergoes *cis-trans* isomerization into tachysterol which accumulates up to 70 % of summary content of the photoisomer mixture in the quasi-photostationary state. The further irradiation leads to the mixture photodegradation owing to accumulation of toxysterols.

Such photoreaction kinetics is reflected by the initial spectrum transformation in the process of UV irradiation (Figure 9): the initial insignificant absorbance decrease caused by formation of previtamin D, is replaced by its appreciable increase caused by accumulation of tachysterol (Figure 9a). Gradual reduction of absorbance during further UV irradiation (Figure 9b) reflects process of irreversible photodegradation.

It should be noted that the same picture of spectral kinetics of both 7-DHC and ergosterol photoisomerization was observed in all investigated solvents.

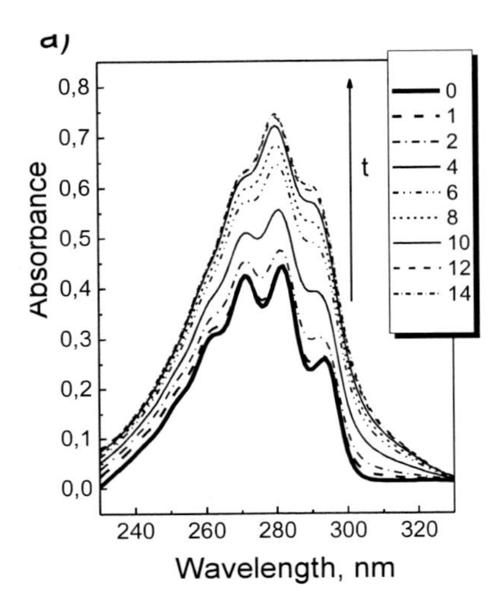

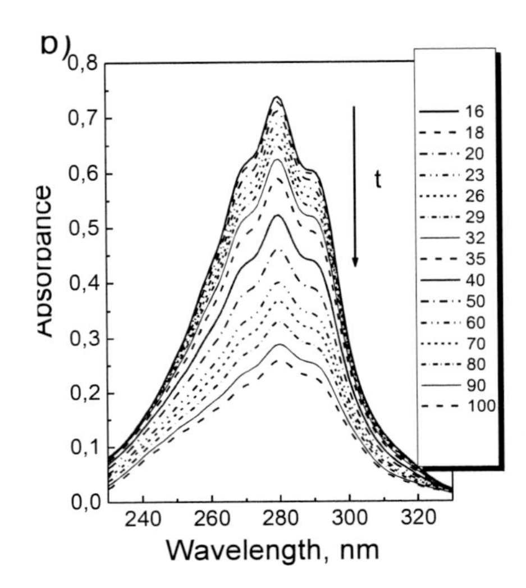

Figure 9. Spectral transformation upon KrF laser irradiation: a) initial stage, b) overirradiation stage. (the numbers of laser pulses are shown)

Concentration kinetics in methanol is shown in Figure 10.

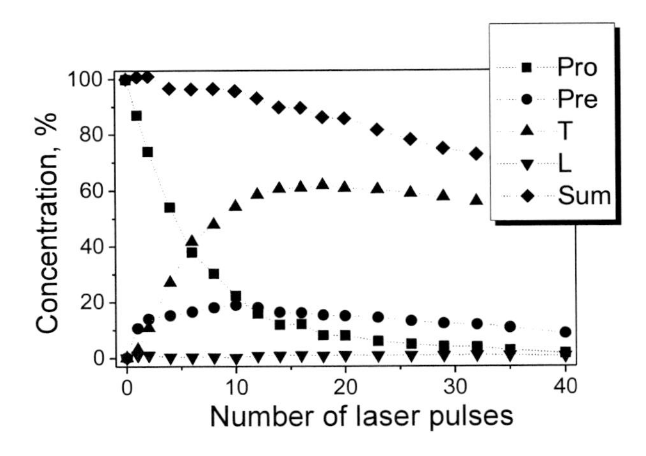

Figure 10. Kinetics of 7-DHC photoisomerization upon laser irradiation at 248 nm.

In addition the photoreaction kinetics was calculated taking into account the channel Pre → Tox. Comparison of the calculated kinetics with the experimentally measured ones shows rather good conformity. Predictably, tachysterol is the basic photoproduct, but by the time of achievement of its maximum concentration the contribution of the irreversible reactions is tangible.

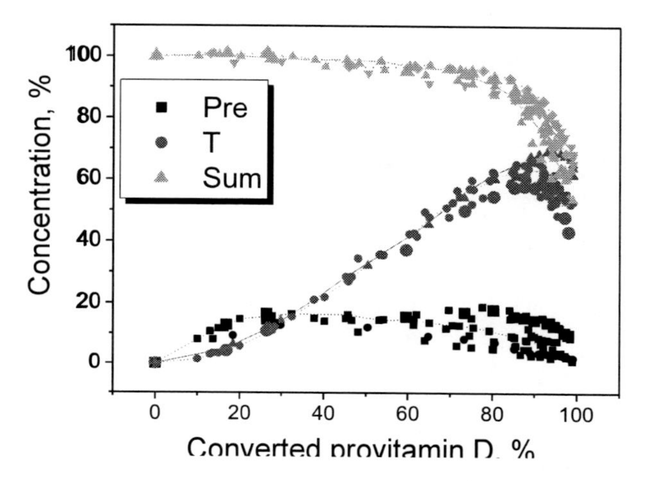

Figure 11. Comparison of calculated and experimentally measured concentrations of Pre, T and Sum in various solvents in relation to the converted provitamin D.

Thus, the results obtained differ essentially from the published data [7] without account for the irreversible photodegradation.

4.2. Concentration Kinetics under Irradiation with the Xecl Laser At 308 Nm

Spectral observation of the photoreaction kinetics in the conditions of irradiation with XeCl laser (λ = 308 nm) has revealed a considerable role of the irreversible photodegradation (Figure 12).

Apparently from a Figure 12, that in the process of UV irradiation the absorption in the spectral range, characteristic for the basic photoisomers, falls almost to zero, and at the same time it increases in the short-wave side of the spectra where toxysterols absorb efficiently.

The typical picture of experimentally measured kinetics of 7-DHC photoisomerization (in methanol) is presented in Figure 13a) from which the qualitative consent is visible with the numerical calculation (Figure 13b) taking into account irreversible channel Pre → Tox.

It is interesting that in spite of appreciable distinction in spectral kinetic in different solvents (Figure 12), the concentration kinetic are in close agreement. It is shown in the figures below where for all investigated solvents the measured dependences are shown of previtamin accumulation (Figure14a) and of reduction of the summary concentration of the basic four photoisomers (Figure 14b) (designated as "Sum") in process of conversion of initial provitamin D.

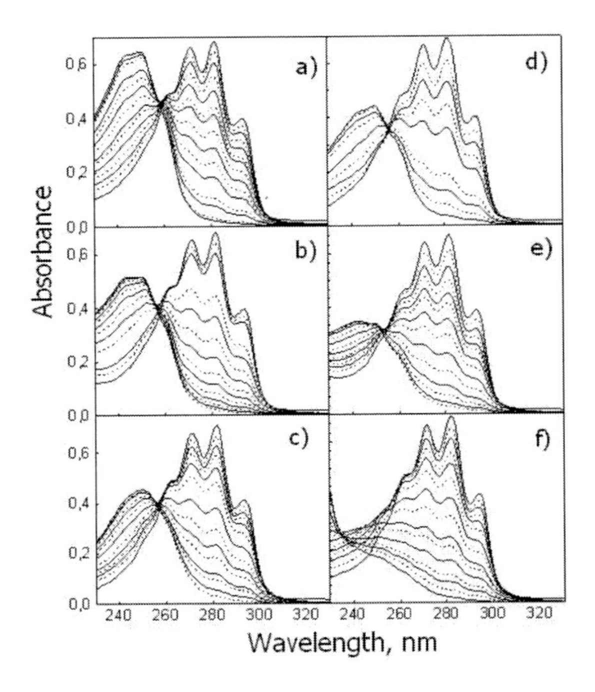

Figure 12. Transformation of the initial absorption spectrum of 7-DHC in methanol (a), ethanol (b), isopropanole (c), deuteromethanol (d), deuteroethanol (e), and cyclohexane (f) during irradiation with the XeCl laser.

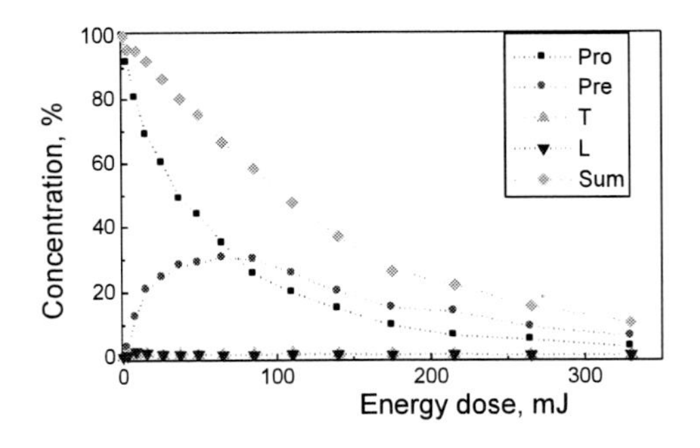

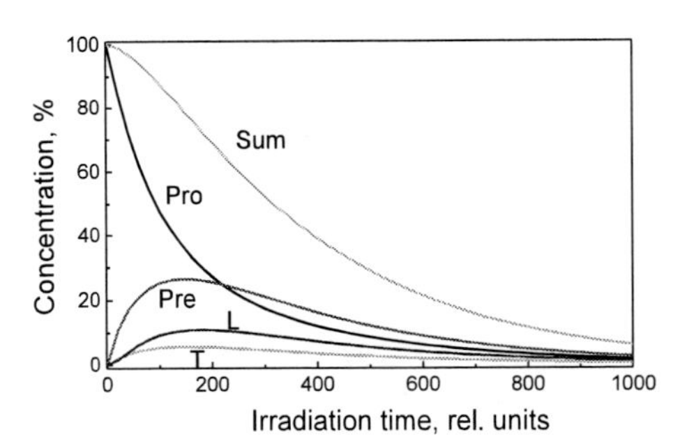

Figure 13. Experimentally measured concentration kinetics (a) and calculated one (b) at irradiation with XeCl laser.

Nevertheless, at good enough consent of experimental data for all tested solvents, the comparison of experimentally measured and calculated concentration dependences of previtamin D accumulation shows appreciable numerical distinction (Figure 14a). There is good reason to think that an actual quantum yield of previtamin D *cis-trans* isomerization is less than used in the calculations. Really, the value $\varphi_{Pre \rightarrow T} = 0.26$ has been defined in [8] in the conditions of neglect by the irreversible photodegradation that, in our opinion, has led to considerable errors in the concentration analysis, and from here - in definition of quantum yields. It counts in favor of this hypothesis that good enough consent was observed between the experimentally measured and the calculated dependences of total concentration of 4 basic photoisomers in relation to the converted provitamin D (Figure 14b).

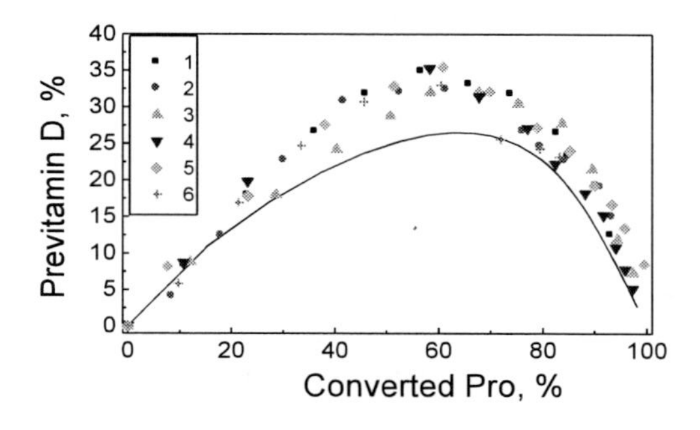

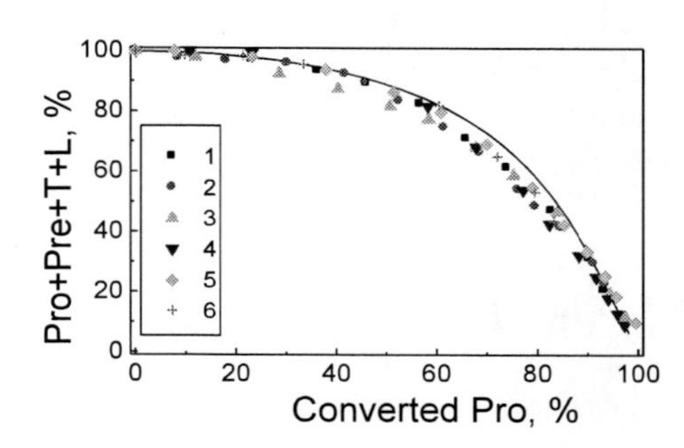

Figure 14. Experimental (symbols) and calculated (line) concentration dependences of previtamin D concentration (a) and summary concentration of the four basic photoisomers (b) on the converted provitamin D: 1 – methanol, 2 – ethanol, 3 – isopropanol, 4 – deuteromethanol, 5 – deuteroethanol, 6 – cyclohexane.

Hence these studies strongly support the necessity of proper account for the irreversible channel Pre → Tox in the concentration analysis of the photoisomer mixtures when the provitamin D photoisomerization is excited by monochromatic irradiation at the wavelengths outside of the range where the PS approximation is available.

5. The Problem of the Photochemical Stage Optimization in the Industrial Production of Vitamin D

The optimizing the photosynthesis process means obtaining the maximum yield of previtamin D at a high conversion of initial provitamin D and a minimal content of side photoproducts. In practice, the increase in the content of side products (mainly, tachysterol) at a high degree of conversion of the initial provitamin D necessitates interruption of the photoreaction at an early stage of irradiation and further separation of the unconverted provitamin D by recrystallization for its repeated irradiation, that leads to the losses and to the increase of the production cost.

Since the concentration of previtamin D in the photoisomeric mixture formed during the irradiation of the initial provitamin D is dependent on the spectral composition of the initiating radiation [27], the conventional methods of optimization of the photosynthesis stage are based on the selection of an irradiation source with the "optimum" spectrum. The maximal yield of previtamin D (about 60%) is observed during the monochromatic irradiation at a wavelength of 295 nm (where the absorption coefficient of provitamin D maximally exceeds the absorption coefficient of the previtamin) [28,29].

Since in the vast majority of papers the photoisomerization scheme is used, which includes only the reversible photoreactions of previtamin D, the main attention is devoted to the prevention of the tachysterol accumulation in view of high quantum yield of *cis-trans* isomerization. It was found [1] that under the shortwave irradiation (254 nm) tachysterol (T) is accumulated in a large amount. In order to reduce the yield of T, the use of a sunlamp whose spectrum lies within the range 290-350 nm has been suggested as the source for industrial synthesis of vitamin D.

However, it was a common practice that in view of the low quantum yield of the Pre → Tox reactions in the analysis of the photoisomeric mixture the presence of Tox was generally neglected and the overall concentration of the main four photoisomers was assumed to be 100%.

5.1. The Method of Two-Stage Irradiation

At the beginning of the 1980s the method of two-stage irradiation of provitamin D [7,8] was proposed, enabling, in the opinion of the authors, obtaining a high yield (more than 80%) of previtamin D at almost complete (more than 95%) conversion of the starting provitamin D and at low content of tachysterol.

The idea behind this method was to achieve a high conversion of provitamin D and to produce a considerable amount of tachysterol at the first stage upon the short-wave irradiation in the 250 nm region, and then at the second stage to subject the photoisomeric mixture to the long-wave irradiation (330-350 nm), which is absorbed by tachysterol alone, and selectively excite the T → P reaction, converting all the tachysterol into previtamin D.

In [7] the two-stage irradiation of 7-DHC was carried out by UV radiation from the laser sources and after the analysis of the irradiated mixture (the HPLC method with UV registration at $\lambda = 282$ nm), the yield of previtamin D after the second stage achieved 80%.

In [8] the two-stage irradiation of 7-DHC was carried out using the lamp sources. The analysis was also carried out by the HPLC without taking into account the Tox amount. The authors pointed out that with irradiation at 254 nm they obtained a result which agreed well with the calculated data: 2.3 % of Pro, 23.5 % of Pre, 70.8 % of T, and 3.4 % of L. When this mixture was then irradiated by a Rayonet lamp ($\lambda_{max} = 350$ nm), the concentrations changed and a mixture was formed with the following composition of the photoproducts: 4% of Pro, 83% of Pre, 11% of T, and 2% of L.

As we have shown, when there is proper concentration analysis and the irreversible photodegradation is duly accounted for, our results of irradiation with KrF laser at $\lambda = 248$ nm significantly differ and testify convincingly the uselessness of the two-step irradiation for optimization of technology of the vitamin D synthesis. This is clearly seen from the comparison of our data with the results from [7] in the Table 1. This conclusion was also supported by the numerical calculations of the photoreaction kinetics and the experiment on the two-step irradiation when the irreversible channel was taken properly into account [31]

Table 1. Comparison of the photoisomer mixture content in the photostationary state with proper account of the irreversible channel with the data from [7]

	Pro, %	Pre, %	T, %	L,%	Sum
Our data	4,5	13,3	63,7	Not detected	84,2
[7]	2,9	25,8	71,3	Not detected	100

5.2. The Role of the Irreversible Channel Under Optically Dense Layer Conditions

The role of the irreversible reactions channel was studied under the conditions of the so-called "film" effect [28] arising during the illumination of concentrated solutions and leading to substantial nonuniformity of the illumination of the solution layer along its thickness [32].

When monochromatic irradiation is used, such screening of deeper layers by the layers located closer to the radiation source, in the absence of effective stirring of the irradiated solution, will thereby lead to different intensity of illumination of different layers of the solution, and, as a result, of different photoreaction rates in the layers.

But when an irradiation source with a broad radiation spectrum is used, the changes may also occur with respect to the spectral composition, as illustrated by Figure 15.

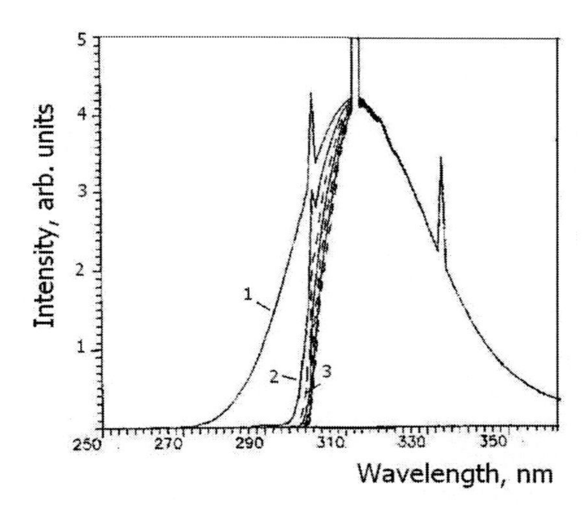

Figure 15. Radiation spectrum of the fluorescent sunlamp El-30 (1) and its transformations after passing through a layer of a 7-DHC solution (C = 0.02%) of 0.05 cm thickness (2), after passing two layers (3), etc.

It is seen from Figure 15 that the part of the lamp radiation spectrum (280-300 nm) which is most suitable for the initiation of the photoreaction is absorbed in a 0.05 cm layer, and only the long-wave part of its radiation spectrum acts on the following layers of the solution.

To illustrate the role of the irreversible channel with a 'film' effect (no stirring of the solution over the cuvette volume) in the mathematical model not only the time dependence of the photoisomer concentrations C_i was taken into account, but also their dependence on the distance x from the front wall of the cuvette. In this case the photoreaction kinetics is described by the system of the two-dimensional equations [32], and in addition to the calculation of the concentration kinetics in each separate layer x the integral (along the cuvette thickness l) concentrations $C_i = \int_0^l C_i(x)dx$ were calculated.

The consequence of the 'film' effect is seen in Figure 16a) where the kinetics of previtamin D formation in separate layers was simulated for monochromatic irradiation at

295 nm. It is seen that in the layers most remote from the radiation source, the maximum values of Pre (equal for each layer) are reached after a much longer period of time. For example, at the moment of time when the concentration of Pre peaks in the 3-rd layer, the overirradiation already occurs in the first layer while the more remote layers are still at the stage of formation of Pre. Hence the highest attainable integrated (over the entire thickness of the solution) concentration of previtamin D decreases significantly.

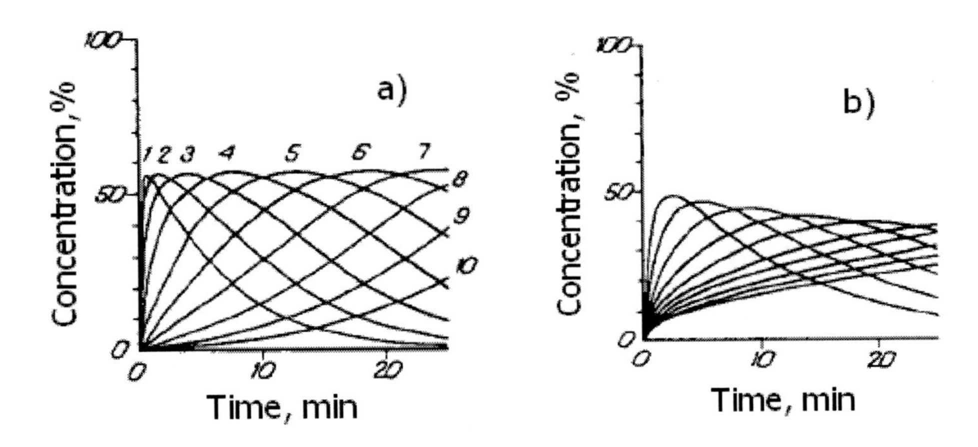

Figure 16. Calculation of the concentrational kinetics of Pre in ten optically thin layers (layer thickness 0.05 cm, cuvette thickness 0.5 cm) in the absence of stirring between the layer: a) monochromatic irradiation at $\lambda = 295$ nm, b) broad-band irradiation with the sunlamp.

The most pronounced negative influence of the irreversible channel with ‘film’ effects appears under the actual conditions of irradiation of concentrated solutions of Pro (C = 0.1-0.2%) by a source with a broad radiation spectrum. As has already been mentioned above, the radiation spectrum of a sunlamp after passing through a 0.05 cm layer of the solution is sharply changed in its composition, so that the remote layers of the solution are irradiated by a source whose spectrum contains only the long-wave components. Thus, the photoreaction kinetics in the front layers is determined by the short-wave part, while in the remote layers it is determined by the long-wave part of the radiation spectrum of the lamp. As a result, not only formation of Pre slows down in each remote layer, but its maximal achievable concentration becomes lower and lower in view of the increase in the rate of the irreversible reactions Pre $\rightarrow$ Tox during long-wave irradiation.

The calculations show that at a higher initial concentration of Pro, the stirring effectiveness of the irradiated solution along the light flux becomes more critical for reaching the maximal possible concentration of Pre. This effect must be taken into consideration during the experimental and theoretical modeling of the photoreaction for the correct prediction of its behavior under actual production conditions. Naturally, with decrease in the layer thickness (or the concentration of the solution) the demands with respect to stirring effectiveness decrease.

The experiments on the broad-band irradiation of the 7-DHC solutions of different optical density confirmed the calculations [30] and clearly demonstrated the destructive influence of the irreversible channel, especially if there is a "film" effect (Figure 17).

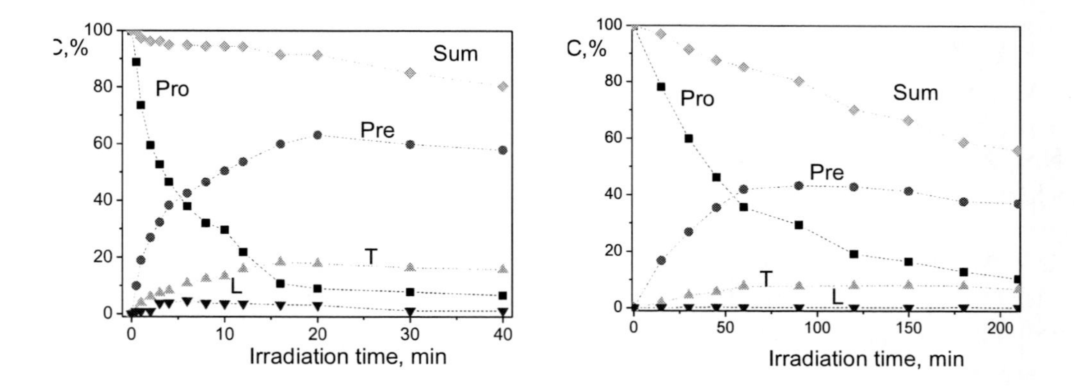

Figure 17. Experimentally measured kinetics of 7-DHC photoisomerization under irradiation with the sunlamp: a) $C = 3 \cdot 10^{-3}\%$, b) $C = 0.075\%$.

Thus we have experimentally shown that the kinetics of provitamin D photoisomerization in optically dense layers depends considerably on the irradiation conditions: different concentration compositions of the mixture and different extents of its degradation may correspond to the same conversion of the provitamin D, and only in optically thin layers does a clear connection exist between the spectral composition of the initiating radiation and the photoreaction kinetics.

Besides, it was shown that under production conditions, upon broad-band irradiation of highly concentrated solutions of provitamin D (C = 0.1-0.2%) the thickness of the layer of the irradiated solution is an extremely important parameter, and according to the estimates made in [32], the optimal thickness should not be greater than 0.5-1 mm.

In conclusion it might be well to point out that significant advantage of the incoherent UV source - XeCl lamp - over a high-pressure mercury lamp used in manufacturing of previtamin D [33] allows us to recommend the XeCl excimer lamp as the most promising source for industrial application.

6. Biological Dosimetry of Solar UV Radiation *In Situ* Using an *In Vitro* Model of Vitamin D Synthesis

6.1. Duality of Solar UV-B Radiation: Vitamin D Synthesis versus Skin Erythema

In the beginning of eighties years of the last century the effect of stratospheric ozone depletion, entailing an increase in the amount of ground-level solar UV-B radiation (280–315 nm), has been established. Concerns of the environmental and health effects of solar ultraviolet radiation penetrating into the biosphere through the depleted stratospheric ozone layer have emphasized the barest necessity for reliable instrumentation and accurate measurements of increased amount of solar UV-B radiation.

Besides, other factors such as cloudiness, aerosols, air pollution and tropospheric ozone also affect penetration of solar UV-B radiation to the Earth's surface and mask the effect of decreasing stratospheric ozone that should be taken into account when estimating the increase in UV-B radiation and its potential consequences on the ecosystem.

As it is known, biological responses to UV radiation begin with light energy penetrating into the skin. Absorption of high-energy UV photons by UV sensitive macromolecules (DNA, RNA, etc.) and photoinduced alterations in their structures lead to subsequent biochemical alterations in the cells that later cause tissue alterations. The details of this sequence of events are rather well studied for vitamin D_3 synthesis in skin, but are less well defined for the skin erythema.

It is significant to note that the UV radiation exhibits two-way biological action depending on the accepted dose. Excessive exposure to sunlight is causing premature aging of the skin and skin cancer but proper amounts of UV-B are beneficial for people and essential in the production of vitamin D_3.

Nevertheless, most attention is focused on raising public awareness and providing information about potential risks of solar UV exposure.

In theory, to determine the biologically effective irradiance, the spectrum measured by a spectroradiometer is to be weighted (integrated over the wavelengths) according to the action spectrum of a specific biological effect [34]. The CIE erythema action spectrum is the most common choice in the so-called 'weighted spectroradiometry' [35].

$$E_{eff} = \int E_{\lambda}(\lambda) * S_{\lambda}(\lambda) d\lambda$$

Here $E_{\lambda}(\lambda)$ - solar spectral irradiance [$Wm^{-2}nm^{-1}$], $S_{\lambda}(\lambda)$ - action spectrum [relative units], λ- wavelength [nm]. Integration of the biologically effective irradiance E_{eff} over the time gives the biologically effective dose H_{eff} [Jm^{-2}]:

$$H_{eff} = \iint E_{\lambda}(\lambda, t) * S_{\lambda}(\lambda) d\lambda dt$$

Just biodosimeters that use the simplest biologic systems (bacteria, spores, biomolecules) directly integrate UV radiation during an exposure according to the action spectrum of the photobiologic effect involved [34].

As a rule biological effectiveness of solar UV radiation is evaluated in relation to the erythemal effect, and the biologically effective doses are expressed in the erythemal units called MEDs (the minimum erythema dose 1 MED = 200 J/m^2 as the upper limit of safety UV dose has been introduced [35]). To date a number of personal dosimeters are widely available to determine the safety limit of sun-tanning exposure depending on the individual skin type.

At the same time only during the last decade monitoring of the vitamin D synthetic capacity of sunlight received the attention it deserved [36-38] in view of the crucial importance of vitamin D for human health [39] and taking into account widespread natural occurrence of endogenous vitamin D synthesis in biosphere.

Comparison of provitamin D absorption spectrum with the CIE erythema action spectrum [35] shows that in view of marked difference between the two spectra the 'erythemal' UV measurements will not directly reflect the 'vitamin D' synthetic capacity of sunlight (Figure.18).

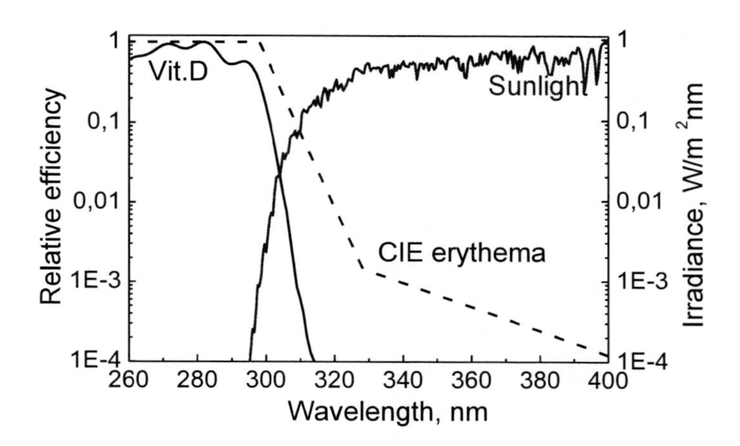

Figure 18. The provitamin D absorption spectrum and the CIE erythema action spectrum in relation to the solar spectrum (SZA = 55^0).

Just taking into account the steep drop in provitamin D absorption spectrum from 280 toward 310 nm one may assume high sensitivity of the vitamin D synthesis to the seasonal and latitudinal changes in solar UV-B radiation. Besides, our observation on the substantial impact of the Pre $\rightarrow$ Tox photoconversions on the previtamin D content at the long-wave irradiation led us to suggest that the provitamin D photoisomerization can be applied for the UVB monitoring [40], and in 1994 an original method based on an *in vitro* model of vitamin D synthesis (the so-called 'D-dosimeter') had been introduced for the UV dosimetry *in situ* [41].

Contrary to the majority of biological dosimeters, for which there is an imperfect understanding of the photobiological process details (especially when such a complex system as living cells is used), in the case of the D-dosimeter the photoreaction mechanism and physical parameters are known in detail, and the adequate mathematical model sets up a correspondence between the physical (J/m^2) and biological (previtamin D concentration) units of the accepted UV dose.

Another advantage of the method suggested is its ability to give the results of UV dose measurements immediately after an exposure owing to original spectrophotometric concentrational analysis. This differs from many biodosimeters that after an UV exposure require further (often time-consuming) laboratory analysis of the dosimeter material to determine the degree of the UV-initiated change.

6.2. Spectral Selectivity of the 'D-Dosimeter'

Most UV biodosimeters measure biologically effective dose by means of only one dose-dependent parameter. A different situation arises with the D-dosimeter, where at least two parameters can serve as an output signal. The first one is the rate of provitamin D concentration decrease upon UV irradiation. This parameter shows the decay of the initial product and is analogous to many biodosimeters that measure the inactivation effects of the UV radiation. Because the decay of provitaminD is exponential over a large range of doses, one can use the logarithm of its concentration as an output signal to increase the linearity range of the dosimeter.

The second parameter is the concentration of previtamin D, which is closely related to vitamin D synthesis and by this means characterizes the antirachitic activity of solar UV radiation. The experimental studies with tunable dye-laser as well as model calculations with monochromatic UV radiation discussed above showed that due to the irreversible photoconversions Pre → Tox the maximum concentration of the previtamin D decreases from 62 to 18 % when the irradiation wavelength was being shifted from 295 nm to 310 nm [40].

It was of interest to check if similar dependence holds true under solar irradiation with different spectral positions of the short-wave edge that could suggest the potential of D-dosimeter for the monitoring of ozone-layer thickness. For this purpose a number of solar spectra were simulated for different thickness of ozone layer (Figure 19a), and previtamin D photosynthesis was calculated using these spectra at the input of the photoreaction model (Figure 19b).

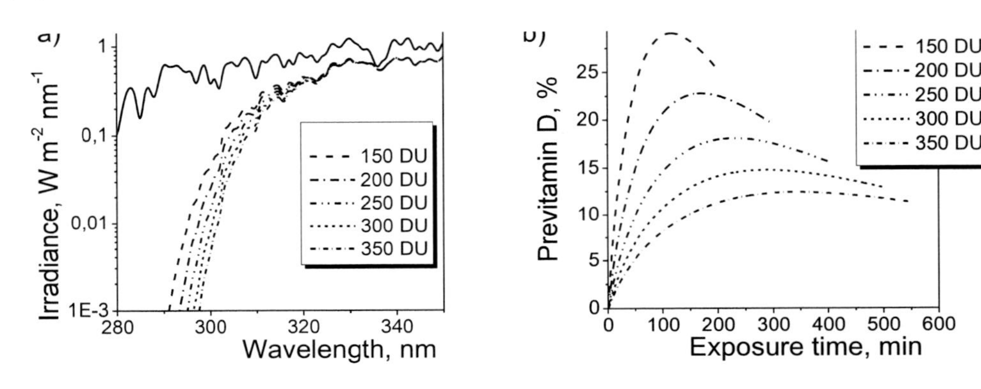

Figure 19. Simulated solar UV spectra for different thickness of ozone layer (SZA = 27^0, clear sky) (a) and calculated kinetics of previtamin D accumulation with the appropriate solar spectra (b).

It follows from these results that the photoreaction rate is controlled by the integral intensity of the initiating UV radiation whereas the maximum achievable concentration of previtamin D is sensitive only to the position of the short-wavelength edge of the solar spectrum [42]. According to preliminary estimations, additional focusing (10:1) of the radiation onto the cuvette with 7-DHC solution can provide a previtamin D maximum within a 1 hour exposure.

It is worthy of note that biodosimeters with only one dose-dependent parameter are not able to reveal ozone-layer depletion in the situation when air pollution (fog, dust, aerosols) attenuates the intensity of UV radiation like a grey filter.

As it is apparent from the history of vitamin D discovery [43], the vitamin D synthesis is very sensitive to air pollution because air pollutants such as sulfur dioxide or nitrogen dioxide hinder the penetration of UV-B radiation to the Earth's surface resulting in a growing incidence of rickets. But the increase of air pollution and cloudiness slows down the rate of previtamin D formation, whereas the maximum attainable concentration of previtamin D is not changed.

6.3. Measurement of the Vitamin D Action Spectrum in Vitro

As it is known, an action spectrum (AS) is defined as spectral dependence of the value of the biological effect initiated by monochromatic radiation of different wavelength with the same dose.

The rate of previtamin D accumulation at the initial stage of UV irradiation is primarily dependent on the efficiency of photon absorption by provitamin D molecule. Thus one would expect correlation of the action spectrum of vitamin D synthesis with the absorption spectrum of provitamin D.

Determination of the vitamin D action spectrum *in vitro* was performed at the Belgian Institute for Space Aeronomy [44] using UV irradiation from the xenon arc lamp with a number of narrow-band filters ($\Delta\lambda = 2.5$ nm). Solution of 7-DHC in ethanol (C = 20 μg/ml, V = 2 ml) was irradiated in standard rectangular cuvette (Hellma) of 0.5 cm thickness.

During the time when the cuvette with the provitamin D solution was exposed, the incident UV radiation was measured by the spectroradiometer Spex Model 1672M.

Absorption spectra of provitamin D before and after several exposures were recorded, and the photoisomer concentrations were derived from the spectra by computer processing with the original program taking into consideration the irreversible photoconversions Pre → Tox.

Table 2.

Wavelength (nm)	260	270	280	290	300	310
Antirachitic dose (%PreD)	5	5	5	5	5	5
Erythemal dose (MED)	0.28	0.21	0.20	0.48	0.70	3.87
Physical dose (J/m^2)	55	42.7	40	95	205	10400

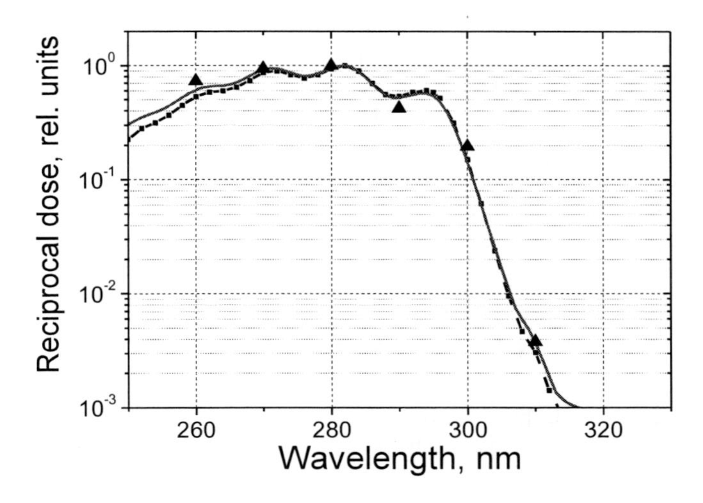

Figure 20. The experimental data (symbols), the 7-DHC absorption spectrum (solid line) and calculated action spectrum of vitamin D synthesis (dash line).

The concentrations of accumulated previtamin D against the irradiation time were plotted, and the required energy doses for 5% of previtamin D were calculated by using linear portion of the calibration curves. The results are presented in Table 2 where the physical and erythemal doses to achieve 5% accumulation of previtamin D are shown for the six wavelengths. The sharp increase in the physical dose and in MEDs required for accumulation of 5% previtamin D at 310nm is of special interest.

Additionally the action spectrum for monochromatic irradiation ($\Delta\lambda = 1$ nm) within a spectral range of 250 - 320 nm using the adequate system of differential equations that described the photoreaction kinetics has been calculated [41]. As may be inferred from the Figure 20, rather good agreement between the experimental data and the calculated action spectrum is observed. Besides close relation between the measured *in vitro* action spectrum of vitamin D synthesis and the absorption spectrum of provitamin D can be seen graphically.

6.4. Inter-Relation between the in Vivo and in Vitro Action Spectra of Vitamin D Synthesis

Earlier studies on the quantitative efficiency of different parts of the ultraviolet spectrum for the albino rats [45] showed that the antirachitic curve follows somewhat the absorption curve of provitamin D.

However, UV irradiation of human skin [46] did not show any correlation of the *in vivo* action spectrum of previtamin D_3 photosynthesis with the absorption spectrum of 7-DHC although close agreement of previtamin D_3 formation from 7-DHC was observed in human epidermis and in solution [46].

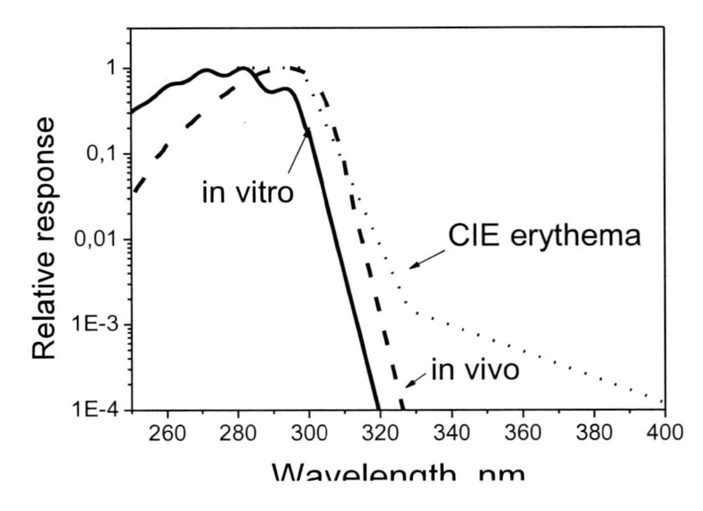

Figure 21. The absorption spectrum of 7-DHC (extrapolated from 300nm) and the CIE Vitamin D action spectrum in human skin [47] in comparison with the CIE erythema action spectrum.

Hence remarkable difference exists between the two measured *in vivo* action spectra of vitamin D synthesis: for the albino rats it closely resembles the absorption spectrum of provitamin D molecule with maximum at 282nm, but the AS for human skin with its maximum at 295nm is completely unrelated to the absorption spectrum of provitamin D (Figure 21).

It may be assumed right away that human skin transmission is responsible for such difference and that taking into account skin transmission could eliminate the difference between the *in vivo* and *in vitro* action spectra.

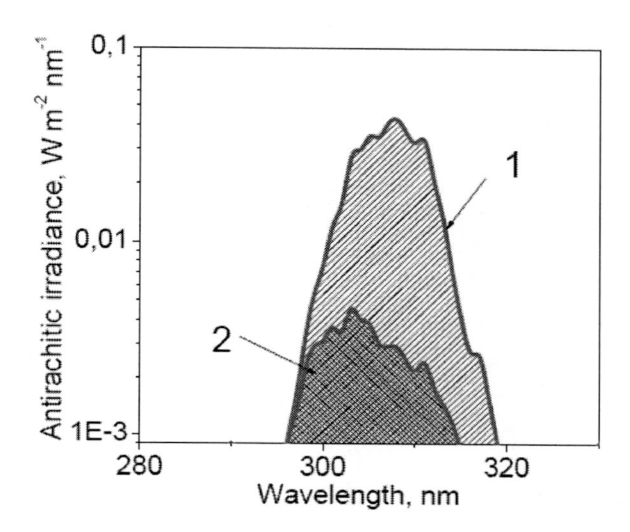

Figure 22. Antirachitic irradiance of sunlight (SZA = 270, clear sky) calculated with the CIE Vitamin D AS *in vivo* (1) and with the skin-corrected AS *in vitro* (2).

To check this assumption the absorption spectrum of 7-DHC was multiplied with the transmission spectra of stratum corneum and human epidermis measured by Bruls et al. [48], and biologic (antirachitic) effectiveness of sunlight using the CIE *in vivo* Vitamin D action spectrum [47] and the skin-corrected AS *in vitro* was calculated [49]. However pronounced difference in the antirachitic irradiance calculated with the two action spectra is still clearly seen from Figure 22.

Notice that significant difference between the *in vivo* AS measured in [46] and the absorption spectrum of provitamin D synthesis is inconsistent with closely related kinetic of previtamin D photosynthesis from 7-DHC observed in human skin and in solution [46].

On the basis of present knowledge there are good grounds to believe that the reason of such self-contradiction is neglect of the irreversible photoconversions of previtamin D in the concentrational analysis fulfilled by MacLaughlin's *et al.* [46] within the limits of generally accepted photostationary approximation assuming total concentration of the four main photoisomers equal to 100%.

Revealed discrepancy between the *in vivo* and *in vitro* action spectra of vitamin D synthesis necessitates detailed consideration to avoid pitfalls in calculations of the vitamin-D-effective UV irradiance of sunlight and artificial UV sources.

Conclusions

The limitations of the generally accepted photostationary state approximation in the photochemistry of provitamin D due to laser excitation and spectral monitoring of the photoreaction kinetics were revealed. The wavelength dependence of the irreversible channel effectiveness using a simplified model was explained, and it was clearly shown that the PS approximation with neglect to the irreversible channel is valid only upon irradiation within the range 266 - 300 nm (between the isosbetic points of provitamin D and previtamin D absorption spectra). Outside of this range and especially at the longer wavelengths the role of the irreversible channel significantly increases, and an adequate concentration analysis should take into account the mixture photodegradation.

In summary it must be emphasized that disregard of the irreversible channel with a low quantum yield in any system of reversible photochemical reactions is not always justified over a wide spectral range

References

[1] Havinga, E. Vitamin D, example and challenge, *Experentia*, 1973, 29, 81-93.

[2] Abillon, E.; Mermet-Bouvier, R. *J. Pharm. Sci*., 1973, 62, 1688-1691.

[3] Jacobs, H. J. C.; Havinga E. *In Advances in photochemistry*, Pitts, J. N. et al. Ed.; Wiley, NY, 1979, 11, 305-373.

[4] Dauben, W. G.; Phillips, R. B. *J. Am. Chem. Soc.,* 1982, 104, 5780-5781.

[5] Boomsma, F.; Jacobs, H. J. C.; Havinga, E., and Van der Gen, *Recuel, J. Royal Netherlands Chem. Soc*., 1977, 96, 104-112.

[6] Hanewald, K.H.; Rappoldt, M.P.; Roborgh, J.R. *Rec. Trav. Chim.* 1961, 80, 1003–1014.

[7] Malatesta, V.; Willis, C.; and Hackett, P. A. *J. Am. Chem. Soc.*, 1981, 103, 6781-6783.

[8] Dauben, W. G. and Phillips, R. B. *J. Am. Chem. Soc.*, 1982, 104, 5780-5781.

[9] Braun, M.; Fuss, W., and Kompa, K. L. *J. Photochem. Photobiol. A*, 1991, 61, 15-26.

[10] Dauben, W. G.; Share, P. E., and Ollmann, R. R Jr. *J. Am. Chem. Soc.*, 1988, 110, 2548-2554.

[11] Dauben, W. G.; Disanyaka, B., Funhoff, D. J. H., et al., *J. Am. Chem. Soc.*, 1991, 113, 8367-8374.

[12] Fuss, W., and Lochbrunner, S. *J. Photochem. Photobiol. A,* 1997, 105, 159-164.

[13] Dmitrenko, O.; Frederick, J. H., and Reischl, W. *J. Photochem. Photobiol. A: Chem.* , 2001, 139, 125-131.

[14] Saltiel, J.; Cires, L., and Turek, A. V. In *Handbook of Organic Photo-chemistry and Photobiology*, Ed.: Horspool, W.M. and Lenci, F., CRC Press, London 2004, Chapter 27, 1-22.

[15] Terenetskaya, I.P.; Gundorov, S. I.; Kravchenko, V. I., and Berik, E. B. *Sov. J. Quantum Electronics*, 1988, 18, 1323-1327.

[16] Bogoslovsky, N.A.; Berik, I.K., Gundorov, S.I., Terenetskaya, I.P. *High Energy Chem.*, 1989, 23, 218-222.

[17] Terenetskaya, I.P.; Gundorov, S. I., and Berik, E. B. *Sov. J. Quantum Electron.*, 1991, 21, 472-474.

[18] Terenetskaya, I. *Theor. Experim. Chem.* 2008, 44, 286-291.
[19] Terenetskaya, I.P., and Repeyev, Yu.A. In "Laser Applications in Life Sciences", Part Two: Lasers in Biophysics and Biomedicine, *SPIE* 1990, 1403, 500-503.
[20] Terenetskaya, I.P., and Repeyev, Yu.A. *High Energy Chemistry*, 1996, 30, 402-406.
[21] Repeyev, Yu.A., and Terenetskaya, I.P. *Sov. J. Quantum Elect*ron., 1996, 23, 765-768.
[22] Sternberg, J. C.; Stillo, H. S., and Schwendeman, R. H. *Anal. Chem.*, 1960, 32, 84-90.
[23] Terenetskaya, I., and Ouchi, A. *Book of Abstracts*, ICP XXI, Nara, Japan, 2003, 452.
[24] Terenetskaya , I. *Talanta,* 2000, Vol.53, No.1, pp. 195-203.
[25] Larichev, O,I,, and Gorvits, G.G. In: The methods of the local extremum search of ravine functions, Moscow, Science, 1990, 93p.
[26] Fletcher, R. *Computer J.*, 1970, 13, 317-322.
[27] Fuss, W.; Hofer, T.; Hering, P.; Kompa, K.L.; Lochbrunner, S.; Schikarski, T.; Schmid, W.E. *J. Phys. Chem.* 1996, 100, 921-927
[28] Havinga, E,; de Kock, R.J., and Rappoldt, M.P. *Tetrahedron*, 1960, 11, 276-284.
[29] Kobayashi, T., and Yasumura, M. *J. Nutr. Sci.-Vitaminol.*, 1973, 19, 123-129.
[30] Terenetskaya, I.P.; Bogoslovsky, N.A., Vysotsky, L.N., and Luknitsky, F.I. *Pharm. Chem. J.*, 1994, 28, 589-596.
[31] Terenetskaya, I.P.; Samojlov, Y.A.; Bogoslovsky, N.A.; Vysotsky, L.N., and Luknitsky, F.I. *Pharm. Chem. J.,* 1993, 27, 797- 803.
[32] Terenetskaya, I.P.; Galkin, O.N., and Bogoslovsky, N.A. *Pharm. Chem. J.*, 1993, 27, 282-288.
[33] Terenetskaya I.; Oppenlander T. *Book of Abstracts* XVIII IUPAC Symposium on Potochemistry, Dresden, 2000, 585-586.
[34] Horneck, G. *Trends in Photochem. Photobiol.*, 1997, 4, 67-78.
[35] McKinlay, and Diffey, A. F. *CIE J.*, 1987, 6, 17-22.
[36] Moan, J., Porojnicu, A.C., Dahlback, A., Setlow, R.B. *Proc.Natl. Acad. Sci. USA*, 2006, 105, 668–673.
[37] Fioletov, V.E., McArthur, L.J.B., Mathews, T.W., and Marrett, L. *J. Photochem. Photobiol. B: Biology*, 2009, 95, 9–16
[38] McKenzie, R.L., Liley, J.B., Björn, L.O. *Photochem. Photobiol.*, 2009, 85, 88–98.
[39] Mayar C., and Norman, A. W. Vitamin D, in*:* Encyclopedia of Human Biology, Academic Press, Inc. 1991, 7, 859-871.
[40] Terenetskaya, I.P. *SPIE Proc.*, 1994, 2134B, 135-140.
[41] Galkin, O. N., and Terenetskaya, I. P. *J. Photochem. Photobiol. B: Biology*, 1999, 53, 12-19.
[42] Terenetskaya, I. *Agricul. Forest Meteorol.*, 2003, 120, 45-50.
[43] M.F. Holick, Vitamin D3 and sunlight: an intimate beneficial relationship, In: M.F. Holick, A.M. Kligman *(*Eds.*)*, Biological Effects of Light, W. de Gruyter, Berlin, 1992, 11–33.
[44] Bolsee, D.; Webb, A.R.; Gillotay, D.; Dorschel, B.; Knuschke, P.; Krins, A., and Terenetskaya, I. *Applied Optics*, 2000, 39, 2813-2822.
[45] Knudsen, A., and Bedford, F. *J. Chem. Biol.*, 1938, 124, 287-299.
[46] MacLaughlin, J.A.; Anderson, R.R., and Holick, M.F. *Science*, 1982, 216, 1001-1003.
[47] Bouillon, R.; Eisman, J.; Garabedian, M.; Holick, M.; Kleinschmidt, J.; Suda, T.; Terenetskaya, I.; and Webb, A. 2006, Action Spectrum for the Production of Previtamin D_3 in Human Skin. UDC: 612.014.481-06, *CIE*, Vienna.

[48] Bruls, W.A.G.; Slaper, G., van der Leun, J.C., and Berrens, L. *Photochem. Photobiol.*, 1984, 40, 485-494.

[49] Terenetskaya, I. *CIE Proceedings* of the 2nd CIE Expert Symposium on Lighting and Health, 7-8 September, 2006, Ottawa, Ontario, Canada, 182-185.

In: Photochemistry
Editors: Karen J. Maes and Jaime M. Willems
ISBN: 978-1-61209-506-6

Chapter 7

EXPLORING ORGANIC PHOTOCHEMISTRY WITH COMPUTATIONAL METHODS: AN INTRODUCTION

Diego Sampedro
Departamento de Química, Universidad de La Rioja,
Grupo de Síntesis Química de La Rioja, Unidad Asociada al CSIC, Logroño, Spain

1. ABSTRACT

Research in photochemistry has substantially changed in the last decades. Some time ago, the focus of research on photochemistry was mainly the effect of light on molecules and the photochemical reactivity or photophysical properties caused by light excitation. More recently, while interest in photochemistry at the molecular level still remains, new and more complex applications to exploit light energy have been developed: photocatalysis, sensing and signaling, photoactivated molecular machines, photoprotection, biomedical and environmental uses and others. The rational design of molecules capable of performing such tasks depends on the knowledge of the mechanisms operating at the molecular level. Thus, much effort has been devoted lately to gather a deeper insight into the way these photoreactions occur.

In recent years, computational chemistry has emerged has an important tool for the detailed investigation of photochemical reactions. With the exponential increase of computer power, together with fundamental breakthroughs in the theoretical descriptions of photochemical reactions the scientific community has now the tools to explore and understand those organic photoreactions.

In this contribution, a brief introduction to the computational methods currently available to study photoreactions of organic substrates will be presented. The main target will be to introduce a non-specialist reader into the features of computational organic photochemistry. This research field has their own methods, concepts and tools, advantages and drawbacks, usually different from those found in relatively closer topics such as ground state computational chemistry or experimental photochemistry. Thus, theoreticians trained to study thermal reactions or experimental photochemists trying to complement their research will find here a starting point. However, it is far beyond the scope of this chapter to comprehensively explore the state-of-the-art in computational photochemistry. Instead, a general overview will be presented and the reader will be invited to follow the references for an in-depth treatment of some sections. Due to the optimum balance between accuracy and computational cost for medium-sized organic compounds, the strategy based in the CASPT2//CASSCF level of theory will be especially explained. Some case studies will also be discussed in order to provide the reader with a general idea of both the importance and capacity of these tools to get a comprehensive picture of relevant processes for photochemical reactions at the molecular-level. Limitations as well as some perspectives into the near future will also be discussed.

2. Introduction

In the past, the only predictive tool available to chemists in general, and photochemists in particular was the use of correlation diagrams, such the Van der Lugt-Oosteroff diagrams,[1] based on avoided crossings between surfaces. However, several experimental observations challenged this vision. For instance, decay at an avoided crossing with a gap between states of some kcal/mol would occur in the same time scale than fluorescence (nanoseconds), while it is known than some photochemical processes are much faster (picoseconds and below) [2,3]. Trying to provide a theoretical model to explain the observations made, Teller proposed [4] that a very fast decay to ground state could happen in special regions of the potential energy surface, *i.e.* conical intersection points. However, for some years conical intersections were thought to be quite exotic points in the energy surface for organic compounds and thus with no practical importance. In fact, the main problem at that time was the lack of adequate methodologies to explore excited state surfaces.

By the end of 1980's a new set of *ab initio* tools were developed in order to explore photochemical reactions. Specifically, these methods allowed to provide a balanced treatment of both the ground and excited states and the possibility of computing gradients that allowed geometry optimizations, chemists then had the tools to explore the reaction path for photochemical reactions.

The first application of this method, *i.e.*, *ab initio* Multiconfigurational Self-Consistent Field (MCSCF) was reported in 1990. [5] Starting from there, conical intersections changed from extremely rare to ubiquitous in organic photochemistry. Lots of different chromophores involved in all types of chemical reactivity were explored [6,7] which allowed to provide a general framework for the computational study of organic photoreactivity.

3. Theoretical Background

It is far beyond the scope of this chapter to offer a comprehensive view of the computational methods in use for the study of photochemical reactivity. Instead, a general overview of some of the most used computational methods will be presented and the reader should refer to excellent general books already available for an in-depth treatment. [8-10]

In order to solve the Schrödinger equation for any non-trivial system, several approximations have been introduced in the development of the different theoretical models. For example, molecular orbital theory uses one-electron functions or orbitals to approximate the full wavefunction. The Hartree-Fock (HF) theory is the simplest version and considers only a single assignment of electrons to orbitals, that is, uses only one electron configuration.

Due to its simplicity HF theory was amply used in the early days of computational chemistry, and it is still in use when the size of the system under study is too big for other methods or only qualitative results are needed. However, the use of only one electron configuration has serious drawbacks, especially in the case of photochemical reactions where several electronic states have to be treated in a balanced way. To correct for such a deficiency, it is necessary to use wavefunctions than represent more than a single configuration as it is done in the so-called post-HF methods.

3.1. The CASSCF Method

The CASSCF (Complete Active Space Self-Consistent Field) method [11] is an extension of the HF method to include diverse electron configurations representing different electronic states. The same molecular orbitals generated in a HF computation are used to generate a multireference wavefunction describing different electronic configurations. Solving the equations provides the energies associated with the ground and different excited states. However, in order to get accurate results, the wavefunction must contain an extremely large number of configurations, which usually makes the calculations very difficult, if not impossible. In order to solve this problem, one can limit the number of configurations effectively considered in the calculation in order to make it viable. Doing so, only a limited number of electron configurations describing a limited number of excited states are included in the computation. To define this subset of configurations, the series of orbitals are grouped in three sets. The core orbitals remain always doubly occupied. The virtual orbitals remain always empty. Finally the so-called active orbitals have occupancies ranging from zero to two electrons. For the CASSCF method, all the configurations describing all possible electronic excitations within the chosen subset (the active space) are used. The choice of this subset of configurations to be used (orbitals and electrons) is critical. It is a function of the molecular system and the reactivity under study. For example, if the subject of study is the excitation process of a chromophore containing a π-system, the active space must include all the orbitals and electrons involved in the absorption process. Similarly, for the study of a chemical reaction, all the orbitals involved in the bond-breaking and bond-forming processes have to be included in the active space. For the CASSCF method, a nomenclature (n,m) is used to define the number of active electrons (n) and active orbitals (m) included in the active space. The choice of active space is always the first decision to make and also the most important for a photochemical calculation. As the computational time increases enormously with the size of the active space, for practical reasons sometimes the number of included orbitals has to be carefully evaluated. A practical upper limit for the active space can be placed in (14,14). Difficulties inherent to the CASSCF method and the choice and size of the active space constitute practical reasons that make this method quite difficult to use. This method is not intended as a black-box and some expertise and chemical intuition are needed for good, reliable results. However, under the proper use, this method can provide the correct shape of the potential energy surface with enough accuracy and minima, transition structures and reaction coordinates can be computed for organic molecules of medium size.

3.2. The CASPT2 Method

As said before, the CASSCF method can provide the correct potential energy surface shape. However, CASSCF energies are usually not accurate enough to reproduce experimental excitation values or reaction barriers. The reason behind this is the fact that CASSCF cannot describe satisfactorily the dynamic correlation, that is, the instantaneous mutual repulsion between two electrons of opposite spins. In order to improve the results, it is necessary to go beyond the CASSCF method and employ any other method capable of treating the dynamic correlation. Several methods can provide such an improved wavefunction, although all of them share the fact of being computationally very expensive. Thus, a common strategy used in the study of photochemical reactions of medium-sized molecules is to first compute the potential surface critical points (minima, transition structures, conical intersections) with a more accessible method (CASSCF, for instance) and then re-evaluate the energy of these points with a more reliable although computationally more expensive method that accounts for the dynamic correlation. One of these methods capable of improving the energetics is the CASPT2 method (Complete Active Space Second-order Perturbation Theory).[12] Basically, the CASSCF result is used as the zeroth-order wavefunction and then a second-order multiconfigurational perturbation is computed. In this way, the initial CASSCF wavefunction is enriched with additional excitations and most of the dynamic correlation is included. As a result, the energies for the calculated points are improved and a better correlation with experimental values is obtained. As happens with the CASSCF method, the practical use of CASPT2 is not exempt from problems. For example, the presence of intruder states can cause the energy corrections to be not reliable. This problem can be overcome by increasing the active space. However, as explained above, there is a practical limit to the size of the active space, so this solution is not always accessible. Another possible solution is the use of a technique so-called level-shift. [13] The level-shift CASPT2 method can remove weak intruder states by addition of a constant, the shift parameter, to the zeroth-order wavefunction and a subsequent back correction to remove it. The CASPT2 energies remain unaltered while the effect of the intruder states is removed. This powerful technique has to be used with caution and proper calibration and analysis of the trends for different values for the shift parameter has to be carried out in order to get reliable results. As with the CASSCF method, good results can be obtained but some expertise in the use of these methods is required, since in their current formulation they are far from being a black-box.

3.3. The TD-DFT Methods

Among the available tools for ground state computations, methods based in the density functional theory (DFT) are nowadays the most popular mainly due to the balance between accuracy and computational cost. They are methods relatively easy-to-use and they can afford relatively reliable results for systems of quite a big size in the fields of biochemistry, solid state and material science. However, DFT is limited to ground state in its basic formulation, which excluded its use in photochemistry. Several attempts have been made to extend DFT to the study of excited states, [14] the most important been the time-dependent density functional theory (TD-DFT). [15] As it happens with the ground state version, TD-DFT methods imply the crucial choice of functional. Both empirical, *i.e.* functionals containing a large number of parameters, and non-empirical functionals can be used. In the case of empirical functionals, a good agreement with experiment is expected for the set of molecules used in parameterization, but they can fail for other systems by far. For non-empirical functionals a more uniform behavior in anticipated. In any case, every functional can offer a certain accuracy beyond which better results are more a matter of fortune than chemical accuracy. TD-DFT methods have been extensively used for calculation of optical properties; especially in the case of systems of big size beyond the reach of *ab initio* methods. Photophysical properties of fullerenes, aromatic compounds, porphyrins, polymers, biomolecules and transition metal compounds have been subject of research. Analytical gradients have been also implemented [16] which allows for the computation of excited states properties, although the treatment of conical intersections, minimum energy paths and dynamics is still under development. Although these methods present serious drawbacks, mainly related to their low accuracy, the ability to treat systems with quite a big size still make them quite useful for some selected applications when qualitative results are enough.

3.4. The Choice of a Method

In the precedent paragraphs a very limited introduction to some of the methods in use for the study of photochemistry and photophysics has been presented. Of course, some other methods are available and the reader is encouraged to refer to excellent books for a deeper treatment of these and other methods for computational photochemistry. [17,18] Recently, the influence of the medium has been also included. For instance, Olivucci *et. al.* have developed [19] a new QM/MM method which allows the computation of a chromophore embedded in a protein cavity. Also, dynamics of photochemical reactions are important in order to determine excited state lifetimes or quantum yields, so an adequate treatment is required to allow comparison with experiment. Different methods have been developed in order to consider the dynamic effects provided by the kinetic energy of the nuclei. High-level quantum dynamics are completely unapproachable when dealing with organic systems. Thus, inclusion of dynamics in nonadiabatic processes has to rely on semi-classical methods such as the Trajectory-Surface-Hopping or Ehrenfest dynamics. A full discussion of these methods can be found elsewhere. [20]

The choice of an adequate method for a certain project implies a previous careful evaluation of the problem, the objectives and the resources available. The accuracy needed for the system and problem under study has to be evaluated before any conclusion can be drawn. In the case that the quality of the results is not enough for our needs, the level of theory should be improved in order to get stable conclusions. As a general and thus coarse approximation, TD-DFT could be used to get a low-level, qualitative general picture of the problem. It should be noted that a low accuracy can be expected from these methods and thus some calibration with experimental data or high level computations could be necessary. However, the size of the systems that can be explored is clearly bigger than in other methods. If the system is of a reasonable size, CASPT2 is probably the method of choice, as it can treat almost all kind of molecular systems and reactivities with considerably accuracy. A detailed literature reviewing would be of much help in order to discriminate which method provides accurate results for any given problem. After the calculation has been finally carried out, a critical evaluation of the results is also important. However, an exact agreement with experiment should not be expected nor needed if the conditions of both types of data are different enough to prevent a comparison.

4. Computing the Photochemical Reaction Path

The properties of materials can be related to the chemical structure of the constituting molecules. In the vast field of photochemical applications, this means that the properties of photoactive polymers, the protection capabilities of sunscreens, the effectiveness of photo-curing paints or the efficacy of photocaged drugs can be related to the molecular photochemical reactivity of each of these compounds. Even more, the properties of materials after light absorption can also be linked to chemical structures and photochemical reaction paths, for instance in the case of luminescent molecules or compounds with strong absorption bands. Thus, it is clear that a detailed knowledge of the photochemical reaction path at the molecular level will not only help to understand the chemical and physical processes after light absorption, but will also allow for a rational design of new molecules and materials with specific properties and functions. In this context, computational photochemistry has emerged as a new and powerful set of tools for the detailed investigation of the photochemical reaction path. Nowadays, it is possible to explore the chemical changes at the molecular level with precisions similar to the experiment. Moreover, the information obtained from careful computational exploration of a compound can be used to tune their properties and behavior and also to design new variants with improved features, for instance, more photostable compounds or faster light-driven molecular motors.

The most common approach used to construct a photochemical reaction path implies the computation of minimum energy paths (MEP) [21] starting from the structure of the reactants and connecting all the relevant points in the potential energy surface to finally end at the structure of the reaction products. This approach has been named the *pathway approach* and it has been extensively used to explain and study lots of photochemical transformations for different chromophores. [22] It is important to note some important features of this approach. First, the information of the system is obtained from the local properties of the potential energy surface, that is, the relative energy of minima, the height of energy barriers, and the crossing points between electronic states. Second, the information obtained from the computation of the reaction path is strictly structural, that is, the path represents the motion of a molecule with no vibrational energy covering the path at infinitesimal speed. Of course, the path calculated this way does not represent the real trajectory that a molecule with at least some vibrational energy would follow. As the computed path implies that the molecule will travel it with no speed, the picture obtained from these calculations is necessarily static, and no quantitative information on the dynamic features of the reaction could be inferred. Yet, some valuable knowledge relevant to chemical reactivity and experiment comparison could be obtained. For instance, the formation of photoproducts or the ratio between different products, the absorption and emission properties of reactants, intermediates and products or the photostability of compounds can be determined by the computation of the photochemical reaction path. As we will see later, if we want to get some dynamic information on our system, other approaches should be used instead.

Among the different levels of theory available for computing reaction paths for organic molecules, clearly the combined *ab initio* CASPT2//CASSCF strategy is the preferred choice. As explained in the precedent section, this is due to the balanced way in which it can describe different electronic states with reasonable accuracy for medium-sized organic molecules. It is not possible to describe here in detail how these computational tools work. Although powerful, this method requires some learning to use and some expertise to get adequate results. Instead, a quick description of the procedure and computational tools will be offered and the reader is invited to turn to the excellent literature available [17,18] for a more in-depth description.

As described in the previous section, computation of photochemical reactivity presents some further complications with respect to thermal reactions. The entire path for a thermal reaction, that is, the complete progression from a reactant structure through the transition state towards the final reaction product, can be computed with considerable accuracy with high level *ab initio* quantum chemical methods. In contrast, the progress of the equivalent computational tools for photochemical reactions was somewhat slower due to the intrinsic difficulties associated with the fact that several electronic states are involved and, thus, the photochemical reaction path includes at least two branches in different potential energy surfaces. The main technical difficulty here is the computation of the region where these surfaces are close enough to allow the molecule to go from the excited state to the ground state. The progression of the molecule through any of these branches of the reaction path (ground or excited state) can be computed using similar tools regardless the electronic nature of the energy surface. However, non-standard computational tools are required to locate the crossing point between surfaces where these two or more branches of the reaction path connect. Once developed these new tools, from the theoretical work to the implementation of algorithms capable of effectively computing these crossings, photochemists and computational chemists have now the possibility to compute photochemical reaction paths with accuracy.

The practical computation of a photochemical reaction path implies to calculate and analyze the reaction coordinates and energies connecting the Franck-Condon region, through the minima and transition structures in the different potential energy surfaces (if present) to the crossings between surfaces and finally the products in the ground state. All the MEPs between these points have to be computed to get a full picture of the photochemical reaction path. As the MEP is computed following the steepest descent in the energy surface region, only the minima or transition structures directly available to the system will be computed. This usually prevents to compute regions of the potential energy surface irrelevant to the reactivity under study. However, as stated above, it should be noted that MEP calculations are done assuming that the system is vibrationally cold and the molecules travel the MEP at infinitesimal speed. This is just a simplistic view of the real process, where the system has in fact some extra energy that may cause some serious deviations from the MEP. Although MEP computations provide some very valuable information on the system under study especially for the directly available reactivity, one should be very careful not to restrict the description of the problem to the static view provided by the MEP. In the case of systems with a large excess of vibrational energy, the real trajectories followed by the molecule along the reaction path could deviate considerably from the MEP. In those cases, other regions in the potential energy surface not considered in the MEP computations can be relevant for the description of the reactivity and dynamic effects should be considered. If this is the case, the mechanistic information provided by the computations can be only qualitatively correct or even completely wrong. However, if the experimental conditions allow for a controlled vibrational energy or thermal equilibration is possible, the MEP provides with a generally good, averaged representation of the real path.

Computation of a photochemical reaction path can be done using the same standard techniques as those used for the exploration of a thermal reaction in the case of stationary points, *i.e.* transition structures and minima, and MEP (for instance the intrinsic reaction coordinate IRC [23] can be used). However these standard tools fail when it comes to computing a region where two or more potential energy surfaces cross. Of course, these regions are of utmost importance in the photochemical reaction path as the decay from the excited state to the ground state will take place there. Optimization of the crossing points between surfaces required the development of new techniques and it will not be described here in detail. Further information on the theoretical background of the exploration of crossing points can be found in the literature. [8,24] These tools are currently available in a number of software packages (see below). Once reached a conical intersection point, evolution of the system in the ground state should also be followed using non-standard techniques, due to the fact that these points are non-stationary and methods such as IRC will fail. Different possibilities are also available to compute these paths starting from the crossing point. For instance, the calculation of the initial relaxation direction (IRD) [25,26] allow to locate different paths starting from a non-stationary point (Franck-Condon region, conical intersection) which eventually end up in valleys in the ground state corresponding to the photoproducts. The practical implementation of this approach consists of computing the relaxation paths from the conical intersection using the IRD and then computing the MEP from the geometries obtained. Further details on the practical use of IRD can be found elsewhere. [27,28] Especially complex is the computation of the relaxation paths starting from a conical intersection. In the ground state, the conical intersection geometry represents a singularity (a cusp) in the potential energy surface. Due to this, the gradient cannot be computed unambiguously, that is, there are an infinite number of different gradient vectors pointing to an infinite number of directions as the system tries to leave the conical intersection point (the cusp) in the ground state. It is obvious however, that for any photochemical reaction there will be a finite number of photoproducts and, thus, some of those directions are preferred to the others. This can be solved using the IRD approach which computes the energy of the system along a circular cross-section centered in the point of interest (conical intersection, Franck-Condon region). Using a small radius for the circular section it is possible to evaluate the energy changes in the vicinity of the starting point. Doing so, a number of relaxation channels from the starting point will be found. Once detected these relaxation paths, standard MEP computations will allow to find the photoproducts (when computing the IRD from a conical intersection point) or the relaxation direction in the excited state (when starting from the Franck-Condon point).

5. Practical Considerations

As explained above, the use of these powerful computational tools for the study of a photochemical reaction requires some knowledge of the theoretical background of the method used (including drawbacks, limitations and systematic errors) and a certain degree of expertise in its use. Of course, chemical comprehension of the system under study is extremely valuable and usually contributes to simplify a lot the planning of the calculations as well as to allow for an early detection of errors or inconsistencies in the computed data. A complete tutorial on how to use these tools is unapproachable here, but in this section a general protocol to study a photochemical reaction will be presented, together with some common technical problems and some hints on how to avoid them. For a deeper treatment of the topic, some references are available. [8,17,18]

5.1. Choice of the Chemical Structure

When performing computations on a photochemical reaction, the first crucial choice to make is the model structure. As explained in the Theoretical Background section, the CASSCF method has some practical limitations with regard to the size of the system. For organic reactions, only medium-sized molecules can be treated with adequate accuracy, *i. e.* using a reasonable active space and basis set. Thus, it is often unavoidable to reduce the size of the system to be computed relative to the system used in the experiments. As a general tip for photochemical reactions, the chemical structure of the chromophore should be completely included. If the chromophore is too big and no simplification can be made, this probably implies that the CASPT2//CASSCF strategy is not a good choice in order to get reliable results. If the complete chromophore can be included in the calculations at least the light absorption process and the initial steps of the photochemical reactivity will probably be qualitatively reproduced in our simplified model. Of course, all atoms directly included in the reaction under study should be included, but also those atoms or groups indirectly involved would be necessary. For instance, a group of atoms which only contributes to stabilize an intermediate through charge donation should be included in the calculations. Although, in this situation this group does not take part in the reactivity, it contributes indirectly as the formation of the intermediate would be prevented if this group is not present in the molecule. Not only electronic aspects are important. Also steric effects can play a major role in the choice of the structure. When some large groups do not affect the reactivity, they can be simply replaced with smaller groups. The chemical effect will be the same, but at much lower computational cost. For instance, a large alky chain in a photochemical reaction can be usually replaced by a smaller methyl group. The alkyl chains are not included in the chromophore and, if the photochemical reaction does not affect the chain (no hydrogen abstraction or fragmentation) they can be removed from the model with no risk. However, it should be noted that sometimes the steric effects are also important for the photochemical reactivity, and thus groups causing these steric effects have to be included in the model. For instance, a large steric hindrance can block a conformation which favors (or prevents) a certain photoreactivity. In this case, the safer alternative is to include the complete set of atoms causing this steric hindrance in order to be sure that this effect will be reproduced in the calculations. If this approach is not possible, a smaller system could be used together with some geometrical constraints which can mimic the effect of the steric hindrance. Other possibility is to compute the whole system with a smaller basis set. It is important to note that these two alternatives imply some serious simplification of the chemical structure and the data obtained should be treated with great care. Results found for these simplified models could be only qualitatively similar to the real systems and no definitive conclusions should be drawn without proper calibration or higher level computations. Another alternative for the study of big systems is to divide the structure in two parts. The central part of the system where the reaction takes place is treated at a higher computational level and the rest of the system which only affects the steric hindrance or the environment of the central part is treated at a lower theoretical level. This approach is termed QM/MM as the important part of the system is computed using a quantum chemical method while the rest of the molecule is only computed using a molecular mechanics parameterized potential. This method has proven to be extremely valuable for the treatment of biological systems of big size. [19]

Although several alternatives for the study of photochemical reactions are currently available, the choice of the structure to be computed is still crucial. Only in few cases the relevant molecule can be completely included in our calculations. Thus, the choice of a simple, but realistic model is necessary in order to reduce the size and computational cost. Even more, the correct choice would also depend on the results (both in quantity and quality) that we want to get for our system. In this case, the chemical knowledge of the system and also the chemical intuition of the researcher are of extreme importance in order to get reliable and accurate results.

5.2. Choice of the Active Space

Even more difficult than the selection of a model structure could be the choice of an appropriate active space for our calculation. Once we have selected a chemical structure suitable for our method of choice and the information on our system that we want to acquire, the next step is to decide the active space that can provide and accurate description of the system within those limits. The choice of the active space is far from trivial and there are a number of issues that we have to consider to make an adequate decision. Even more, the right choice can vary a lot depending on the system and the questions we want to answer.

When selecting and active space for a CASSCF calculation it is strongly advisable to gather as much information on the chemical behavior of the system as possible. The choice of the orbitals and electrons relevant to the processes under consideration might be simplified with a strong background on the photochemical and photophysical behavior of the molecule under study and/or its analogues.

A key factor to bear in mind is the fact that the active space should be kept along the entire reaction path. This means that the same set of orbitals and electrons should be valid for both the reactants and products, together with all possible intermediates, critical points and conical intersection points. Obviously, this requires to have some chemical knowledge on the system or, at least, to be able to provide some adequate hypothesis.

As a first guideline, one should include in the active space every orbital and electron that could possible contribute to the reactivity of the system in any part of the reaction path to be considered. This implies that if we know (or expect/suspect) that a bond will be broken on the reaction path the bonding and antibonding orbitals describing this bond should be included in the active space. Also the complete subset of orbitals and electrons responsible for the light absorption (the chromophore) should be included. For instance, in an aromatic ring we should include both the bonding and antibonding π orbitals, together with the π electrons. In the case we want to study a particular excited state, we also have to include the orbitals that could affect the energetics of this particular state. Heteroatom lone pairs could contribute to stabilize or destabilize intermediates along the reaction path. Thus, when selecting the active space we should include all the orbitals and electrons that could possible affect the structure or energy of all the points along the reaction path. However, we should also bear in mind that the bigger the active space, the slower the calculation will be. Therefore, we want to keep the active space to the minimum necessary to adequately represent our system, without including additional orbitals or electrons that not contribute to an improvement in the description but to slow down our calculations. Regarding the computational cost of our calculations, it is important to know that the time employed in a CASSCF calculation increases factorially with the size of the active space. This means that the inclusion, for instance, of an additional orbital with a couple of electrons will lengthen our computation to a big extent. Even more, with big active spaces we could reach a limit where our computation would take a time so long to be unpractical or even to exceed the capacity of our computer. A practical limit can be placed in an active space including 14 orbitals with 14 electrons, although this value depends on some other variables such as the available computer power, the number of atoms, the basis set and the symmetry of the system. Therefore, it might happen that once decided the orbitals and electrons to be included in the active space, the size of the system prevents a practical computation. In these cases a first approximation could be to exclude some of the orbitals trying to reduce the size of the active space. This should be done with care in order not to exclude relevant orbitals. Some test calculations could help in the decision of which orbitals to exclude without a loss of accuracy. For instance, orbitals with occupancy very close to 2 or to 0 might, in principle, be transferred out of the active space into the occupied orbitals and virtual orbitals subsets, respectively. However, one must be sure that the occupancy of those orbitals will remain the same along the reaction path, that is, the orbitals will remain fully occupied or empty in all the computed structures. Again, some test calculations could help in this decision. Another possibility to carry out calculations with a big active space is a modification of the CASSCF method, the restricted active space, RASSCF. [29] This method implies the computation of only a subset of the configurations generated by the orbitals and electrons of the active space. Thus, the active space is not complete anymore (such in CASSCF, complete active space) but it allows for the computation of systems with bigger active spaces at the price of being less accurate. The selection of the more relevant orbitals imply some previous calculation to get a picture of the energy of the different orbitals that allow for a critical evaluation of the most relevant orbitals, a decision often also hard to make.

The practical evaluation and selection of the orbitals to be included in the active space can be done by evaluation of the occupation numbers, energy and location on the molecule of low-level computations. Although all this information is usually available in the output of the computation, it is strongly advisable to employ visualization software to display graphically the form and location of the orbitals. This allows for a quick selection of the active space that should be further tested to assure the right decision before a real set of high-level calculations would be performed.

5.3. Computing a Photochemical Reaction

Once assessed the right choice for both the structure to be computed and the active space included in the calculation, the real calculation of the photophysical and photochemical processes can start. The complete protocol will depend on the question we want to answer. For instance, in order to get information on the absorption or emission spectra of a molecule we will carry out some calculations that differ from those needed to completely map a reaction path from reactants to photoproducts. Also, the approach to compute an excited singlet state reactivity is different from a triplet state. In this section we will deal with singlet excited states. The computation of triplet states is affected by the way these electronic excited states are populated (sensitized irradiation or direct irradiation with intersystem crossing) and usually implies the calculation of the spin-orbit coupling. [30] In any case, it is advisable to validate the model (chemical structure, active space, basis set,…) by comparing some of our theoretical results with available experimental data. This can be done by comparing the computed structure of reactants or products with available X-ray diffraction structures or any other property. As we are interested in the photochemical behavior of the system, it is usually convenient to compare the theoretical absorption spectrum with the experimental one. If no experimental data is available one should rely on data obtained for similar compounds, other properties or high-level computations. In order to do this, CASPT2 corrected energies of the vertical excitation to different states should be computed, together with the oscillator strength (*f*) values for each of the relevant electronic states. This allows for the construction of a theoretical absorption spectrum that can be compared with the experimental one. If both spectra match to a reasonable extent we can be confident on the validity of the model we have chosen. It is important to note that the experimental and theoretical data are not measured in the same conditions (for example, gas phase *vs.* solution). Thus, a perfect match of both types of spectra is neither expected nor desired. If big differences rise after the comparison in either the band maximum absorption or the intensity of the absorptions (represented by the *f* value in the theoretical data) some errors in the choice of the model have to be considered. Perhaps the chemical structure was simplified too much and the chromophore does not represent the real molecule. Or the active space was improperly chosen, or some relevant orbitals were not considered. A big influence of the medium effect has also to be considered. If not included in the calculation, the solvent can modify the absorption spectrum. Comparing spectra obtained in the gas phase with those measured in solution has to be done with caution and the polarity of the solvent has to be considered. If the solvent effect is important it should be included in the calculation in order to get reliable results.

Calculation of the absorption spectra can be valuable information on its own apart from just a mere validation of the method. The effect on the band maxima or band intensities of different modifications of the molecule (solvents, substituents, conformations,…) can allow for the design of new compounds with the desired spectral properties. Besides, the calculation of vertical excitations provides information on the electronic nature of the excited states. For instance, a careful analysis of the values obtained for the configuration interaction allows to describe the electronic excited state with the electronic configurations with higher values. Thus, one can know where are mainly located the electrons after the excitation and compare it with the ground state electron density. A related piece of information can be obtained from the charge distribution of the excited states. This data allow to determine the ionic or covalent nature of the state. All these data provide some useful information on the features of the electronic excited state after vertical excitation, but can also be used to anticipate some of the photochemical features after relaxation on the excited state potential energy surface.

If we are only interested in the absorption properties of our system and we are not concerned about any photoreactivity, this is the end of the road. However, if we want to explore the fate of our system in the excited state (radiative or non-radiative decay, reactivity,...) the evolution of the molecule beyond the Franck-Condon region has to be considered. Once the nature of the different electronic states has been assessed, the study of a photochemical reaction path implies the relaxation of the system on the potential energy surface of the state of interest using CASSCF. Different possibilities may rise depending on the features of the energy surface. For instance, several electronic excited states may be involved and crossing points among them have to be computed, the presence of intermediates along the path in the excited state may be responsible for the luminescent properties of our system, the absence of any intermediate suggests a very fast non-radiative decay to the ground state. Thus, the specific protocol will vary a lot according to the photophysical and photochemical properties of the system under study. In any case, the more information relevant for our system we have before the calculations, the simpler this process would be. For instance, a fluorescence spectrum of our molecule would not only suggest the presence of an intermediate in the potential energy surface, but it would also approximately demarcate its energy. For any photochemical reaction, all the tools briefly described in section 4 can be used, together with standard geometry optimizations to minima or transition structures. Starting from the Franck-Condon region, IRD and IRC could lead to a minimum or a crossing point. Optimizations of minima, transition structures and conical intersection points have to be carried out with the corresponding algorithms. After the decay to the ground state, the reaction path leading to the products has to be also computed. When the decay point is a conical intersection, all the different paths departing from it have to be followed until the minima corresponding to the photoproducts (and perhaps also the reactants).

Once the ground state reaction path is mapped, a complete view of the full process can be obtained from the CASSCF calculations. However, as said before, CASSCF does not account for the dynamic correlation. This means that, although the geometries obtained at the CASSCF level are reasonably good, the energy computed for those structures not allow a direct comparison with experiment. Thus, the next step is to re-compute the energy profile at the CASPT2 level. The CASPT2 energy correction may give rise to a problem when two or more states are in close vicinity. In a nutshell, this problem arises when the order in the states is changed due to the different effect of dynamic correlation in both states which causes a state swapping. The energy order of the states at the CASSCF level is different than the energy order at the CASPT2 level. This means that the photochemical reaction path computed at the lower level (CASSCF) is not valid at the higher level (CASPT2). The extent of this problem very much depends on the region of the potential energy surface where takes place but usually the solution implies to compute selected points along the MEP at the CASPT2 level in order to map the region where the states swap. By doing this, a corrected potential energy surface can be obtained and both the right reaction path and energetics would be found. Usually this state swapping entails the disappearance (or at least the displacement) of a crossing point found at the CASSCF level when correcting the energy at the CASPT2 level or the other way round, the appearance of a crossing point in a region of the surface where the states were not too close to each other.

5.4. Software

In previous sections a brief introduction on the general aspects of computing a photochemical reaction path has been presented together with some practical considerations. The specific steps to tackle a computational study of a photochemical reaction will depend on the type of system and the processes under study, but also on the software available. In this section, a concise presentation of some of the software packages capable of performing computations on a photochemical reaction will be shown. It should be noted that most of the programs discussed below are under continuous development. Thus, some differences or improvements can be expected with every new version available.

GAUSSIAN (http://www.gaussian.com/): [31] It has versions for different operating systems (including Unix/Linux, Windows and MacOS). This software package allows for the calculation of very different systems and properties beyond the photochemical reactions. Regarding excited state processes, spectral properties can be computed using TD-DFT methods and photochemical reaction paths can be studied using CASSCF. Search for conical intersections among states can also be performed. Standard methods such as minima and transition state structures optimization and IRC computations at the CASSCF level allow for a complete mapping of the potential energy surface. Solvent effects can also be included when computing excited states by using the ONIOM (Our own N-layered Integrated molecular Orbital and molecular Mechanics) approach. [32] This allows to perform calculations using a molecular mechanics method for the system as a whole (including solvent) and an *ab initio* one for the site of interest (usually the chromophore).

MOLCAS (http://www.teokem.lu.se/molcas/introduction.html): [33] It is a software developed at Lund (Sweden) as a set of programs that will allow an accurate *ab initio* treatment of very general electronic structure problems for molecular systems in both ground and excited states. It is available for Unix/Linux systems, MacOS and Windows under a Cygwin environment. *MOLCAS* contains codes for general and effective multiconfigurational SCF calculations at the Complete Active Space (CASSCF) level, but also employing more restricted MCSCF wave functions (RASSCF). It is also possible, at this level of theory, to optimize geometries for equilibrium and transition states using gradient techniques and to compute force fields and vibrational energies. Also of interest for an accurate description of a photochemical reaction is the possibility to carry out CASPT2 computations. The CASPT2 approach has become especially important in studies of excited states and spectroscopic properties of large molecules, where no other *ab initio* method has, so far, been applicable. It also includes a program (RASSI) that can be used to compute transition dipole moments in spectroscopy. It is also possible to model solvent effects by adding a reaction field Hamiltonian (PCM).

COLUMBUS (http://www.univie.ac.at/columbus/): [34] It is a collection of programs for high-level *ab initio* molecular electronic structure calculations. The programs are designed primarily for extended multi-reference calculations on electronic ground and excited states of atoms and molecules. It allows the computation of automatic searches for minima on the crossing seam (conical intersections) and the topography of conical intersections. This program can be obtained free of charge.

NWChem (http://www.emsl.pnl.gov/capabilities/computing/nwchem/) and TURBO MOLE (http://www.turbomole-gmbh.com/) also allow for some computations of excited states, specially using TD-DFT methods.

In addition to the before mentioned computer packages, visualization software is not only advisable but sometimes necessary. The progress of critical points optimization can be easily followed by any of the visualization programs available. These programs allow to track the convergence to transition structures and minima, but in the case of a photochemical reaction path it is also useful to represent graphically other types of data. The software in this subsection can in any case perform a graphical representation of common data. Thus, only relevant information regarding the computation of photochemical reactions will be discussed.

MOLDEN (http://www.cmbi.ru.nl/molden/): It reads directly the GAUSSIAN or MOLCAS output and it is capable of displaying molecular orbitals which is very useful for the selection of the appropriate orbitals for the active space. Molden has also a powerful Z-matrix editor which gives full control over the geometry and allows to build molecules from scratch. This program is free for academic use and it is available as source or executable for a variety of operating systems.

GaussView (http://www.gaussian.com/g_prod/gv5.htm): It is a graphical interface for GAUSSIAN. It is sold separately and it can be used to prepare the input, control the calculation and read the output. Among different functionalities for other types of calculations, GaussView can be used to prepare the active space by graphically selecting the orbitals and to follow a conical intersection optimization.

Molekel (http://molekel.cscs.ch/wiki/pmwiki.php/Main/HomePage): It can be used to represent orbitals and follow optimizations. It can prepare animations of atoms, molecules, trajectories and surfaces and export them in different formats with high resolutions. It is free of charge and available for Linux, Windows and MacOS.

Moplot (http://www-chem.unifr.ch/tb/moplot/moplot.html): It can provide fast 2D representations of molecular orbitals. Although not as detailed nor as graphically attractive as other software, this program allows for a fast visualization of orbitals which is useful when a rapid inspection of a list or orbitals (for instance preparing an active space) is needed. Gradient and derivative coupling vectors from a conical intersection calculation can also be graphically represented. It is available free of charge for Linux, Windows and MacOS.

6. Case Study: Photocycloaddition of Imines to Alkenes

In this section, some examples of computations of photochemical reactions will be shown. The scope of this section is not to provide an in-depth treatment of the calculation technical issues (the reader can refer to the works cited in previous sections) nor to describe the photochemical results (see cited articles below) but to illustrate the process of computing a photochemical process. Some relevant features of the calculations will be discussed together with the relationship between computational and experimental data.

While the [2+2] photocycloaddition of alkenes to C=C, C=O and C=S double bonds are well known, the related cycloaddition of olefins to imines is less common and scarcely used in organic synthesis. In order to shed some light on the reaction mechanism and in an effort to turn this reactivity into a more general tool, a theoretical exploration of the cycloaddition of imines to alkenes was carried out. The first factor to consider is the choice of an appropriate chemical structure. Successful photocycloadditions of imines to alkenes usually require the imine moiety to be constrained in a five or six-membered ring. Thus, pyrroline was chosen as the simplest system to be computed. On the other hand, isoxazoline allows to consider the effect of an electron donor group within the system. Finally two substituted isoxazolines were also computed in order to consider the effect of substitution. Calculated structures are show in Figure 1.

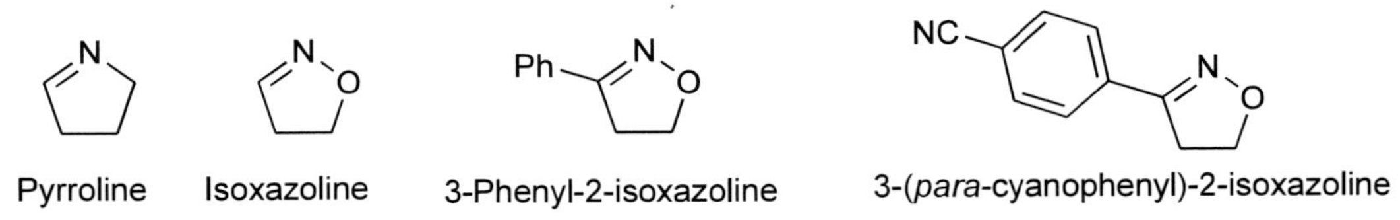

Pyrroline Isoxazoline 3-Phenyl-2-isoxazoline 3-(*para*-cyanophenyl)-2-isoxazoline

Figure 1. Model structures.

As described in the previous section, the next step is to decide which orbitals should be included within the active space. In this case, four different structures were computed. Thus, the active space should reflect the modifications in the chemical structure to correctly evaluate the inclusion of the oxygen atom in isoxazoline and the phenyl rings in the substituted isoxazolines. For every compound, the general advice was followed including all the orbitals that could potentially affect the photochemical reaction course. In the case of pyrroline an active space of 6 electrons in 5 orbitals was chosen (π and π^*, σ and σ^* for the C-N bond and the nitrogen lone pair). Test calculations revealed that, not surprisingly, the C-N σ bond remained fully occupied and unaltered along the reaction and could be removed from the active space. Thus, for isoxazoline, the set of orbitals included the π, π^* and nitrogen atom n orbitals (4 electrons in 3 orbitals). Oxygen orbitals showed no relevant effects in the active space selection. For the isoxazoline derivatives, the active space included 10 electrons in 9 orbitals (4 π, 4 π^* and n orbital for nitrogen). In all cases, two more electrons and orbitals were included for the reaction with ethylene, which in the larger case gave an active space of (12, 11). Initial calculations to prepare the active space were done by using the Gaussian 03 program package and MOLDEN as the visualization software.

After selecting the chemical structures and active spaces, the vertical absorption was computed. As said before, dynamic correlation has to be included in order to obtain reliable energetics. This was done by computing the oscillator strengths and energies associated with the transitions among states using the CASPT2 methodology implemented in MOLCAS. Pyrroline is not experimentally accessible, so comparison had to be done with related compounds including the pyrroline moiety. Once the method was validated, critical points, *i.e.* ground and excited state minima and conical intersection points were computed at the CASSCF level by using GAUSSIAN and energy was re-evaluated at the CASPT2 level using MOLCAS.

Comparison with experimental data allowed to rule out the presence of relevant species in the triplet state along the reaction course, so the discussion could be limited to states of singlet multiplicity. In every case, vertical absorption leads to the S_2 state with oscillator strength values of *ca.* 0.3, while S_1 is an excited dark state. Thus, at least two different conical intersection points could be expected, S_2/S_1 and S_1/S_0. For the structures under study, the S_2/S_1 conical intersection point can be easily reached after relaxation in the S_2 potential energy surface with minimal structural changes. S_1 is then populated and a minimum could be found for every compound and a relatively close (both in energy and structure) S_1/S_0 conical intersection point. The energy difference between the S_1 minimum and the conical intersection point is the key factor controlling the possibility of fast deactivation to the ground state. When these two points have similar energies and structures, the system can reach the conical intersection from the excited state minimum quite easily. Thus, fast decay can take place and the molecule won't have the chance to react in its excited state with other reactants (alkenes in this case). However, if an energy barrier separates the minimum from the conical intersection point, the system will spend some time on the excited state (depending on the value of the energy difference) and a photoreaction could happen (photocycloaddition to alkenes).

The potential energy surface for the system including the imine compounds (see Figure 1) and ethylene (as the simplest alkene) was also carried out. This is important in order to check the presence of a reaction path connecting the reactants (imine and alkene) with the expected photoproducts (azetidine cycloadducts). Relevant conical intersection points were found in every case, thus assessing the possibility of forming the corresponding cycloadducts for all the imines. The vectors defining the conical intersection points were graphically represented and evaluated using MOplot. This allowed to ensure the relevance of the conical intersection points found.

Inspection of the relevant portions of the potential energy surface shows a main difference between these compounds. While for the pyrroline and the 3-phenyl-isoxazoline the conical intersection is located very close in energy to the minimum in the excited state, in the case of isoxazoline and 3-(*para*-cyanophenyl)-2-isoxazoline the conical intersection point is higher in energy than the minimum. This means that in the first two cases the system will rapidly decay without possibility to photoreact, while in the last two cases, before the decay to the ground state, the system will take some time in the exited state. Thus, only isoxazoline and 3-(*para*-cyanophenyl)-2-isoxazoline will be able to photocycloadd to alkenes.

A detailed description of the results can be found [35] but, in summary, the different rates between two possible reactions should explain the experimental data. In the case of 3-phenyl-2-isoxazoline the aromatic ring stabilizes the CI point and fast deactivation may not allow the cycloaddition to take place – thus the azetidine product was not detected experimentally. However, when deactivation is not that fast due to the presence of an energy barrier, cycloaddition is competitive and the cycloadduct becomes the main product. This is the case for isoxazoline (where an aromatic ring is not present) and 3-(*para*-cyanophenyl)-2-isoxazoline (where the cyano group counteracts the effect of the aromatic ring). The available experimental data agree well with this observation in that only *p*-cyano- or *p*-carboxymethyl-substituted isoxazolines have been reported to give cycloaddition, while other types of substituents render the isoxazoline inactive toward alkenes. This result could be helpful for synthetic chemists since it provides an explanation for the experimental data together with some predictions about potentially useful substrates.

In order to fully understand the process and the structural features needed for a successful photocycloaddition of imines to alkenes, the geometries of the relevant conical intersection points were further analyzed. [36] The key feature of these geometries is the triangular shape of the C=N-O moiety, which is planar in the ground state. This deformation causes the ground-state energy to increase while the excited state is stabilized, a situation that leads to the conical intersection. The only distinction between 3-phenyl-isoxazoline and 3-(*para*-cyanophenyl)-2-isoxazoline is the cyano group. The electron-withdrawing effect of the CN transmitted by the phenyl ring causes a partial positive charge to appear on the carbon atom of the C=N-O moiety in 3-(*para*-cyanophenyl)-2-isoxazoline. This partial positive charge interacts with the O atom, causing the C-O distance to shorten from 1.830 Å in 3-phenyl-isoxazoline to 1.818 Å in 3-(*para*-cyanophenyl)-2-isoxazoline. This deformation contributes to an increase in the energy of the conical intersection point, thus allowing the system to spend some time in the excited state and yield photocycloaddition. To check that the enlargement of the active space is not essential, further calculations on 3-(*para*-cyanophenyl)-2-isoxazoline including the π orbitals of the cyano group (CAS(14,11)) were carried out. Both geometries and relative energies remained almost unaltered. Thus, the orbitals of the cyano group can be safely left out of the active space.

The regio- and stereochemistry of the reaction using a model compound for the imine and different olefins was also studied. Methyl vinyl ether, dimethylvinyl amine and acryladehyde were chosen to study the regiochemistry of this photocycloaddition and the effect of the electronic nature of the olefin. It was found that the regiochemistry is completely dependent on the alkene used, that is, the electron nature of its substituents, as illustrated by the fact that the regioisomer obtained is different when olefins substituted by electron withdrawing (aldehyde) or electron-releasing groups (amine, ether) are involved. This was shown by the relative energies of the different conical intersection points which lead to the formation of the corresponding photoproducts in the ground state. A preference for a particular stereoisomer was not found, and as a consequence, mixtures of stereoisomers were computationally predicted. A high correlation was found between data obtained in the calculations and the equivalent experiments, a finding that supports the mechanistic proposals. These results also provide an explanation for previous experimental data and could be used as a guide for the generalization of the photocycloaddition of imines to alkenes.

Conclusion

Throughout this chapter a basic introduction to the computational tools currently available for the study of organic photochemical reactions has been presented. A particular emphasis has been placed in the use of the CASPT2//CASSCF methodology as far as it provides a balance between accuracy and the size of the systems to be treated. By including dynamic correlation after the CASPT2 correction, energies comparable with experimental data can be obtained. As a drawback, the dimension of the system, both in terms of number of atoms and size of the active space, can reach a practical limit with relative ease. In those cases, both RASSCF and TD-DFT methods can provide an alternative, although less accurate results are expected. It is hoped that in the near future, the rapid evolution of hardware capabilities would allow for the computation of bigger systems and bigger active spaces at the high-level CASPT2 method. At the same time, the development in theories and their implementation in new computer codes will also contribute to improved software with increased capabilities and features. Hopefully soon enough, computational photochemistry will be able to treat with accuracy the same systems that have already proven valuable in the experimental photochemical reactions. New and exciting opportunities will emerge from the collaborative effort between both disciplines.

ACKNOWLEDGMENT

Continuous support from the Ministerio de Ciencia e Innovación, the Comunidad Autónoma de La Rioja and Universidad de La Rioja is gratefully acknowledged.

REFERENCES

[1] Van der Lugt, W. T. A. M.; Oosteroff, L. J. *J. Am. Chem. Soc.* 1969, *91*, 6042.
[2] Kandori, H.; Katsuta, Y.; Ito, M.; Sasabe, H. *J. Am. Chem. Soc.* 1995, *117*, 2669.
[3] Wang, Q.; Schoenlein, R. W.; Petenu, L. A.; Mathies, R. A.; Shank, C. V. *Science* 1994, *226*, 422.
[4] Teller, E. *Isr. J. Chem.* 1969, *7*, 227.
[5] Bernardi, F.; De, S.; Olivucci, M.; Robb, M. A. *J. Am. Chem. Soc.* 1990, *112*, 1737.
[6] Bernardi, F.; Olivucci, M.; Robb, M. A. *Chem. Soc. Rev.* 1996, *25*, 321.
[7] Migani, A.; Olivucci, M. In *Conical Intersections: Electronic Structure, Dynamics and Spectroscopy*; Domcke, W., Yarkony, D., Köppel, H., Eds.; World Scientific: Singapore, 2004.
[8] Foresman, J. B.; Frisch, A. *Exploring Chemistry with Electronic Structure methods*; Gaussian, Inc.: Pittsburg, 1996.
[9] Hehre, W. J.; Radom, L.; Schleyer, P. v. R.; Pople, J. A. *Ab initio Molecular Orbital Theory*; Wiley: New York, 1986.
[10] Szabo, A.; Ostlund, N. S. *Modern Quantum Chemistry: Introduction to Advanced Electronic Structure Theory*; Dover Publications: Mineola, 1996.
[11] Roos, B. O. In *Adv. Chem. Phys. (Ab Initio Methods in Quantum Chemistry-II)*; Lawley, K. P., Ed.; Wiley: New York, 1987; Vol. 69.
[12] Roos, B. O.; Linse, P.; Siegbahn, P. E. M.; Blomberg, M. R. A. *Chem. Phys.* 1981, *66*, 197.
[13] Roos, B. O.; Andersson, K.; Fülscher, M. P.; Serrano-Andrés, L.; Pierloot, K.; Merchán, M.; V., M. *J. Mol. Struct.* 1996, *388*, 257.
[14] Görling, A. *Phys. Rev. Lett.* 1999, *59*, 3359.
[15] Furche, F.; Burke, K. In *Ann. Rep. Comp. Chem.*; David, C. S., Ed.; Elsevier: 2005; Vol. 1.
[16] Hutter, J. *Chem. Phys.* 2003, *118*, 3928.
[17] *Computational Photochemistry*; Olivucci, M., Ed.; Elsevier: Amsterdam, 2005.
[18] *Computational Methods in Photochemistry*; Kutateladze, A. G., Ed.; Taylor & Francis: Boca Raton, 2005.
[19] Melloni, A.; Rossi Paccani, R.; Donati, D.; Zanirato, V.; Sinicropi, A.; Parisi, M. L.; Martin, E.; Ryazantsev, M.; Ding, W. J.; Frutos, L. M.; Basosi, R.; Fusi, S.; Latterini, L.; Ferré, N.; Olivucci, M. *J. Am. Chem. Soc.* 2010, *132*, 9310.
[20] Hack, M. D.; Truhlar, D. G. *J. Phys. Chem A* 2000, *104*, 7917.
[21] Trulhar, D. G.; Gordon, M. S. *Science* 1990, *249*, 491.
[22] Robb, M. A.; Garavelli, M.; Olivucci, M.; Bernardi, F. *Rev. Comp. Chem.* 2000, *15*, 87.
[23] González, C.; Schlegel, H. B. *J. Phys. Chem.* 1990, *94*, 5523.

[24] Schlegel, H. B. In *Ab initio Methods in Quantum Chemistry*; Lawley, K. P., Ed.; Wiley: New York, 1987.

[25] Garavelli, M.; Celani, P.; Fato, M.; Bearpark, M. J.; Smith, B. R.; Olivucci, M.; Robb, M. A. *J. Phys. Chem. A* 1997, *101*, 2023.

[26] Celani, P.; Robb, M. A.; Garavelli, M.; Bernardi, F.; Olivucci, M. *Chem. Phys. Lett.* 1995, *243*, 1.

[27] Garavelli, M.; Bernardi, F.; Cembran, A. in *Computational Photochemistry,* Olivucci, M., Ed.; Elsevier: Amsterdam, 2005.

[28] Blancafort, L.; Ogliaro, F.; Olivucci, M.; Robb, M. A.; Bearpark, M. J.; Sinicropi, A. in *Computational Methods in Photochemistry,* Kutateladze, A. G., Ed.; Taylor & Francis: Boca Raton, 2005.

[29] Olsen, J.; Roos, B. O.; Jorgenssen, P.; Jensen, H. J. A. *J. Chem. Phys.* 1998, *89*, 2185.

[30] Marian, C. M. In *Spin-Orbit Coupling in Molecules*; Lipkowitz, K. B., Boyd, D. B., Eds.; Wiley-VCH: New York, 2001; Vol. 17.

[31] Frisch, M. J.; Trucks, G. W.; Schlegel, H. B.; Scuseria, G. E.; Robb, M. A.; Cheeseman, J. R.; Scalmani, G.; Barone, V.; Mennucci, B.; Petersson, G. A.; Nakatsuji, H.; Caricato, M.; Li, X.; Hratchian, H. P.; Izmaylov, A. F.; Bloino, J.; Zheng, G.; Sonnenberg, J. L.; Hada, M.; Ehara, M.; Toyota, K.; Fukuda, R.; Hasegawa, J.; Ishida, M.; Nakajima, T.; Honda, Y.; Kitao, O.; Nakai, H.; Vreven, T.; Montgomery, Jr., J. A.; Peralta, J. E.; Ogliaro, F.; Bearpark, M.; Heyd, J. J.; Brothers, E.; Kudin, K. N.; Staroverov, V. N.; Kobayashi, R.; Normand, J.; Raghavachari, K.; Rendell, A.; Burant, J. C.; Iyengar, S. S.; Tomasi, J.; Cossi, M.; Rega, N.; Millam, N. J.; Klene, M.; Knox, J. E.; Cross, J. B.; Bakken, V.; Adamo, C.; Jaramillo, J.; Gomperts, R.; Stratmann, R. E.; Yazyev, O.; Austin, A. J.; Cammi, R.; Pomelli, C.; Ochterski, J. W.; Martin, R. L.; Morokuma, K.; Zakrzewski, V. G.; Voth, G. A.; Salvador, P.; Dannenberg, J. J.; Dapprich, S.; Daniels, A. D.; Farkas, .; Foresman, J. B.; Ortiz, J. V.; Cioslowski, J.; Fox, D. J. *Gaussian 09; Gaussian, Inc.*: Wallingford CT, 2009.

[32] Dapprich, S.; Komáromi, I.; Byun, K. S.; Morokuma, K.; Frisch, M. J. *THEOCHEM* 199, *462*, 1.

[33] Karlström, G.; Lindh, R.; Malmqvist, P.-Å.; Roos, B. O.; Ryde, U.; Veryazov, V.; Widmark, P.-O.; Cossi, M.; Schimmelpfennig, B.; Neogrady, P.; Seijo, L. *Computational Material Science* 2003, *28*, 222.

[34] H. Lischka, R. Shepard, I. Shavitt, R. M. Pitzer, M. Dallos, Th. Müller, P. G. Szalay, F. B. Brown, R. Ahlrichs, H. J. Böhm, A. Chang, D. C. Comeau, R. Gdanitz, H. Dachsel, C. Ehrhardt, M. Ernzerhof, P. Höchtl, S. Irle, G. Kedziora, T. Kovar, V. Parasuk, M. J. M. Pepper, P. Scharf, H. Schiffer, M. Schindler, M. Schüler, M. Seth, E. A. Stahlberg, J.-G. Zhao, S. Yabushita, Z. Zhang, M. Barbatti, S. Matsika, M. Schuurmann, D. R. Yarkony, S. R. Brozell, E. V. Beck, and J.-P. Blaudeau, *COLUMBUS, an ab initio electronic structure program, release* 5.9.1, 2006.

[35] Sampedro, D. *ChemPhysChem* 2006, *7*, 2456.

[36] Sampedro, D.; Soldevilla, A.; Campos, P. J.; Ruiz, R.; Rodríguez, M. A. *J. Org. Chem.* 2008, *73*, 8331.

In: Photochemistry
Editors: Karen J. Maes and Jaime M. Willems

ISBN: 978-1-61209-506-6

Chapter 8

RECENT PROGRESS IN CHIRALITY RESEARCH USING CIRCULARLY POLARIZED LIGHT

Tsubasa Fukue*

National Astronomical Observatory of Japan, Tokyo, Japan

ABSTRACT

We review recent studies of molecular chirality using circularly polarized light, along with the birth and evolution of life and planetary systems. Terrestrial life consists almost exclusively of one enantiomer, *left-handed* amino acids and *right-handed* sugars. This characteristic feature is called homochirality, whose origin is still unknown. The route to homogeneity of chirality would be connected with the origin and development of life on early Earth along with evolution of the solar system. Detections of enantiomeric excess in several meteorites support the possibility that the seed of life was injected from space onto Earth, considering the possible destruction and racemization in the perilous environment on early Earth. Circularly polarized light could bring the enantiomeric excess of prebiotic molecules in space. Recent experimental works on photochemistry under ultraviolet circularly polarized light are remarkable. Asymmetric photolysis by circularly polarized light can work for even amino acid leucine in the solid state. Amino acid precursors can be asymmetrically synthesized by circularly polarized light from complex organics. Astronomical observations by imaging polarimetry of star-forming regions are now revealing the distribution of circularly polarized light in space. Enantiomeric excess by photochemistry under circularly polarized light would be small. However, several mechanisms for amplification of the excess into almost pure enantiomers have been shown in experiments. When enantiomeric excess of amino acids appears in the prebiotic environment, it might initiate the homochirality of sugars as a catalyst. Astrobiological view on chirality of life would contribute to understanding of the origin and development of life, from the birth to the end of stars and planetary systems in space. Deep insights on terrestrial life, extrasolar life, and the origin of life in the universe, would be brought by consideration of both the place where life is able to live, *the habitable zone*, and the place where life is able to originate, what we shall call *the originable zone*.

* National Astronomical Observatory of Japan, 2-21-1 Osawa, Mitaka, Tokyo 181-8588, Japan, E-mail: tsubasa.fukue@nao.ac.jp

1. Introduction

1.1. Astrobiology

There are a lot of astronomical objects that human beings have been observing so far. Findings of new type of celestial bodies have been encouraging us to make a step toward new idea. In the end of 20th century, we faced the new historical situation. Mayor and Queloz discovered a planet candidate around a solar-like star, 51 Pegasi, in 1995, for the first time [1]. They observed spectrum of a star, not a planet, and detected the oscillation of the spectrum due to Doppler effect because both a star and its planet orbit their barycenter. This planet is thought to be one of hot Jupiters. The separation between the host star and the planet is about 1/20 of that between our Sun and Earth, and the mass of the planet is about a half of Jupiter in our solar system. While life such as us seems not to live on this reckless planet, this discovery encouraged us to consider the existence of planets around stars other than our Sun, extrasolar planets (ESPs). Following this success, Charbonneau and co-workers observed a star by photometry in 2000, whose ESP candidate had been indicated by Doppler effect [2]. They discovered that the star transiently dims because the planet occults the host star when the planet transits between the host star and us. These detections of both Doppler effect and transit on a star indicate the existence of ESPs strongly. Astronomers have also tried to directly detect ESPs by imaging, suffering from darkness of ESPs very close to the bright host star, uncertainties of mass of detected objects, and uncertainties whether the ESP candidate accompanies the star. In 2008, Kalas and colleagues reported the direct detection of an ESP using Hubble Space Telescope [3]. Remarkably, they compared direct images of an ESP observed in 2004 and 2006, and showed the orbital movement of the ESP. In September 2010, the number of all the ESP candidates is approaching 500. These discoveries in 15 years have impacted us. Now, it would be thought that 8 planets in our solar system and ESPs are not special in space. It is possible that another Earth would exist around another Sun. Astronomers are searching ESPs, and also biomarkers which indicate the existence of extrasolar life [4]. When observational data of a lot of ESPs [5] and their circumstances are available in future, it would lead to understanding their many evolutional stages, from their birth to end, along with life on them. Astronomical view now involves origins of life on ESPs, which might involve the hint of the origin of the life on Earth. Naturally, the new academic field, *Astrobiology*, has been launched in face of the new interdisciplinary stage involving extraterrestrial life and origins of various life including us in space.

1.2. Habitability and Originability

The discoveries of ESPs have led to deep consideration of the possible area where life is able to live, *a habitable zone* [6, 7]: a habitable zone in a stellar system [8], a habitable zone in a galaxy [9], a habitable zone in the universe, involving planets [10] and moons [11], and various type of the host star [12, 13]. In view of habitability, the spatial distribution of temperatures, and the spatial distribution of elements for terrestrial planets and terrestrial life, around the host star, along with the evolution of the planetary atmosphere, are sometimes focused on.

However, it is still unknown whether the origin of life is always generated in a habitable zone. It is important to consider not only *a habitable zone*, where life is able to live, but also the place where life is able to originate, what we shall call *the originable zone*. A habitable zone and an originable zone would not be equal. However, they would have partial overlap. When biomarkers of ESPs are directly observed in future, consideration of both *habitability* and *originability* for life in space would contribute to confirmation of the biomarkers.

Generally, we do not know extrasolar life on ESPs, even extrasolar life on earth-type planets around sun-like stars. There are also ESPs around stars whose types are different from our Sun. There might be various type of life, and various route to life in space. In this chapter, originability is focused on, rather than habitability.

1.3. Enantiomer and Homochirality of the Terrestrial Life

Life on Earth consists almost exclusively of one enantiomer, *left-handed* amino acids and *right-handed* sugars. An enantiomer is one of a pair of stereoisomers. Atoms of an enantiomer are bonded similarly to that of another enantiomer, and then chemical property of an enantiomer resembles with that of another enantiomer. On the other hand, the spatial configuration of atoms of an enantiomer is set as a mirror image of that of another enantiomer. Enantiomers of amino acids consist of left-handed amino acids and right-handed amino acids. Usual products of amino acids are racemic, which include almost the same quantity of left- and right-handed amino acids. However, the terrestrial life uses almost only left-handed amino acids. This characteristic feature of biomolecular is called *homochirality*, whose origin is still unknown. Some right-handed amino acids work for life, and some right-handed amino acids would be involved with aging and disease [14, 15, 16, 17]. Enantiomers are essential for terrestrial life.

1.4. Early Earth and Early Life

The route to homogeneity of chirality, which is characteristic of terrestrial life, would be involved with the origin and development of life [18, 19, 20, 21]. The age of our solar system is thought to be about 4.6 billion years, indicated by analysis of meteorites [22]. When we look back at the initiation of life, we would watch primeval Earth and our solar system, where life and its material were formed. Earth is thought to be formed by accumulation of planetesimals and protoplanets [23], as well as these events seem to occur in extrasolar planetary systems [24]. The accumulation is thought to have brought thermal energy from gravitational energy to early Earth, about 4.6 billion years ago. There could be a molten magma ocean on early Earth [25, 26], high temperature, and bombardment from comets and asteroids [19], as well as magma oceans were present on other bodies in the early solar system [27]. In particular, the giant impact [28], the impact between protoplanets accreted from planetesimals [29] which would yield Moon [30, 31], would bring a completely molten magma ocean [32, 33], which would yield the core and mantle of Earth [34, 35]. Such a heat environment would destroy enantiomers, even if homochirality occurred on start of Earth. Therefore, the origin of life with homochirality would be initiated after Earth cooled down, in short time before the first terrestrial life appeared (probably before about 3.8 billion years ago) [36, 18, 19]. The mechanism of production of homochirality is very controversial, involving many chemical routes on Earth, and in space [18, 19].

1.5. The Birth Place of Our Solar System

In space, low mass stars such as our Sun can be formed in two-type star-forming regions: a massive star-forming region where both high-mass and low-mass stars are formed, or a relatively isolated region where only low-mass stars are formed [37, 38]. The decay products of short-lived radionuclides in meteorites indicate the birth place of our solar system [37]. ^{60}Fe has a half-life of ~1.49 million years. This half-life is very short, in comparison with the age of our solar system (~4.6 billion years). ^{60}Fe is a neutron-rich isotope that is formed exclusively in stars, e.g., in core collapse supernova which is the end of a massive star. The presence of ^{60}Fe in primitive meteorites is confirmed, suggesting that a supernova explosion occurred near our Sun [39, 40, 41, 42]. This indicates that the birth place of our solar system was located in a massive star-forming region. The nearest massive star-forming region is the Orion nebula, and then it is a familiar and important target in astronomy.

1.6. Linear and Circular Imaging Polarimetry in Astronomy

Because we can not directly explore astronomical objects far from Earth, it is essential to investigate the property of light from the astronomical objects. The property of light involves wavelength, and polarization [43, 44]. When the light is scattered, the light is often polarized depending on the scattering angle and the optical property of scattering body. In many cases in the universe, the light from stars is scattered at the circumstellar matter and polarized. In particular, during the formation of star/stars and its planetary system, dust grains around the host star/stars evolve and polarize the stellar light. The polarization provides useful information for the circumstellar structures and the property of dust grains.

The polarization status generally consists of linear and circular polarizations, which are obtained with linear and circular polarimetry in astronomy, respectively. In partially polarized light, circular polarization (CP) can not be derived from linear polarization (LP), vice versa. To obtain clear view of polarization status, it is preferable to perform linear and circular polarimetry with the same telescope. The 1.4-m IRSF telescope, which is set up with the SIRIUS camera [45] and its polarimeter (SIRPOL) [46], brought the first performance of wide-field imaging linear and circular polarimetry in near-infrared (NIR). The IRSF telescope is located at the South African Astronomical Observatory. The observations by IRSF/SIRPOL are reviewed in section 4.

In the star-forming region, the cloud harbors the young forming star or stars. When a cavity exists around the forming star, the light from the forming star preferably propagates in the cavity. The scattered light at the wall of the cavity will be linearly polarized. In addition, the cavity decreases opacity in the line of sight. Therefore, when we observe the star-forming region from the outside of the region by imaging polarimetry, the observed spatial distribution of LP indicates the existence of a cavity in the cloud [44]. In fact, on star formation, a bipolar cavity around the forming star is often produced in a parent cloud, by outflows and jets from the forming star. As numerical simulations show [47], the LP region can appear depending on inclination of the system. Further, the direction of LP tends to be centrosymmetric around the major light source [44]. The star-forming region often has many young stars. The direction map of observed LP indicates the dominant light source of the observed region.

Although discussion of many proposed mechanisms leading to homochirality of the terrestrial life is useful for consideration of the terrestrial life on our Earth and various extrasolar life on ESPs around nearby stars which would be directly observed in future [48], in this chapter, we focus on the idea of the extraterrestrial origin leading to homochirality, considering circular polarization in space. It would be important to consider a consistent theory with the birth and evolution of Earth in the solar system. In section 2, the organic materials and water in space is reviewed, as possible necessary material for life and chirality. In section 3, enantiomeric excess (EE) in meteorites and asymmetric photochemistry by circularly polarized light (CPL), and amplification of EE are reviewed. In section 4, the source of CPL in space is reviewed. These are summarized in section 5.

2. Organic Materials and Water in Space

2.1. Extraterrestrial Organic Matter, Amino Acid, Amino Acid Precursor, and Sugar

Amino acids or amino acid precursors, molecules that provide amino acids after acid hydrolysis, are thought to exist in space [49]. Organic matters seem to be popular in space, considering detections in our solar and interstellar materials [50, 49].

In our solar system, the carbonaceous chondrite meteorites, which are thought to be the most ancient meteorites indicating the starting materials of the presolar molecular cloud [51, 52], present organic matters including amino acids [53, 54, 55, 56, 57]. Interplanetary dust particles, which are thought to be samples of primitive objects [58, 59], also show organics [60, 61, 62, 63]. The recent returned samples by spacecraft from comet 81P/Wild 2, which can be the accreted materials during the formation of our solar system, have showed organics [64, 65, 66, 67] including amino acid [68], glycine confirmed by the stable carbon isotopic ratios [69].

Glycine (NH_2CH_2COOH) is the simplest and achiral (i.e., not chiral) amino acid, which is used in life. Since 1979 [70], glycine has been investigated in interstellar medium involving controversy [71, 49]. The detection of interstellar glycine was proposed in several interstellar clouds including the Orion KL [72], followed by different perspectives [73].

Amino acetonitrile (NH_2CH_2CN), which is a possibly direct amino acid precursor of glycine, was detected in one of Galactic major center sources of activity, Sagittarius B2(N), indicating of formation by grain surface chemistry [74]; Sagittarius B2 is a very massive (several million solar masses) and extremely active region of massive-star formation near the Galactic center (projected distance of about 100 pc from the Galactic center).

Organic molecules have been detected in interstellar and circumstellar medium [49], for example, in high mass star forming region such as the Orion KL [75, 76], low mass star forming region such as IRAS 16293-2422 [77, 78], and protoplanetary disks [79, 80].

Interstellar glycolaldehyde (CH_2OHCHO), the simplest possible aldeyde sugar, was detected in the Sagittarius B2(N) [81]. Toward the Sagittarius B2(N-LMH), interstellar ethylene glycol ($HOCH_2CH_2OH$), the sugar alcohol of glycolaldehyde, was detected [82]; The three-carbon keto ring, cyclopropenone (c-H_2C_3O) was detected, with no detection of the three-carbon sugar glyceraldehyde [83].

2.2. Extraterrestrial Water

Water, which could contribute to development of chirality and life, seems to be broadly distributed in the universe [84], apart from its phase. In our solar system, our Earth has solid, liquid, and vapor water. High resolution images of Martian surface by the Mars Global Surveyor suggested the liquid water seepage and surface runoff on Mars [85]. The ultraviolet imaging spectrograph of the Cassini space craft revealed a water vapor plume in the south polar region of Saturn's moon, Enceladus [86]. The observation of the disk-averaged light of another moon of Saturn, Titan, indicated water icy bedrock [87]. The solid water ice deposits were directly detected on the surface of comet 9p/Tempel 1 on the Deep Impact mission, suggesting that the surface deposits are loose aggregates [88]. In our outer solar system, the presence of crystalline water ice was reported on the Kuiper belt object (50000) Quaoar in the Kuiper belt, which is consist of solid bodies beyond Neptune and yields comets [89].

In outside of our solar system, the water vapor was detected in the atmosphere of an ESP, a transiting hot Jupiter [90]. Water vapor and organic molecules were indicated in the inner protoplanetary disk around a classical T Tauri star, which is thought to be a young sun-like star, using the Spitzer Space Telescope [91]. The water ice grains were detected in the circumstellar disk around a Herbig Ae star, which is thought to be a young massive star, using Subaru Telescope [92]. H_2O maser emission is detected in massive star-forming regions [93] including the Orion KL [94]. The ortho-H_2O emission was detected in molecular cloud cores [95]. Numerical simulations of terrestrial planet formation indicate the possibility of broadly distributed water in planetary system, depending on conditions [96].

3. Enantiomeric Excess and Asymmetric Photochemistry

3.1. Enantiomeric Excess in Meteorites and the Late Heavy Bombardment

Detections of enantiomeric excesses (EEs) of amino acids in several meteorites (Murchison, Murray, Orgueil) have been reported, with small EEs of the same handedness as terrestrial life [97, 98, 99, 100, 101, 56]. The detections of EEs from meteorites which have fallen down to Earth indicate that EEs in meteorites survive through infall to Earth's atmosphere, and EEs in meteorites survive through long time after EEs were formed, as the stability of chiral amino acids against radionuclides decay in comets and asteroids in 4.6 billion years is investigated by experiments [102]. The detections of EEs support the hypothesis of the extraterrestrial origin of life which was seeded by delivery of organics from outer space.

If meteorites (and small objects in our solar system) brought seeds of life to Earth, the efficiency and the period is to be considered. The first terrestrial life appeared on Earth probably before about 3.8 billion years ago, indicated by sedimentary protolith [103, 104] and sedimentary rocks [105] from the Isua supracrustal belt in west Greenland. About 3.9 billion years ago, a lot of meteorites fell down on Earth's moon, and then Earth [106], as also indicated by sedimentary rocks from Isua greenstone belt in west Greenland [107]. This is called *the late heavy bombardment phase*, which would occur throughout the inner solar system [108]. In this late moment from the birth of Earth (about 4.6 billion years ago), many drops of meteorites with peculiar EEs could bring peculiar EEs over Earth before the emergence of life. Coincidentally, the ocean of Earth probably appeared before about 3.8 billion years ago [109]. It could be the period for the appearance of terrestrial life. The late heavy bombardment is thought to be brought due to the migration of giant planets [110, 111]. If so, the late heavy bombardment could be connected with evolution of some type of planetary systems and could occur in another planetary system.

3.2. Amino Acid and Amino Acid Precursor by Photochemistry

The extraterrestrial origin leading to homochirality requires the material and the mechanism of production of EEs of amino acids, *in space*. Regarding the material, as denoted in section 2, organic materials involving amino acids, amino acid precursors, or their elements can be available in space, apart from chirality.

Apart from EE, amino acids (or amino acid precursors) have been obtained by photochemistry in laboratory experiments for interstellar chemistry. In 2002, the analogues of icy interstellar grains (an ice film consisting of amorphous H_2O, NH_3, CH_3OH, and HCN) at 15 K were irradiated by ultraviolet (UV) light in vacuum. After warming the ices to room temperature, and after hydrolysis, racemic amino acids (glycine, alanine, and serine) were obtained [112]. In another experiment, an interstellar ice analogue (an ice mixture containing H_2O, CH_3OH, NH_3, CO, and CO_2) was irradiated by UV light at 12 K in vacuum. After warming the system to room temperature, 16 amino acids were identified [113]. Only after acid hydrolysis, amino acids were detected at considerable amounts. Results in these two experiments were also confirmed in experiments in 2007 [114]. In 2005, an interstellar ice analogue containing H_2O, CH_3OH, NH_3, CO, CO_2 were irradiated by UV light, at 12 K in vacuum. After warming up to room temperature, N-heterocyclic molecules and amines were detected in water extracts [115]. Carboxylic acid salts as part of the refractory products were shown using infrared spectroscopy in 2003 [116].

In 2007, an ice mixture containing H_2O, CO_2, and NH_3 was irradiated with UV light at 16 K in vacuum. This starting ice mixture did not contain any organic compound such as methanol (CH_3OH) and methane (CH_4). After 6 times repeat of warming up to room temperature, cooling down, and irradiation, finally, the proteinaceous amino acids were identified in the production [117]. A detailed analysis of amino acids which was produced by the UV irradiation of interstellar ice analogues was reported in 2008, highlighting the contribution of acid hydrolysis to yield amino acids [118]. The experiment using naphthalene ($C_{10}H_8$), the smallest polycyclic aromatic hydrocarbon (so called, PAH), was also performed with UV in vacuum at 15 K, producing amino acids. The naphthalene was mixed in an ice mixture of H_2O and NH_3 [119].

The mechanism for the formation of the amino acids glycine, serine, and alanine in interstellar ice analogs was investigated using isotopic labeling techniques in 2007, indicating the multiple pathways to amino acid formation [120]. In 2009, the structures of the products and their formation pathways were investigated using deuterium-labeling experiments, indicating the initial photochemical cleavages of C-H and N-H bonds, for glycine [121]. The efficiency of photochemical synthesis of glycine on the ice surfaces and steady-state equilibrium between photosynthesis and photodestruction of glycine were pointed out.

Photostability of amino acids against UV photodestruction were investigated [122], as well as other small biomolecules were investigated [123]. Amino acids may be preferably destructed in UV, so some protections [122] or escapes during accretion [124] may be necessary. On the other hand, aminoacetonitrile (H_2NCH_2CN), which is an amino acid precursor to the amino acid glycine, is more stable than amino acid against UV photolysis [125].

3.3. Enantiomeric Excess by Asymmetric Photochemistry

The laboratory experiments show that EE can be yielded by asymmetric photochemistry using CPL [126, 127]: asymmetric photolysis, asymmetric synthesis, and asymmetric photoisomerization.

In asymmetric photolysis [18, 19, 128], one of a pair of enantiomers is preferentially destructed under CPL, and then the other is enriched. The handedness of EE depends on the handedness of the CPL. Even elliptically-polarized light can induce asymmetric photolysis with a lesser degree than CPL [129]. In 2005, the amino acid leucine in the solid state was photolysed by UV CPL in vacuum, while other experiments were often performed in solutions [130]. The experiment showed the highest gain of ~2.6% in D-leucine.

Remarkably, several recent laboratory experiments have been performed with the interstellar analogues, although the interpretation of the results for effective mechanisms is more complex than in simple experiments. In 2006, the interstellar ice analogues (gas mixtures) were irradiated by UV CPL under interstellar-like conditions (80 K, ~10^{-7} mbar). However, very small EEs (at most ~1%) were produced, comparable with the detection limit (~1%) of the chromatography used in the experiment (GC-MS) [131].

In 2007, the possibility of asymmetric synthesis of amino acid precursors in interstellar complex organic under CPL was demonstrated in laboratory [132]. In their experiment, initially, gas mixtures (at room temperature) of carbon monoxide, ammonia and water, which are identified in the interstellar medium, were irradiated with high energy protons, yielding complex organic compounds. A liquid portion of the proton-irradiated sample was irradiated with UV CPL. Following acid hydrolysis, alanine showed EEs of +0.44% and -0.65% by right-handed CPL and left-handed CPL, respectively. Comparing an unhydrolyzed fraction with a product following acid hydrolysis, they speculated that combined amino acid analogs, rather than free amino acids, were present in the CPL-irradiated samples.

3.4. Asymmetric Amplification of Tiny Enantiomeric Excess for Amimo Acid and Sugar

Even if the EE is small as described in the previous section 3.3, the EE can be amplified by some process: asymmetric amplification [133, 134]. Experiments have shown that low EEs can be amplified by asymmetric autocatalysis (autocatalysed reactions) [135, 136]. In this reaction, a chiral product serves as a chiral catalyst for its own formation in the reaction. Asymmetric autocatalysis of 2-(*tert*-butylethynyl)-5-pyrimidyl alkanol showed that low EE (~0.00005%) was amplified to large EE (>99.5%) [137].

The small EEs can be amplified into solutions, because the solubility of an exclusive enaniomer is higher than that of the racemic compound crystal [138, 139, 140, 141]. This solubility-based amplification (solid-liquid phase behavior) is asymmetric aldol reaction and can occur in aqueous systems. Small EE (~1%) of serine was amplified to large EE (>99%) under solid-liquid equilibrium conditions [138].

Homogeneity of right-handed sugars may be initiated by low EEs of amino acids as a catalyst [142, 143, 144, 145]. The solubility-based amplification for amino acids in water [138] can also work for nucleosides, which would lead to the RNA world [146]. These reactions support that exogenous injection of low EEs of amino acids on early Earth yields the homochirality of amino acid and sugar, and then the terrestrial life.

4. Circularly Polarized Light Source in Space

4.1. Possible Sources of Circular Polarization

Neutron stars and magnetic white dwarfs seems not to be a CP source yielding of EEs for early solar system, considering no detection of CP in optical or few encounters with a molecular cloud or star-forming region [147, 124]. On the other hand, CP of young stellar objects (YSOs), which are thought to be young stars, has been detected by circular polarimetry. In table 1, the degree of CP of YSOs is summarized from previous circular polarimetry.

Table 1. The degree of CP (%) of young stellar objects in previous observations

Mass	Object name	The degree of CP (*color band*)	Reference
high	OMC-1 in Orion NGC 6334-V	17(K_n), 5(H), 2(J) 23(K)	[147], [150], [148] [151]
intermediate	HH 135-136 R CrA	8 (K_n), 3 (H), 2.5 (J) 5 (H)	[149] [152]
low	GSS 30 Cha IRN	1.7 (K_n), 0.8 (H) 1 (H)	[153] [154]
high	R Mon	0.4 (R)	[155]
low	HL Tau	no detection, <0.5 (J, H, K)	[156]
intermediate	PV Cep V633 Cas	< 1 (V, I) < 1 (V, I)	[157] [157]
low/intermediate	L1551 IR5	< 3 (V, I)	[157]
high	GL 2591	< 1 (I)	[157]
low	~hundreds point-like sources in Orion	<~1.5% (H, K_s)	[148]

The table 1 in [157] is updated in September 2010 (see also [158]). The nebulae in the first three columns (denoted by bold face) were detected by imaging polarimetry.

In September 2010, the *imaging* circular polarimetry of YSOs is still scarce, although hundreds of *point-like sources* in the core of the Orion nebula were performed in [148]. According to the previous observations, more massive YSOs appear to have larger CP. The possibility of the contribution of the stronger magnetic field with the formation of higher-mass stars was pointed out, which leads to more efficient alignment of dust grains [124, 149]. The Orion nebula is one of YSOs which have the largest CP.

Previous imaging circular polarimetry of YSOs (in Table 1) and numerical simulations producing CP in YSOs [159, 160, 47, 161, 162, 149] indicate that a YSO will usually produce a low net CP because it will have regions of right-handed and left-handed CP that cancel globally. Such patchy spatial distribution of right-handed and left-handed CP could yield patchy distribution of right-handed and left-handed EEs, so that meteorites in late heavy bombardment could bring inefficient EEs. To inject efficient EEs on early Earth, the entire irradiation on early solar system or its materials by one-handed CP would be necessary.

4.2. Wide Extension of Linear and Circular Polarization in the Orion Nebula

LP images of the Orion nebula in NIR using IRSF/SIRPOL revealed the large LP region, which extends over about 0.7 pc, in 2006 [163]. The three color bands were used: J (1.25μm), H (1.63μm), K_s (2.14μm). The nebulae emitting LP are located around the massive star-forming region, the BN/KL region. This extension of LP indicates the existence of large cavities around the young massive stars in the BN/KL region. Moreover, the observation showed the linearly polarized Orion bar, several small linearly polarized nebulae, and the low LP near the Trapezium. The Trapezium is a group of massive young stars and is located near the BN/KL region.

CP images of the Orion nebula in K_s band using IRSF/SIRPOL revealed the large CP region, which extends over about 0.4 pc, in 2010 (see Figure 1 in [148]). The observed CP extends over a region of about 400 times the size of the solar system, when the size of the solar system is assumed to be about 200 AU in diameter. This extension of the CP region is almost comparable to that of the LP region. This CP region is located around the BN/KL region. The degrees of CP range from +17% to -5%. The linearly polarized Orion bar in linear polarimetry shows no significant CP in circular polarimetry. The small linearly polarized nebulae in linear polarimetry show no significant CP. The aperture circular polarimetry of the 353 point-like sources, many of which are low-mass young stars, showed that the degree of CP for each source is generally small, less than ~1.5% (see Figure 2 in [148]).

Although the spatial distribution of CP within point-like sources is not spatially resolved in aperture polarimetry, the result in [148] showed that the point-like sources do not have generally large degree of CP. In other words, a point-like source of them does not emit large one-handed CP from its entire face to us. Even if a point-like source of them has inherent CP locally within it, the CP degree would be low as spatially resolved nebulae showed low CP degree in previous observations (see Table 1), or/and the spatial distribution of right-handed and left-handed CP would be patchy, in which the right-handed and left-handed CP are cancelled leading to the appearance of low CP degree when the spatial distribution of CP is not spatially resolved. Such spatial distribution of CP would lead to inefficient EEs on early Earth, as described in section 4.1.

The major contribution to production of the observed CP in the Orion nebula is thought to be dichroic extinction [164, 165]. The light from the central star/stars propagates inside the cavity around the central star/stars in the cloud. The light is scattered at the wall of the cavity, linearly polarized, and goes through the surrounding cloud. When the dust grains in the cloud are non-spherical and somewhat aligned, the incident linearly polarized light is (partially) circularly polarized. In other words, the aligned dust grains in the cloud behave like 1/4 wave plate for the incident LP. This situation can be connected with a simple relation between LP, CP and the color excess in the imaging observation. In fact, the observed correlation between LP, CP and the color excess agrees with the relation [165].

UV light in star-forming regions can not be directly observed. The dust grains drifting in star-forming regions prevent us to observe by UV light. Therefore, observations of star-forming regions are often performed in near-infrared. Numerical simulations to produce CP in a modeled space are helpful to investigate UV CP in star-forming regions [147, 162].

CONCLUSION

Detections of EEs in several meteorites support the possibility that the seed of life was injected from space onto Earth. CPL could bring the EE of prebiotic molecules in space, as shown in experimental works on photochemistry under UV CPL. EE by photochemistry under CPL would be small. However, several mechanisms for amplification of the EE into almost pure enantiomers have been shown in experiments.

Astronomical observations by imaging circular polarimetry of star-forming regions are now revealing the distribution of CPL in space. The observed significant CP in the core of the Orion nebula extends over a region of about 400 times the size of the solar system. If a solar system would be formed in such nebula and be irradiated by CP, EE would be produced by asymmetric photochemistry and be brought with meteorites and small objects onto an early planet/moon. This could result in life with homochirality.

The observed CP in the Orion nebula showed widely extended regions of both right-handed and left-handed CP. This result might indicate that the handednesses of possible EEs produced by CP in the Orion nebula are different among those CP regions, since the handedness of EE yielded by CP is dependent on the handedness of CP.

For deep insights on terrestrial life, extrasolar life, and the origin of life in the universe, it would be important to consider both the place where life is able to live, *the habitable zone*, and the place where life is able to originate, what we shall call *the originable zone*. A habitable zone and an originable zone would not be equal. However, they would have partial overlap.

The Orion nebula seems to be located in the galactic habitable zone. The Orion nebula harbors a lot of stellar systems, some of which would harbor the circumstellar habitable zone. When the CPL in the massive star forming region, the BN/KL region, contributes to the EEs and leads to the homochirality for the life with homochirality, the Orion nebula would harbors the originable zone for the life with homochirality. The Orion nebula, a young star-forming region, whose age is about one million years, seems to yield the extraterrestrial life in future. In billions of years, some extraterrestrial intelligences on some ESPs from the Orion nebula might watch the end of our Earth and us, who might watch the birth of them.

ACKNOWLEDGMENT

The author thanks Jun Fukue for useful comments and encouragements.

REFERENCES

[1] Mayor, M.; Queloz, D. *Nature* 1995, 378, 355-359.
[2] Charbonneau, D.; Brown, T. M.; Latham, D. W.; Mayor, M. *Astrophys J* 2000, 529, 45-48.
[3] Kalas, P.; Graham, J. R.; Chiang, E.; Fitzgerald, M. P.; Clampin, M.; Kite, E. S.; Stapelfeldt, K.; Marois, C.; Krist, J. *Science* 2008, 322, 1345-1348.
[4] Arnold, L. *Space Sci Rev* 2008, 135, 323-333.
[5] Kaltenegger, L.; Traub, W. A.; Jucks, K. W. *Astrophys J*, 2007, 658, 598-616.
[6] Gonzalez, G. *Orig Life Evol Biosph* 2005, 35, 555-606.
[7] Javaux, E. J.; Dehant, V. *Astron Astrophys Rev*, 2010, 18, 383-416.
[8] Kasting, James F.; Whitmire, Daniel P.; Reynolds, Ray T. *Icarus*, 1993, 101, 108-128.
[9] Gonzalez, G.; Brownlee, D.; Ward, P. *Icarus*, 2001, 152, 185-200.
[10] Kasting, J. F.; Catling, D. *Annu Rev Astron Astrophys*, 2003, 41, 429-463.
[11] Kaltenegger, L. *Astrophys J L*, 2010, 712, L125-L130.
[12] Buccino, A. P.; Lemarchand, G. A.; Mauas, P. J. D. *Icarus*, 2007, 192, 582-587.
[13] Kaltenegger, L.; Eiroa, C.; Ribas, I.; Paresce, F.; Leitzinger, M.; Odert, P.; Hanslmeier, A.; Fridlund, M.; Lammer, H.; Beichman, C.; Danchi, W.; Henning, T.; Herbst, T.; Léger, A.; Liseau, R.; Lunine, J.; Penny, A.; Quirrenbach, A.; Röttgering, H.; Selsis, F.; Schneider, J.; Stam, D.; Tinetti, G.; White, G. J. *Astrobiology* 2010, 10, 103-112.
[14] Fujii, N. *Orig Life Evol Biosph* 2002, 32, 103-127.
[15] Fujii, N.; Saito, T. *The Chemical Record* 2004, 4, 267-278.
[16] Fujii, N. *Biol Pharm Bull* 2005, 28, 1585-1589.
[17] Fuchs, S. A.; Berger, R.; Klomp, L. W. J.; de Koning, T. J. *Mol Genet Metab* 2005, 85, 168-180.
[18] Bonner, W. A. *Orig Life Evol Biosph* 1991, 21, 59-111.
[19] Bonner, W. A. *Orig Life Evol Biosph* 1995, 25, 175-190.
[20] Meierhenrich, U. J.; Thiemann, W. H.-P. *Orig Life Evol Biosph* 2004, 34, 111-121.
[21] Barron, L. D. *Space Sci Rev* 2008, 135, 187-201.
[22] Bouvier, A.; Wadhwa, M. *Nature Geoscience* 2010, 3, 637-641.
[23] Morishima, R.; Stadel, J.; Moore, B. *Icarus*, 2010, 207, 517-535.
[24] Moro-Martín, A.; Malhotra, R.; Bryden, G.; Rieke, G. H.; Su, K. Y. L.; Beichman, C. A.; Lawler, S. M. *Astrophys J* 2010, 717, 1123.
[25] Matsui, T.; Abe, Y. *Nature*, 1986, 319, 303-305.
[26] Abe, Y. *Physics of The Earth and Planetary Interiors*, 1997, 100, 27-39.
[27] Greenwood, R. C.; Franchi, I. A.; Jambon, A.; Buchanan, P. C. *Nature*, 2005, 435, 916-918.
[28] Wetherill, G. W. *Science*, 1985, 228, 877-879.
[29] Kokubo, E.; Ida, S. *Icarus*, 2000, 143, 15-27.
[30] Canup, R. M.; Asphaug, E. *Nature*, 2001, 412, 708-712.

[31] Canup, R. M. *Annu Rev Astron Astrophys*, 2004, 42, 441-475.
[32] Melosh, H. J. In Origin of the Earth; Newsom, H. E.; Jones, J. H.; Ed.; Oxford University Press: NY, 1990, 69-83.
[33] Cameron, A. G. W. *Icarus*, 1997, 126, 126-137.
[34] Wood, B. J.; Halliday, A. N. *Nature*, 2005, 437, 1345-1348.
[35] Wood, B. J.; Walter, M. J.; Wade, J. *Nature*, 2006, 441, 825-833.
[36] Oberbeck, V. R.; Fogleman, G. *Orig Life Evol Biosph* 1989, 19, 549-560.
[37] Hester, J. J.; Desch, S. J. In: Krot, A. N. et al (ed) Chondrites and the protoplanetary disk, ASP, San Francisco, 2005, 341, 107-130.
[38] McKee, C. F.; Ostriker, E. C. *Annual Review of Astronomy & Astrophysics*, 2007, 45, 565-687.
[39] Tachibana, S.; Huss, G. R. *Astrophys J*, 2003, 588, 41-44.
[40] Mostefaoui, S.; Lugmair, G. W.; Hoppe, P. *Astrophys J*, 2005, 625, 271-277.
[41] Tachibana, S.; Huss, G. R.; Kita, N. T.; Shimoda, G.; Morishita, Y. *Astrophys J*, 2006, 639, 87-90.
[42] Dauphas, N.; Cook, D. L.; Sacarabany, A.; Fröhlich, C.; Davis, A. M.; Wadhwa, M.; Pourmand, A.; Rauscher, T.; Gallino, R. *Astrophys J*, 2008, 686, 560-569.
[43] Hough, J. *Astronomy & Geophysics*, 2006, 47, 31-35.
[44] Hough, J. H. *Journal of Quantitative Spectroscopy and Radiative Transfer*, 2007, 106, 122-132.
[45] Nagayama, T.; Nagashima, C.; Nakajima, Y.; Nagata, T.; Sato, S.; Nakaya, H.; Yamamuro, T.; Sugitani, K.; Tamura, M. *Proceedings of the SPIE*, 2003, 4841, 459-464.
[46] Kandori, R.; Kusakabe, N.; Tamura, M.; Nakajima, Y.; Nagayama, T.; Nagashima, C.; Hashimoto, J.; Hough, J.; Sato, S.; Nagata, T.; Ishihara, A.; Lucas, P.; Fukagawa, M. *Proceedings of the SPIE*, 2006, 6269, 159.
[47] Whitney, B. A.; Wolff, M. J. *Astrophys J*, 2002, 574, 205-231.
[48] Kaltenegger, L.; Eiroa, C.; Fridlund, C. V. M. *Astrophysics and Space Science*, 2010, 326, 233-247.
[49] Herbst, E.; van Dishoeck, E. F. *Annu Rev Astron Astrophys*, 2009, 47, 427-480.
[50] Ehrenfreund, P.; Charnley, S. B. *Annu Rev Astron Astrophys*, 2000, 38, 427-483.
[51] Busemann, H.; Young, A. F.; O'D. Alexander, C. M.; Hoppe, P.; Mukhopadhyay, S.; Nittler, L. R. *Science*, 2006, 312, 727-730.
[52] Nakamura-Messenger, K.; Messenger, S.; Keller, L. P.; Clemett, S. J.; Zolensky, M. E. *Science*, 2006, 314, 1439-1442.
[53] Kvenvolden, K.; Lawless, J.; Pering, K.; Peterson, E.; Flores, J.; Ponnamperuma, C. *Nature*, 1970, 228, 923-926.
[54] Botta, O.; Bada, J. L. *Surveys in Geophysics*, 2002, 23, 411-467.
[55] Cody, G. D.; Alexander, C. M. O.; Tera, F. *Geochim Cosmochim Acta*, 2002, 66, 1851-1865.
[56] Sephton, M. A. *Nat Prod Rep*, 2002, 19, 292-311.
[57] Schmitt-Kopplina, P.; Gabelicab, Z.; Gougeonc, R. D.; Feketea, A.; Kanawatia, B.; Harira, M.; Gebefuegia, I.; Eckeld, G.; Hertkorna, N. *PNAS*, 2010, 107, 2763-2768.
[58] Messenger, S. *Nature*, 2000, 404, 968-971.
[59] Duprat, J.; Dobrică, E.; Engrand, C.; Aléon, J.; Marrocchi, Y.; Mostefaoui, S.; Meibom, A.; Leroux, H.; Rouzaud, J.-N.; Gounelle, M.; Robert, F. *Science*, 2010, 328, 742-745.

[60] Maurette, M.; Duprat, J.; Engrand, C.; Gounelle, M.; Kurat, G.; Matrajt, G.; Toppani, A. *Planetary and Space Science*, 2000, 48, 1117-1137.

[61] Flynn, G. J.; Keller, L. P.; Feser, M.; Wirick, S.; Jacobsen, C. *Geochim Cosmochim Acta*, 2003, 67, 4791-4806.

[62] Messenger, S.; Stadermann, F. J.; Floss, C.; Nittler, L. R.; Mukhopadhyay, S. *Space Sci Rev*, 2003, 106, 155-172.

[63] Floss, C.; Stadermann, F. J.; Bradley, J. P.; Dai, Z. R.; Bajt, S.; Graham, G.; Lea, A. S. *Geochim Cosmochim Acta*, 2006, 70, 2371-2399.

[64] Sandford, S. A.; Aléon, J.; Alexander, C. M. O.'D.; Araki, T.; Bajt, S.; Baratta, G. A.; Borg, J.; Bradley, J. P.; Brownlee, D. E.; Brucato, J. R.; Burchell, M. J.; Busemann, H.; Butterworth, A.; Clemett, S. J.; Cody, G.; Colangeli, L.; Cooper, G.; D'Hendecourt, L.; Djouadi, Z.; Dworkin, J. P.; Ferrini, G.; Fleckenstein, H.; Flynn, G. J.; Franchi, I. A.; Fries, M.; Gilles, M. K.; Glavin, D. P.; Gounelle, M.; Grossemy, F.; Jacobsen, C.; Keller, L. P.; Kilcoyne, A. L. D.; Leitner, J.; Matrajt, G.; Meibom, A.; Mennella, V.; Mostefaoui, S.; Nittler, L. R.; Palumbo, M. E.; Papanastassiou, D. A.; Robert, F.; Rotundi, A.; Snead, C. J.; Spencer, M. K.; Stadermann, F. J.; Steele, A.; Stephan, T.; Tsou, P.; Tyliszczak, T.; Westphal, A. J.; Wirick, S.; Wopenka, B.; Yabuta, H.; Zare, R. N.; Zolensky, M. E. *Science*, 2006, 314, 1720-1724.

[65] Cody, G. D.; Ade, H.; O'D. Alexander, C. M.; Araki, T.; Butterworth, A.; Fleckenstein, H.; Flynn, G.; Gilles, M. K.; Jacobsen, C.; Kilcoyne, A. L. D.; Messenger, K.; Sandford, S. A.; Tyliszczak, T.; Westphal, A. J.; Wirick, S.; Yabuta, H. *Meteoritics & Planetary Science*, 2008, 43, 353-365.

[66] Wirick, S.; Flynn, G. J.; Keller, L. P.; Nakamura-Messenger, K.; Peltzer, C.; Jacobsen, C.; Sandford, S. A.; Zolensky, M. E. *Meteoritics & Planetary Science*, 2009, 44, 1611-1626.

[67] De Gregorio, B. T.; Stroud, R. M.; Nittler, L. R.; Alexander, C. M. O'D.; Kilcoyne, A. L. D.; Zega, T. J. *Geochim Cosmochim Acta*, 2010, 74, 4454-4470.

[68] Glavin, D. P.; Dworkin, J. P.; Sandford, S. A. *Meteoritics & Planetary Science*, 2008, 43, 399-413.

[69] Elsila, J. E.; Glavin, D. P.; Dworkin, J. P. *Meteoritics & Planetary Science*, 2009, 44, 1323-1330.

[70] Brown, R. D.; Godfrey, P. D.; Storey, J. W. V.; Bassez, M.-P.; Robinson, B. J.; Batchelor, R. A.; McCulloch, M. G.; Rydbeck, O. E. H.; Hjalmarson, A. G. *Mon Not R Astron Soc*, 1979, 186, 5-8.

[71] Snyder, L. E. *Orig Life Evol Biosph*, 1997, 27, 115-133.

[72] Kuan, Y.-J.; Charnley, S. B.; Huang, H.-C.; Tseng, W.-L.; Kisiel, Z. *Astrophys J*, 2003, 593, 848-867.

[73] Snyder, L. E.; Lovas, F. J.; Hollis, J. M.; Friedel, D. N.; Jewell, P. R.; Remijan, A.; Ilyushin, V. V.; Alekseev, E. A.; Dyubko, S. F. *Astrophys J*, 2005, 619, 914-930.

[74] Belloche, A.; Menten, K. M.; Comito, C.; Müller, H. S. P.; Schilke, P.; Ott, J.; Thorwirth, S.; Hieret, C. *Astron Astrophys*, 2008, 482, 179-196.

[75] Sutton, E. C.; Peng, R.; Danchi, W. C.; Jaminet, P. A.; Sandell, G.; Russell, A. P. G. *Astrophys J Suppl Ser*, 1995, 97, 455-496.

[76] Schilke, P.; Groesbeck, T. D.; Blake, G. A.; Phillips, T. G. *Astrophys J Suppl*, 1997, 108, 301-337.

[77] van Dishoeck, E. F.; Blake, G. A.; Jansen, D. J.; Groesbeck, T. D. *Astrophys J*, 1995, 447, 760-782.
[78] Cazaux, S.; Tielens, A. G. G. M.; Ceccarelli, C.; Castets, A.; Wakelam, V.; Caux, E.; Parise, B.; Teyssier, D. *Astrophys J*, 2003, 593, 51-55.
[79] Thi, W.-F.; van Zadelhoff, G.-J.; van Dishoeck, E. F. *Astron Astrophys*, 2004, 425, 955-972.
[80] Öberg, K. I.; Qi, C.; Fogel, J. K. J.; Bergin, E. A.; Andrews, S. M.; Espaillat, C.; van Kempen, T. A.; Wilner, D. J.; Pascucci, I. *Astrophys J*, 2010, 720, 480-493.
[81] Hollis, J. M.; Jewell, P. R.; Lovas, F. J.; Remijan, A. *Astrophys J*, 2004, 613, 45-48.
[82] Hollis, J. M.; Lovas, F. J.; Jewell, P. R.; Coudert, L. H. *Astrophys J*, 2002, 571, 59-62.
[83] Hollis, J. M.; Remijan, Anthony J.; Jewell, P. R.; Lovas, F. J. *Astrophys J*, 2006, 642, 933-939.
[84] Encrenaz, T. *Annu Rev Astron Astrophys*, 2008, 46, 57-87.
[85] Malin, M. C.; Edgett, K. S. *Science*, 2000, 288, 2330-2335.
[86] Hansen, C. J.; Esposito, L.; Stewart, A. I. F.; Colwell, J.; Hendrix, A.; Pryor, W.; Shemansky, D.; West, R. *Science*, 2006, 311, 1422-1425.
[87] Griffith, C. A.; Owen, T.; Geballe, T. R.; Rayner, J.; Rannou, P. *Science*, 2003, 300, 628-630.
[88] Sunshine, J. M.; A'Hearn, M. F.; Groussin, O.; Li, J.-Y.; Belton, M. J. S.; Delamere, W. A.; Kissel, J.; Klaasen, K. P.; McFadden, L. A.; Meech, K. J.; Melosh, H. J.; Schultz, P. H.; Thomas, P. C.; Veverka, J.; Yeomans, D. K.; Busko, I. C.; Desnoyer, M.; Farnham, T. L.; Feaga, L. M.; Hampton, D. L.; Lindler, D. J.; Lisse, C. M.; Wellnitz, D. D. *Science*, 2006, 311, 1453-1455.
[89] Jewitt, D. C.; Luu, J. *Nature*, 2004, 432, 731-733.
[90] Tinetti, G.; Vidal-Madjar, A.; Liang, M.-C.; Beaulieu, J.-P.; Yung, Y.; Carey, S.; Barber, R. J.; Tennyson, J.; Ribas, I.; Allard, N.; Ballester, G. E.; Sing, D. K.; Selsis, F. *Nature*, 2007, 448, 169-171.
[91] Carr, J. S.; Najita, J. R. *Science*, 2008, 319, 1504-1506.
[92] Honda, M.; Inoue, A. K.; Fukagawa, M.; Oka, A.; Nakamoto, T.; Ishii, M.; Terada, H.; Takato, N.; Kawakita, H.; Okamoto, Y. K.; Shibai, H.; Tamura, M.; Kudo, T.; Itoh, Y. *Astrophys J L*, 2009, 690, 110-113.
[93] Beuther, H.; Walsh, A.; Schilke, P.; Sridharan, T. K.; Menten, K. M.; Wyrowski, F. *Astron Astrophys*, 2002, 390, 289-298.
[94] Wright, M. C. H.; Plambeck, R. L.; Wilner, D. J. *Astrophys J*, 1996, 469, 216-237.
[95] Snell, R. L.; Howe, J. E.; Ashby, M. L. N.; Bergin, E. A.; Chin, G.; Erickson, N. R.; Goldsmith, P. F.; Harwit, M.; Kleiner, S. C.; Koch, D. G.; Neufeld, D. A.; Patten, B. M.; Plume, R.; Schieder, R.; Stauffer, J. R.; Tolls, V.; Wang, Z.; Winnewisser, G.; Zhang, Y. F.; Melnick, G. J. *Astrophys J*, 2000, 539, 101-105.
[96] Raymond, S. N.; Quinn, T.; Lunine, J. I. *Icarus*, 2004, 168, 1-17.
[97] Cronin, J. R.; Pizzarello, S. *Science*, 1997, 275, 951-955.
[98] Pizzarello, S.; Cronin, J. R. *Geochim Cosmochim Acta*, 2000, 64, 329-338.
[99] Pizzarello, S.; Zolensky, M.; Turk, K. A. *Geochim Cosmochim Acta*, 2003, 67, 1589-1595.
[100] Pizzarello, S.; Huang, Y.; Alexandre, M. R. *PNAS*, 2008, 105, 3700-3704.
[101] Glavin, D. P.; Dworkin, J. P. *Proc Natl Acad Sci USA*, 2009, 106, 5487-5492.
[102] Iglesias-Groth, S.; Cataldo, F.; Ursini, O.; Manchado, A. 2010, arXiv:1007.4529

[103] Mojzsis, S. J.; Arrhenius, G.; McKeegan, K. D.; Harrison, T. M.; Nutman, A. P.; Friend, C. R. L. *Nature*, 1996, 384, 55-59.
[104] Mojzsis, S. J.; Harrison, T. M. *Earth and Planetary Science Letters*, 2002, 202, 563-576.
[105] Rosing, M. T. *Science*, 1999, 283, 674-676.
[106] Cohen, B. A.; Swindle, T. D.; Kring, D. A. *Science*, 2000, 290, 1754-1756.
[107] Schoenberg, R.; Kamber, B. S.; Collerson, K. D.; Moorbath, S. *Nature*, 2002, 418, 403-405.
[108] Kring, D. A.; Cohen, B. A. *Journal of Geophysical Research*, 2002, 107, 4-1 - 4-6
[109] Harald F.; Maarten de W.; Hubert S.; Minik R.; Karlis M. *Science*, 2007, 315, 1704-1707.
[110] Gomes, R.; Levison, H. F.; Tsiganis, K.; Morbidelli, A. *Nature*, 2005, 435, 466-469.
[111] Strom, R. G.; Malhotra, R.; Ito, T.; Yoshida, F.; Kring, D. A. *Science*, 2005, 309, 1847-1850.
[112] Bernstein, M. P.; Dworkin, J. P.; Sandford, S. A.; Cooper, G. W.; Allamandola, L. J. *Nature*, 2002, 416, 401-403.
[113] Muñoz Caro, G. M.; Meierhenrich, U. J.; Schutte, W. A.; Barbier, B.; Arcones Segovia, A.; Rosenbauer, H.; Thiemann, W. H.-P.; Brack, A.; Greenberg, J. M. *Nature*, 2002, 416, 403-406.
[114] Nuevo, M.; Meierhenrich, U. J.; D'Hendecourt, L.; Muñoz Caro, G. M.; Dartois, E.; Deboffle, D.; Thiemann, W. H.-P.; Bredehöft, J.-H.; Nahon, L. *Advances in Space Research*, 2007, 39, 400-404.
[115] Meierhenrich, U. J.; Muñoz Caro, G. M.; Schutte, W. A.; Thiemann, W. H.; Barbier, B.; Brack, A. *Chemistry*, 2005, 11, 4895-4900.
[116] Muñoz Caro, G. M.; Schutte, W. A. *Astron Astrophys*, 2003, 412, 121-132.
[117] Nuevo, M.; Chen, Y.-J.; Yih, T.-S.; Ip, W.-H.; Fung, H.-S.; Cheng, C.-Y.; Tsai, H.-R.; Wu, C.-Y. R. *Advances in Space Research*, 2007, 40, 1628-1633.
[118] Nuevo, M.; Auger, G.; Blanot, D.; D'Hendecourt, L. *Origins of Life and Evolution of Biospheres*, 2008, 38, 37-56.
[119] Chen, Y.-J.; Nuevo, M.; Yih, T.-S.; Ip, W.-H.; Fung, H.-S.; Cheng, C.-Y.; Tsai, H.-R.; Wu, C.-Y. R. *Mon Not R Astron Soc*, 2008, 384, 605-610.
[120] Elsila, J. E.; Dworkin, J. P.; Bernstein, M. P.; Martin, M. P.; Sandford, S. A. *Astrophys J*, 2007, 660, 911-918.
[121] Lee, C.-W.; Kim, J.-K.; Moon, E.-S.; Minh, Y. C.; Kang, H. *Astrophys J*, 2009, 697, 428-435.
[122] Ehrenfreund, P.; Bernstein, M. P.; Dworkin, J. P.; Sandford, S. A.; Allamandola, L. J. *Astrophys J*, 2001, 550, 95-99.
[123] Schwell, M.; Jochims, H.-W.; Baumgärtel, H.; Dulieu, F.; Leach, S. *Planetary and Space Science*, 2006, 54, 1073-1085.
[124] Bailey, J. *Orig Life Evol Biosph*, 2001, 31, 167-183.
[125] Bernstein, M. P.; Ashbourn, S. F. M.; Sandford, S. A.; Allamandola, L. J. *Astrophys J*, 2004, 601, 365-370.
[126] Feringa, B.L.; Delden, R.A. van *Angew Chem Int Ed*, 1999, 38, 3418-3438.
[127] Griesbeck, A. G.; Meierhenrich, U. J. *Angew Chem Int Ed*, 2002, 41, 3147-3154.
[128] Cerf, C.; Jorissen, A. *Space Sci Rev*, 2000, 92, 603-612.
[129] Bonner, W. A.; Bean, B. D. *Orig Life Evol Biosph*, 2000, 30, 513-517.

[130] Meierhenrich, U. J.; Nahon, L.; Alcaraz, C.; Bredehöft, J. H.; Hoffmann, S. V.; Barbier, B.; Brack, A. *Angew Chem Int Ed*, 2005, 44, 5630-5634.
[131] Nuevo, M.; Meierhenrich, U. J.; Muñoz Caro, G. M.; Dartois, E.; D'Hendecourt, L.; Deboffle, D.; Auger, G.; Blanot, D.; Bredehöft, J.-H.; Nahon, L. *Astron Astrophys*, 2006, 457, 741-751.
[132] Takano, Y.; Takahashi, J.; Kaneko, T.; Marumo, K.; Kobayashi, K. *Earth and Planetary Science Letters*, 2007, 254, 106-114.
[133] Melchiorre, P.; Marigo, M.; Carlone, A.; Bartoli, G. *Angew Chem Int Ed Engl,* 2008, 47, 6138-6171.
[134] Soai, K.; Kawasaki, T. *Chirality*, 2006, 18, 469-478.
[135] Soai, K.; Shibata, T.; Morioka, H.; Choji, K. *Nature*, 1995, 378, 767-768.
[136] Shibata, T.; Yamamoto, J.; Matsumoto, N.; Yonekubo, S.; Osanai, S.; Soai, K. *J Am Chem Soc*, 1998, 120, 12157-12158.
[137] Kawasaki, T.; Soai, K. *Journal of Fluorine Chemistry*, 2010, 131, 525-534.
[138] Klussmann, M.; Iwamura, H.; Mathew, S. P.; Wells, D. H.; Pandya, U.; Armstrong, A.; Blackmond, D. G. *Nature*, 2006, 441, 621-623.
[139] Breslow, R.; Levine, M. S. *PNAS*, 2006, 103, 12979-12980.
[140] Dziedzica, P.; Zoua. W.; Ibrahema, I.; Sundéna, H.; Córdova, A. *Tetrahedron Letters*, 2006, 47, 6657-6661.
[141] Aratake, S.; Itoh, T.; Okano, T.; Usui, T.; Shoji, M.; Hayashi, Y. *Chem Commun*, 2007, 2524-2526.
[142] Weber, A. L. *Orig Life Evol Biosph* 2001, 31, 71-86.
[143] Pizzarello, S.; Weber, A. L. *Science* 2004, 303, 1151.
[144] Córdova, A.; Engqvist, M.; Ibrahem, I.; Casas, J.; Sunde´n, H. *Chem Commun* 2005, 2047–2049.
[145] Córdova, A.; Zou, W.; Dziedzic, P.; Ibrahem, I.; Reyes, E.; Xu, Y. *Chem Eur J* 2006, 12, 5383–5397.
[146] Breslow, R.; Cheng, Z.-L. *PNAS*, 2009, 106, 9144-9146.
[147] Bailey, J.; Chrysostomou, A.; Hough, J. H.; Gledhill, T. M.; McCall, A.; Clark, S.; Menard, F.; Tamura, M. *Science*, 1998, 281, 672-674.
[148] Fukue, T.; Tamura, M.; Kandori, R.; Kusakabe, N.; Hough, J. H.; Bailey, J.; Whittet, D. C. B.; Lucas, P. W.; Nakajima, Y.; Hashimoto, J. *Origins of Life and Evolution of Biospheres*, 2010, 40, 335-346.
[149] Chrysostomou, A.; Lucas, P. W.; Hough, J. H. *Nature*, 2007, 450, 71-73.
[150] Chrysostomou, A.; Gledhill, T. M.; Ménard, F.; Hough, J. H.; Tamura, M.; Bailey, J. *Mon Not R Astron Soc*, 2000, 312, 103-115.
[151] Ménard, F.; Chrysostomou, A.; Gledhill, T.; Hough, J. H.; Bailey, J. In: Lemarchand G, Meech K (ed) Bioastronomy 99: a new era in the search for Life in the Universe, San Francisco, ASP Conf. 2000, 213, 355-358.
[152] Clark, S.; McCall, A.; Chrysostomou, A.; Gledhill, T.; Yates, J.; Hough, J. *Mon Not R Astron Soc*, 2000, 319, 337-349.
[153] Chrysostomou, A.; Menard, F.; Gledhill, T. M.; Clark, S.; Hough, J. H.; McCall, A.; Tamura, M. *Mon Not R Astron Soc*, 1997, 285, 750-758.
[154] Gledhill, T. M.; Chrysostomou, A.; Hough, J. H. *Mon Not R Astron Soc*, 1996, 282, 1418-1436.
[155] Menard, F.; Bastien, P.; Robert, C. *Astrophys J*, 1988, 335, 290-294.

[156] Takami, M.; Gledhill, T.; Clark, S.; Mnard, F; Hough, J. H. in Star Formation 1999, Ed. T. Nakamoto, 1999, 205.

[157] Clayton, G. C.; Whitney, B. A.; Wolff, M. J.; Smith, P.; Gordon, K. D. In: Adamson, A. et al. (ed) Astronomical polarimetry: current status and future directions. ASP, San Francisco, 2005, ASP Conf. Ser. 343, 122-127.

[158] Fukue, T. "Polarimetric Study of Star/Planet-Forming Regions", PhD Thesis, Kyoto University, 2009.

[159] Fischer, O.; Henning, T.; Yorke, H. W. *Astron Astrophys*, 1996, 308, 863-885.

[160] Wolf, S.; Voshchinnikov, N. V.; Henning, Th. *Astron Astrophys*, 2002, 385, 365-376.

[161] Lucas, P. W.; Fukagawa, M.; Tamura, M.; Beckford, A. F.; Itoh, Y.; Murakawa, K.; Suto, H.; Hayashi, S. S.; Oasa, Y.; Naoi, T.; Doi, Y.; Ebizuka, N.; Kaifu, N. *Mon Not R Astron Soc*, 2004, 352, 1347-1364.

[162] Lucas, P. W.; Hough, J. H.; Bailey, J.; Chrysostomou, A.; Gledhill, T. M.; McCall, A. *Origins of Life and Evolution of Biospheres*, 2005, 35, 29-60.

[163] Tamura, M.; Kandori, R.; Kusakabe, N.; Nakajima, Y.; Hashimoto, J.; Nagashima, C.; Nagata, T.; Nagayama, T.; Kimura, H.; Yamamoto, T.; Hough, J. H.; Lucas, P.; Chrysostomou, A.; Bailey, J. *Astrophys J*, 2006, 649, 29-32.

[164] Buschermöhle, M.; Whittet, D. C. B.; Chrysostomou, A.; Hough, J. H.; Lucas, P. W.; Adamson, A. J.; Whitney, B. A.; Wolff, M. J. *Astrophys J*, 2005, 624, 821-826.

[165] Fukue, T.; Tamura, M.; Kandori, R.; Kusakabe, N.; Hough, J. H.; Lucas, P. W.; Bailey, J.; Whittet, D. C. B.; Nakajima, Y.; Hashimoto, J.; Nagata, T. *Astrophys J L*, 2009, 692, 88-91.

In: Photochemistry
Editors: Karen J. Maes and Jaime M. Willems

ISBN: 978-1-61209-506-6

Chapter 9

PHOTOINDUCED TRANSFORMATION PROCESSES IN SURFACE WATERS

Davide Vione[a,b]**, Claudio Minero**[a]**, Valter Maurino**[a]**,***
***Romeo-Iulian Olariu**[c]**, Cecilia Arsene**[c]** and Khan M. G. Mostofa**[d]*

[a] Dipartimento di Chimica Analitica, Università di Torino, Torino, Italy.
[b] Centro Interdipartimentale NatRisk,
Università degli Studi di Torino, Grugliasco (TO), Italy.
[c] Department of Chemistry, "Al. I. Cuza" University of Iasi, Iasi, Romania.
[d] Key Laboratory of Environmental Geochemistry, Institute of Geochemistry,
Chinese Academy of Sciences, Guiyang, P. R. China.

ABSTRACT

Photochemical processes are important pathways for the transformation of biologically refractory organic compounds, including harmful pollutants, in surface waters. They include the direct photolysis of sunlight-absorbing molecules, the transformation photosensitised by dissolved organic matter, and the reaction with photochemically generated transient species (*e.g.* $^{\bullet}OH$, $CO_3^{-\bullet}$, $^1O_2$). This chapter provides first an overview of the main photoinduced processes that can take place in surface waters, leading to the transformation of the primary compounds but also to the production of harmful intermediates. The second part is devoted to the modelling of the main photochemical processes, such as direct photolysis and reaction with hydroxyl and carbonate radicals, singlet oxygen and the excited triplet states of chromophoric dissolved organic matter.

* Tel. +39-011-6707838, Fax +39-011-6707615, E-mail: davide.vione@unito.it,
http://naturali.campusnet.unito.it/cgi.bin/docenti.pl/Show?_id=vione

1. INTRODUCTION

The persistence in surface water bodies of dissolved organic compounds, including both natural organic molecules and man-made xenobiotics and pollutants strongly depends on their transformation kinetics due to abiotic and biological processes. Transformation by micro-organisms can be very important for readily biodegradable molecules, including most notably nutrients during the phytoplankton blooming period [1].

The abiotic transformation processes include hydrolysis, oxidation mediated by dissolved oxidising species or by metal oxides, and light-induced reactions. Many organic pollutants such as polycyclic aromatic hydrocarbons, pesticides, pharmaceuticals and their transformation intermediates are refractory to biological degradation. In such cases the abiotic transformation processes represent the main removal pathways from surface waters. Within the abiotic transformation reactions of xenobiotics, those induced by sunlight are receiving increasing attention nowadays [2] because of their potential importance in the removal of the parent molecules and for the possible production of harmful secondary pollutants [3]. The present chapter will be dedicated to the description of the photochemically induced reactions. They include the direct photolysis upon absorption of sunlight, the transformation photo-sensitised by coloured or chromophoric dissolved organic matter (CDOM), predominantly by fulvic and humic acids (humic substances), and the reaction with photogenerated, reactive transients (e.g. $^{\bullet}OH$, $CO_3^{-\bullet}$ and $^1O_2$) that often have the ability of oxidising the organic substrates [4].

It is noteworthy that the direct photolysis alone, although very important in defining the lifetime of many photolabile compounds in surface waters, will seldom lead to their complete removal. The process, which is sometimes very efficient, would rather yield a number of transformation intermediates. In some cases they are even more harmful than the parent molecule [5]. The complete removal of the xenobiotic species, *e.g.* by mineralization, usually requires the reaction with oxidising transients such as $^{\bullet}OH$, or microbial processing. In many cases the mineralization is much slower than the primary step of phototransformation of the parent molecule [6].

2. AN OVERVIEW OF PHOTOINDUCED REACTIONS IN SURFACE WATERS

2.1. Direct Photolysis Processes

The direct photolysis of a molecule usually consists of the break of a chemical bond, as a consequence of photon absorption. The photon flux of sunlight absorbed by a given compound, P_a can be expressed in [einstein L^{-1} s^{-1}], where 1 einstein = 1 mole of photons [7]. The quantity P_a depends on the intensity of sunlight, the ability of the compound to absorb it, and the absorption of radiation by the other dissolved components. For a certain compound at the wavelength λ it is defined the absorbance $A_\lambda = \varepsilon_\lambda\ b\ c$, where ε_λ is the molar absorption coefficient (decadic), b the optical path length of the solution, and c the concentration of the compound. The quantities b and c are usually expressed in cm and in mol L^{-1}, respectively, in which case the unit of ε_λ is [L mol^{-1} cm^{-1}], and A_λ is dimensionless.

Assume $i_o(\lambda)$ as the spectral photon flux density of the incident radiation in solution, which is usually expressed in [einstein cm^{-2} s^{-1} nm^{-1}]. In a surface water body the intensity of the incident radiation is not constant over the whole water column, because absorption and scattering phenomena will reduce the radiation intensity as the depth increases. As a consequence, the spectral photon flux density of the incident radiation at the depth b will be a function of both depth and wavelength, as $i_o(\lambda,b)$. Under the simplified assumption that the scattering of radiation is negligible compared to absorption, the Lambert-Beer relationship applies as:

$$i_o(\lambda,b) = i_o(\lambda)\cdot 10^{-A_1(\lambda)\cdot b} \tag{1}$$

where $i_o(\lambda)$ is the spectral photon flux density of incident radiation on top of the water column and $A_1(\lambda)$ is the attenuation coefficient [8]. The coefficient $A_1(\lambda)$ depends on the absorption spectrum of water and is mainly accounted for by CDOM absorption [9]. The measure unit of $A_1(\lambda)$ is cm^{-1} if b (optical path length) is in cm. In the most general case, for $\lambda_1 \leq \lambda \leq \lambda_2$ and $0 \leq x \leq b$, P_a is given by equation (2), where 2.3 = ln 10 and 10^3 is the conversion factor between cm^{-3} and L^{-1} [3]:

$$P_a = \frac{2.3\cdot 10^3}{b}\cdot \int_{\lambda_1}^{\lambda_2}\int_0^b i_o(\lambda)\cdot 10^{-A_1(\lambda)\cdot x}\cdot \varepsilon_\lambda \cdot c\, dx\, d\lambda \tag{2}$$

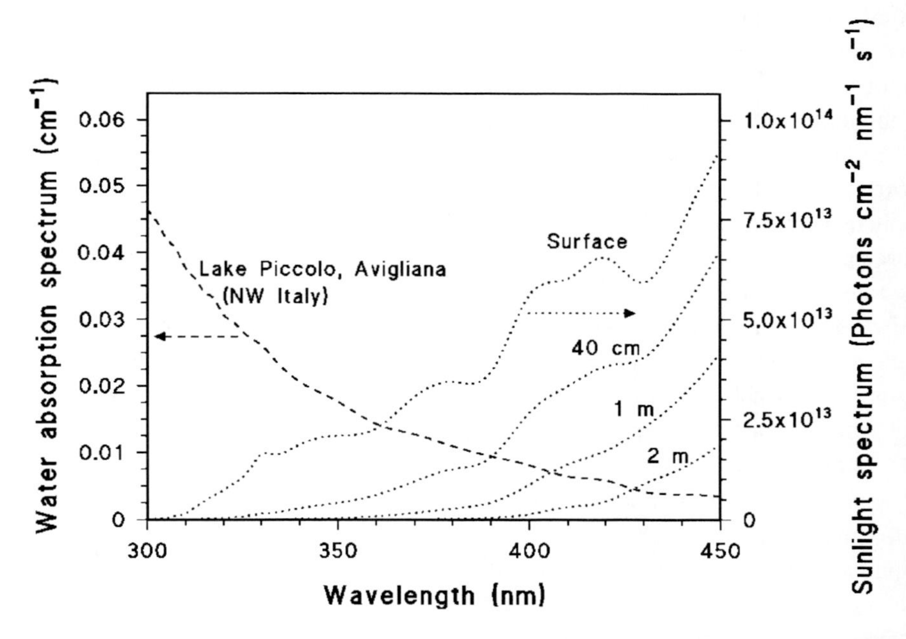

Figure 1. Absorption spectrum $A_1(\lambda)$ of water from Lake Piccolo in Avigliana (NW Italy), and spectral fluence rate density of sunlight at different depths of the water column (surface to 2 m depth), calculated according to the Lambert-Beer approximation [8]. Note that the spectral fluence rate density [photons cm^{-2} nm^{-1} s^{-1}] is the spectral photon flux density $i_o(\lambda)$ [einstein cm^{-2} nm^{-1} s^{-1}] times the Avogadro's number.

Figure 1 reports the spectral fluence rate density of sunlight at different values of the water column depth, for a given absorption spectrum $A_1(\lambda)$ of water [8,10,11]. It is apparent the rapid decrease of the fluence density with increasing depth of the water column, in particular at the shorter wavelengths.

The incident spectral photon flux density at a given depth is a key factor in defining the rate R of the direct photolysis, which is directly proportional to absorbed photon flux P_a. However, not all of the absorbed photons are able to induce photolysis. This is reflected by the photolysis quantum yield, which measures the probability that a photolysis process is induced by the absorption of a photon. The photolysis quantum yield is usually indicated as Φ, and it is $0 \le \Phi \le 1$. In the most general case the quantum yield of photolysis is not constant with wavelength. The absorption spectra of the dissolved molecules are often characterised by the overlapping of different absorption bands, each corresponding to a different transition with a peculiar quantum yield for photolysis. For irradiation at the wavelength λ, the rate of photolysis is $r(\lambda) = \Phi(\lambda)\, p_a(\lambda)$, and the rate R of photolysis for $\lambda_1 \le \lambda \le \lambda_2$ and $0 \le x \le b$ is [11]:

$$R = \frac{2.3 \cdot 10^3}{b} \cdot \int_{\lambda_1}^{\lambda_2}\int_0^b \Phi(\lambda) \cdot i_o(\lambda) \cdot 10^{-A_1(\lambda) \cdot x} \cdot \varepsilon_\lambda \cdot c \, dx \, d\lambda \qquad (3)$$

Direct photolysis processes have received much attention in the context of the degradation of xenobiotic compounds of high environmental concern, such as polycyclic aromatic hydrocarbons, haloaromatic compounds, pesticides and their metabolites, and more recently pharmaceuticals [2, 3].

Chlorophenols are a class of chlorinated aromatic compounds of considerable environmental concern because of their potential carcinogenic effect in humans and micro-organisms. They can be released either as by-products from various industrial activities or as secondary pollutants upon the environmental transformation of various pesticides, mainly the chlorophenoxy-acetic and propionic acids [12], and of the antimicrobial agent triclosan [13]. The absorption of radiation brings chlorophenols to their first excited singlet state, which can then be transformed into the first excited triplet state by inter-system crossing (ISC) [3]. In the case of *ortho*-chlorophenols [14], the first excited singlet state is sufficiently long-lived to allow chemical reactivity as an alternative to ISC. Accordingly, the excited singlet state can undergo ring contraction and loss of HCl to form a cyclopentadienyl carboxyaldehyde (Figure 2). The ring-contraction process would be particularly significant for the phenolate anions [15]. In contrast, the first excited triplet state would mainly react by dechlorination [16], either reductive (with $HO_2^{\bullet}$) to give the corresponding phenol, or involving oxygen with the final formation of dihydroxyphenols and quinones (Figure 2). In the case of *para*-chlorophenols, only the triplet state is sufficiently long-lived to undergo chemical reactions. Therefore, the direct photolysis of these compounds yields the corresponding phenols, dihydroxyphenols and quinones [16].

Many xenobiotic compounds of environmental concern undergo different photolysis processes in their protonated or deprotonated forms, such as 2-methyl-4-chlorophenoxyacetic acid (MCPA). The photolysis pathways are, therefore, strongly dependent on pH. In the case of MCPA, the protonated form undergoes photoisomerisation, while the deprotonated one follows a dechlorination pathway [17]. MCPA is also interesting because its direct photolysis quantum yield decreases in the presence of dissolved organic compounds, which can have important consequence for the MCPA fate in surface water bodies [18].

In the case of triclosan, it has been found that its photocyclisation produces a dichlorodibenzodioxin (Figure 3) [19]. This is a good example of a photolysis process, which yields an intermediate that is more harmful than the parent compound. It is even more significant because the direct photolysis is the most likely sink of triclosan in surface waters [20]. Indeed the photodegradation of a pollutant is not always beneficial to the environment, and the environmental and health impact of the transformation intermediates is to be considered as well [21]. Another interesting example is the direct photolysis of the anti-epileptic drug carbamazepine that yields, among the other intermediates, the mutagenic acridine [5,22].

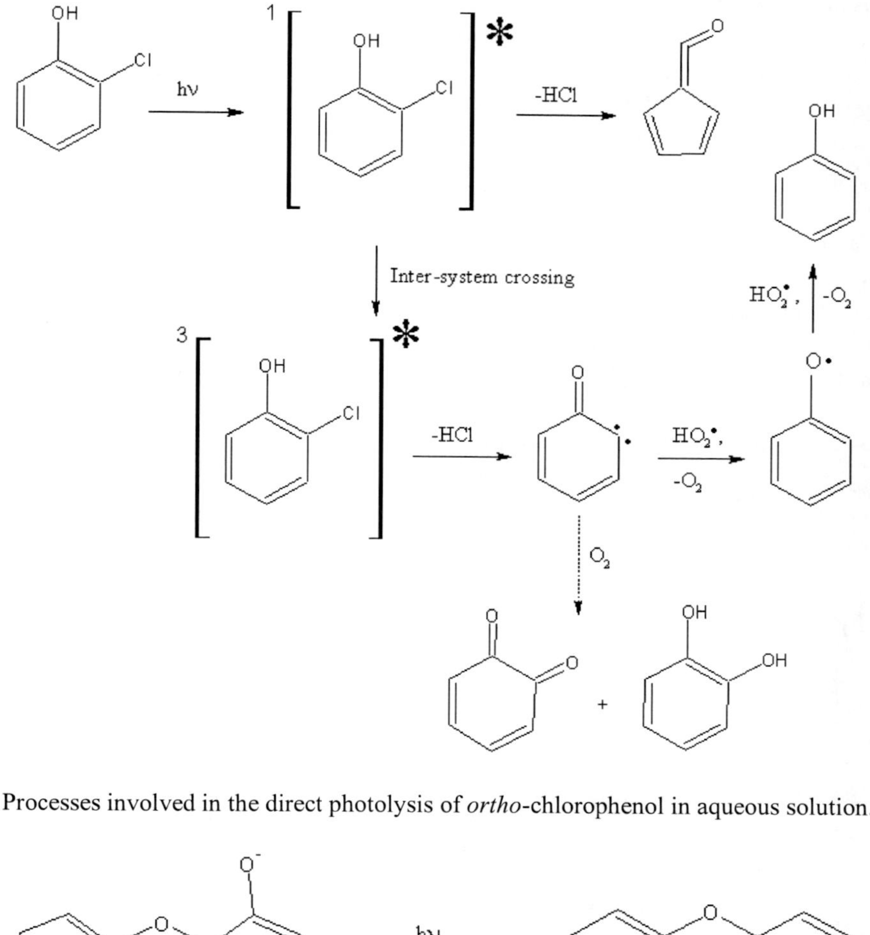

Figure 2. Processes involved in the direct photolysis of *ortho*-chlorophenol in aqueous solution.

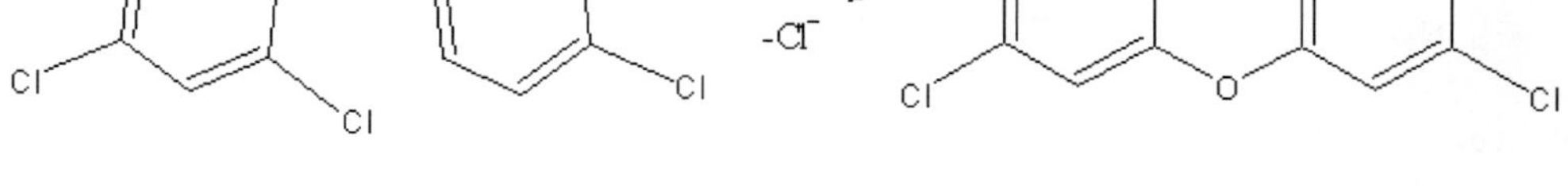

Figure 3. Processes involved in the direct photolysis of triclosan in aqueous solution.

2.2. Transformation Photosensitised by Chromophoric Dissolved Organic Matter (CDOM)

Natural dissolved organic matter in surface waters consists of both autochthonous (aquagenic) material, mainly composed of autochthonous fulvic acid or marine humic-like substances, polysaccharides, complex carbohydrates, peptides, proteins, and of allochthonous compounds, mostly fulvic and humic acids that derive from soil runoff. Most of the autochthonous and allochthonous compounds contain a significant amount of functional groups or chromophores that absorb sunlight [1, 23].

The functional groups or chromophores in CDOM constitute the fraction of organic matter that is susceptible to absorb radiation in surface waters [9]. A very interesting issue is that CDOM contains, among other constituents, quinonoid substances, aromatic carbonyls, methoxylate and phenolic groups, as well as O, N, S, and P-atom-containing functional groups, which are known photosensitisers. This means that the excitation of CDOM by sunlight can cause the degradation of other dissolved molecules, which do not need to absorb sunlight to be transformed [24]. The photochemical activity of CDOM depends on the reactivity of the excited triplet states of the photosensitising functional groups it contains [3].

The absorption of radiation by a sensitiser S causes the transition from the ground state to the first excited singlet state ($^1S^*$). Various alternative pathways are then possible. The sensitiser can reach its ground state via thermal loss of energy, *e.g.* by collision with the solvent. Some molecules lose energy by radiation, emitting fluorescence photons. A further alternative can be the so-called inter-system crossing (ISC) to the excited triplet state ($^3S^*$). The ground vibrational state of $^1S^*$ can have the same energy as an excited vibrational state of $^3S^*$, which allows the transition to occur [11]. Deactivation of $^3S^*$ to the ground state can take place via thermal energy loss, by combination of vibration and collision. Solid systems might also emit phosphorescence radiation. A further possibility is the chemical reactivity. The excited triplet states of the sensitisers are sufficiently long-lived to transfer energy (*e.g.* to molecular oxygen to produce $^1O_2$), electrons or atoms (usually H ones) to or from other molecules. The following reactions can take place, where M is a generic dissolved molecule [11]:

$$S + h\nu \rightarrow {}^1S^* \text{— ISC} \rightarrow {}^3S^* \quad (4)$$

$${}^3S^* + M \rightarrow {}^1S + M^* \quad (5)$$

$$M^* \rightarrow \text{Products} \quad (6)$$

$${}^3S^* + M \rightarrow S^{-\bullet} + M^{+\bullet} \quad (7)$$

$${}^3S^* + M \rightarrow (S{+}H)^\bullet + (M{-}H)^\bullet \quad (8)$$

The sensitiser would often undergo reduction when reacting with M. The reduced sensitiser ($S^{-\bullet}$ or $(S{+}H)^\bullet$) could undergo further transformation or be recycled back to S by dissolved oxygen (reactions 9,10) [25]:

$$S^{-\bullet} + O_2 \rightarrow S + O_2^{-\bullet} \quad (9)$$

$$(S{+}H)^{\bullet} + O_2 \rightarrow S + HO_2^{\bullet} \quad (10)$$

When reactions (9,10) take place, limited or no transformation of S is observed in the process. Reactions (4-8) suggest that photoexcited CDOM sensitisers can induce the transformation of dissolved organic compounds. The transformation processes photosensitised by CDOM play a substantial role in the degradation of important classes of pollutants such as phenols, phenylurea herbicides and sulfonamide antibiotics [3]. In the case of the herbicides isoproturon and diuron, the modelling of CDOM-sensitised photoreactions has predicted with precision their vertical profiles in the Greifensee lake (Switzerland), thereby confirming the important role of CDOM in inducing the phototransformation of these compounds [26].

2.3. Reactions induced by the Hydroxyl Radical, $^{\bullet}OH$

The hydroxyl radical is one of the most reactive transients that are formed in natural waters. The high reactivity implies that this species can be involved into the degradation of refractory pollutants, and some of them (including most notably alkanes, chloroalkanes such as butyl chloride, benzene, toluene, nitrobenzene, benzoic acid) would almost exclusively be degraded by $^{\bullet}OH$ in surface waters. The cited molecules can also be used as probes to quantify $^{\bullet}OH$ in surface-water samples, because they undergo to a limited or null extent direct photolysis or side reactions with $^3CDOM^*$, $CO_3^{-\bullet}$ and $^1O_2$ [27]. However, the very high reactivity of $^{\bullet}OH$ is a drawback for its overall importance as an oxidant in surface waters. The very vast majority of $^{\bullet}OH$ radicals, which are formed upon irradiation of photoactive precursors, would in fact be scavenged by dissolved organic matter (DOM), carbonate/bicarbonate, nitrite and bromide ion (Br^-), the latter particularly in sea waters. Therefore, only a small fraction would be available for the degradation of the xenobiotic compounds depending on their concentration levels [28].

The main photochemical sources of hydroxyl radicals in surface waters are the photolysis of nitrate, nitrite and H_2O_2 (reactions (11-13) and the photo-Fenton reaction (Fe^{2+}/H_2O_2) [29]. Photoexcited $^3CDOM^*$ states could be able to oxidise water. An alternative pathway for the production of $^{\bullet}OH$ from $^3CDOM^*$ is the photo-induced generation of H_2O_2 that is then able to undergo photolysis (reactions 13,14) [30], as well as photo-Fenton reactions in the presence of dissolved Fe^{2+}/Fe^{3+} or Fe^{III}-complexes (reactions 15,16, where L is an organic ligand of Fe) [31].

$$NO_3^- + h\nu + H^+ \rightarrow {}^{\bullet}OH + {}^{\bullet}NO_2 \quad (11)$$

$$NO_2^- + h\nu + H^+ \rightarrow {}^{\bullet}OH + {}^{\bullet}NO \quad (12)$$

$$H_2O_2 + h\nu \rightarrow 2\ {}^{\bullet}OH \quad (13)$$

$$2\ CDOM + O_2 + 2\ H^+ + 2\ h\nu \rightarrow H_2O_2 + 2\ CDOM^{\bullet +} \quad (14)$$

$$Fe^{III}\text{-}L + h\nu \rightarrow Fe^{2+} + L^{+\bullet} \quad (15)$$

$$Fe^{2+} + H_2O_2 \rightarrow Fe^{3+} + {}^{\bullet}OH + OH^- \quad (16)$$

$$FeOH^{2+} + h\nu \rightarrow Fe^{2+} + {}^{\bullet}OH \quad (17)$$

Fe(III) could also produce $^{\bullet}OH$ directly, upon photolysis of its hydroxocomplexes (mainly $FeOH^{2+}$, reaction 17). However, $FeOH^{2+}$ is present at significant levels only under acidic conditions (typically at pH < 5), which are little representative of surface waters [32]. A major exception could be acidic mine drainage water, where $FeOH^{2+}$ photolysis could be an important source of $^{\bullet}OH$ [33]. Hydroxyl groups on the surface of Fe(III) (hydr)oxide colloids (=Fe^{III}-OH) would also be able to yield $^{\bullet}OH$ upon photolysis, but different studies suggest that the efficiency of the heterogeneous process is very low [22,32]. Accordingly, if Fe plays a significant role in the generation of $^{\bullet}OH$ radicals under non-acidic conditions, it is mainly through its possible interaction with DOM (reactions 15,16).

The relative role of nitrate, nitrite and CDOM as $^{\bullet}OH$ sources is in the order CDOM > NO_2^- > NO_3^- [34]. The main scavengers of $^{\bullet}OH$ radicals in surface waters are, in the order, DOM (obviously also including the non-absorbing compounds), inorganic carbon (carbonate and bicarbonate), and nitrite [28,35]. From experimental data on natural water irradiation and the literature second-order rate constants for reaction with $^{\bullet}OH$ [36], it is possible to predict the steady-state [$^{\bullet}OH$] as a function of the concentration values of nitrate, nitrite, carbonate, bicarbonate, and the amount of organic matter [8]. The dissolved organic matter in surface waters is measured as NPOC (Non-Purgeable Organic Carbon). The following equation holds for a sunlight irradiance of 22 W m^{-2} in the UV, as can be found in the water surface layer during the morning at mid-July and mid-latitude. The contribution of organic matter as $^{\bullet}OH$ source and sink has been obtained as the average result of irradiation experiments of lake water samples collected in NW Italy [8,35]:

$$[{}^{\bullet}OH] = \frac{1.7\cdot 10^{-7}[NO_3^-] + 2.6\cdot 10^{-5}[NO_2^-] + 5.7\cdot 10^{-12}\,NPOC}{8.5\cdot 10^{6}[HCO_3^-] + 3.9\cdot 10^{8}[CO_3^{2-}] + 1.0\cdot 10^{10}[NO_2^-] + 2.0\cdot 10^{4}\,NPOC} \quad (18)$$

The numerator takes into account the $^{\bullet}OH$ sources, and the sinks are at the denominator. To simplify the equation, it is possible to introduce IC [mg C L^{-1}] = ([HCO_3^-] + [CO_3^{2-}])/12000 and to consider that on average in surface waters it is [HCO_3^-] ≈ 10^2 [CO_3^{2-}] [8]. In the neutral to basic conditions that characterise most surface water bodies it is possible to neglect [H_2CO_3], also because of its low reaction rate constant with $^{\bullet}OH$ [36].

Figure 4 reports the steady-state [$^{\bullet}OH$] in the presence of 3×10^{-6} M nitrate and 2×10^{-8} M nitrite, as a function of IC and NPOC. At elevated IC and low NPOC, inorganic carbon can be the main sink of $^{\bullet}OH$ and, if nitrate and nitrite are not too elevated, the contribution of dissolved organic matter as a $^{\bullet}OH$ source can be more important than that as a sink. This explains why, at elevated IC, [$^{\bullet}OH$] increases with increasing NPOC. In contrast, at low IC, DOM would soon become the main $^{\bullet}OH$ sink. Because nitrate and nitrite would be non-negligible $^{\bullet}OH$ sources under such conditions, the role of DOM as $^{\bullet}OH$ scavenger would prevail. As a consequence, [$^{\bullet}OH$] decreases with increasing NPOC.

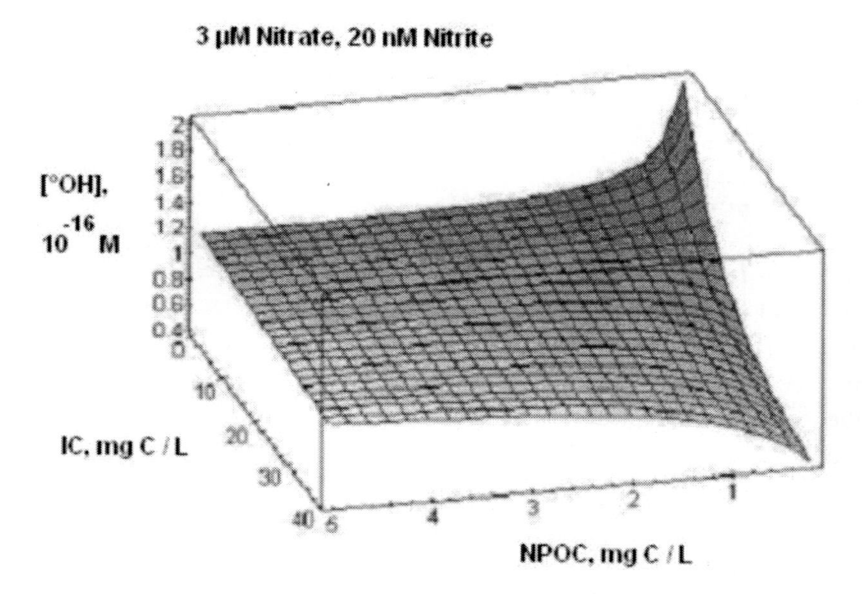

Figure 4. Steady-state [$^{\bullet}$OH], according to equation (18), as a function of IC and NPOC, in the presence of 3 μM nitrate and 20 nM nitrite. Sunlight irradiance: 22 W m^{-2} in the UV.

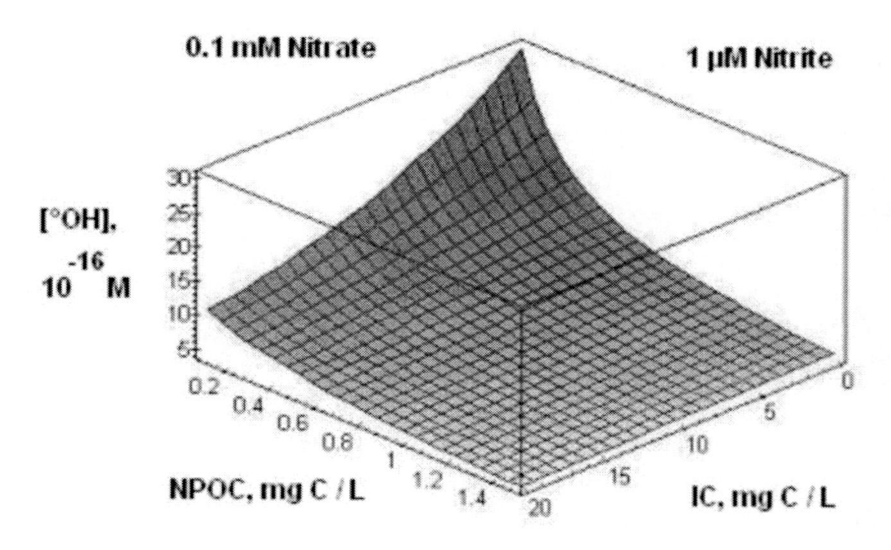

Figure 5. Steady-state [$^{\bullet}$OH], according to equation (18), as a function of IC and NPOC, in the presence of 0.1 mM nitrate and 1 μM nitrite. Sunlight irradiance: 22 W m^{-2} in the UV.

Figure 5 reports the steady-state [$^{\bullet}$OH], vs. IC and NPOC, for more elevated concentrations of nitrate and nitrite (0.1 mM and 1 μM, respectively). In this case [$^{\bullet}$OH] decreases with increasing NPOC, even at high IC. Indeed, at sufficiently high [NO_3^-] and [NO_2^-] there is no possibility for organic matter to be the main source of $^{\bullet}$OH without being at the same time the main sink.

The radical $^{\bullet}OH$ often has elevated reaction rate constants with organic compounds. When the rate constants are in the range of 10^{9}-10^{10} M^{-1} s^{-1} [36], the reaction with $^{\bullet}OH$ can play an important role in the degradation of the relevant substances in surface waters.

2.4. Reactions Induced by the Carbonate Radical, $CO_3^{-\bullet}$

The carbonate radical anion is a fairly reactive transient with relatively elevated reduction potential (E^0 = 1.59 V), although its oxidising capability cannot be compared with the hydroxyl radical (E^0 = 2.59 V) [3, 27]. The radical $CO_3^{-\bullet}$ is formed in surface waters upon reaction between $^{\bullet}OH$ and CO_3^{2-} or HCO_3^{-}. An additional pathway for the generation of $CO_3^{-\bullet}$ is the reaction between carbonate and ^{3}CDOM*, but its weight could be at most 10% of the $^{\bullet}OH$-mediated one [37].

$$^{\bullet}OH + CO_3^{2-} \rightarrow OH^- + CO_3^{-\bullet}\ [k_{21} = 3.9\times10^8\ M^{-1}\ s^{-1}] \quad (19)$$

$$^{\bullet}OH + HCO_3^{-} \rightarrow H_2O + CO_3^{-}\ [k_{22} = 8.5\times10^6\ M^{-1}\ s^{-1}] \quad (20)$$

$$^3CDOM^* + CO_3^{2-} \rightarrow DOM^{-\bullet} + CO_3^{-\bullet}\ [k_{23} = 1\times10^5\ M^{-1}\ s^{-1}] \quad (21)$$

The reaction with DOM is the main sink of the carbonate radical in surface waters. Literature sources do not agree for the rate constant of such a reaction (40 or 280 L (mg C)$^{-1}$ s^{-1}) [37], but a value of 100 L (mg C)$^{-1}$ s^{-1} would not be very far from reality. By neglecting for simplicity reaction (21), and applying the steady-state approximation to [$CO_3^{-\bullet}$], one obtains the following result:

$$\left[CO_3^{-\bullet}\right] = \left[^{\bullet}OH\right]\cdot\frac{8.5\cdot10^6\cdot\left[HCO_3^-\right]+3.9\cdot10^8\cdot\left[CO_3^{2-}\right]}{10^2\cdot NPOC} \quad (22)$$

The steady-state [$^{\bullet}OH$] can be obtained by the model equation (18). In such a case equation (22) would be valid for 22 W m^{-2} sunlight UV irradiance as already discussed. The main problem with equation (22) is that the concentration values of carbonate and bicarbonate are not always determined together in surface water samples, and the value of the inorganic carbon (IC [mg C L^{-1}] = ([HCO_3^-] + [CO_3^{2-}])/12000) is more often available. An additional, reasonable hypothesis is that [HCO_3^-] $\approx 10^2$ [CO_3^{2-}] [8]. Figure 6 shows the trend of [$CO_3^{-\bullet}$] vs. IC and NPOC, in the presence of 10 μM nitrate and 50 nM nitrite. The steady-state [$CO_3^{-\bullet}$] increases with increasing IC, which is reasonable because carbonate and bicarbonate are the immediate precursors of $CO_3^{-\bullet}$. [$CO_3^{-\bullet}$] decreases with NPOC, because DOM is the main sink of both $CO_3^{-\bullet}$ and $^{\bullet}OH$ that is involved in the formation of the carbonate radical. Figure 6 indicates that [$CO_3^{-\bullet}$] would reach values up to 10^{-14} M under conditions that are typical of surface waters, in agreement with field data [38]. This means that $CO_3^{-\bullet}$ would be one-two orders of magnitude more concentrated than $^{\bullet}OH$, which in some cases could compensate for its lower reactivity.

The rate constants of the reactions that involve $CO_3^{-\bullet}$ and organic compounds are highly variable. The upper limit is around 10^9 M^{-1} s^{-1} [39], which is easier to be reached by phenolates, anilines and some sulphur-containing molecules. The bimolecular rate constants in aqueous solution have a diffusion-controlled upper limit of around 2×10^{10} M^{-1} s^{-1}, which is also the upper limit for the reactions that involve $^{\bullet}OH$ [36]. The carbonate radical can be a more important sink than $^{\bullet}OH$ for compounds that have a bimolecular rate constant of around 10^9 M^{-1} s^{-1} for reaction with $CO_3^{-\bullet}$.

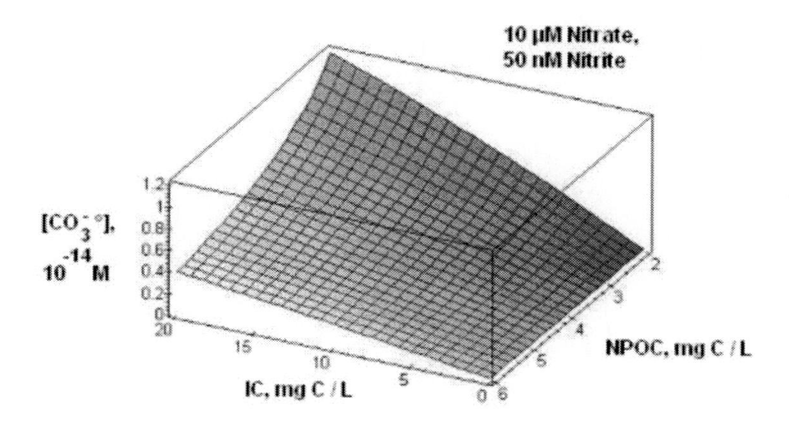

Figure 6. Trend of $[CO_3^{-\bullet}]$ as a function of IC and NPOC, in the presence of 10 µM nitrate and 50 nM nitrite, according to equations (22) and (18). Sunlight intensity: 22 W m^{-2} in the UV.

2.5. Other Reactions

CDOM can induce phototransformation processes directly, via the reactions of its excited triplet states ($^3CDOM^*$) with dissolved compounds, or indirectly upon generation of $^{\bullet}OH$ and singlet oxygen ($^1O_2$, reaction 23). In reaction (25), S represents a dissolved organic molecule. The light-induced transformation reactions of CDOM can yield peroxy radicals ($ROO^{\bullet}$), which could show a specific reactivity [3].

$$^3CDOM^* + O_2 \rightarrow CDOM + {}^1O_2 \quad (23)$$

$$^1O_2 \rightarrow O_2 \quad (24)$$

$$S + {}^1O_2 \rightarrow \text{Products} \quad (25)$$

Singlet oxygen can lose the surplus energy because of collisions with the solvent. Such a deactivation process (reaction 24), with rate constant $k_{24} = 2.5\times10^5$ s^{-1}, is in competition with the reaction (25) that involves dissolved compounds. There is evidence that reaction (24) prevails to a large extent over (25) as a sink of $^1O_2$ in surface waters [27]. As a consequence, $^1O_2$ cannot be accumulated in solution and its ability to induce the degradation of organic substrates depends on their reactivity toward $^1O_2$. It is likely that $^1O_2$ plays an important role in inducing the degradation of photolabile aminoacids [40]. Very little is known regarding the importance of $ROO^•$ in surface-water photochemistry. The relevant reactions could simulate those of $^3CDOM^*$, but it is generally accepted that $^3CDOM^*$ is more important than $ROO^•$ in the phototransformation processes that take place in surface waters [3].

An interesting class of photochemical reactions is initiated by Fe(III) (hydr)oxide colloids. They are semiconductor oxides, and the absorption of visible radiation promotes electrons to the conduction band of the semiconductor, leaving electron vacancies (holes) in the valence band. Conduction-band electrons are reducing species and can for instance interact with oxygen. In contrast, valence-band holes are oxidising and can extract electrons from a number of compounds adsorbed on the semiconductor surface [41]. Figure 7 shows the main processes that take place upon irradiation of the Fe(III) (hydr)oxides.

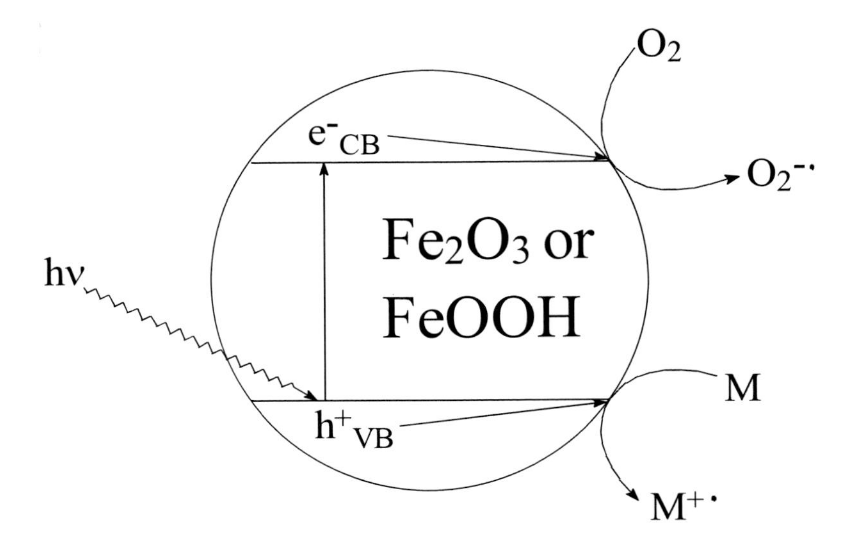

Figure 7. Schematic of the processes that follow the absorption of visible radiation by Fe(III) (hydr)oxides. M represents a dissolved molecule, e^-_{CB} an electron of the conduction band, h^+_{VB} a hole of the valence band. Note that the absorption of UV radiation would also induce photolysis of the surface Fe^{III}-OH groups, albeit with low quantum yield.

Among the species that can be oxidised at the surface of Fe(III) (hydr)oxides there are organic ligands that can form surface complexes [3], and dissolved anions such as carbonate [42], nitrite [43-45], chloride [22,46], and bromide [46]. We have shown that the oxidation of CO_3^{2-} by irradiated hematite yields the oxidising agent $CO_3^{-•}$ [42], and the oxidation of nitrite yields $^•NO_2$ that is a nitrating agent in the aqueous solution. The radical $^•NO_2$ is also produced by irradiation of nitrate and nitrite [15,34,42-45]. The irradiation of Fe(III) (hydr)oxides in the presence of chloride and bromide results into the generation of the radicals $Cl^•$ and $Br^•$, soon transformed into $Cl_2^{-•}$ and $Br_2^{-•}$ upon reaction with the corresponding halogenide ions [22,46,47].

The radical $Cl_2^{-•}$ is an oxidising and chlorinating agent, and is for instance able to chlorinate aromatic hydrocarbons [22,46]. The radical $Br_2^{-•}$ is a less powerful oxidant, but a very efficient brominating agent for aromatic compounds [46,47].

3. MODELLING OF THE MAIN PHOTOCHEMICAL REACTIONS IN SURFACE WATERS

This section presents an approach to model the phototransformation kinetics of dissolved compounds as a function of their photochemical reactivity, of the water chemical composition and the column depth.

3.1. Surface-Water Absorption Spectrum

It is possible to find a reasonable correlation between the absorption spectrum of surface waters and their content of dissolved organic matter, expressed as NPOC. The following equation holds for the water spectrum [48] (error bounds represent $\pm \sigma$):

$$A_1(\lambda) = (0.45 \pm 0.04) \cdot NPOC \cdot e^{-(0.015 \pm 0.002) \cdot \lambda} \tag{26}$$

Equation (26) can be adopted when the absorption spectrum of water is not available.

3.2. Direct Photolysis

The calculation of the photon flux absorbed by a molecule S requires taking into account the mutual competition for sunlight irradiance between S and the other water components (mostly CDOM, which is the main sunlight absorber in the spectral region of interest, around 300-500 nm).

Under the Lambert-Beer approximation, at a given wavelength λ, the ratio of the photon flux densities absorbed by two different species is equal to the ratio of the respective absorbances. The same is also true of the ratio of the photon flux density absorbed by species to the total photon flux density absorbed by the solution ($p_a^{tot}(\lambda)$) [49]. Accordingly, the photon flux absorbed by S in a water column of depth d (expressed in cm) can be obtained by the following equations (note that $A_1(\lambda)$ is the specific absorbance of the surface water sample over a 1 cm optical path length, $A_{tot}(\lambda)$ the total absorbance of the water column, $p°(\lambda)$ the spectrum of sunlight, $\varepsilon_S(\lambda)$ the molar absorption coefficient of S, in units of M^{-1} cm^{-1}, and $p_a^S(\lambda)$ its absorbed spectral photon flux density; it is also $p_a^S(\lambda) \ll p_a^{tot}(\lambda)$ and $A_S(\lambda) \ll A_{tot}(\lambda)$ in the very vast majority of the environmental cases):

$$A_{tot}(\lambda) = A_1(\lambda) \cdot d \tag{27}$$

$$A_S(\lambda) = \varepsilon_S(\lambda) \cdot d \cdot [S] \tag{28}$$

$$p_a^{tot}(\lambda) = p°(\lambda) \cdot (1 - 10^{-A_{tot}(\lambda)}) \tag{29}$$

$$p_a^S(\lambda) = p_a^{tot}(\lambda) \cdot A_S(\lambda) \cdot [A_{tot}(\lambda)]^{-1} \quad (30)$$

Note that the sunlight spectrum $p°(\lambda)$ in the calculations will be referred to a sunlight UV irradiance of 22 W m^{-2} (see Figure 8). Finally, the absorbed photon flux P_a^S is the integral over wavelength of the absorbed photon flux density:

$$P_a^S = \int_\lambda p_a^S(\lambda)\, d\lambda \quad (31)$$

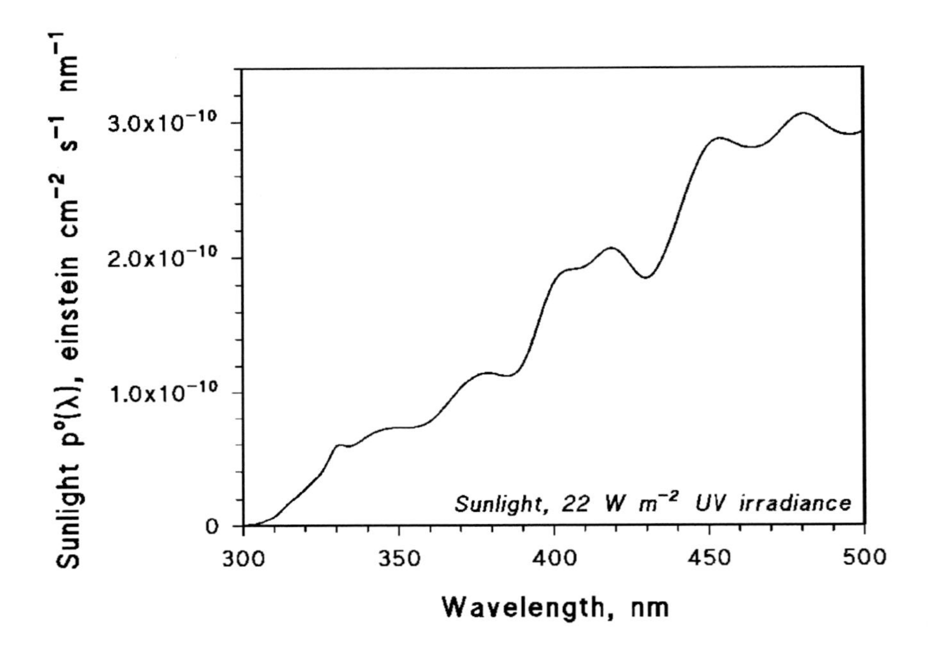

Figure 8. Spectral photon flux density of sunlight ($p°(\lambda)$), corresponding to 22 W m^{-2} UV irradiance [10], as can be found on 15 July at 45°N latitude, under clear-sky conditions, at 10 am or 2 pm.

The sunlight spectrum $p°(\lambda)$ is referred to a unit surface area (units of einstein s^{-1} nm^{-1} cm^{-2}, see Figure 8), thus P_a^S (units of einstein s^{-1} cm^{-2}) represents the photon flux absorbed by S inside a cylinder of unit area (1 cm^2) and depth d. The rate of photolysis of S, expressed in M s^{-1}, can be approximated as $Rate_S = 10^3\ \Phi_S\ P_a^S\ d^{-1}$, where Φ_S is the polychromatic photolysis quantum yield of S in the relevant wavelength interval, and d is expressed in cm (also note that 1 L = 10^3 cm^3). This approximated expression of $Rate_S$ can be adopted if the detailed wavelength trend of Φ_S is not known, provided that Φ_S is referred to the same wavelength interval where the spectra of S and sunlight overlap. The pseudo-first order degradation rate constant of S is $k_S = Rate_S\ [S]^{-1}$, which corresponds to a half-life time $t_S = ln\ 2\ (k_S)^{-1}$.

The time t_S is expressed in seconds of continuous irradiation under sunlight, at 22 W m^{-2} UV irradiance. It has been shown that the sunlight energy reaching the ground in a summer sunny day (SSD) such as 15 July at 45°N latitude corresponds to 10 h = 3.6×10^4 s continuous irradiation at 22 W m^{-2} UV irradiance [34]. Accordingly, the half-life time expressed in SSD units would be given by:

$$\tau^{SSD}_S = (3.6\times10^4)^{-1}\ ln\ 2\ (k_S)^{-1} = 1.9\times10^{-5}\ [S]\ d\ 10^{-3}\ (\Phi_S\, P_a^S)^{-1} =$$
$$= 1.9\times10^{-5}\ [S]\ d\ 10^{-3}\ (\Phi_S \int_\lambda p_a^S(\lambda)\, d\lambda\,)^{-1} =$$
$$= 1.9\times10^{-5}\ [S]\ d\ 10^{-3}\ (\Phi_S \int_\lambda p_a^{tot}(\lambda)\cdot A_S(\lambda)\cdot[A_{tot}(\lambda)]^{-1}\, d\lambda\,)^{-1} =$$
$$= \frac{1.9\times10^{-8}\ d}{\Phi_S \int_\lambda p°(\lambda)\,(1-10^{-A_1(\lambda)\,d})\,\frac{\varepsilon_S(\lambda)}{A_1(\lambda)}\,d\lambda} \qquad (32)$$

3.3. Reactions with ^{3}CDOM*

The formation of the excited triplet states of CDOM (^{3}CDOM*) in surface waters is a direct consequence of the radiation absorption by CDOM. In aerated solution, ^{3}CDOM* could undergo thermal deactivation or reaction with O_2, and a pseudo-first order quenching rate constant $k_{3CDOM*} \approx 5\times10^5$ s^{-1} has been observed. The quenching of ^{3}CDOM* would be in competition with the reaction between ^{3}CDOM* and S, with rate constant $k_{S,3CDOM*}$ [3,26].

In the Rhône delta waters it has been found that the formation rate of ^{3}CDOM* is $R_{3CDOM*} = 1.28\times10^{-3}\ P_a^{CDOM}$ [50], where $P_a^{CDOM} = 10^3\ d^{-1} \int_\lambda p_a^{CDOM}(\lambda)\, d\lambda$ is the photon flux absorbed by CDOM. Considering the competition between ^{3}CDOM* deactivation and the reaction with S, the following expression for the degradation rate of S by ^{3}CDOM* is obtained:

$$R_S^{^3CDOM*} = R_{^3CDOM*}\cdot\frac{k_{S,^3CDOM*}\cdot[S]}{k_{^3CDOM*}} \qquad (33)$$

In a pseudo-first order approximation, the degradation rate constant is $k_S = R_S^{3CDOM*}\ [S]^{-1}$ and the half-life time is $t_S = ln\ 2\ k_S^{-1}$. Considering the usual conversion ($\approx$ 10 h) between a constant 22 W m^{-2} sunlight UV irradiance and a SSD unit, one gets the following expression for $\tau^{SSD}_{S,^3CDOM*}$ (remembering that $P_a^{CDOM} = 10^3\ d^{-1} \int_\lambda p_a^{CDOM}(\lambda)\, d\lambda$):

$$\tau^{SSD}_{DCNP,^3CDOM*} = \frac{7.52\cdot d}{k_{DCNP,^3CDOM*}\cdot \int_\lambda p_a^{CDOM}(\lambda)\, d\lambda} \qquad (34)$$

Note that 7.52 = (ln 2) $k_{3CDOM*}\,(1.28\cdot10^{-3}\cdot 3.60\cdot10^4\cdot 10^3)^{-1}$.

3.4. Reactions with $^{\bullet}$OH.

In natural surface waters under sunlight illumination, the main $^{\bullet}$OH sources are (in order of average importance) CDOM, nitrite and nitrate. It can be noted that CDOM can photochemically produce H_2O_2, which is then responsible for the occurrence of the photo-Fenton reaction (photo-Fe^{2+}/H_2O_2) in surface waters.

At the present state of knowledge it is reasonable to hypothesise that these three sources generate $^{\bullet}$OH independently, with no mutual interactions. Therefore, the total formation rate of $^{\bullet}$OH ($R_{\bullet OH}^{tot}$) is the sum of the contributions of the three species:

$$R_{\bullet OH}^{tot} = R_{\bullet OH}^{CDOM} + R_{\bullet OH}^{NO2-} + R_{\bullet OH}^{NO3-} \quad (35)$$

Various studies have yielded useful correlation between the formation rate of $^{\bullet}$OH by the photoactive species and the respective absorbed photon fluxes of sunlight (P_a^{CDOM}, P_a^{NO2-}, P_a^{NO3-}). In particular, it has been found that [48,51]:

$$R_{\bullet OH}^{CDOM} = (3.0 \pm 0.4) \cdot 10^{-5} \cdot P_a^{CDOM} \quad (36)$$

$$R_{\bullet OH}^{NO2-} = (7.2 \pm 0.3) \cdot 10^{-2} \cdot P_a^{NO2-} \quad (37)$$

$$R_{\bullet OH}^{NO3-} = (4.3 \pm 0.2) \cdot 10^{-2} \cdot \frac{[IC] + 0.0075}{2.25\,[IC] + 0.0075} \cdot P_a^{NO3-} \quad (38)$$

where $[IC] = [H_2CO_3] + [HCO_3^-] + [CO_3^{2-}]$ is the total amount of inorganic carbon. Error bounds represent $\pm\ \sigma$. The calculation of the photon fluxes absorbed by CDOM, nitrate and nitrite requires taking into account the mutual competition for sunlight irradiance, also considering that CDOM is the main absorber in the UV region where also nitrite and nitrate absorb radiation. From the already cited proportionality between the absorbance and the absorbed photon flux density [49], one gets the following equations:

$$A_{tot}(\lambda) = A_1(\lambda) \cdot d \quad (39)$$

$$A_{NO3-}(\lambda) = \varepsilon_{NO3-}(\lambda) \cdot d \cdot [NO_3^-] \quad (40)$$

$$A_{NO2-}(\lambda) = \varepsilon_{NO2-}(\lambda) \cdot d \cdot [NO_2^-] \quad (41)$$

$$A_{CDOM}(\lambda) = A_{tot}(\lambda) - A_{NO3-}(\lambda) - A_{NO2-}(\lambda) \approx A_{tot}(\lambda) \quad (42)$$

$$p_a^{tot}(\lambda) = p°(\lambda)\cdot(1-10^{-A_{tot}(\lambda)}) \quad (43)$$

$$p_a^{CDOM}(\lambda) = p_a^{tot}(\lambda)\cdot A_{CDOM}(\lambda)\cdot[A_{tot}(\lambda)]^{-1} \approx p_a^{tot}(\lambda) \quad (44)$$

$$p_a^{NO2-}(\lambda) = p_a^{tot}(\lambda)\cdot A_{NO2-}(\lambda)\cdot[A_{tot}(\lambda)]^{-1} \quad (45)$$

$$p_a^{NO3-}(\lambda) = p_a^{tot}(\lambda)\cdot A_{NO3-}(\lambda)\cdot[A_{tot}(\lambda)]^{-1} \quad (46)$$

To express the formation rates of $^{\bullet}OH$ in M s^{-1}, the absorbed photon fluxes P_a^i (i = nitrate, nitrite or CDOM) should be expressed in einstein L^{-1} s^{-1}. Integration of $p_a^i(\lambda)$ over wavelength would give units of einstein cm^{-2} s^{-1} that represent the moles of photons absorbed per unit surface area and unit time. Assuming a cylindrical volume of unit surface area (1 cm^2) and depth d (expressed in cm), the absorbed photon fluxes in einstein L^{-1} s^{-1} units would be expressed as follows (note that 1 L = 10^3 cm^3):

$$P_a^{CDOM} = 10^3\ d^{-1} \int_\lambda p_a^{CDOM}(\lambda)\, d\lambda \quad (47)$$

$$P_a^{NO2-} = 10^3\ d^{-1} \int_\lambda p_a^{NO2-}(\lambda)\, d\lambda \quad (48)$$

$$P_a^{NO3-} = 10^3\ d^{-1} \int_\lambda p_a^{NO3-}(\lambda)\, d\lambda \quad (49)$$

Accordingly, having as input data d, $A_1(\lambda)$, $[NO_3^-]$, $[NO_2^-]$ and $p°(\lambda)$ (the latter referred to a 22 W m^{-2} sunlight UV irradiance, see Figure 8), it is possible to model the expected $R_{\bullet OH}^{tot}$ of the sample. The photogenerated $^{\bullet}OH$ radicals could react either with S or with the natural scavengers present in surface water (mainly organic matter, bicarbonate, carbonate and nitrite). The natural scavengers have a $^{\bullet}OH$ scavenging rate constant (in s^{-1}) of $\Sigma_i k_{Si} [S_i] = 2\times10^4\ NPOC + 8.5\times10^6\ [HCO_3^-] + 3.9\times10^8\ [CO_3^{2-}] + 1.0\times10^{10}\ [NO_2^-]$ (NPOC is expressed in mg C L^{-1} and the other concentration values are in molarity) [35,36]. Accordingly, the reaction rate between S and $^{\bullet}OH$ can be expressed as follows [48]:

$$R_S^{\bullet OH} = R_{\bullet OH}^{tot} \frac{k_{S,\bullet OH}\ [S]}{\sum_i k_{Si}\ [S_i]} \quad (50)$$

where $k_{S,\bullet OH}$ is the second-order reaction rate constant between S and $^{\bullet}OH$ and [S] is a molar concentration. Note, that in the vast majority of the environmental cases it would be $k_{S,\bullet OH} [S] \ll \Sigma_i k_{Si} [S_i]$. The pseudo-first order degradation rate constant of S is $k_S = R_S^{\bullet OH} [S]^{-1}$, and the half-life time is $t_S = ln\ 2\ k_S^{-1}$. By shifting as usual to the SSD units one gets the following equation:

$$\tau_{S,^{\bullet}OH}^{SSD} = \frac{\ln 2 \sum_i k_{Si}\,[S_i]}{3.6\cdot 10^4\; R_{^{\bullet}OH}^{tot}\; k_{S,^{\bullet}OH}} = 1.9\cdot 10^{-5}\, \frac{\sum_i k_{Si}\,[S_i]}{R_{^{\bullet}OH}^{tot}\; k_{S,^{\bullet}OH}} \quad (51)$$

Note that $1.9\cdot10^{-5} = \ln 2\ (3.6\cdot10^4)^{-1}$.

3.5. Reactions with $CO_3^{-\bullet}$

The radical $CO_3^{-\bullet}$ can be produced in reactions (19-21) and is mainly scavenged by dissolved organic matter, with a reaction rate constant of $\approx 10^2$ (mg C)$^{-1}$ s^{-1}.

The formation rate of $CO_3^{-\bullet}$ in reactions (19,20) is given by the formation rate of $^{\bullet}OH$ times the fraction of $^{\bullet}OH$ that reacts with carbonate and bicarbonate, as follows [52]:

$$R_{CO3-\bullet(\bullet OH)} = R_{^{\bullet}OH}^{tot} \cdot \frac{8.5\cdot10^6\cdot[HCO_3^-]+3.9\cdot10^8\cdot[CO_3^{2-}]}{2.0\cdot10^4\cdot NPOC+1.0\cdot10^{10}\cdot[NO_2^-]+8.5\cdot10^6\cdot[HCO_3^-]+3.9\cdot10^8\cdot[CO_3^{2-}]} \quad (52)$$

The formation of $CO_3^{-\bullet}$ in reaction (22) is given by [52]:

$$R_{CO3-\bullet(CDOM)} = 6.5\cdot10^{-3}\cdot[CO_3^{2-}]\cdot P_a^{CDOM} \quad (53)$$

The total formation rate of $CO_3^{-\bullet}$ is $R_{CO3-\bullet}^{tot} = R_{CO3-\bullet(\bullet OH)} + R_{CO3-\bullet(CDOM)}$. The transformation rate of S by $CO_3^{-\bullet}$ is given by the fraction of $CO_3^{-\bullet}$ that reacts with S, in competition with the reaction between $CO_3^{-\bullet}$ and DOM:

$$R_{DCNP,CO3-\bullet} = \frac{R_{CO3-\bullet}^{tot}\cdot k_{S,CO3-\bullet}\cdot[S]}{10^2\cdot NPOC} \quad (54)$$

where $k_{S,CO3-\bullet}$ is the second-order reaction rate constant between S and $CO_3^{-\bullet}$. In a pseudo-first order approximation, the rate constant is $k_S = R_{S,CO3-\bullet}\,[S]^{-1}$ and the half-life time is $t_S = \ln 2\ k_S^{-1}$. Considering the usual conversion ($\approx$ 10 h) between a constant 22 W m^{-2} sunlight UV irradiance and a SSD unit, the following expression for $\tau_{S,CO3-\bullet}^{SSD}$ is obtained:

$$\tau_{S,CO3-\bullet}^{SSD} = 1.9\cdot10^{-5}\cdot\left(\frac{10^2\cdot NPOC}{R_{CO3-\bullet}^{tot}\cdot k_{S,CO3-\bullet}}\right) \quad (55)$$

Note that $1.9\cdot10^{-5} = \ln 2\ (3.6\cdot10^4)^{-1}$.

3.6. Reactions with $^1O_2$

The formation of singlet oxygen in surface waters arises from the energy transfer between ground-state molecular oxygen and the excited triplet states of CDOM ($^3CDOM^*$). Accordingly, irradiated CDOM is practically the only source of $^1O_2$ in the aquatic systems. In contrast, the main $^1O_2$ sink is the energy loss to ground-state O_2 by collision with the water molecules, with a pseudo-first order rate constant $k_{1O2} = 2.5\times10^5$ s^{-1} [3, 27]. The dissolved species, including the dissolved organic matter that is certainly able to react with $^1O_2$, would play a minor to negligible role as sinks of $^1O_2$ in the aquatic systems. The main processes involving $^1O_2$ and S in surface waters would be the following [3, 27]:

$$^3CDOM^* + O_2 \rightarrow CDOM + {}^1O_2 \qquad (56)$$

$$^1O_2 + H_2O \rightarrow O_2 + H_2O + \text{heat} \qquad (57)$$

$$^1O_2 + S \rightarrow \text{Products} \qquad (58)$$

In the Rhône delta waters it has been found that the formation rate of $^1O_2$ by CDOM is $R_{1O2}^{CDOM} = 1.25\cdot10^{-3}\ P_a^{CDOM}$ [50]. Considering the competition between the deactivation of $^1O_2$ by collision with the solvent (reaction 57) and the reaction (58) with S, one gets the following expression for the degradation rate of S by $^1O_2$:

$$R_S^{^1O_2} = R_{^1O_2}^{CDOM} \cdot \frac{k_{S,^1O_2} \cdot [S]}{k_{^1O_2}} \qquad (59)$$

where $k_{S,1O2}$ is the second-order reaction rate constant between S and $^1O_2$. In a pseudo-first order approximation, the transformation rate constant is $k_S = R_S^{1O2}\ [S]^{-1}$ and the half-life time is $t_S = \ln 2\ k_S^{-1}$. Considering the usual conversion ($\approx$ 10 h) between a constant 22 W m^{-2} sunlight UV irradiance and a SSD unit, the following expression for $\tau_{S,1O2}^{SSD}$ is obtained (remembering that $R_{1O2}^{CDOM} = 1.25\cdot10^{-3}\ P_a^{CDOM}$ and $P_a^{CDOM} = 10^3\ d^{-1} \int_\lambda p_a^{CDOM}(\lambda)\, d\lambda$):

$$\tau_{S,^1O_2}^{SSD} = \frac{4.81}{R_{^1O_2}^{CDOM}\ k_{S,^1O_2}} = \frac{3.85\cdot d}{k_{S,^1O_2} \cdot \int_\lambda p_a^{CDOM}(\lambda)\, d\lambda} \qquad (60)$$

Note that $3.85 = (\ln 2)\ k_{1O2}\ (1.25\cdot10^{-3} \cdot 3.60\cdot10^4 \cdot 10^3)^{-1}$.

3.7. Application of the Model to the Lifetime of 2,4-Dichloro-6-Nitrophenol (DCNP)

DCNP is a toxic and genotoxic chloronitroaromatic compound that arises from the photoinduced nitration of 2,4-dichlorophenol (DCP) by $^{\bullet}NO_2$. DCP is, in turn, a transformation intermediate of the herbicide dichlorprop that is adopted against broad-leaf weeds, *e.g.* in flooded rice farming [45]. The dissociated, anionic form of DCNP, which prevails in surface waters, has a polychromatic photolysis quantum yield of 4.5×10^{-6} and second-order reaction rate constants of 2.8×10^{9} M^{-1} s^{-1} with $^{\bullet}OH$, 3.7×10^{9} M^{-1} s^{-1} with $^{1}O_2$, and 1.4×10^{8} M^{-1} s^{-1} with $^{3}CDOM^{*}$ [53]. It has a negligible reactivity with the carbonate radical. Enough data are available for the modelling of the lifetime of DCNP as a function of the chemical composition of surface waters and of the water column depth. Table 1 reports as an example the data for the Rhône delta lagoons, where sufficient field data are available to check the model predictions [45].

Figure 9 shows the trend of the DCNP half-life time (SSD units) as a function of *d* and NPOC, the other water parameters being as for Table 1. The reactions with $^{\bullet}OH$ and $^{1}O_2$ and the direct photolysis are the main transformation processes for DCNP, while the reaction with $^{3}CDOM^{*}$ plays a less important role. Obviously, the most important processes are those that yield the lower values of the half-life time.

Table 1. Water parameters for the Rhône delta lagoons

	Rhône delta (S France)
Water depth d, m	1.0
Nitrate, M	5.1×10^{-5}
Nitrite, M	3.2×10^{-6}
Bicarbonate, M	2.1×10^{-3}
Carbonate, M	2.6×10^{-5}
pH	7.5
NPOC, mg C L^{-1}	4.5

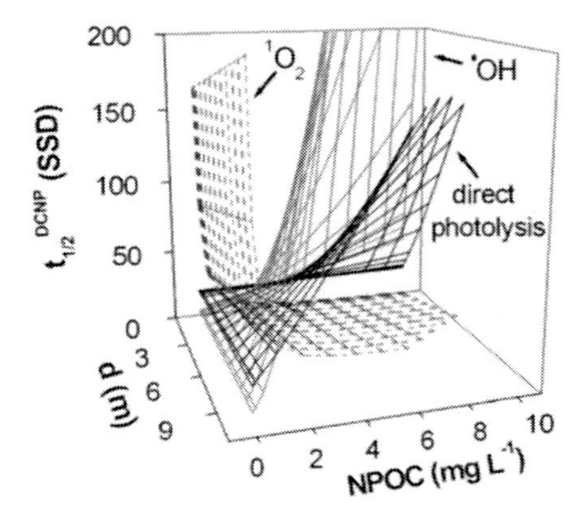

Figure 9. Modelled half-life time of DCNP (SSD units) as a function of water column depth *d* and the NPOC values. The model results are reported for the most important pathways (direct photolysis and reactions with $^{\bullet}OH$ and $^{1}O_2$).

It can be seen that all the half-life time values increase with increasing *d*, which is reasonable because the photochemical processes are more important in shallow waters. Interestingly, direct photolysis and the reactions with $^{\bullet}OH$ and $^{1}O_2$ would all play a similar role for NPOC ≈ 2 mg C L^{-1}; $^{\bullet}OH$ would prevail at lower NPOC values, $^{1}O_2$ at higher NPOC (the prevailing pathway is that yielding the lowest half-life time). In particular, in the Rhône delta lagoons (NPOC = 4.5 mg C L^{-1}) the reaction with $^{1}O_2$ would be the main removal process for DCNP. For the water composition data reported in Table 1, the model foresees a half-life time of 6.3±2.3 days [53], which compares quite well with the 8.3 days observed in the field [45].

Conclusion

Photochemical processes can play an important role in the transformation of biorefractory organic compounds, and most notably of organic pollutants, in surface waters. They include direct photolysis, phototransformation sensitised by CDOM, and reactions with transient species generated by irradiation of photoactive solutes and colloids. The main photoactive compounds in surface waters are CDOM, nitrate, nitrite and Fe(III). Interestingly, nitrite and CDOM are both sources and sinks of reactive species such as the hydroxyl radical.

The relative importance of direct photolysis and of indirect photodegradation is highly variable depending on the molecule under consideration and on the composition of water. Direct photolysis is not operational with molecules that do not absorb sunlight, in which case indirect photodegradation has to prevail. However, the situation is more complicated for molecules that both absorb sunlight and are reactive toward $^{3}CDOM^{*}$, $^{\bullet}OH$, $CO_3^{-\bullet}$ and other transients. It should also be considered that the direct photolysis is more energy-effective than the indirect photodegradation. Indeed, the generation of the excited triplet states of CDOM, followed by reaction with the substrate, and the production of reactive radicals that can either be scavenged or react with the relevant compounds cause a much higher dissipation of radiation energy than the direct photolysis [3]. The latter, however, requires radiation absorption by the substrate to be operational.

A model is presented to predict the lifetime of a dissolved compound as a function of its reactivity toward direct photolysis and with the transients $^{\bullet}OH$, $^{1}O_2$, $CO_3^{-\bullet}$ and $^{3}CDOM^{*}$, of the water chemical composition and the column depth. The described approach was able to predict the lifetime of DCNP in the Rhône delta lagoons, with good agreement with the field data.

References

[1] J.L. Oliveira *et al.*, *Acta Hydrochim. Hydrobiol.*, 2006, 34, 608.

[2] M. Caselli, *Chim. Ind. (Milan)*, 2009, 91, 144.

[3] P. Boule *et al.* (eds.), Environmental Photochemistry Part II (The Handbook of Environmental Chemistry, Vol. 2M), Springer: Berlin, 2005.

[4] M. Czaplicka, *J. Haz. Mat.*, 2006, 134, 45.

[5] D. Vogna *et al.*, *Chemosphere*, 2004, 54, 497.

[6] S. Rafqah *et al.*, *Environ. Chem. Lett.*, 2006, 4, 213.
[7] S.E. Braslavsky, *Pure Appl. Chem.*, 2007, 79, 293.
[8] C. Minero *et al.*, *Ann. Chim. (Rome)*, 2007, 97, 685.
[9] S.A. Loiselle *et al.*, *Limnol. Oceanogr.*, 2009, 54, 590.
[10] R. Frank, W. Klöpffer, *Chemosphere*, 1988, 17, 985.
[11] J.G. Calvert, J.N. Pitts, Photochemistry, J. Wiley, New York, 1966.
[12] M.A. Crespin *et al.*, *Environ. Sci. Technol.*, 2001, 35, 4265.
[13] H.C. Zhang, C.H. Huang, *Environ. Sci. Technol.*, 2003, 37, 2421.
[14] C. Guyon *et al.*, *New J. Chem.*, 1984, 8, 685.
[15] S. Chiron *et al.*, *Chemosphere*, 2009, 74, 599.
[16] F. Bonnichon *et al.*, *Chem. Commun.*, 2001, 73.
[17] A. Zertal *et al.*, *J. Photochem. Photobiol. A: Chem.*, 2001, 146, 37.
[18] D. Vione *et al.*, *Wat. Res.*, 2010, 44, 6053-6062.
[19] D.E. Latch *et al.*, *J. Photochem. Photobiol. A: Chem.*, 2003, 158, 63.
[20] C. Tixier *et al.*, *Environ. Sci. Technol.*, 2002, 36, 3482.
[21] M. Cermola *et al.*, *Environ. Chem. Lett.*, 2005, 3, 43.
[22] S. Chiron *et al.*, *Environ. Sci. Technol.* 2006, 40, 5977.
[23] P.G. Coble, *Chem. Rev.*, 2007, 107, 402.
[24] S. Halladja *et al.*, *Environ. Sci. Technol.*, 2007, 41, 6066.
[25] V. Maurino *et al.*, *Photochem. Photobiol. Sci.*, 2008, 7, 321.
[26] A. C. Gerecke *et al.*, *Environ. Sci. Technol.*, 2001, 35, 3915.
[27] W. Stumm (ed.), Aquatic Chemical Kinetics, J. Wiley, New York, 1990, 43.
[28] P.L. Brezonik, J. Fulkerson-Brekken, *Environ. Sci. Technol.*, 1998, 32, 3004.
[29] J. Mack, J.R. Bolton, *J. Photochem. Photobiol. A: Chem.*, 1999, 128, 1.
[30] K. M. G. Mostofa, H. Sakugawa, *Environ. Chem.*, 2009, 6, 524.
[31] B.M. Voelker, B. Sulzberger, *Environ. Sci. Technol.*, 1996, 30, 1106.
[32] P. Mazellier *et al.*, *New J. Chem.*, 1997, 21, 389.
[33] J.M. Allen *et al.*, *Environ. Toxicol. Chem.*, 1996, 15, 107.
[34] C. Minero *et al.*, *Aquat. Sci.*, 2007, 69, 71.
[35] D. Vione *et al.*, *Environ. Sci. Technol.*, 2006, 40, 3775.
[36] G.V. Buxton *et al.*, *J. Phys. Chem. Ref. Data*, 1988, 17, 513.
[37] S. Canonica *et al.*, *Environ. Sci. Technol.*, 2005, 39, 9182.
[38] J.P. Huang, S.A. Mabury, *Environ. Toxicol. Chem.*, 2000, 19, 2181.
[39] P. Neta *et al.*, *J. Phys. Chem. Ref. Data*, 1988, 17, 1027.
[40] A. L. Boreen *et al.*, *Environ. Sci. Technol.*, 2008, 42, 5492-98.
[41] B.C. Faust *et al.*, *J. Phys. Chem.*, 1989, 93, 6371.
[42] S. Chiron *et al.*, *Photochem. Photobiol. Sci.*, 2009, 8, 91.
[43] D. Vione *et al.*, *Environ. Sci. Technol.*, 2002, 36, 669.
[44] D. Vione *et al.*, *Environ. Sci. Technol.*, 2003, 37, 4635.
[45] S. Chiron *et al.*, *Environ. Sci. Technol.*, 2007, 41, 3127.
[46] P. Calza *et al.*, *J. Photochem. Photobiol. A: Chem.*, 2005, 170, 61.
[47] D. Vione *et al.*, *ChemSusChem*, 2008, 1, 197.
[48] D. Vione *et al.*, *Intern. J. Environ. Anal. Chem.*, 2010, 90, 258-273.
[49] S. E. Braslavsky, *Pure Appl. Chem.*, 2007, 79, 293-465.
[50] F. Al-Housari *et al.*, *Photochem. Photobiol. Sci.*, 2010, 9, 78-86.
[51] D. Vione *et al.*, *Wat. Res.*, 2009, 43, 4718-4728.

[52] D. Vione *et al.*, *C. R. Chimie*, 2009, 12, 865-871.
[53] P.R. Maddigapu *et al.*, *Environ. Sci. Technol.*, in press. DOI: 10.1021/es102458n.

In: Photochemistry
Editors: Karen J. Maes and Jaime M. Willems

ISBN: 978-1-61209-506-6

Chapter 10

PHOTOCHEMICAL PROCESSES IN NEEDLES OF OVERWINTERING EVERGREEN CONIFERS

Piotr Robakowski*

Poznan University of Life Sciences, Department of Forestry, Poznan, Poland

ABSTRACT

Overwintering conifer tree species differ in ability to adapt photochemical processes to light environments from other taxonomic groups of plants. Contrastingly to deciduous trees, they maintain leaves for many years and they adapted to conduct photochemical processes in highly varying irradiance. An ability to photosynthesize in different light environments depends on a species' light requirements and its successional status. Moreover, shade-intolerant species which appear in succession early are characterized by greater efficiency of photochemical processes in high irradiance compared with shade-tolerant, late-successional species with a better photochemical performance in shade. Evergreen shade tolerant conifers are adapted to winter photoinhibition, however, their response to winter stress also depends on the light environment of growth. Seedlings growing in high irradiance show a greater winter photoinhibition and shade-acclimated ones only a small, fast-recovering decline in maximum quantum yield of photosystem II photochemistry (F_v/F_m). On the other hand, high light acclimated seedlings develop more efficient photoprotective mechanisms for excess energy dissipation.

In natural conditions under the canopy of trees or in artificial shading the parameters characterising photochemical processes such as effective quantum yield of PSII photochemistry (Φ_{PSII}) and apparent electron transfer rate (ETR) are modified by light regimes. Tree seedlings of the same species acclimated to high irradiance have lower F_v/F_m indicating the PSII down-regulation compared with shade-acclimated ones. This temporarily occurring phenomenon protecting photosynthetic apparatus against permanent damage in PSII is associated with dissipation of excess energy as heat in the xanthophyll cycle. This protective mechanism can be indirectly estimated using non-photochemical quenching of fluorescence (NPQ) which is greater in leaves exposed to high irradiance compared with shade leaves. Down-regulation of PSII photochemistry decreases ETR and overall intensity of photosynthesis, but at the same time a risk of

* Poznan University of Life Sciences, Department of Forestry, Wojska Polskiego 69 St., PL 60-625 Poznan, Poland, e-mail: pierrot@up.poznan.pl

irreversible photoinhibition is diminished. However, when a reduction of F_v/F_m is significantly below 0.8 and it is long-lasting, the photoinhibition of PSII may lead to a permanent dysfunction of the photosynthetic apparatus. PSII down-regulation was observed in a daily scale at midday, but also in a seasonal scale during winter and in early spring at chilling or freezing temperatures. In these situations the reversible photoinhibition can play a photoprotective role unless a cumulative stress of high light and low temperature prolongs and permanently damages the structural proteins in PSII.

Overwintering evergreen shade-tolerant conifers have the ability to adapt the photosynthetic apparatus to changing irradiance which involves some plasticity of their photochemistry and developing of photoprotective mechanisms. Due to these photochemical traits shade-tolerant species can efficiently compete with fast-growing, shade-intolerant pioneer species. High photochemical capacity and well functioning photoprotective mechanisms may give advantage in natural selection and be of critical importance in achieving an evolutionary success.

Abbreviations: ETR – apparent electron transfer rate (μmol m^{-2} s^{-1}); ETR_{max} – maximum value of ETR; F_v/F_m – maximum quantum yield of photosystem II photochemistry; F_m – maximum fluorescence yield; F'_m – maximum fluorescence in the light; F_0 – minimum fluorescence yield; F_s – steady-state fluorescence; NPQ – non-photochemical quenching of fluorescence; PPFD – photosynthetic photon flux density (μmol m^{-2} s^{-1}); $PPFD_{sat}$ - saturating level of photosynthetic photon flux density; PSII – photosystem II; Φ_{PSII} – effective quantum yield of PSII photochemistry; $\Phi_{PPFDsat}$ - quantum yield of PSII photochemistry at the saturation level of PPFD.

INTRODUCTION

Trees adapt their photosynthetic apparatus to function in different light regimes which substantially change within their ontogenesis. Usually, seedlings develop in shade of mature trees' canopy, but with a growing age, crowns of trees are exposed to higher and higher irradiance. Thus, adult trees live in an entirely different light environment than young trees. Light requirements of trees are genetically determined and depend on species. In contrast to conservative shade tolerant species, pioneer species can grow from seed germination to maturity in the open. In Europe, in moderate climate zone, most tree species can grow as a seedling in some shade, but there are only few overwintering evergreen tree species (e.g. *Abies alba* Mill. and *Taxus baccata* L.) which can survive and slowly grow for many years at around 5% of full sun irradiance (Ellenberg et al. 1991; Brzeziecki 1995). It appears evident that the photochemical processes play an important role in trees' adaptation to a varying light environment. In particular, overwintering evergreen conifers are considered to be an interesting object for research into the photochemical processes. Long-leaved needles adapted their photochemical machinery to seasonally strongly varying irradiance (Öquist, Huner 2003).

A photosynthetic response to a changing irradiance depends on the genetic background and also on the ability to acclimate the photosynthetic apparatus to light and temperature conditions of growth. Overwintering evergreen conifers maintain leaves for all seasons and many years, which suggest their high plasticity in response to changing irradiance. Tree species can create ecotypes adapted to a light environment, e.g. photoperiodic ecotypes of

Pinus sylvestris L. (Oleksyn et al. 1992). Ecotypic differences in sensitivity of the photochemistry to freezing and cold stress were identified between the two populations of *Quercus virginiana*, corresponding to their climates of origin (Cavender-Bares 2007).

Young trees growing under a canopy of adults develop in dispersed light of low energy filtered through leaves of higher trees' crowns. Their photosynthetic apparatus is able to use light of low energy, especially in red and blue wavelengths which to a great extent is absorbed by canopy foliage. They also consume light from sunflecks, which may provide from about 30 to 80% of energy absorbed by leaves (Küppers et al. 1996). For shade-tolerant, late-successional species light patches are from one hand a source of energy for photosynthesis, and from the other hand a danger of being exposed to high illumination which can cause a down-regulation of PSII photochemistry (Pearcy et al. 1994; Küppers et al. 1996; Lovelock et al. 1998). In contrast, pioneer trees growing in gaps or clearings adapted their photosynthetic functions to high irradiance and even if their seeds germinate, they will not be able to grow under a dense canopy of mature trees (e.g. *Pinus sylvestris* L.). There is evidence that an ability of seedlings to use energy from sunflecks for photochemistry is important to grow in shade of a canopy of mature trees (Robakowski, Antczak 2008).

Plants require light for photosynthesis and photomorphogenetic processes, but growing in high light they are not able to use all available energy for photochemistry and developed many different morpho-anatomical traits and biochemical mechanisms to protect their photosynthetic apparatus from excess light energy (Adams, Demmig-Adams 1994; Verhoeven et al. 1996; Niyogi 1999; Adams et al. 2004). Due to these adaptations they maintain photostasis which consists in sustaining the balance between the absorption and use of light energy even in low winter temperatures drastically decreasing an 'energy sink' for photosynthesis (Öquist, Huner 2003).

Light absorbed by leaf is mainly captured by chlorophylls located in light harvesting complexes which act as antennae and may transfer their excitation energy to the reaction centres of photosystems PSI and PSII to be used in photochemistry. In PSII an electron derived from the splitting of water is transferred to the first electron acceptor of the photosynthetic electron-transport chain. Simultaneously, protons are transported across the membrane into the thylacoid lumen which thus becomes acidified. Electrochemical gradient across the thylacoid membrane is used to phosphorylate ADP and produce ATP. In non-cyclic electron transport NADP is the terminal acceptor of electrons from PSI and NDPH is produced. In cyclic electron transport, electrons are transferred from PSI to cytochrome in the thylacoid lumen, an extrusion of protons occurs and ATP is synthesised in the reaction catalyzed by an ATPase (Lambers et al. 1998). The products of the light-depended reactions of photosynthesis i.e. ATP and NADPH are subsequently used in carbon-reduction cycle. Nevertheless, the energy absorbed by leaf can also be either converted into heat or re-emitted as chlorophyll fluorescence. Most fluorescence is emitted by chlorophyll *a* of PSII. The wavelength of fluorescence is longer than that of the absorbed light and prompt fluorescence is emitted at a peak of 685 nm (Schreiber et al. 1998; Lawlor 2001). Fluorescence increases under different stress conditions: excessive light, high or low temperatures, drought, and other stressors (Havaux 1993). The parameters based on chlorophyll *a* fluorescence allow to assess the functioning of the photosynthetic apparatus, in particular the state of PSII, electron transfer rate, and efficiency of photochemical processes including mechanisms of defence from excessive energy (Schreiber et al. 1998).

A surplus of high light energy can be dissipated as heat in xanthophyll cycle that is associated with an increase in non-photochemical quenching of fluorescence (NPQ). This biochemical mechanism protects photosynthetic apparatus against photodamage (Adams, Demmig-Adams 1994; Niyogi 1999; Adams et al. 2004). However, it does not ensure a sufficient protection when the balance between the sink and the delivery of light energy is strongly upset, for example at high light or at high light occurring together with low temperature in winter or in early spring. In such circumstances maximum quantum yield of PSII photochemistry (F_v/F_m) can be permanently decreased below 0.8 which indicates photoinhibition (Örgen 1988; Maxwell and Johnson 2000).

In contrast to deciduous tree species losing their leaves seasonally, overwintering evergreen conifers maintain leaves also in winter. Their needles are threatened by low temperatures and high light. To avoid damage resulted from excess light energy, conifers developed photoprotective mechanisms e.g. energy-dependent down-regulation of PSII (recovering to the dark-adapted state within seconds to minutes). Temporal photoinhibition (recovering over several hours to days) can also be accounted for the protective plant behaviour (Adams et al. 2004). However, long-lasted photoinhibition connected with a prolonged stress of high light and low temperature may cause permanent damage to PSII, needless discolouration, and fall. The energy-dependent quenching of chlorophyll *a* fluorescence is mediated by energization of the thylakoid membrane: under bright light, protons are accumulated in the thylakoid lumen. A lower pH of the lumen induces a reversible enzymatic conversion of the carotenoid violaxanthin into zeaxanthin, which efficiently de-excites chlorophyll (Adams, Demmig-Adams 1994; Verhoeven et al. 1996; Adams et al. 2004). In contrast, photoinhibition is connected with photooxidative damage to D1 proteins of PSII. This damage may be repaired by the resynthesis of the proteins and lead to a slow recovery of Φ_{PSII} (Kitao et al. 2000; Ebbert et al. 2005). In realistic ecological situations, the difference between these two groups might not be obvious (Robakowski, Wyka 2004).

This review is focused on the photochemical processes in needles of overwintering evergreen conifer tree species of different light requirements, successional status, and susceptibility to photoinhibition. The main objective was to determine the ability of conifers to adapt or acclimate their photochemistry to irradiance. The photochemical capacity of moderately shade-tolerant and shade-intolerant, pioneer species which appear in succession early characterized by higher efficiency of photochemical processes in high irradiance were compared with shade-tolerant, late-successional species showing a better photochemical performance in shade. Moreover, susceptibility to photoinhibition induced by high light and low temperature among the pioneer and shade-tolerant species was assessed in the different light environments. The relationship between quantum yield of PSII photochemistry, ambient air temperature and light was studied in daily and seasonal scale to show a role of PSII down-regulation and winter photoinhibition for the photosynthetic apparatus. Leaf aging effect on effective quantum yield of PSII photochemistry and non-photochemical quenching of fluorescence was determined in *Abies alba* needles. At the end, the question about the role of photochemical processes in between-species competition and in evolutionary success of overwintering evergreen conifers in harsh climate was addressed.

The results shown in this review were chosen from different experiments conducted with young conifer trees in controlled conditions or in forest using the measurements of chlorophyll *a* fluorescence. This method has often been applied in photochemical research providing the information of the functioning of photochemistry *in vivo*.

Use of Chlorophyll *A* Fluorescence Measurements to Investigate the Photochemical Processes

The measurement of chlorophyll *a* fluorescence is a non-invasive and rapid technique widely used in plant ecophysiology for studying primary photochemical events and energy transfer (Krause and Weis 1984; Havaux 1993; Strasser et al. 2000). The method is usually combined with the measurements of net CO_2 assimilation rates. Yet, it is noteworthy that the fluorescence parameters alone allow to predict photosynthesis (Genty et al. 1989; Öquist and Chow 1992). There is an apparent linear relationship between quantum yield of PSII photochemistry related to inhibition of PSII electron transport and the quantum yield of CO_2 uptake, at least in laboratory (Genty et al. 1989). The measurements of chlorophyll *a* fluorescence were applied to investigate the photochemical processes occurring in leaves of conifer trees *in vivo* and *in situ* or in laboratory. Two types of fluorometers were used: 1. Plant Efficiency Analyser (PEA, Hansatech, Norfolk, UK) inducing fluorescence with a continuous light emitted by red diodes, and 2. Fluorescence Monitoring System (FMS 2, Hansatech, Norfolk, UK) which measure pulse-modulated fluorescence (see Schreiber et al. 1986 for comparison of these techniques).

Measurements of Chlorophyll A Fluorescence

The terminology used here for chlorophyll fluorescence parameters was based on the review article by Maxwell and Johnson (2000) and the experimental protocols were earlier given in detail in Robakowski (2005a, b) and Wyka et al. (2007).

Prior to the measurements of maximum quantum yield of photosystem II photochemistry (F_v/F_m), leaves were dark adapted for 30 minutes. When FMS 2 was used, the fiberoptics encased in a light-tight chamber was inserted onto the leaf clip and the leaves were exposed to modulated measuring light of 0.05 μmol m^{-2} s^{-1}. After reading minimum fluorescence F_0, a saturating 0.7 s pulse of light (PPFD = 15.3 mmol quanta m^{-2} s^{-1}) was delivered to induce a maximum fluorescence (F_m). Maximum quantum yield of PSII photochemistry was calculated according to the formula: maximum quantum yield = F_v/F_m, where $F_v = F_m - F_0$ is variable fluorescence.

To measure F_v/F_m with PEA a pulse of red light was given to induce F_m in a dark adapted leaf. F_0 was derived from the initial part of the fluorescence kinetics curve and F_v/F_m was calculated by the in-built software. The value of photosynthetic photon flux density (PPFD) inducing F_m was experimentally determined and usually ranged between 2400 to 3200 μmol m^{-2} s^{-1}.

The photochemical processes in leaves were described using light response curves of fluorescence determined with FMS 2. To generate light response curves of effective PSII quantum yield (Φ_{PSII}) leaves in the clip were illuminated with actinic light using an inbuilt halogen lamp. The intensity of actinic light corresponding to values indicated by the software was measured prior to the experiment using a light sensor inserted in a leaf-clip in the position of needles. Up to 12 levels of actinic light were applied in the order of increasing intensity, and for each level, after a stable steady state fluorescence (F_s) was reached, 0.7 s saturating pulse was delivered and maximum light-adapted fluorescence (F'_m) was determined. Quantum yield of PSII was calculated by the built-in software as: $\Phi_{PSII} = (F'_m - F_s) / F'_m$ (Genty et al. 1989). The course of fluorescence and all of the measured parameters were monitored on the computer screen. This was particularly important after changing actinic illumination levels, when stable F_s values had to be reached before a saturating pulse was applied. At each actinic light level, non-photochemical quenching of fluorescence (NPQ) was calculated according to the formula: $NPQ = (F_m - F'_m) / F'_m$ (Maxwell and Johnson 2000).

For each light level, apparent rates of photosynthetic electron transport (ETR) were also calculated following the formula $ETR = 0.84 * \Phi_{PSII} * PPFD * 0.5$ (Maxwell and Johnson 2000; Lüttge et al. 2003). The assumptions were made that the excitation energy is partitioned equally between the two photosystems (hence the factor 0.5; Maxwell and Johnson 2000) and that 84% of the incident radiation is absorbed by the photosystems (Rascher et al. 2000; Lüttge et al. 2003, but see von Caemmerer 2000). However, leaf absorptance may differ among plants depending on species and adaptation to microclimate conditions. Moreover, to verify the appropriateness of the factor 0.84 leaf absorptance was calculated for each species using the model given by Evans (1993), which is based on total chlorophyll content in the leaf. The measurements were taken at ambient temperature, which was monitored during the fluorescence measurement using a thermocouple installed in the leaf clip and the leaf clip was shaded to avoid warming up due to solar irradiation.

Derivation of Cardinal Values from Light Curves of Fluorescence

Cardinal points of the light curves of Φ_{PSII} and ETR were derived by fitting exponential functions to the data (Rascher et al. 2000; Robakowski 2005a). To estimate the maximum apparent rate of photosynthetic electron transport of PSII (ETR_{max}), quantum yield of PSII photochemistry at saturating PPFD ($\Phi_{PPFDsat}$) and the saturating level of photosynthetic photon flux density ($PPFD_{sat}$), an exponential rise to maximum function

$$ETR = ETR_{max} (1-e^{-b*PPFD}), \text{ where b is an independent parameter} \quad (1)$$

was fitted to PPFD and calculated ETR values.

$PPFD_{sat}$ was assumed to be reached at $0.9*ETR_{max}$. $\Phi_{PPFDsat}$ was calculated from the equation:

$$\Phi_{PPFDsat} = m + ae^{-b*PPFDsat} + ce^{-d*PPFDsat}, \quad (2)$$

where a, b, c, d, m are independent parameters.

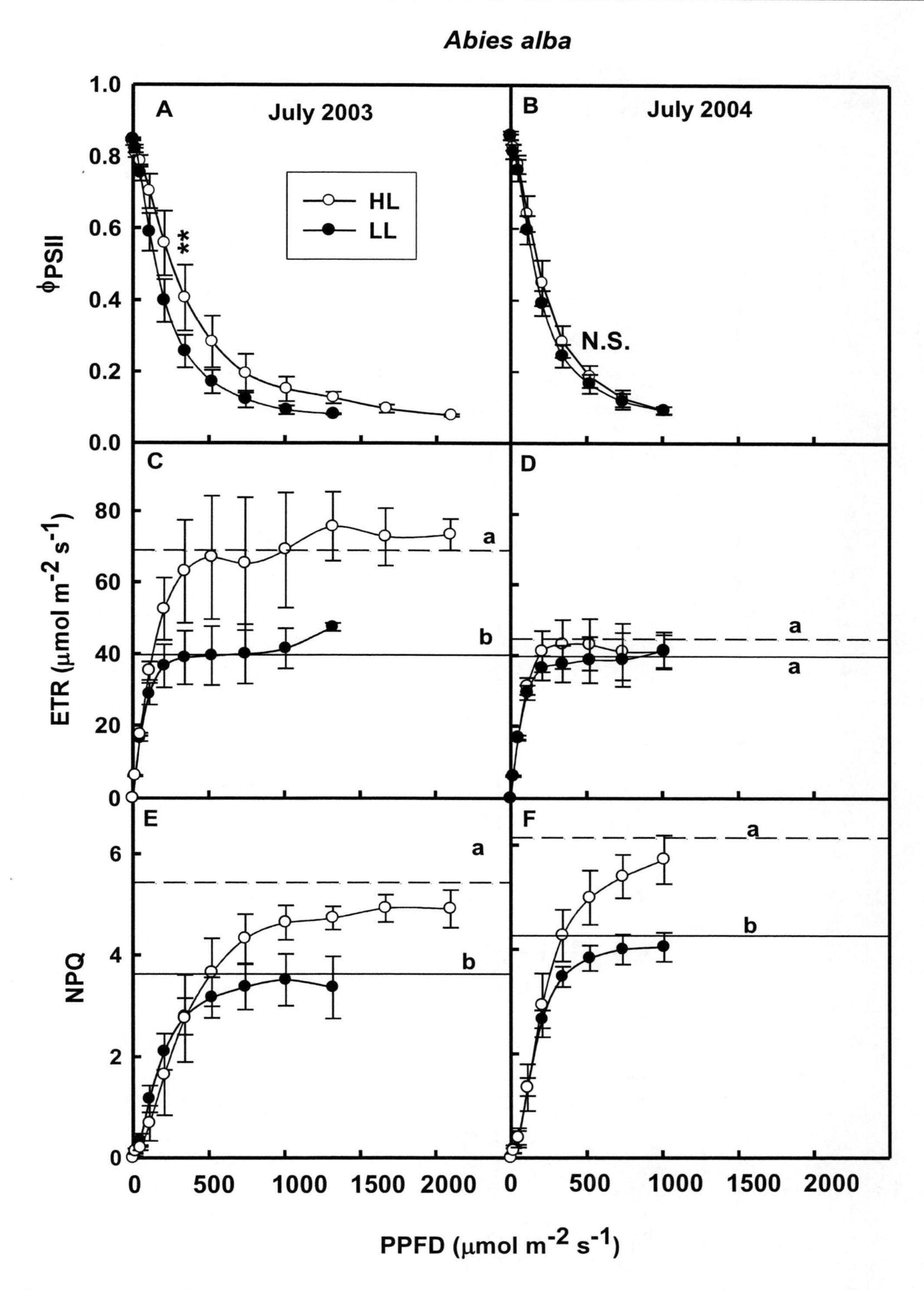

Figure 1. (continued)

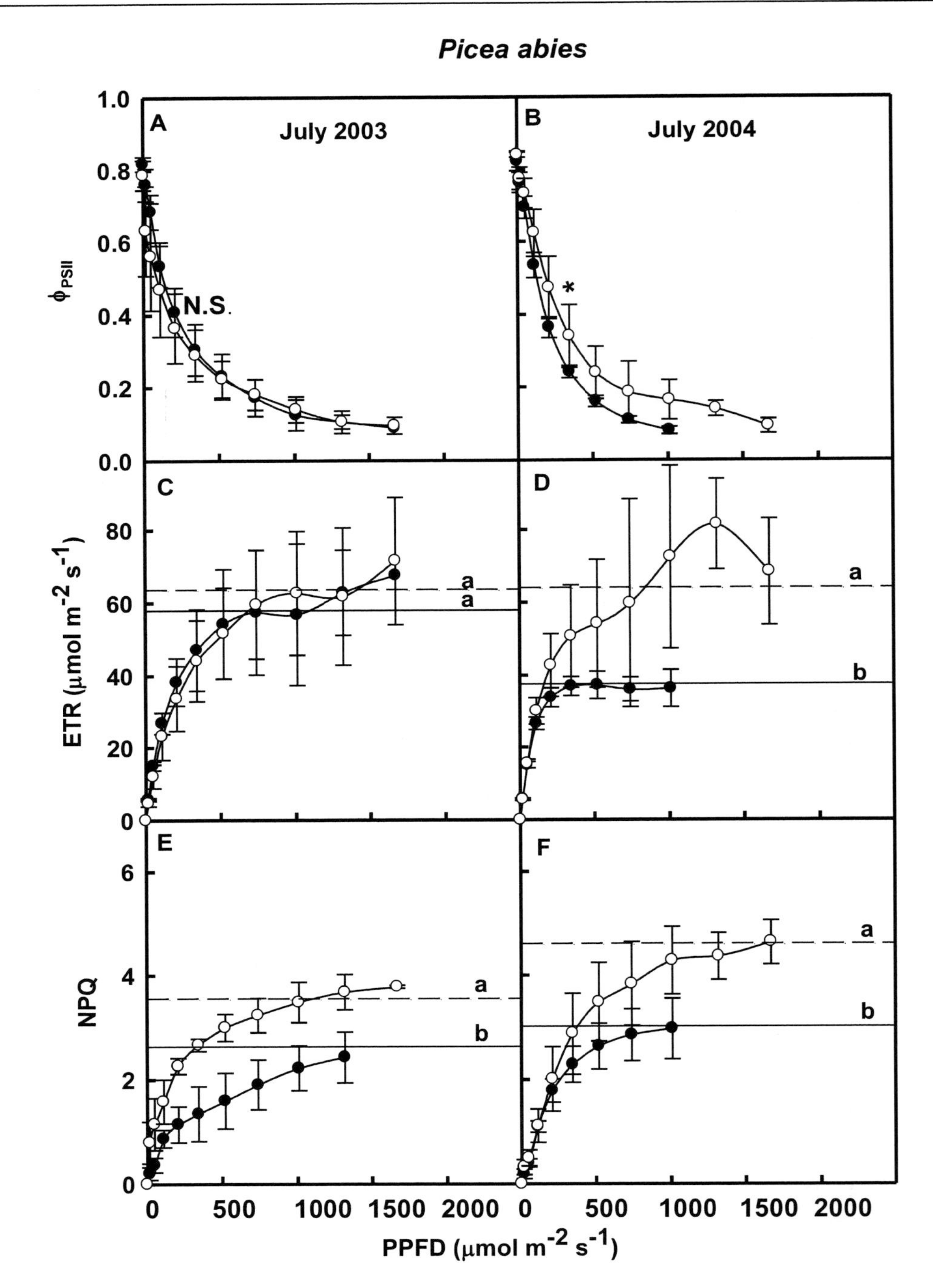

Figure 1. Light response curves of quantum efficiency of PSII (Φ_{PSII}; A, B), apparent electron transport rate (ETR; C, D), and nonphotochemical quenching (NPQ; E, F) in current season leaves of *Abies alba* and *Picea abies* acclimated to HL (open symbols) or LL (closed symbols) for one (A, C, E) or two (B, D, F) growing seasons. In panels A, B asterisks indicate significant contrasts between treatments for Φ_{PSII} corresponding to PPFD = 343 µmol m^{-2} s^{-1} (* *P*<0.05, ** *P*<0.01, *** *P*<0.001). In panels C-F, horizontal lines indicate maximal values of ETR or NPQ derived through curve fitting (dashed line-HL, solid line - LL). Different letters at these lines indicate significant contrasts between treatments (Wyka et al. 2007).

ACCLIMATION OF PHOTOCHEMICAL PROCESSES TO DIFFERENT LIGHT ENVIRONMENTS

Extremely shade tolerant conifer seedlings of *Abies alba* and moderately shade tolerant *Picea abies* acclimated their photochemistry to full solar irradiance (HL – high light, 100% of full solar irradiance) or low light (LL – low light, 5% of full solar irradiance) to different extent. In the first season of acclimation to the light conditions 3-year-old seedlings of *Abies* showed lower Φ_{PSII} with increasing light in LL grown plants compared with HL plants, but in the second season the differences between HL and LL were not significant for *Abies*, but only for *Picea* (Fig. 1-2A, B). ETR was increasing to a saturation level of ETR_{max} with a growing level of actinic light and was higher in HL than LL seedlings, except for *Picea* in the first and *Abies* in the second season (Fig 1-2C, D). Non-photochemical quenching was greater in HL plants independently of the species and year of experiment (Fig. 1-2E, F). At the same actinic light values *Abies alba* showed higher NPQ than *Picea abies*, which suggested a more dynamic defensive response of the former against excessive light energy (Wyka et al. 2007). The fact that in the second season statistical differences were consequently found between HL and LL *Picea* seedlings for all characteristics based on chlorophyll *a* fluorescence indicated that the photochemical response to shade of this moderately shade-tolerant pioneer mid- or late-successional species was more plastic than that of *Abies*. On the other hand, higher NPQ values in *Abies alba* needles compared with *Picea abies* suggested that its great tolerance to shade did not exclude an acclimation to high light conditions due to an efficient mechanism of energy dissipation.

PHOTOINHIBITION AND PHOTOPROTECTIVE MECHANISMS IN DAILY AND SEASONAL SCALE

Daily PSII Down-Regulation

Depression of net CO_2 assimilation rate observed at midday can be caused by stomatal closure and temporary photoinhibition that occur concomitantly with the increase in light and temperature. Sun-adapted leaves of *Citrus* showed a deeper decrease in F_v/F_m at midday, which recovered longer within a day compared with shade-adapted leaves (Jifon, Syversten 2003). *Quercus suber* leaves had the ability to dissipate excessive energy at midday by a non-photochemical mechanism (Faria et al. 1996). In the diurnal study of quantum yield of PSII in *Abies alba* seedlings growing under a dense canopy of *Picea abies* or a more open canopy of *Larix decidua*, the hourly mean values of Φ_{PSII} were lower and down-regulation of PSII significantly higher under the larch canopy when sunny days were compared. This tendency was absent on cloudy days. In both stands a full recovery of F_v/F_m was noticed in the afternoon (Robakowski, Wyka 2004). Additionally, under the *Larix* canopy instantaneous deep depressions of F_v/F_m in sunflecks were more frequently observed than under the *Picea* crowns. The midday depression of F_v/F_m was due to an increase in F_0 and decrease in F_m together with an excess energy dissipation as heat in xanthophyll cycle (Adams and Demmig-Adams1994; Verhoeven et al. 1996; Robakowski et al. 2004). It can be stated that the midday

PSII down-regulation efficiently protected the photosynthetic apparatus of young fir seedlings against excess energy when they grew under a mature trees' canopy. However, a light level under a canopy of trees is always lower than in the open where this photoprotective mechanism does not need to be sufficiently effective.

Winter Photoinhibition

Overwintering evergreen conifer trees sustain photosynthetic capacity within winter. Nonetheless, the energy that is absorbed by needles at low winter temperature cannot often be used through photosynthesis (Adams et al. 2004). To protect the photosynthetic apparatus against excessive energy chlorophyll concentration can be decreased in needles to reduce light absorption. On the other hand, in winter, xanthophyll carotenoids concentration increases and excessive energy can be, at least to some extent, dissipated in xanthophyll cycle as heat. However, winter decrease in F_v/F_m was also correlated positively with the decrease in non-phosphorylated D1 protein and the degradation of this protein (Ebbert et al. 2005).

It was hypothesised that pioneering, light demanding conifers should be more tolerant to winter photoinhibition than late-successional and shade-tolerant species. The seasonal course of F_v/F_m in needles of shade intolerant *Pinus mugo* Turra, moderately shade tolerant *Picea abies* Karst., and shade intolerant *Abies alba* Mill. did not support this hypothesis (Fig. 2). The differences among the species in response to winter photoinhibition with regard to F_v/F_m, $\Phi_{PPFDsat}$, ETR_{max}, and $PPFD_{sat}$, did not allow to range them according to the successional status. It seems that shade-tolerant *Abies alba* was able to dissipate chlorophyll excitation pressure by sustaining a high photochemical capacity and by energy dissipation in xanthophyll cycle (high NPQ). In contrast, pioneer *Pinus mugo* and moderately shade-tolerant *Picea abies* decreased their photochemical capacity. It was suggested that the differences between species resulted from their adaptation to temperature at low (*Abies alba*) or high (*Pinus mugo*, *Picea abies*) altitudes (Robakowski 2005b).

Seasonal course of F_v/F_m was monitored in current needles of *Taxus baccata* seedlings acclimated for one of three light treatments: high light (HL, full irradiance), moderate light (ML, 18%), low light (LL, 5%) during one vegetative season. From the beginning of the study F_v/F_m was generally lowest in needles of HL seedlings (Fig. 3). It decreased especially sharply with falling temperature in mid-January and February and started to recover from around the beginning of March. The lowest level of F_v/F_m (0.37) in HL seedlings occurred in February. Minima of F_v/F_m in both ML and LL seedlings (0.666 and 0.750, respectively) were noted on January 25 when the temperature values on previous day were T_{mean} = - 2.8 °C, T_{max} = - 2.0 °C, and T_{min} = - 3.9 °C. Full recovery of F_v/F_m to the summer level in ML and LL leaves took place in March while in HL leaves full recovery was observed as late as in the end of May. Similar trend of the F_v/F_m changes during winter was observed for many conifer species, e.g. in *Pinus ponderosa* Laws., *Pseudotsuga menziesii* Beissn. Franco, *Picea pungens* (Adams, Demmig-Adams 1994; Nippert et al. 2004), *Cryptomeria japonica* (Han et al. 2003), *Thuja plicata* Donn ex D. Donn, *Picea Engelmanii* Parry, *Abies grandis* Dougl., *Pinus contorta* Dougl. (Nippert et al. 2004), *Pinus sylvestris* (Porcar-Castell et al. 2008b).

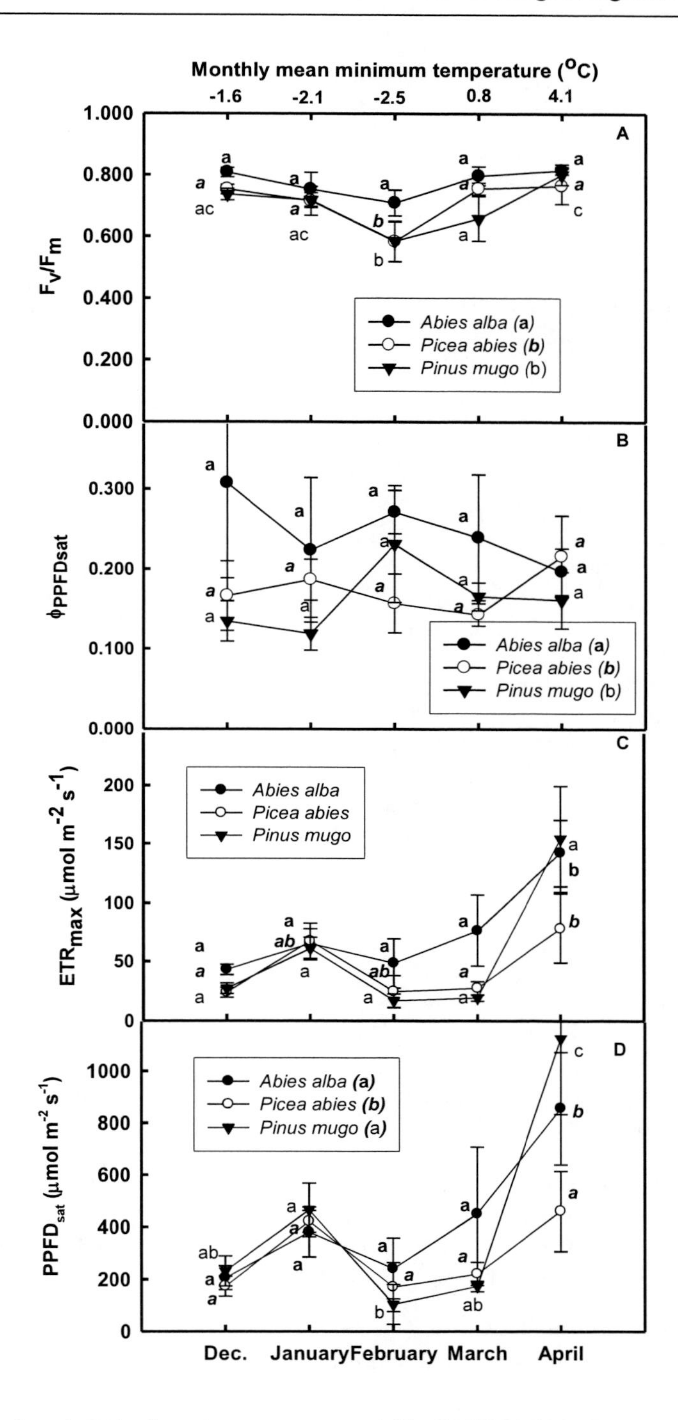

Figure 2. Mean values (±SD) of maximum quantum yield of PSII (F_v/F_m), $\Phi_{PPFDsat}$, apparent maximum electron transport rate (ETR_{max}), and $PPFD_{sat}$ in the needles of *Abies alba*, *Picea abies* and *Pinus mugo* (A, B, C, D). Two-way MANOVA with interaction and Tukey's *post-hoc* test at $P < 0.05$ were applied to compare the mean values of the determined parameters between the species and between months in each species. The shared letters indicate a lack of statistically significant differences. Letters in bold are for *Abies alba*, bold and italic – *Picea abies*, normal font – *Pinus mugo* (Robakowski 2005a).

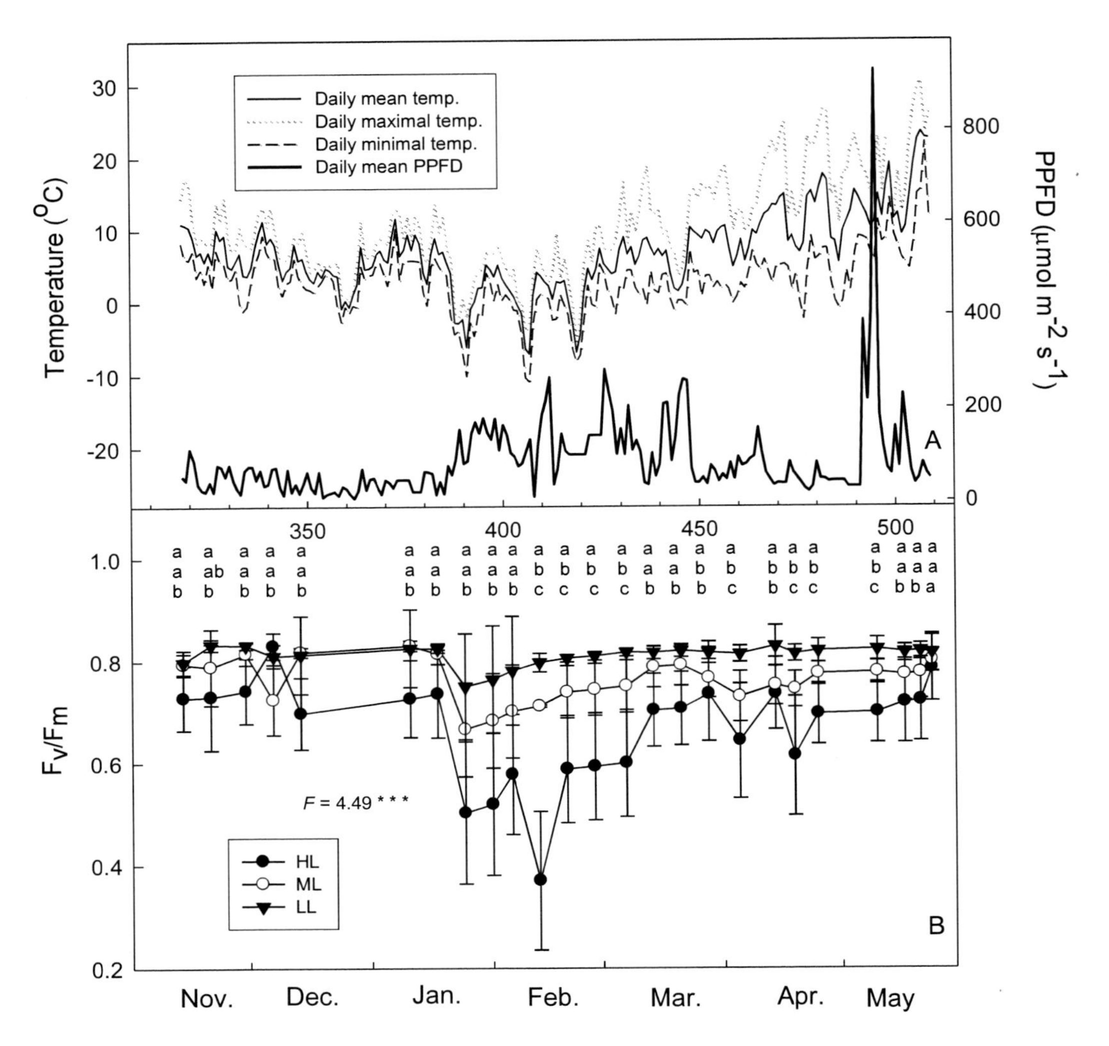

Figure 3. A. Daily mean, minimal and maximal temperature and daily mean PPFD at 2 m above ground level from Nov. 16, 2006 (DOY – 318) to May 24, 2007 (DOY – 509). B. Time course of F_v/F_m from 16 November 2006 to 24 May 2007 in needles of *Taxus baccata* seedlings acclimated to high light (100 % of full irradiance), mid light (18 %) or low light (5 %). Asterisks indicate statistically significant differences between the light treatments (*** $P < 0.001$) according to MANOVA ($P < 0.05$). The same letters indicate that there are not statistically significant differences between light treatments in *post-hoc* Tukey's test ($P < 0.05$; $n = 5$) (Robakowski, Wyka 2009).

Within each light environment, F_v/F_m in *Taxus* needles was positively correlated with air temperature. The strength and statistical significance of this correlation depended, however, on the specific temperature parameter (daily mean, maximum or minimum temperature) and the width of time window prior to the measurement used for the estimation of the temperature moving average. Of the three temperature measures studied, daily minimum temperatures were the best predictors of F_v/F_m in HL and ML plants, but this was not the case in LL plants. The closest fit in all light treatments was achieved when temperatures were averaged over 3 or 5-day time-windows (Robakowski, Wyka 2009). In contrast, Nippert et al. (2004) found that maximum temperature was a better predictor of F_v/F_m in needles of montane conifers than minimum temperature.

Seasonal changes in F_v/F_m can be caused by adjustments in the photochemical capacity in PSII, the capacity of thermal dissipation in PSII, or both. Porcar-Castell et al. (2008) complemented and extended F_v/F_m measurements introducing two new parameters: rate constant of sustained thermal energy dissipation (k_{NPQ}) and of photochemistry (k_p), which allowed to separate changes in F_v/F_m caused by adjustments either the photochemical capacity in PSII, or the capacity of thermal dissipation in PSII (Porcar-Castell et al. 2008a, b).

Leaf Age Effect on Photochemistry

Overwintering evergreen conifers which maintain their leaves for several years have often been an object for the study of the effects of leaf aging on photosynthesis (Freeland 1952; Kayama et al. 2007; Han et al. 2008; Wyka et al. 2008). Generally, photosynthetic capacity decreased with a greater needle age which resulted from morpho-anatomical needle modification and physiological changes in the photosynthetic apparatus. Shade-tolerant, late-successional *Abies alba* is able to maintain needles up to 8 (11) years. These needles differ in photosynthetic capacity showing changes in Φ_{PSII} in function of actinic light intensity and needle age (Robakowski and Bielinis, unpubl.). Maximum quantum yield of PSII photochemistry measured in dark (F_v/F_m) did not vary and attained optimal values independently of needle age. However, quantum yield of PSII photochemistry in light (Φ_{PSII}) decreased with a greater actinic light level and non-linearly changed with needle age (Fig. 4A). At actinic light = 111 and 208 $\mu mol\ m^{-2}\ s^{-1}$ the highest Φ_{PSII} was detected in three-year-old needles. This parameter was lower in the youngest needles and decreased with a greater age from four-year-old to seven-year-old needles. Inversely to Φ_{PSII}, NPQ increased with a greater actinic light level and was lowest in three-year-old needles (Fig. 4B). NPQ decreased from the youngest (one- and two-year-old) needles and then increased from three-year-old needles up to seven-year-old ones. Net CO_2 assimilation rate together with nitrogen concentration decreased exponentially with the needles' age (Robakowski, Bielinis, unpubl.) The content of macroelements N, P, K, Mg decreased and Ca, inversely, increased with the *Abies alba* needles' aged from one to six year (Szymura 2009). Changes in the content of elements in function of needles' age, especially the decrease in nitrogen concentration, which is an important constituent of enzymes, affect the photochemistry and the overall photosynthetic performance. Nevertheless, it is unclear why the three-year-old needles had the highest Φ_{PSII} and, simultaneously, the lowest NPQ. It can be suggested that young (1- or 2-year-old) needles were exposed to higher irradiance and were less protected by self-shading which induced photoinhibition. On the other hand, the needles older than the three-year-old ones were strongly shaded and nitrogen was plausibly retranslocated from them to the younger needles (Szymura 2009). Therefore, high NPQ in old needles might be caused by nitrogen deficit and the lower energy 'sink' for photosynthesis.

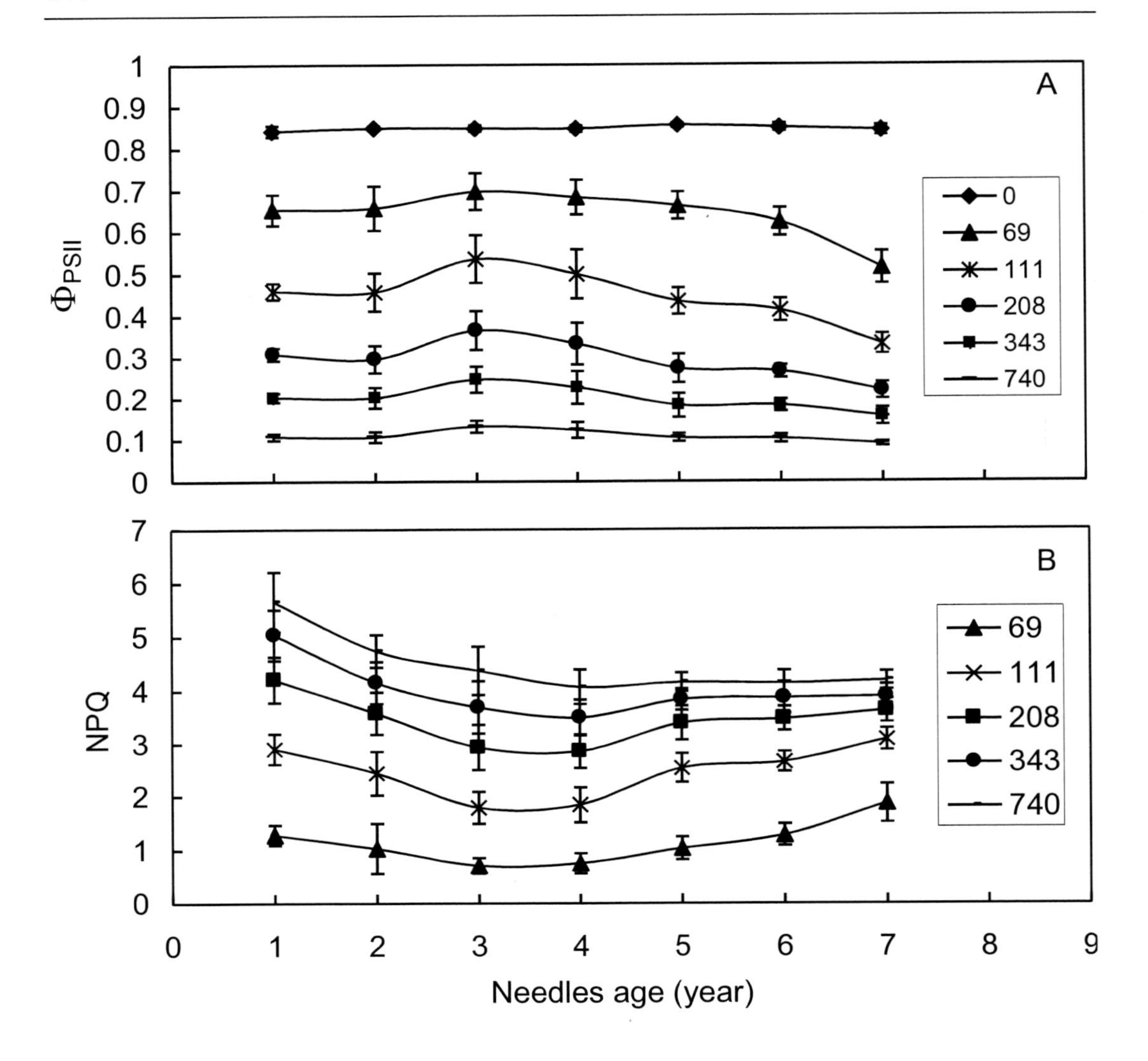

Figure 4. Mean (±SD) quantum yield of PSII photochemistry (Φ_{PSII}) (A) and non-photochemical quenching (NPQ) (B) in *Abies alba* needles differing in age (n = 5). Needles of different age were acclimated to increasing actinic light (0 – 740 μmol m^{-2} s^{-1}). The PPFD values in μmol m^{-2} s^{-1} with corresponding symbols are given in the legends of each plot.

Evolutionary Importance of an Ability to Adapt Photochemistry to a Changing Light Environment

Trees developed different strategies to increase or sustain the photosynthetic capacity and to protect the photosynthetic apparatus (Adams et al. 2004). The strategy which consists in shedding leaves before winter and production of their new set each year seem to be more costly compared to maintaining leaves. On the other hand, conifers use more resources to sustain needles for many years (e.g. in *Abies alba* up to 8(11) years) and protect them against photoinhibition and other abiotic and biotic stressors (Öquist, Huner 2003). When altitudinal ranges of overwintering evergreen conifers with deciduous tree species are compared, it appears that conifers, for example *Picea abies*, *Pinus mugo* in central European Mts. are more tolerant to severe climate and nutrient deficit at the forest limit at high altitudes. Geographical distribution of conifers indicates that they are adapted to cold, but also some deciduous species co-exist with them (Öquist, Huner 2003). The photochemical processes along with different photoprotective mechanisms play an important role in plants' adaptation to climate and in some circumstances may give an advantage in the between-species competition. It has often been suggested that species equipped with the photosynthetic apparatus more tolerant to warming will be better adapted to future climate than cold-adapted species. What is more, high thermostability and an effective regulating mechanism in energy partitioning to minimize potential damage in PSII can give advantage to an invasive species over its indigenous competitors (Song et al. 2010).

Plants growing in land protect photochemical processes crucial for photosynthesis as electron transfer associated with ATP and NADP production through morpho-anatomical changes in leaf and also physiological and biochemical mechanisms. It appears that an evolutionary success of terrestrial plants would have not been possible if they had not developed the photoprotective mechanisms such as excess energy dissipation in xanthophyll cycle, synthesis of screening pigments (carotenoids, flavonoids), switching off some PSII reaction centres or reduction of chlorophyll concentration in high light and low temperature conditions. A temporal down-regulation of PSII in daily scale or within a winter, which involves a depression of net CO_2 assimilation rates, may protect the photosynthetic machinery against permanent photoinhibition. All these protective processes allow a plant to grow in different light environments; however, each species has its own optimum of photosynthesis. Pioneer species can grow in the open, whereas in the same light conditions late-successional trees can be more sensitive to photoinhibition.

Differences in successional status and light requirements do not always imply the expected photochemical responses, for example shade tolerant *Abies alba* showed a higher tolerance to the photoinhibitory stress and faster recovered F_v/F_m than less shade tolerant, mid-successional *Picea abies* (Fig. 1). Photochemical responses of early and late-successional species can be strongly modified by light conditions of growth. Acclimation to shade diminished photoinhibition expressed as a decrease in F_v/F_m and fastened a recovery of the photochemical functions in shade-tolerant *Taxus baccata* and shade intolerant *Pinus sylvestris* compared with seedlings exposed to full sun irradiance (Porcar-Castell et al. 2008b; Robakowski, Wyka 2009). Seasonal changes in F_v/F_m can be caused by adjustments in either photochemical capacity in PSII, the capacity of thermal dissipation in PSII (constitutive and sustained thermal dissipation), or both (Porcar-Castell et al. 2008a).

At present the question arises which strategy can ensure the evolutionary success of different taxonomic groups of trees under global climate changes. Photosynthesis is sensitive to both low and high temperatures; however, photochemistry is more stable than biochemical processes in response to temperature changes (Berry, Björkman 1980). It appears of great importance to know the values of optimal, minimal and maximal temperatures at which the different photochemical processes occur. Moreover, we should address the question how far are the critical temperatures of the photochemical processes from the current and projected temperatures (Saxe et al. 2001). The projected global warming, and any other naturally occurring climate changes, directly can influence efficiency of the photochemical processes, net CO_2 assimilation rates and plants' productivity.

References

Adams, III, W. W., and B. Demmig-Adams 1994. Carotenoid composition and down regulation of photosystem II in three conifer species during the winter. *Physiologia Plantarum*, 92: 451-458.

Adams, III, W. W., Zarter, C. R., Ebbert, V., Demmig-Adams, B. 2004. Photoprotective strategies of overwintering evergreens. *BioScience*, 54(1): 41-49.

Berry, J. A., Björkman, O. 1980. Photosynthetic response and adaptation to temperature in higher plants. *Annual Review of Plant Physiology*, 31: 491-543.

Brzeziecki, B. 1995. Skale nominalne wymagań klimatycznych gatunk ów drzew leśnych. [Scales of climatic requirements of forest tree species]. *Sylwan*, 3: 53-65. (in Polish)

Cavender-Bares, J. 2007. Chilling and freezing stress in live oaks (*Quercus* section *Virentes*): intra- and inter-specific variation in PSII sensitivity corresponds to latitude of origin. *Photosynthesis Research*, 94: 437-453.

Ebbert, V., Adams III, W. W., Mattoo, A. K., Sokolenko, A., Demmig-Adams, B. 2005. Up-regulation of a photosystem II core protein phosphatase inhibitor and sustained D1 phosphorylation in zeaxanthin-retaining, photoinhibited needles of overwintering Douglas fir. *Plant, Cell and Environment*, 28: 232-240.

Ellenberg, H. 1991. Zeigerwerte der Gefäβpflanzen (ohne *Rubus*). In*: Zeigerwerte von Pflanzen in Mitteleuropa*. Eds. R.D.H. Ellenberg, V. Wirth, W. Werner and D. Pauliβen. Erich Goltze KG, Göttingen, pp. 9-166.

Evans, R. J. 1993. Photosynthetic acclimation and nitrogen partitioning within a lucerne canopy. II Stability through time and comparison with a theoretical optimum. *Australian Journal of Plant Physiology*, 20: 69-82.

Faria, T., Garcia-Plazaola, I. J., Abadia, A., Cerasoli, S., Pereira, S.J., Chaves, M.M. 1996. Diurnal changes in photoprotective mechanisms in leaves of cork oak (*Quercus suber*) during summer. *Tree Physiology*, 16: 115-123.

Freeland, R. O. 1952. Effect of age of leaves upon the rate of photosynthesis in some conifers. *Plant Physiology*, 27: 685-690.

Genty B., Briantais J-M., Baker N.R. 1989. The relationship between the quantum yield of photosynthetic electron transport and quenching of chlorophyll fluorescence. *Bioch. Biophys. Acta*, 990: 87-92.

Han, Q., Shinohara, K., Kakubari, Y., Mukai, Y. 2003. Photoprotective role of rhodoxanthin during cold acclimation in *Cryptomeria japonica. Plant, Cell and Environment, 26: 715-723.*

Han, Q., Kawasaki, T., Nakano, T., Chyba, Y. 2008. Leaf-age effects on seasonal variability in photosynthetic parameters and its relationships with leaf mass per area and leaf nitrogen concentration within a *Pinus densiflora* crown. *Tree Physiology*, 28: 551-558.

Havaux, M. 1993. La fluorescence de la chlorophylle in vivo: Quelques concepts appliqués à l'étude de la résistance de la photosynthèse aux contraintes de l'environment. In: Ed. INRA, Paris 1993. Tolérance à la sécheresse des céréales en zone méditerranéenne. *Diversité génétique et amélioration variétale*. Montpellier (France), 15-17 décembre 1992, pp. 19-29.

Jifon, L. J., Syversten, P. J. 2003. Moderate shade can increase net gas exchange and reduce photoinhibition in citrus leaves. *Tree Physiology*, 23: 119-127.

Kayama, M., Kitaoka, S., Wang, W., Choi, D., Koike, T. 2007. Needle longevity, photosynthetic rate and nitrogen concentration of eight spruce taxa planted in northern Japan. *Tree Physiology*, 27: 1585-1593.

Kitao, M., Lei, T. T., Koike, T., Tobita, H., Maruyama, Y. 2000. Susceptibility to photoinhibition of three deciduous broadleaf tree species with different successional traits raised under various light regimes. *Plant, Cell and Environment*, 23: 81-89.

Krause, G. H., Weis, E. 1984. Chlorophyll fluorescence as a tool in plant physiology. II. Interpretation of fluorescence signals. *Photosynthesis Research*, 5: 139-157.

Küppers, M., Timm, H., Orth, F., Stegemann, J., Stöber, R., Schneider, H., Paliwal, K., Karunaichamy, K. S. T. K., Oritz, R. 1996. Effects of light environment and successional status on lightfleck use by understory trees of temperate and tropical forests. *Tree Physiology*, 16: 69-80.

Lambers, H., Stuart Chapin III, F., Pons, L. T. 1998. *Plant Physiological Ecology*. Springer-Verlag New York, Inc.

Lawlor, D. W. 2001. *Photosynthesis*. BIOS Scientific Publishers Ltd., UK, pp.: 79-112.

Lovelock E. C., Kursar A. T., Skillman B. J., Winter K. 1998. Photoinhibition in tropical forest understorey species with short- and long-lived leaves. *Functional Ecology*, 12: 553-560.

Lüttge, U., Berg, A., Fetene, M., Nauke, P., Dirk, P., Beck, E. 2003. Comparative characterization of photosynthetic performance and water relations of native trees and exotic plantation trees in Ethiopian forest. *Trees*, 17(1): 40-50.

Maxwell, K., Johnson, N. G. 2000. Chlorophyll fluorescence - a practical guide. *Journal of Experimental Botany*, 51: 659-668.

Nippert, J. B., Duursma, R. A., Marshall, D. 2004. Seasonal variation in photosynthetic capacity of montane conifers. *Functional Ecology*, 18: 876-886.

Niyogi, K. K. 1999. Photoprotection revisited: genetic and molecular approaches. *Annual Review of Plant Physiology*, 50: 333-359.

Oleksyn, J., Tjoelker, M.G., Reich, P. B. 1992. Growth and biomass partitioning of populations of European *Pinus sylvestris* L. under simulated 50° and 60° N daylengths: evidence for photoperiodic ecotypes. *New Phytologist*, 120(4): 561–574.

Öquist, G., Chow, W. S. 1992. On the relationship between the quantum yield of Photosystem II electron transport, as determined by chlorophyll fluorescence and the quantum yield of CO_2-dependent O_2 evolution. *Photosynthesis Research*, 33: 51-62.

Öquist, G., Huner, A. P. N. 2003. Photosynthesis of overwintering evergreen plants. *Annual Review of Plant Biology*, 54: 329-355.

Örgen, E. 1988. Photoinhibition of photosynthesis in willow leaves. *Planta*, 175: 229-236.

Pearcy, R. W., Chazdon, R. L., Gross, L. J., Mott, K.A. 1994. Photosynthetic utilization of sunflecks: a temporally patchy resource on a time scale of seconds to minutes. In: *Exploitation of environmental heterogeneity by plants*. Eds. M.W. Caldwell and R.W. Pearcy. Academic Press, pp. 175-208.

Porcar-Castell, A., Juurola, E., Nikinmaa, E., Berninger, F., Ensminger, I., Hari, P. 2008a. Seasonal acclimation of photosystem II in Pinus sylvestris. I. Estimating the rate constants of sustained thermal energy dissipation and photochemistry. *Tree Physiology*, 28: 1475-1482.

Porcar-Castell, A., Juurola, E., Ensminger, I., Berninger, F., Hari, P. Nikinmaa, E., 2008b. Seasonal acclimation of photosystem II in Pinus sylvestris. II. Using the rate constants of sustained thermal energy dissipation and photochemistry to study the effect of the light environment. *Tree Physiology*, 28: 1483-1491.

Rascher, U., Liebieg, M., Lüttge, U. 2000. Evaluation of instant light-response curves of chlorophyll fluorescence parameters obtained with a portable chlorophyll fluorometer on site in the field. *Plant, Cell and Environment*, 23: 1397-1405.

Robakowski, P. 2005a. Susceptibility to low-temperature photoinhibition in three conifer tree species differing in successional status. *Tree Physiology*, 25: 1091-1100.

Robakowski, P. 2005b. Species-specific acclimation to strong shade modifies susceptibility of conifers to photoinhibition. *Acta Physiologiae Plantarum*, 27(3A): 255-263.

Robakowski, P., Antczak, P. 2008. Ability of silver fir and European beech saplings to acclimate photochemical processes to the light environment under different canopies of trees. *Polish Journal of Ecology*, 56(1): 3-16.

Robakowski P., Wyka T. 2004. Down-regulation of PSII in needles of silver fir (*Abies alba* Mill.) seedlings growing under the canopy of European larch and Norway spruce. *Zeszyty Problemowe Postępów Nauk Rolniczych*, 496: 421-431.

Robakowski, P., Wyka, T. 2009. Winter photoinhibition in needles of *Taxus baccata* seedlings acclimated to different light levels. *Photosynthetica*, 47(4): 527-535.

Robakowski, P., Wyka, T., Samardakiewicz, S. and Kierzkowski, D. 2004. Growth, photosynthesis and needle structure of silver fir (*Abies alba* Mill.) seedlings under different canopies. *Forest Ecology and Management*, 201(2-3): 211-227.

Saxe, H., Cannell, R.G.M, Johnsen, B, Ryan, G. M., Vourlitis, G. 2001. Tree and forest functioning in response to global warming. *New Phytologist*, 149: 369-400.

Schreiber, U. Schliwa, U. Bilger, W. 1986. Continuous recording of photochemical and non-photochemical chlorophyll fluorescence quenching with a new type of modulation fluorometer. *Photosynthesis Research*, 10: 51-62.

Schreiber, U., Bilger, W., Hormann, H., Neubauer, C. 1998. *Chlorophyll fluorescence as a diagnostic tool: the basic and some aspects of practical relevance*. Photosynthesis (Ed. A. Raghavendia). Cambridge University Press.

Song, L., Chow, W.S., Sun, L., Li, Ch., Peng, Ch. 2010. Acclimation of photosystem II to high temperature in two *Wedelia* species from different geographical origins: implications for biological invasions upon global warming. *Journal of Experimental Botany*, doi: 10.1093/jxb/erq220.

Strasser, R.J., Srivastava, A., Tsimilli-Michael, M. 2000. *The fluorescence transient as a tool to characterize and screen photosynthetic samples*. Taylor & Francis, London.

Szymura, T.H. 2009. Concentration of elements in silver fir (*Abies alba* Mill.) needles as a function of needles' age. *Trees*, 23: 211-217.

Verhoeven, S.A., Adams III, W. W., Demming-Adams, B. 1996. Close relationship between the state of xanthophyll cycle pigments and photosystem II efficiency during recovery from winter stress. *Physiologia Plantarum*, 96: 567-576.

Von Caemmerer, S. 2000. *Biochemical models of leaf photosynthesis*. CSIRO Publishing, pp. 73.

Wyka, T. Robakowski, P., Żytkowiak, R. 2007. Leaf acclimation to contrasting irradiance in juvenile evergreen and deciduous trees. *Tree Physiology*, 27: 1293-1306.

Wyka, T., Robakowski, P., Żytkowiak, R. 2008. Leaf age as a factor in anatomical and physiological acclimative responses of *Taxus baccata* L. needles to contrasting irradiance environments. *Photosynthesis Research*, 95: 87-99.

In: Photochemistry
Editors: Karen J. Maes and Jaime M. Willems
ISBN: 978-1-61209-506-6

Chapter 11

THEORIES IN PAM CHLOROPHYLL FLUORESCENCE MEASUREMENT OF PLANT LEAVES

Ichiro Kasajima[1], Maki Kawai-Yamada[1,2] and Hirofumi Uchimiya[1,3]
[1]Institute for Environmental Science and Technology, Saitama University, Saitama City, Saitama, Japan
[2]Graduate School of Science and Engineering, Saitama University, Saitama City, Saitama, Japan
[3]Iwate Biotechnology Research Center, Kitakami, Iwate, Japan

SUMMARY

Chlorophyll in chloroplasts is the light-harvesting pigment of plant leaves. Chlorophyll is excited by light energy to reach excitation state. This excitation energy is de-excited by several processes *in planta*. Of the two photochemical apparatus of photosynthetic electron transport (photosystem I and photosystem II), rate constants of photochemical reactions around photosystem II can be measured with chlorophyll fluorescence, because it is mainly emitted from photosystem II. Fluorescence itself is one of the de-excitation processes. Internal conversion and intersystem crossing also dissipates excitation energy. These three processes are called as 'basal dissipation'. In addition to basal dissipation, excitation energy is consumed to drive 'photochemistry' (photosynthetic electron transport). Under illumination, a mechanism called as 'non-photochemical quenching' is induced by biochemical mechanisms and dissipates excessive energy as heat.

Based on the Stern-Volmer relationship between fluorescence intensity and de-excitation rate constants, relative sizes of rate constants for these de-excitation pathways (basal dissipation, photochemistry and non-photochemical quenching) can be calculated by the technique called as Pulse Amplitude Modulation (PAM). In the PAM analysis of chlorophyll fluorescence, yield of fluorescence is measured with weak pulses of illumination. PAM technique enables measurement of fluorescence yield under both dark and illuminated conditions. Illumination of the leaf results in decrease of photochemistry and increase of non-photochemical quenching. Such changes are measured by several fluorescence parameters.

The principle equation to calculate relative sizes of rate constants from chlorophyll fluorescence intensities was presented by Kitajima and Butler. Since Kitajima and Butler, PAM was introduced and non-photochemical quenching was discovered, thus the principle equation was updated. This updated principle equation will be referred to as 'Kitajima-Butler equation' in this review. Many fluorescence parameters have been proposed from calculation of Kitajima-Butler equation. These parameters were first represented by complex formulas, and recent finding even showed that relative sizes of rate constants of all de-excitation pathways are easily described as the comparison between inverse values of fluorescence intensities.

PAM fluorescence analysis provides high-throughput method to estimate photosynthetic rate, damage by stressful conditions (photoinhibition) and size of non-photochemical quenching in living plant leaves. PAM fluorescence measurement is routinely used to estimate and visualize the effects of plant genes or stress treatments on these processes.

WHAT IS CHLOROPHYLL FLUORESCENCE?

Chlorophyll is light-harvesting green pigment contained in chloroplasts. Chlorophyll molecules are excited by light. This energy is used to excite electrons to drive photosynthetic electron transport. NADPH is the final acceptor of the excited electron. Proton gradient is also created across the thylakoid membrane in chloroplast by photosynthetic electron transport. This proton gradient drives H^+-ATPase to generate ATP. NADPH and ATP are then used to drive Calvin cycle to fix CO_2. Electron is excited twice in photosynthetic electron transport. Excitations occur at photosystem II (PSII) and photosystem I (PSI). Both PSII and PSI are protein complexes and bind chlorophyll molecules. Chlorophyll fluorescence is mostly emitted from PSII at room temperature. We will focus on the de-excitation processes of excited chlorophyll around PSII in this text.

De-excitation processes of chlorophyll around PSII are illustrated in Figure 1. De-excitation of excitation energy of chlorophyll molecule by photosynthetic electron transport is called as photochemistry. Excitation energy of chlorophyll molecules are also de-excited by other processes. Three kinds of de-excitation processes take place within the chlorophyll molecule. These are chlorophyll fluorescence, internal conversion (IC), and intersystem crossing (IS). These processes are collectively called as basal dissipation. Yield of fluorescence is calculated to be 1~2% of absorbed light in leaves. Yield of fluorescence is higher in extracted chlorophyll. Chlorophyll fluorescence can be visualized by extracting chlorophyll from leaves with ethanol and illuminating with UV light. Non-photochemical quenching (NPQ) is also a de-excitation process. NPQ mainly occurs by transfer of chlorophyll excitation energy to xanthophyll molecules. These de-excitation processes are explained by a simplified Jablonski diagram of chlorophyll molecule (Figure 2). Chlorophyll molecule at ground state (S_0) is excited to first singlet state (S_1) or second singlet state (S_2) by red or blue light. Second singlet state readily reaches first singlet state by IC. IC is a heat dissipation process. Thus excited chlorophylls reach first singlet state directly or indirectly via second singlet state. Five de-excitation processes (photochemistry, fluorescence, IC, IS, NPQ) take place from the first singlet state, and these de-excitation processes are competitive to each other. First triplet state (T_1) is reached after IS from the first singlet state. First triplet states of organic molecules easily react with molecular oxygen and create highly active singlet oxygen ($^1O_2$). Such case is also believed to take place in plant chloroplasts. Singlet oxygen destroys PSII component, especially D1 protein. This process is called as photoinhibition. PSII activity is lowered as a result of photoinhibition.

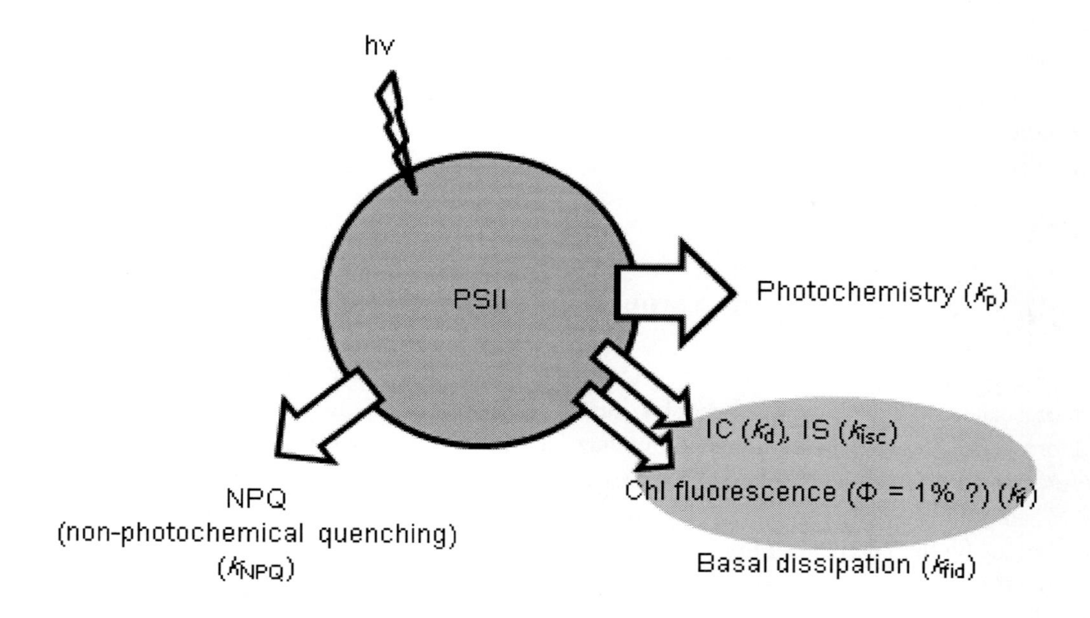

Figure 1. De-excitation processes of chlorophyll excitation energy.

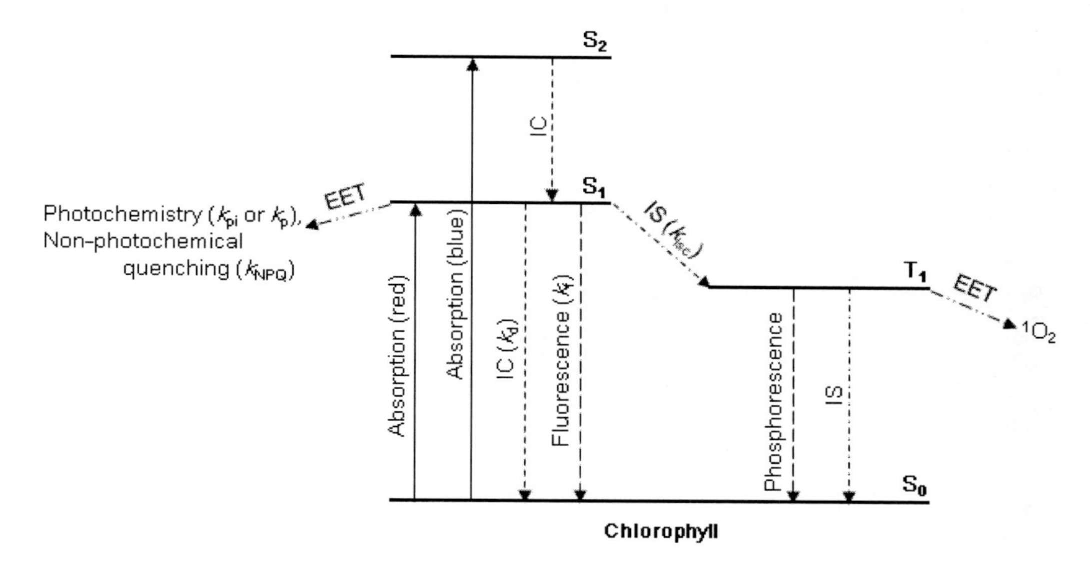

Figure 2. Simplified Jablonski diagram of chlorophyll molecule.

Activity of de-excitation processes can be expresses by rate constants (k). Rate constant of photochemistry become lower by illumination. This is because components of photosynthetic electron transport are partly reduced as the result of electron transport processes. Reduced portion of the components cannot accept additional electron. This chemical change lowers rate constant of photochemistry during illumination. Rate constant of basal dissipation does not change according to light conditions, because components of basal dissipation (fluorescence, IC, IS) are physicochemical processes within chlorophyll molecules. Rate constant of NPQ is 0 (does not exist) in the dark. During illumination, NPQ is induced by biochemical changes. Because de-excitation processes are competitive to each other, yield of chlorophyll fluorescence changes according to the changes in the rate constants of photochemistry and NPQ. Such changes in the yield of chlorophyll fluorescence can be observed by the changes in chlorophyll fluorescence intensities or yields. Rate constants for photochemistry, chlorophyll fluorescence, IC, IS, basal dissipation and NPQ are called as k_p, k_f, k_d, k_{isc}, k_{fid} and k_{NPQ} each.

PULSE AMPLITUDE MODULATION

Pulse Amplitude Modulation (PAM) is a technique used to measure chlorophyll fluorescence. This technique enables direct comparison of the yields of chlorophyll fluorescence under both dark and illuminated conditions. Originally, Amplitude Modulation (AM) and PAM are techniques to express original data by electronic signals (Figure 3). In AM, amplitude of vibration signal is modulated to express the original data. Original data are expressed by the intensities of pulse signals in PAM.

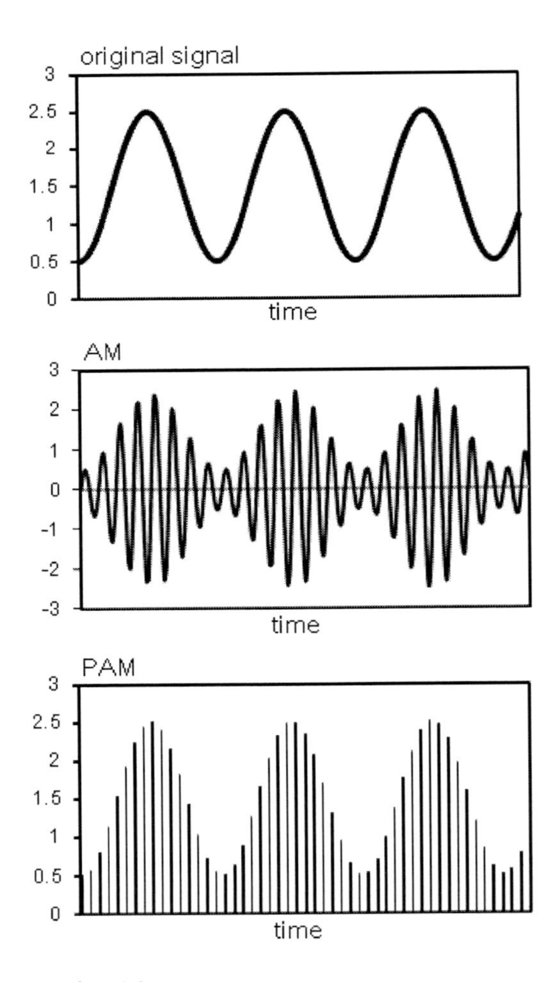

Figure 3. Illustration of AM and PAM.

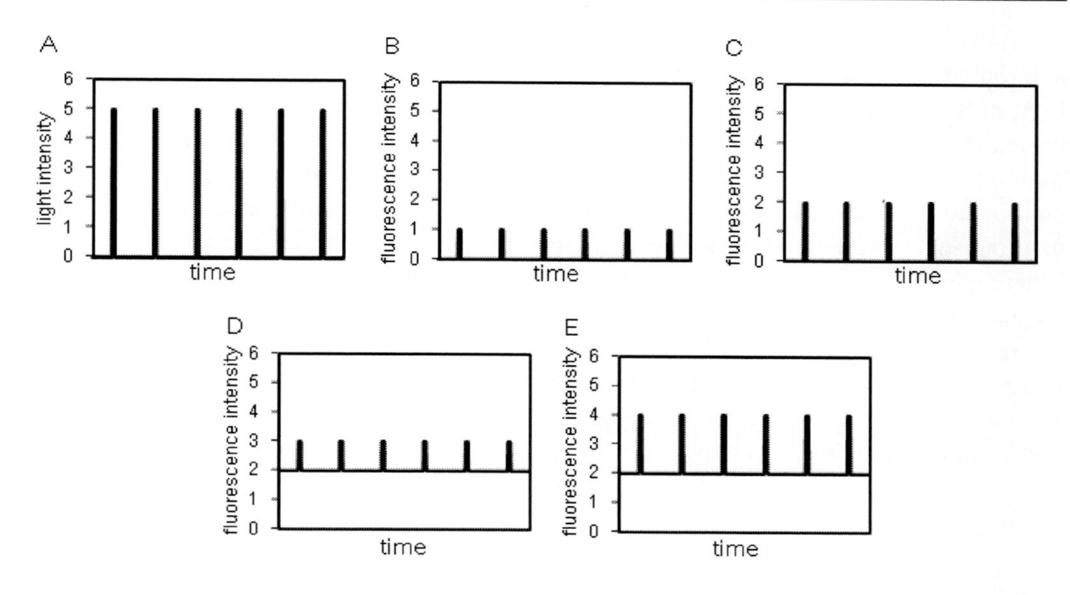

Figure 4. Illustration of PAM fluorescence. Measuring pulses are supplemented as illustrated in A. Four examples of fluorescence signals are shown (B,C,D,E). B and C illustrate fluorescence signals under dark condition. D and E illustrate fluorescence signals under light condition. Basal signal of the intensity 2 are observed under light conditions (D,E). Intensities of pulse signals are 1, 2, 1 and 2 in B, C, D and E each. This represents that relative fluorescence yields are B:C:D:E = 1:2:1:2.

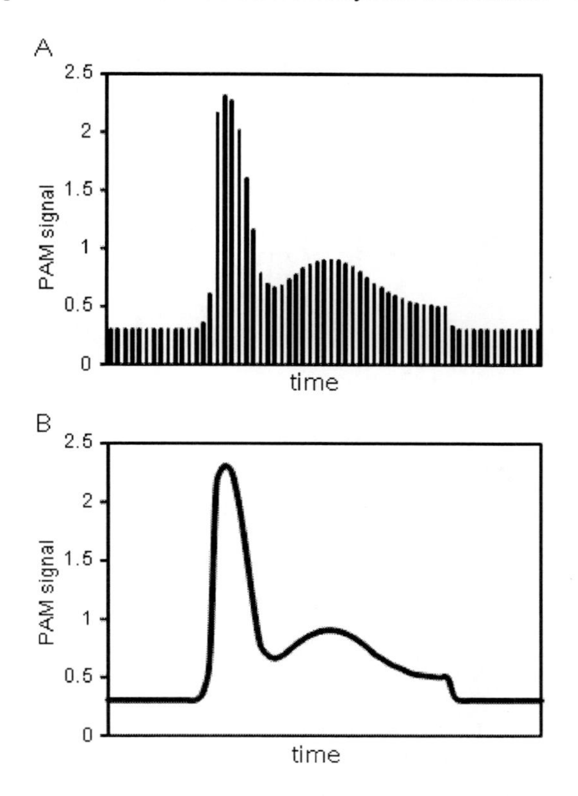

Figure 5. Plot of pulse signals. PAM fluorescence data are measured as pulse signals (A). These pulse signals are plotted as shown in B.

PAM fluorescence analysis is something different from original PAM. In PAM chlorophyll fluorescence measurement, modulation is made by plants (Figure 4). Even short light pulses (measuring pulse) are supplemented to leaves (Figure 4A). In response to the measuring pulse, fluorescence is emitted from leaves (Figure 4A,B,C,D). Intensity of fluorescence varies according to physicochemical properties of the leaves. In the dark environment, fluorescence intensity reflects yield of fluorescence. In illuminated environments, there exists basal level of chlorophyll fluorescence caused by continuous illumination (actinic light). Increase of chlorophyll fluorescence as the result of supplementation of measuring pulse reflects yield of fluorescence in this case. Increase of fluorescence can be compared among any intensity of actinic lights, because intensity of measuring pulse is constant. Yield of fluorescence changes within a measurement of the same leaf according to the changes of light intensities (Figure 5). Pulse data of relative fluorescence yields is plotted in the time course.

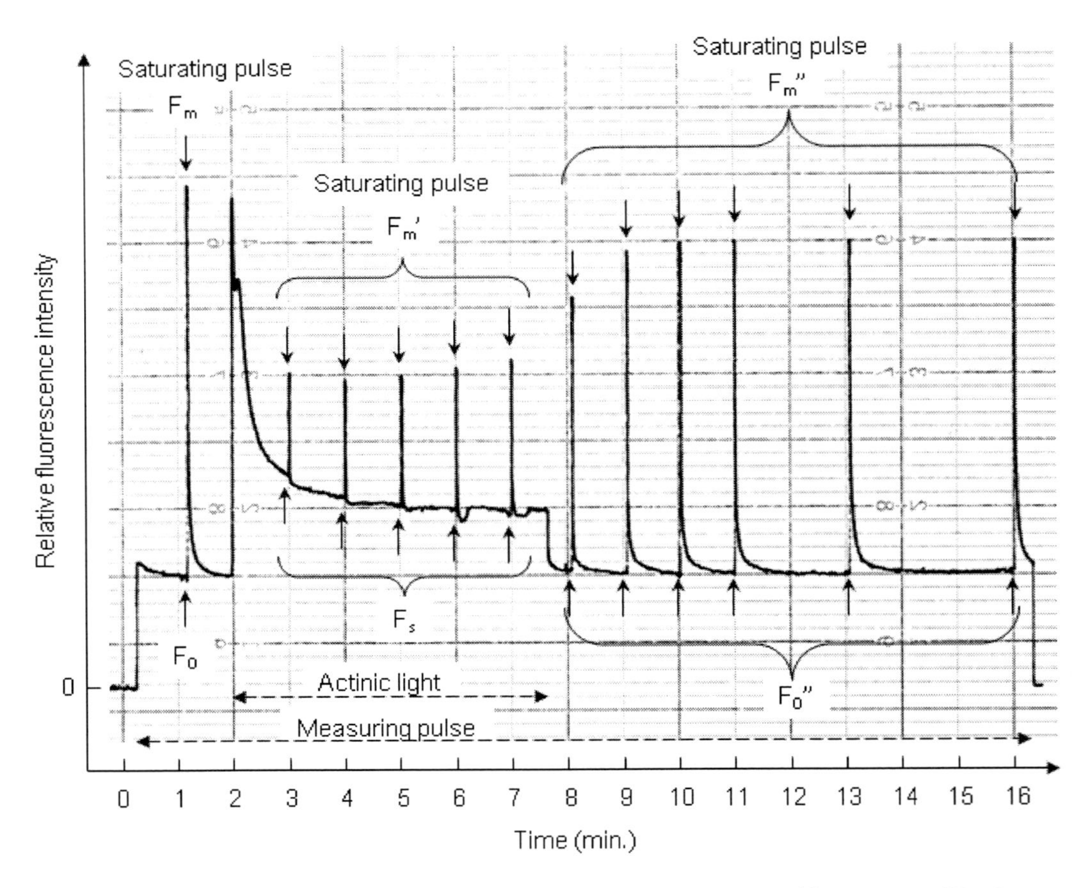

Figure 6. A typical PAM fluorescence measurement. (This data was measured in courtesy of Prof. Ichiro Terashima and co-workers.)

A typical pattern of PAM chlorophyll fluorescence measurement is shown in Figure 6. Fluorescence was measured with a leaf of a model dicot plant *Arabidopsis thaliana* (thale-cress, accession Col-0). In this measurement, measuring pulse was turned on at 0.2 min and turned off at 16.3 min, thus PAM fluorescence signals are measured within these time points.

Actinic light (continuous light) is supplemented from 2.0 to 7.6 min. Saturating pulses are also supplemented at 1.0, 3.0, 4.0, 5.0, 6.0, 7.0, 8.1, 9.1, 10.0, 11.0, 13.1 and 16.0 min. Saturating pulses are short (typically for 1 sec) and very strong light (typically PPFD = 10,000 μ mol m^{-2} s^{-1}). PPFD (photosynthetic photon flux density) is a unit of light intensity which measures the number of photons of wavelength from 400 nm to 700 nm, which can be absorbed by chlorophyll and used for photosynthesis. Light intensity is around PPFD = 2,000 μ mol m^{-2} s^{-1} at the maximum in the field and around PPFD = 15 μ mol m^{-2} s^{-1} in the room in our laboratory lit with fluorescence lamp. Photochemistry is transiently and completely reduced during supplementation of saturating pulses. Rate constant of NPQ is not affected by supplementation of saturating pulses, because the span of illumination is short. After supplementation of actinic light, fluorescence intensity first increases and then reduced with time. This change in fluorescence intensity is called as Kautsky effect.

Specific names are given to fluorescence intensity. Fluorescence intensity in the dark before supplementation of actinic light is called as Fo. Fluorescence intensity before supplementation of actinic light with supplementation of saturating pulse is called as Fm. Fluorescence intensity with supplementation of actinic light is called as Fs. Fluorescence intensity with supplementation of both actinic light and saturating pulse is called as Fm'. Fluorescence intensity after turning off the actinic light is called as Fo''. Fluorescence intensity after turning off the actinic light and with supplementation of saturating pulse is called as Fm''.

As observed in Figure 6, complex changes in fluorescence yield occur in response to alterations in light conditions. These changes can be explained by the changes in rate constants of de-excitation processes. Rate constants of de-excitation processes are illustrated in Figure 7 for each light condition. Under the light condition of Fo measurement, basal dissipation and dark-adapted photochemistry exist. The capacity of photochemistry is at the maximum in the dark. The rate constant of photochemistry in the dark is called as k_{pi}. Under Fm condition, only basal dissipation exists, so total de-excitation size is lower under Fm condition compared to under Fo condition. This reduction in total size of de-excitation results in the increase in the portion of excitation energy which is de-excited by chlorophyll fluorescence. Thus, Fm value is higher than Fo value (Figure 6). Under Fs condition, all basal dissipation, photochemistry and NPQ exist. Photochemistry is partly reduced by supplementation of actinic light, thus k_p is smaller than k_{pi}. k_{NPQ} is induced by supplementation of actinic light. Under Fm' condition, k_p disappears by supplementation of saturating pulse compared with Fs condition. This results in higher yield of fluorescence, and thus higher value of Fm' than Fs. Under Fo'' condition, photosynthetic electron transport is completely reduced, so rate constant of photochemistry should return to k_{pi} again. But actually it is lower than k_{pi} (k_{pi}'), because PSII is partly destroyed by supplementation of actinic light (photoinhibition). NPQ is induced by illumination and relaxed in the dark. Complete relaxation of NPQ needs hours, because slow-relaxing component of NPQ remains. NPQ of higher plants typically consists of fast-relaxing NPQ and slow-relaxing NPQ. Fast-relaxing NPQ is relaxed after several minutes in the dark. Complete relaxation of slow-relaxing NPQ needs hours. Thus rate constant of slow-relaxing NPQ (k_{slow}) exists under Fo'' condition. Under Fm'' condition, k_{pi}' disappears compared to Fo'' condition. Fo' will be explained later in this text.

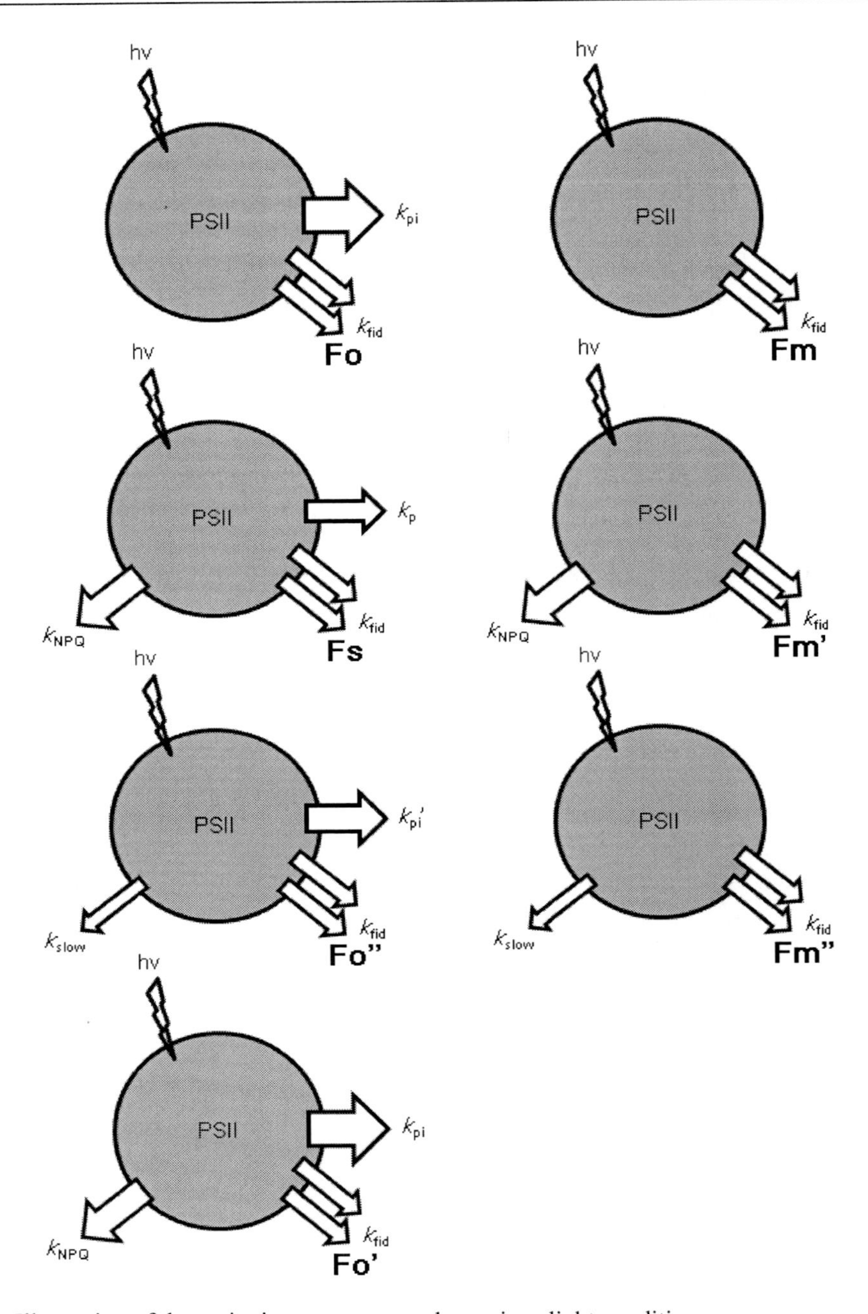

Figure 7. Illustration of de-excitation processes under various light conditions.

Changes in rate constants of photochemistry and NPQ during measurement can be deduced from the changes in fluorescence intensities. For example, difference between Fo and Fm reflects the size of k_{pi}, and difference between Fs and Fm' reflects the size of k_p. As a result, (Fm' – Fs)/(Fm – Fo) may be a good indicator of the portion of open PSII. As to the evaluation of NPQ size, difference between Fm and Fm' reflects NPQ size. Thus (Fm – Fm')/Fm may be a good indicator of NPQ size. These kinds of formulas used to estimate rate constants of de-excitation processes are called as fluorescence parameters. Many fluorescence parameters have been generated by researchers. Of these parameters, several parameters can calculate relative sizes of rate constants between different de-excitation processes.

KITAJIMA-BUTLER EQUATION AND FV/FM

As in the case of Stern-Volmer equation in inorganic photochemistry, rate of de-excitation of excited chlorophyll can be calculated from intensity of chlorophyll fluorescence. Kitajima and Butler demonstrated mathematical relationship between rate constants of de-excitation processes from first singlet chlorophyll and fluorescence intensities in their paper published in 1975 [1]. Their formula was simple compared to the formula used today [2,3]. This is partly because NPQ was discovered later. Kitajima-Butler equation is expressed in the modern form as follows:

$$F = S \cdot k_f /(k_{fid} + k_{NPQ} + k_p) \text{ (generally)} \quad (1)$$

F is fluorescence intensity. S is called as sensitivity factor and includes light intensity (I) and several factors such as fraction of supplemented light which is absorbed by leaf (A_{leaf}), fraction of absorbed light that is received by PSII ($fraction_{PSII}$), and instruments response (*Resp*). These factors are supposed to be constant throughout the measurement. Light intensity represents the intensity of measuring pulse in PAM measurement, which is constant. Thus S and rate constants of basal dissipations (k_f, k_{isc}, k_d and k_{fid}) are unknown but constant values.

From their equation, Kitajima and Butler derived a fluorescence parameter which calculates maximal quantum yield of photochemistry (the quantum yield of photochemistry in the dark). Following general equation (1), the specific equations below are derived for Fo and Fm:

$$Fo = S \cdot k_f /(k_{fid} + k_{pi}) \quad (2)$$

$$Fm = S \cdot k_f /k_{fid} \quad (3)$$

From these equations, (Fm – Fo)/Fm gives:

$$(Fm - Fo)/Fm = k_{pi} /(k_{fid} + k_{pi}) \quad (4)$$

Fm – Fo is also called as Fv, so (Fm – Fo)/Fm is called as Fv/Fm. Importance of this calculation is that $S \cdot k_f$ is deleted from original equations (2) and (3) by division of two different fluorescence values. Fv/Fm calculates the portion of photochemistry within the total rate constants in the dark, which is equal to the quantum yield of photochemistry in the dark (in other words, excitation energy of the first singlet state is de-excited by photochemistry at the ratio of Fv/Fm under the dark condition). Kitajima and Butler observed perfect match between the rate of photochemistry and the theoretical rate of photochemistry calculated from Fv/Fm in their observation. This discovery triggered the application of chlorophyll fluorescence for estimation of the rate of photochemistry and other de-excitation processes in living plant leaves. The value of Fv/Fm is typically from 0.80 to 0.85 in leaves of higher plants.

VARIOUS CHLOROPHYLL FLUORESCENCE PARAMETERS

Quantum yield (Φ) of photochemistry is lower under illumination. Genty et al. found a fluorescence parameter which estimates the quantum yield of photochemistry under supplementation of actinic light in 1989 [4]. This fluorescence parameter is called as Φ_{II}. Formula of Φ_{II} is derived from the Kitajima-Butler equations for Fs and Fm':

$$Fs = S \bullet k_f / (k_{fid} + k_{NPQ} + k_p) \quad (5)$$

$$Fm' = S \bullet k_f / (k_{fid} + k_{NPQ}) \quad (6)$$

Φ_{II} calculates quantum yield of photochemistry under illumination as follows:

$$\Phi_{II} = (Fm' - Fs)/Fm' = k_p / (k_{fid} + k_{NPQ} + k_p) \quad (7)$$

Excellent linear correlation was observed between Φ_{II} and quantum yield of CO_2 fixation. These successfulness of Fv/Fm and Φ_{II} guarantees validity of Kitajima-Butler equation and leads to application of Kitajima-Butler equation to wider range of comparison between rate constants of de-excitation processes.

Fluorescence intensity Fo' is also measured in some cases. Immediately after turning off the actinic light, infra-red ray is supplemented to leaves to immediately and fully oxidize photosynthetic electron transport. This is possible because absorbance spectrum of chlorophyll is different between PSII and PSI. Only PSI can absorb infra-red rays, and electrons harbored by the components of photosynthetic electron transport are immediately consumed by PSI. Fo' is measured after infra-red illumination, thus photochemistry is fully oxidized (open) and NPQ is not relaxed under Fo' condition. Kitajima-Butler equation for Fo' is:

$$Fo' = S \bullet k_f / (k_{fid} + k_{NPQ} + k_{pi}) \quad (8)$$

Using Fo', Kramer et al. found a formula to estimate the ratio of open PSII in 2004 [5]. This parameter is called as qL. qL calculates the ratio of open PSII as follows:

$$qL = (Fm' - Fs)/(Fm' - Fo') \bullet Fo'/Fs = k_p/k_{pi} \quad (9)$$

Interestingly, qL can also be calculated using Fo instead of Fo'. Miyake et al. found the following alternative equation to calculate qL in 2009 [6]:

$$qL = [\Phi_{II}/(1 - \Phi_{II})] \bullet [(1 - Fv/Fm)/(Fv/Fm)] \bullet (NPQ + 1)$$
$$= k_p / k_{pi} \quad (10)$$

Fo is included in Fv, and Fo' is not included in this equation. As is evident from these formulas, Fo' can be calculated from Fo, or vice versa. Mathematical relationship between Fo' and Fo was described by Oxborough and Baker in 1997 as follows [7]:

$$Fo' = Fo/(Fv/Fm + Fo/Fm') \quad (11)$$

All these equations are certainly consistent with Kitajima-Butler equation.

Another parameter NPQ was originally proposed as a qualitative parameter estimating NPQ size. Although, NPQ was recently found to be a quantitative parameter estimating the relative size of k_{NPQ} compared with the size of k_{fid}. NPQ calculates as follows:

$$NPQ = Fm/Fm' - 1 = k_{NPQ}/k_{fid} \quad (12)$$

NPQ is becoming the most popular parameter that estimates the size of NPQ, instead of another popular parameter qN. Some representative fluorescence parameters and their formulas are summarized in Table 1.

Table 1. Chlorophyll fluorescence parameters

Parameter	Formula	Formula by Inverse	Meaning	Rate Constant
Fv/Fm	(Fm – Fo)/Fm	$(Fo^{-1} - Fm^{-1})/Fo^{-1}$	maximal quantum yield of PSII photochemistry	k_{pi}/k_{si}
Φ_{II}	(Fm' – Fs)/Fm'	$(Fs^{-1} - Fm'^{-1})/Fs^{-1}$	quantum yield of PSII photochemistry under illumination	k_p/k_s
qL	(Fm' – Fs)/(Fm' – Fo') • Fo'/Fs	$(Fs^{-1} - Fm'^{-1}) \cdot (Fo'^{-1} - Fm'^{-1})$	ratio of open PSII (calculated with Fo')	k_p/k_{pi}
qL	$[\Phi_{II}/(1 - \Phi_{II})] \cdot [(1 - Fv/Fm)/(Fv/Fm)] \cdot (NPQ + 1)$	$(Fs^{-1} - Fm'^{-1}) \cdot (Fo^{-1} - Fm^{-1})$	ratio of open PSII	k_p/k_{pi}
NPQ	Fm/Fm' – 1	$(Fm'^{-1} - Fm^{-1})/Fm^{-1}$	NPQ size relative to basal dissipation	k_{NPQ}/k_{fid}
qP	(Fm' – Fs)/(Fm' – Fo')	–	ratio of open PSII under the puddle model	–
qN	1 – (Fm' – Fo')/(Fm – Fo)	–	NPQ size (qualitative parameter)	–
Φ_{NO}	$1/[NPQ + 1 + qL \cdot (Fm/Fo - 1)]$	–	quantum yield of basal dissipation under illumination (calculated with Fo')	k_{fid}/k_s
Φ_{NO}	Fs/Fm	Fm^{-1}/Fs^{-1}	quantum yield of basal dissipation under illumination	k_{fid}/k_s
Φ_{NPQ}	$1 - \Phi_{II} - \Phi_{NO}$	–	quantum yield of NPQ under illumination (calculated with Fo')	k_{NPQ}/k_s
Φ_{NPQ}	Fs/Fm' – Fs/Fm	$(Fm'^{-1} - Fm^{-1})/Fs^{-1}$	quantum yield of NPQ under illumination	k_{NPQ}/k_s
qPI	'change of 1/Fo – 1/Fm'	$(Fo''^{-1} - Fm''^{-1})/(Fo^{-1} - Fm^{-1})$	ratio of functioning PSII	k_{pi}'/k_{pi}
qSlow	–	$(Fm''^{-1} - Fm^{-1})/Fm^{-1}$	size of slow-relaxing NPQ	k_{slow}/k_{fid}
qS	–	Fs^{-1}/Fo^{-1}	size of sum de-excitation under illumination	k_s/k_{si}
ETR	$\Phi_{II} \cdot I \cdot 0.84 \cdot 0.5$	–	flux of photochemistry	–

The parameter ETR (electron transport rate) is used to approximately estimate the flux of photochemistry as follows:

$$ETR= \Phi_{II} \cdot I \cdot 0.84 \cdot 0.5 \quad (13)$$

The unit of ETR is μmol m^{-2} s^{-1}. 0.84 in equation (13) is a typical value of A_{leaf} and 0.5 is a typical value of $fraction_{PSII}$ in plant leaves. ETR is a useful parameter, but A_{leaf} and $fraction_{PSII}$ are not necessarily the same between different samples. We have to be also aware that ETR is the sum of the fluxes of photosynthesis, photorespiration and water-water cycle.

Calculation by Inverse Fluorescence Intensities

Quantitative fluorescence parameters above were derived through tremendous efforts of many researchers. But recently a way of calculation was found to easily calculate relative sizes of rate constants for all de-excitation processes. This is done by changing the shape of general Kitajima-Butler equation [equation (1)] as follows [2]:

$$k_{fid} + k_{NPQ} + k_p = S \cdot k_f \cdot F^{-1} \text{ (generally)} \quad (14)$$

This equation only exchanged the left side and the denominator of right side of original Kitajima-Butler equation, but this inverse Kitajima-Butler equation (14) shows the important fact; sum of rate constants of all de-excitation processes is proportional to inverse values of fluorescence intensities under any condition. For example, when fluorescence intensity becomes two-fold by some treatment, we can say that the sum of rate constants became the half by that treatment. Sum of rate constants is called as 'sum de-excitation'. Rate constant of sum de-excitation under dark condition is k_{si} and rate constant of sum de-excitation under illumination is k_s:

$$k_{si}= k_{fid} + k_{pi} \quad (15)$$

$$k_s= k_{fid} + k_{NPQ} + k_p \quad (16)$$

Inverse Kitajima-Butler equation is written for each specific fluorescence intensity as follows:

$$k_{si}= S \cdot k_f \cdot Fo^{-1} \quad (17)$$

$$k_{fid}= S \cdot k_f \cdot Fm^{-1} \quad (18)$$

$$k_s= S \cdot k_f \cdot Fs^{-1} \quad (19)$$

$$k_{fid} + k_{NPQ}= S \cdot k_f \cdot Fm'^{-1} \quad (20)$$

$$k_{fid} + k_{NPQ} + k_{pi}= S \cdot k_f \cdot Fo'^{-1} \quad (21)$$

Subtractions between these formulas also give following equations:

$$k_{pi}= S \cdot k_f \cdot (Fo^{-1} - Fm^{-1}) \quad (22)$$

$$k_p= S \cdot k_f \cdot (Fs^{-1} - Fm'^{-1}) \quad (23)$$

$$k_{pi}= S \cdot k_f \cdot (Fo'^{-1} - Fm'^{-1}) \quad (24)$$

$$k_{NPQ}= S \cdot k_f \cdot (Fm'^{-1} - Fm^{-1}) \quad (25)$$

From these equations, it is quite easy to derive all quantitative fluorescence parameters which were explained above. For example, parameter Fv/Fm is given by the division of equation (22) by equation (17) as follows:

$$k_{pi}/k_{si}= (Fo^{-1} - Fm^{-1})/ Fo^{-1} \quad (26)$$

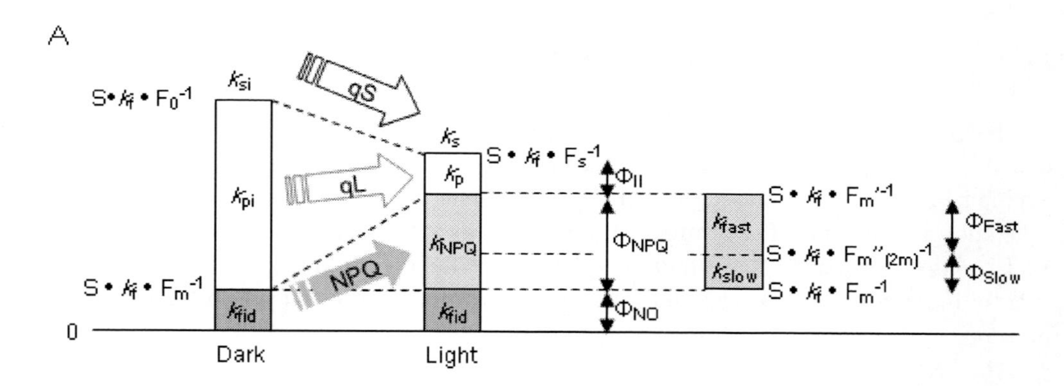

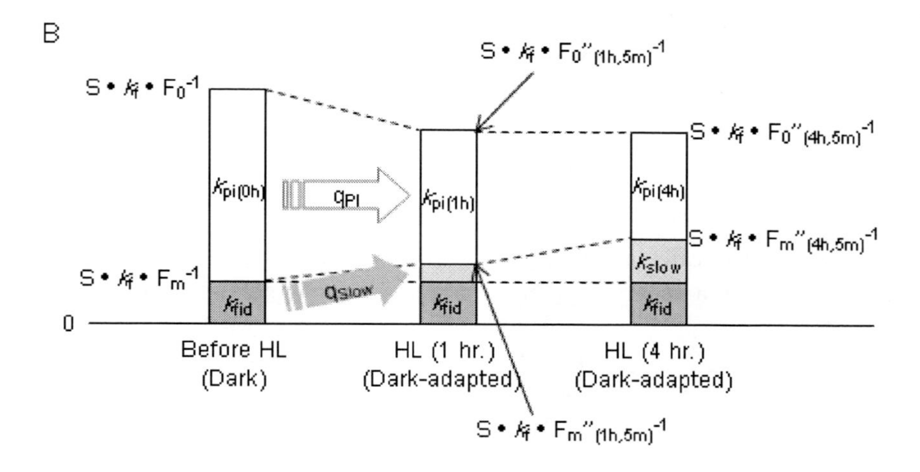

Figure 8. Relationship between rate constants of de-excitation processes and chlorophyll fluorescence intensity. Rate constants (k), fluorescence intensity (F) and chlorophyll fluorescence parameters are shown for typical fluorescence measurement (A) and measurement of photoinhibition (B).

This is the same formula with the original formula for Fv/Fm, presented in a different form. This is also true of the other fluorescence parameters. The shapes of formulas derived by inverse Kitajima-Butler equations are also summarized in Table 1. In addition to these formulas, relationship between Fo and Fo' which was originally shown by the equation (11) is also derived from inverse Kitajima-Butler equation. Because left sides of the equations (22) and (24) are the same (k_{pi}), right sides of these equations are also same. As a result, relationship between Fo and Fo' is expresses by inverse fluorescence intensities as follows:

$$Fo^{-1} - Fm^{-1} = Fo'^{-1} - Fm'^{-1} \quad (27)$$

Relationship between rate constants of de-excitation processes and inverse fluorescence intensities is visually summarized in Figure 8A. In this figure, qS is a parameter estimating the size of sum de-excitation under illumination:

$$qS = k_s / k_{si} = Fs^{-1}/Fo^{-1} \quad (28)$$

COMPONENTS OF NPQ

NPQ consists of at least two components in higher plants. These NPQ components are called as qE quenching and qI quenching. As described above, NPQ is induced by supplementation of actinic light. NPQ components are discriminated by their speed of relaxation in the dark, after turning off the actinic light (Figure 9) [8]. This analysis is called as relaxation analysis. After turning off the actinic light, Fo'' and Fm'' are measured. Fm'' rapidly increases from the value of Fm' for a few minutes after turning off the actinic light, but does not recover to Fm level for hours. From this fact, we can know that there are fast-relaxing NPQ and slow-relaxing NPQ. The fast-relaxing NPQ is called as qE and slow-relaxing NPQ is called as qI.

To estimate sizes of fast-relaxing NPQ and slow-relaxing NPQ, total NPQ size can be divided into two components [2]. For example, NPQ can be divided into fast-relaxing NPQ (k_{fast}) which relaxes within two minutes in the dark and the rest slow-relaxing NPQ (k_{slow}) as follows:

$$k_{NPQ} = k_{fast} + k_{slow} \quad (29)$$

Sizes of k_{fast} and k_{slow} are calculated using $Fm''_{(2m)}$, which is measured two minutes after turning off the actinic light and under supplementation of saturating pulse as follows:

$$qFast = k_{fast} / k_{fid} = (Fm'^{-1} - Fm''_{(2m)}{}^{-1})/Fm^{-1} \quad (30)$$

$$qSlow = k_{slow} / k_{fid} = (Fm''_{(2m)}{}^{-1} - Fm^{-1})/Fm^{-1} \quad (31)$$

qT quenching is a major NPQ instead of qE in unicellular photosynthetic organisms such as cyanobacteria and chlamydomonas. qT quenching is caused by state transition. In state transition, antenna complex of PSII dissociates from PSII core complex and then binds to PSI core complex. This results in reduction in the portion of absorbed light energy which goes to PSII core complex, and thus reduction in the fluorescence intensity. From this mechanism, qT may not be a de-excitation process, but fluctuation of the sensitivity factor (S fluctuation). S fluctuation is one of the topics which should be clarified in the future [2].

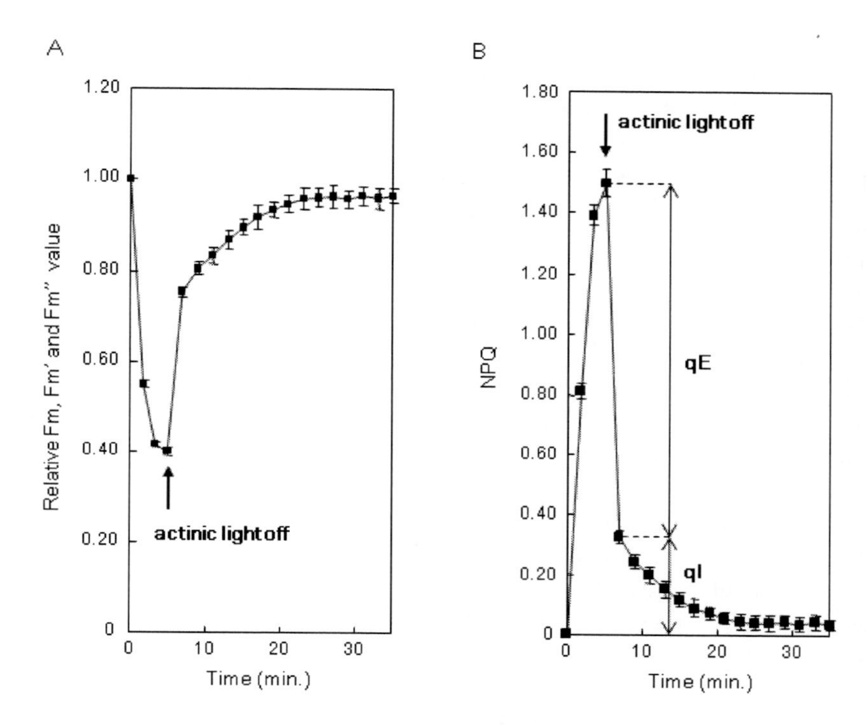

Figure 9. Relaxation analysis. Fm, Fm' and Fm'' values are plotted during supplementation of actinic light for five minutes and following relaxation analysis for 30 minutes of rice leaves (A). qE relaxes within a few minutes in the dark, whereas qI is not completely relaxed even after 30 minutes in the dark (B).

Molecular Mechanism of qE Quenching

The mechanism of qI quenching is not clear. Size of qI quenching is empirically correlated with the degree of photoinhibition, so qI quenching is believed to be caused by photoinhibition.

On the other hand, mechanism of qE quenching is becoming clear, especially after the genes which regulate the size of qE quenching were cloned in model plant *Arabidopsis thaliana*. These genes were isolated by forward genetics by Niyogi and co-workers [9,10]. *Arabidopsis thaliana* (thale cress) is a widely used model plant by plant biologists especially in the field of genetics. *A. thaliana* is a reasonable model plant for genetics, because the genome size is small and researchers can grow many plants in limited spaces and gather many seeds from one plant. Whole genomic DNA sequence is read and can be obtained from database. There are also additional genetic materials available, such as genetic markers, many accessions (varieties) gathered from all over the world, and accumulation of high-throughput methods for genetics.

Genomic DNA is mutagenized by treatment with mutagen such as ethylmethane sulfonate (EMS). *A. thaliana* has more than 20,000 genes. Several genes are mutagenized (DNA sequence is changed) and inactivated in a mutant, and other several genes are mutagenized in another mutant. To isolate genes which regulate qE quenching, population of such mutants were grown and screened by parameter NPQ. As expected qE quenching was deficient in some mutants. *npq1* and *npq4* were among these mutants.

NPQ1 (the mutated gene in *npq1* mutant) encoded violaxanthin de-epoxidase (VDE), which localizes in chloroplast lumen and catalyzes the conversion of violaxanthin (Vx) to anthlaxanthin (Ax) and further to zeaxanthin (Zx) (Figure 10) [9]. On the other hand, Zx is converted to Ax and then to Vx by zeaxanthin epoxidase (ZE). This conversion of xanthophyll is called as xanthophyll cycle. Vx mainly accumulates under normal conditions. Interestingly, VDE is activated under acid environment. H^+ pumping across thylakoid membrane is activated when there is excessive excitation energy. This results in lumen acidification and activation of VDE. Part of xanthophyll is bound to PSII. In PSII, excitation energy of chlorophyll is passed to xanthophyll through resonance and safely dissipated as the form of heat. Zx can more efficiently dissipate excitation energy as heat than Vx. Thus, xanthophyll cycle created by VDE is necessary to improve the size of qE quenching and safely dissipate the harmful excitation energy of chlorophyll when there is excessive light energy than can be consumed by photochemistry. Another pigment lutein also dissipates chlorophyll excitation energy. Mutant lacking both xanthophyll cycle and lutein accumulation completely loses qE quenching [11].

NPQ4 (the mutated gene in *npq4* mutant) encoded PsbS [10]. PsbS is one of the components of PSII protein complex. PsbS is an essential component of qE, because qE quenching is completely lost in *npq4* mutant. The precise role of PsbS in the induction of qE quenching is not clear yet, but there is hypothesis that PsbS is protonated after lumen acidification and this PsbS protonation affects geometric structure of PSII to improve efficiency of excitation energy transfer from chlorophyll to xanthophyll [12].

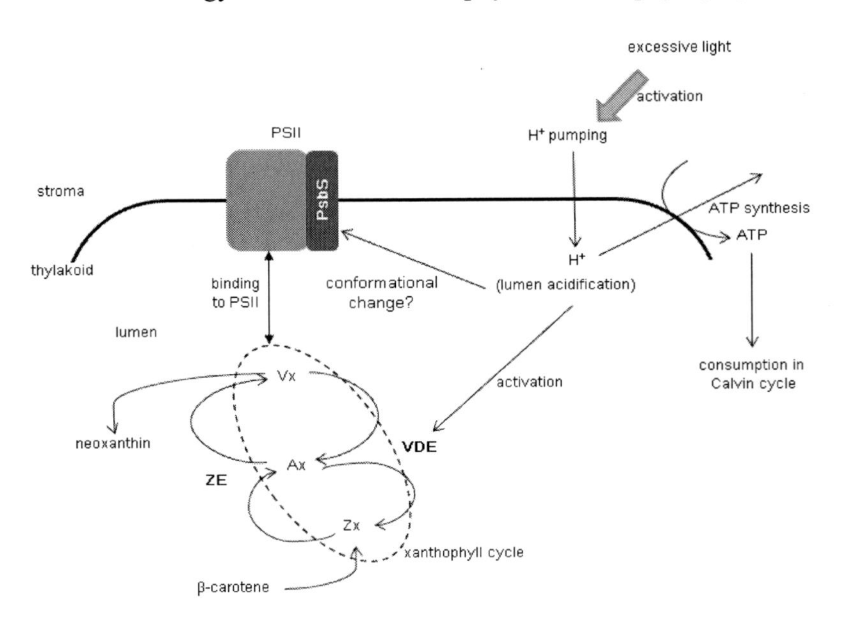

Figure 10. Role of VDE and PsbS in induction of qE quenching.

PHOTOINHIBITION

During photoinhibition, PSII is partly destroyed and size of photochemistry (k_{pi}) is lowered compared with the original state (Figure 8B). Changes in the size of k_{pi} after photoinhibition are estimated by the following equation [2]:

$$qPI= k_{pi}'/k_{pi} = (Fm''^{-1} - Fo''^{-1})/(Fm^{-1} - Fo^{-1}) \quad (31)$$

qPI becomes lower and qSlow (estimate of qI quenching) becomes higher by heavier photoinhibition. As a result, yield of photochemistry in the dark (Fv/Fm) is also lowered by photoinhibition. Damage by photoinhibition is estimated by the decrease of Fv/Fm most frequently, because measurement of Fv/Fm is the easiest. Changes in the size of k_{pi} by photoinhibition were originally estimated by 'changes in the values of 1/Fo – 1/Fm'. This calculation is the same as qPI.

Because PSII core complex, especially D1 protein is constantly destroyed by photoinhibition, D1 protein is rapidly renewed. To observe rate of photoinhibition, leaves are treated with lincomycin to dysfunction this repair mechanism. Photoinhibition occurs even under low light conditions, because some part of chlorophyll excitation energy drives intersystem crossing and singlet oxygen is generated. Because qS is about 0.5 to 0.6 under medium and high light conditions [2], ratio of photoinhibition per illumination is expected to be higher under medium and high light conditions compared with low light condition.

In addition to intersystem crossing, direct destruction of Mn cluster is also discussed to be one of the reasons of photoinhibition. This is based on the observation that ultra-violet and blue light cause photoinhibition more efficiently per quanta than red light [13,14].

LAKE MODEL AND PUDDLE MODEL

Exchanging capacity of chlorophyll excitation energy among PSII units should be considered to strictly calculate quantitative fluorescence parameters. The two major models about the exchangeability of chlorophyll excitation energy among PSII units are lake model and puddle model (Figure 11). In puddle model, it is hypothesized that there is not exchange of chlorophyll excitation energy among PSII units. Excitation energy stays within one PSII unit until it is de-excited by de-excitation processes. On the other hand, in lake model, it is hypothesized that chlorophyll excitation energy can freely travel to any neighboring PSII unit. The virtue of lake model over puddle model is that photochemistry can function more efficiently.

The formulas for quantitative fluorescence parameters may differ between lake model and puddle model. For example, parameter qP calculates the ratio of open PSII under puddle model. qP has older history than qL, which calculates the ratio of open PSII under lake model [5]. But it is now believed that lake model well approximates the actual state in plant leaves [5]. An intermediate state between puddle model and lake model seems to operate in plant leaves. All calculations in this review are based on lake model. Parameters such as Fv/Fm and Φ_{II} are quantitative under both lake model and puddle model.

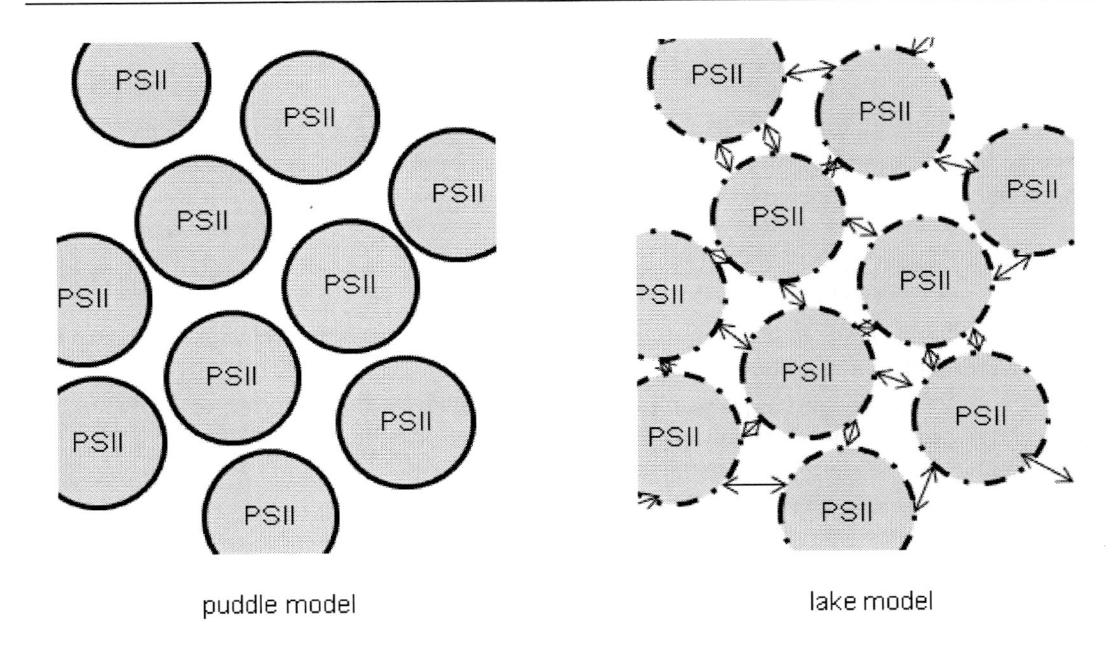

Figure 11. Concepts of lake model and puddle model.

Fluorescence Lifetime Analysis

Fluorescence lifetime analysis (FL analysis) is becoming a popular technique to estimate components of de-excitation processes. In contrast to PAM fluorescence analysis which gives relative values of rate constants for de-excitation processes, FL analysis gives absolute values of lifetime components. Principle of the FL analysis is illustrated in Figure 12. Very short laser light pulses are supplemented to samples to induce fluorescence. Fluorescence (emission of photons from samples) occurs shortly after illumination of the laser and its intensity exponentially decays with time. Fluorescence lifetime (τ) is the inverse value of the rate constant (k) of de-excitation of the excitation energy. In Figure 12A, fluorescent material 2 has a shorter lifetime compared with that of fluorescent material 1. Fluorescence intensity of the mixture of materials of 1 and 2 by the ratio of 1:1 is the average of the fluorescence intensities of materials 1 and 2. Fluorescence intensities are frequently plotted by single logarithmic graph (Figure 12B). In single logarithmic graph, slope of pure materials (material 1 and material 2) is linear. On the other hand, slope of the mixture is the average of those of materials 1 and 2 at time 0 and then approaches that of material 1 with time. From the pattern of fluorescence decay, de-excitation components and their contents are calculated for the sample.

Several lifetime components are typically detected within a few nano seconds in FL analysis of plant chlorophyll. For example, short lifetime components appear under condition of qE quenching, and these components does not appear in *npq4* mutant [15]. FL analysis is in fact a potential tool to quantify de-excitation process. Although, values of lifetimes vary between papers even for the same plant species and the identity of lifetime component is nothing more than speculation in many cases. In addition, only results of model calculations are shown in many cases. FL analysis seems to be an additional tool to PAM analysis at the moment. Technical advances are waited for to overcome these problems in the future. As described in the explanation of lake model, PSII complexes exchange excitation energy each other. Such communication between PSII complexes will be making the FL measurement more complex.

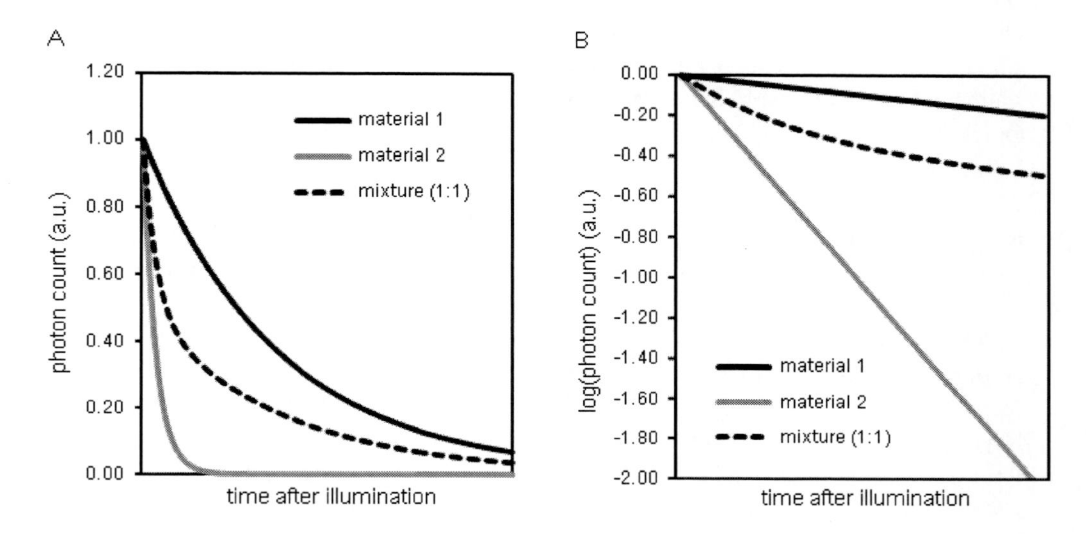

Figure 12. Model of fluorescence lifetime analysis.

REFERENCES AND THEIR CONTENTS

[1] Kitajima M, Butler WL (1975) Quenching of chlorophyll fluorescence and primary photochemistry in chloroplasts by dibromothymoquinone. *Biochim Biophys Acta* 376, 105-115. Original Kitajima-Butler equation. Equation of Fv/Fm. Lake model and puddle model.

[2] Kasajima I, Takahara K, Kawai-Yamada M, Uchimiya H (2009) Estimation of the relative sizes of rate constants for chlorophyll de-excitation processes through comparison of inverse fluorescence intensities. *Plant Cell Physiol* 50, 1600-1616. Calculation of chlorophyll fluorescence parameters by inverse values of fluorescence intensities. Nomenclature and theoretical backgrounds which this text is based on.

[3] Baker NR (2008) Chlorophyll fluorescence: A probe of photosynthesis in vivo. *Annu Rev Plant Biol* 59, 89-113. General review of chlorophyll fluorescence measurement by PAM

[4] Genty B, Briantais JM, Baker NR (1989) The relationship between the quantum yield of photosynthetic electron transport and quenching of chlorophyll fluorescence. *Biochim Biophys Acta* 990, 87-92. Discovery and characterization of the parameter Φ_{II}.

[5] Kramer DM, Johnson G, Kiirats O, Edwards GE (2004) New fluorescence parameters for the determination of Q_A redox state and excitation energy fluxes. *Photosynth Res* 79, 209-218. Formulas of qL, Φ_{NO} and Φ_{NPQ} with Fo'. Linearity of the parameter NPQ. Discussion on lake model and puddle model.

[6] Miyake C, Amako K, Shiraishi N, Sugimoto T (2009) Acclimation of tobacco leaves to high light intensity drives the plastoquinone oxidation system – relationship among the fraction of open PSII centers, non-photochemical quenching of Chl fluorescence and the maximum quantum yield of PSII in the dark. *Plant Cell Physiol* 50, 730-743. Formula of qL without Fo'

[7] Oxborough K, Baker NR (1997) Resolving chlorophyll *a* fluorescence images of photosynthetic efficiency into photochemical and non-photochemical components – calculation of *qP* and *Fv'/Fm'* without measuring *Fo'*. *Photosynth Res* 54, 135-142. Mathematical relationship between Fo' and Fo

[8] Quick WP, Stitt M (1989) An examination of factors contributing to non-photochemical quenching of chlorophyll fluorescence in barley leaves. *Biochim Biophys Acta* 977, 287-296 Relaxation analysis of NPQ in barley leaves

[9] Niyogi KK, Grossman AR, Björkman O (1998) Arabidopsis mutants define a central role for the xanthophyll cycle in the regulation of photosynthetic energy conversion. *Plant Cell* 10, 1121-1134. Analysis of *A. thaliana npq1* mutant

[10] Li XP, Björkman O, Shih C, Grossman AR, Rosenquist M, Jansson S, Niyogi KK (2000) A pigment-binding protein essential for regulation of photosynthetic light harvesting. *Nature* 403, 391-395. Analysis of *A. thaliana npq4* mutant

[11] Niyogi KK, Shih C, Chow WS, Pogson BJ, DellaPenna D, Björkman O (2001) Photoprotection in a zeaxanthin- and lutein-deficient double mutant of *Arabidopsis*. *Photosynth Res* 67, 139-145. *A. thaliana* mutant lacking both xanthophyll cycle and lutein accumulation

[12] Niyogi KK, Li XP, Rosenberg V, Jung HS (2005) Is PsbS the site of non-photochemical quenching in photosynthesis? *J Exp Bot* 56, 375-382. Working model of PsbS function

[13] Oguchi R, Terashima I, Chow WS (2009) The involvement of dual mechanisms of photoinactivation of photosystem II in *Capsicum annuum L. Plants*. *Plant Cell Physiol* 50, 1815-1825. Quantum efficiency of photoinhibition is different among light wave lengths.

[14] Takahashi S, Milward SE, Yamori W, Evans JR, Hillier W, Badger MR (2010) The solar action spectrum of photosystem II damage. *Plant Physiol* 153, 988-993. UV light efficiently causes photoinhibition.

[15] Li XP, Müller-Moulé P, Gilmore AM, Niyogi KK (2002) *PsbS-dependent enhancement of feedback de-excitation protects photosystem II from photoinhibition*. Proc Natl Acad Sci USA 99, 15222-15227. Improvement of NPQ by *PsbS* overexpression in *A. thaliana*. FL analysis of *npq4*.

FURTHER STUDIES

Heldt HW (2005) The use of energy from sunlight by photosynthesis is the basis of life on earth. *In Plant Biochemistry*, 3rd edn. pp. 52-56. Elsevier Academic Press, London. Textbook of photosynthesis in general

Hendrickson L, Furbank RT, Chow WS (2004) A simple alternative approach to assessing the fate of absorbed light energy using chlorophyll fluorescence. *Photosynth Res* 82, 73-81. Formulas of Φ_{NO} and Φ_{NPQ} without Fo'

Hieber AD, Kawabata O, Yamamoto HY (2004) Significance of the lipid phase in the dynamics and functions of the xanthophyll cycle as revealed by PsbS overexpression in tobacco and in-vitro de-epoxidation in monogalactosyldiacylglycerol micelles. *Plant Cell Physiol* 45, 92-102. Overexpression of *VDE* and *PsbS* in tobacco

Külheim C, Ågren J, Jansson S (2002) Rapid regulation of light harvesting and plant fitness in the field. *Science* 297, 91-93. Growth of *A. thaliana npq* mutants under natural condition

Sonoike K Koh-Goh-Sei No Mori (in Japanese) *http://sunlight.k.u-tokyo.ac.jp/fluo1.html*. On-line textbook of chlorophyll fluorescence and photosynthesis

INDEX

C

D

E

F

G

H

I

J

K

L

M

N

O

P

Q

R

S

T

U

V

W

X

Y

Z